D1243675

BUILDING A FOUNDATION IN MATHEMATICS

2e

NJATC

BUILDING A FOUNDATION IN MATHEMATICS

"Mathematics is the queen of the sciences."
Carl Friedrich Gauss

Building a Foundation in Mathematics, 2nd Edition is intended to be an educational resource for the user and contains procedures commonly practiced in industry and the trade. Specific procedures vary with each task and must be performed by a qualified person. For maximum safety, always refer to specific manufacturer recommendations, insurance regulations, specific job site and plant procedures, applicable federal, state, and local regulations, and any authority having jurisdiction. The *electrical training ALLIANCE* assumes no responsibility or liability in connection with this material or its use by any individual or organization.

ACKNOWLEDGMENTS

Technical Editors

Jim Paladino, IBEW/NECA Omaha

John C. Peterson, Chattanooga State Technical Community College (retired)

ADDITIONAL ACKNOWLEDGMENTS

This material is continually reviewed and evaluated by curriculum groups who are also members of the NJATC Inside Education Committee. The invaluable input provided by these individuals allows for the development of instructional material that is of the absolute highest quality. At the time of this printing the Inside Education Committee was comprised of the following members:

Kathleen Barber

Byron Benton

John Biondi

Richard Brooks

Eric Davis

Lawrence Hidalgo

Greg Hojdila

Tony Lewis

David McCraw

David Milazzo Jr.

Tom Minder

Tony Naylor

Jim Patterson

Janet Skipper

Jim Sullivan

Dennis Williamson

Chapter 7

Logarithms 151

Chapter 8

Units and Measurements 175

Chapter 16

Solid Geometry 405

Chapter 17

Trigonometry 427

Chapter 18

Vectors 469

This text has been strengthened from top to bottom with many new features and enhancements to existing content. All-new chapter features provide structure and guidance for learners. Chapter Overviews and Chapter Objectives have been introduced to guide the student's learning and to reinforce the expanded Review Questions and Practice Problems. Step-by-step calculator examples walk students through the process of solving each problem. Throughout each chapter, concepts are explained from their theoretical roots to their application principles, with reminders and tips relating mathematical concepts to on-the-job electrical work and electrical concepts.

Building a Foundation in Mathematics, Second Edition, has been expanded to more fully explore a number of concepts. Tips on using scientific calculators are included, as well as step-by-step visuals to guide students through calculator functions while solving each type of problem. Mathematical concepts are introduced through more elementary topics, such as whole numbers, fractions, and decimals, and build on these basics through more advanced topics, such as vector operations, linear equations, and Boolean algebra. All of the topics are introduced with an emphasis on practicality and application in the electrical field, thereby emphasizing the relationship between mathematics and electrical concepts.

See the following pages for examples of these new features.

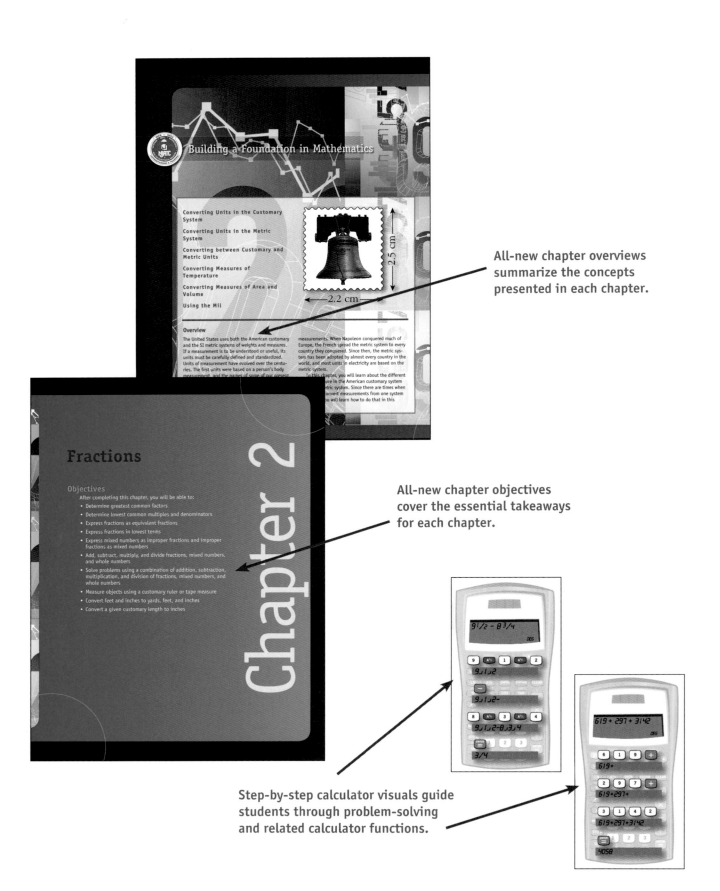

All-new chapter overviews summarize the concepts presented in each chapter.

All-new chapter objectives cover the essential takeaways for each chapter.

Step-by-step calculator visuals guide students through problem-solving and related calculator functions.

New sidebars and Key Ideas highlight need-to-know concepts.

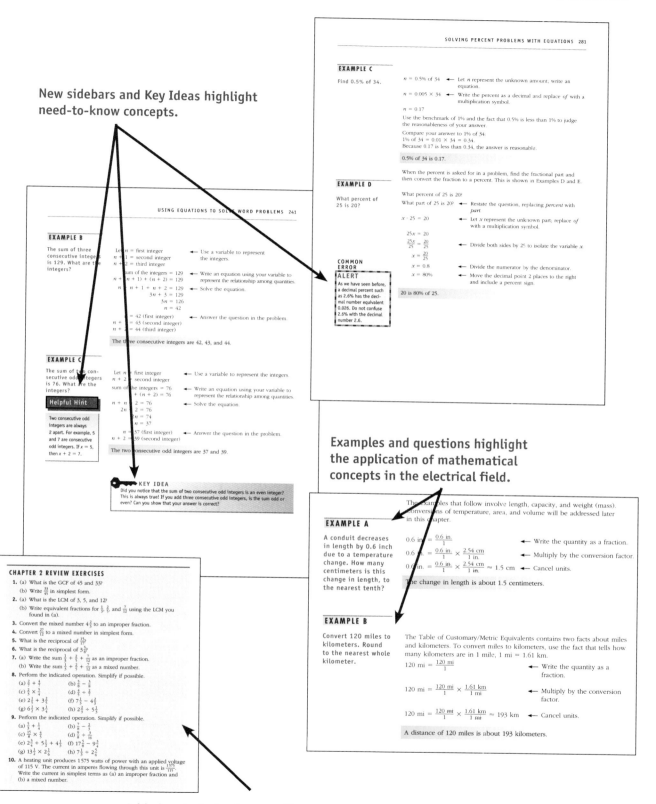

SOLVING PERCENT PROBLEMS WITH EQUATIONS 281

EXAMPLE C

Find 0.5% of 34.

$n = 0.5\%$ of 34   ◄— Let n represent the unknown amount; write an equation.

$n = 0.005 \times 34$ ◄— Write the percent as a decimal and replace *of* with a multiplication symbol.

$n = 0.17$

Use the benchmark of 1% and the fact that 0.5% is less than 1% to judge the reasonableness of your answer.

Compare your answer to 1% of 34:
1% of 34 = $0.01 \times 34 = 0.34$.
Because 0.17 is less than 0.34, the answer is reasonable.

0.5% of 34 is 0.17.

When the percent is asked for in a problem, find the fractional part and then convert the fraction to a percent. This is shown in Examples D and E.

EXAMPLE D

What percent of 25 is 20?

What part of 25 is 20?   ◄— Restate the question, replacing *percent* with *part*.

$x \cdot 25 = 20$ ◄— Let x represent the unknown part; replace *of* with a multiplication symbol.

$25x = 20$

$\frac{25x}{25} = \frac{20}{25}$ ◄— Divide both sides by 25 to isolate the variable x.

$x = \frac{20}{25}$

$x = 0.8$ ◄— Divide the numerator by the denominator.

$x = 80\%$ ◄— Move the decimal point 2 places to the right and include a percent sign.

20 is 80% of 25.

COMMON ERROR ALERT

As we have seen before, a decimal percent such as 2.6% has the decimal number equivalent 0.026. Do not confuse 2.6% with the decimal number 2.6.

USING EQUATIONS TO SOLVE WORD PROBLEMS 241

EXAMPLE B

The sum of three consecutive integers is 129. What are the integers?

Let n = first integer   ◄— Use a variable to represent the integers.
$n + 1$ = second integer
$n + 2$ = third integer

sum of the integers = 129   ◄— Write an equation using your variable to represent the relationship among quantities.
$n + (n + 1) + (n + 2) = 129$
$n + n + 1 + n + 2 = 129$ ◄— Solve the equation.
$3n + 3 = 129$
$3n = 126$
$n = 42$

$n = 42$ (first integer)   ◄— Answer the question in the problem.
$n + 1 = 43$ (second integer)
$n + 2 = 44$ (third integer)

The three consecutive integers are 42, 43, and 44.

EXAMPLE C

The sum of two consecutive odd integers is 76. What are the integers?

Helpful Hint

Two consecutive odd integers are always 2 apart. For example, 5 and 7 are consecutive odd integers. If $x = 5$, then $x + 2 = 7$.

Let n = first integer   ◄— Use a variable to represent the integers.
$n + 2$ = second integer

sum of the integers = 76   ◄— Write an equation using your variable to represent the relationship among quantities.
$n + (n + 2) = 76$
$n + n + 2 = 76$ ◄— Solve the equation.
$2n + 2 = 76$
$2n = 74$
$n = 37$

$n = 37$ (first integer)   ◄— Answer the question in the problem.
$n + 2 = 39$ (second integer)

The two consecutive odd integers are 37 and 39.

KEY IDEA

Did you notice that the sum of two consecutive odd integers is an even integer? This is always true! If you add three consecutive odd integers, is the sum odd or even? Can you show that your answer is correct?

Examples and questions highlight the application of mathematical concepts in the electrical field.

The examples that follow involve length, capacity, and weight (mass). Conversions of temperature, area, and volume will be addressed later in this chapter.

EXAMPLE A

A conduit decreases in length by 0.6 inch due to a temperature change. How many centimeters is this change in length, to the nearest tenth?

$0.6 \text{ in} = \frac{0.6 \text{ in.}}{1}$ ◄— Write the quantity as a fraction.

$0.6 \text{ in.} = \frac{0.6 \text{ in.}}{1} \times \frac{2.54 \text{ cm}}{1 \text{ in.}}$ ◄— Multiply by the conversion factor.

$0.6 \text{ in.} = \frac{0.6 \text{ in.}}{1} \times \frac{2.54 \text{ cm}}{1 \text{ in.}} \approx 1.5 \text{ cm}$ ◄— Cancel units.

The change in length is about 1.5 centimeters.

EXAMPLE B

Convert 120 miles to kilometers. Round to the nearest whole kilometer.

The Table of Customary/Metric Equivalents contains two facts about miles and kilometers. To convert miles to kilometers, use the fact that tells how many kilometers are in 1 mile, 1 mi = 1.61 km.

$120 \text{ mi} = \frac{120 \text{ mi}}{1}$ ◄— Write the quantity as a fraction.

$120 \text{ mi} = \frac{120 \text{ mi}}{1} \times \frac{1.61 \text{ km}}{1 \text{ mi}}$ ◄— Multiply by the conversion factor.

$120 \text{ mi} = \frac{120 \text{ mi}}{1} \times \frac{1.61 \text{ km}}{1 \text{ mi}} \approx 193 \text{ km}$ ◄— Cancel units.

A distance of 120 miles is about 193 kilometers.

CHAPTER 2 REVIEW EXERCISES

1. (a) What is the GCF of 45 and 33?
   (b) Write $\frac{33}{45}$ in simplest form.
2. (a) What is the LCM of 3, 5, and 12?
   (b) Write equivalent fractions for $\frac{1}{3}$, $\frac{2}{5}$, and $\frac{7}{12}$ using the LCM you found in (a).
3. Convert the mixed number $4\frac{2}{3}$ to an improper fraction.
4. Convert $\frac{27}{12}$ to a mixed number in simplest form.
5. What is the reciprocal of $\frac{15}{17}$?
6. What is the reciprocal of $3\frac{3}{7}$?
7. (a) Write the sum $\frac{1}{3} + \frac{2}{5} + \frac{7}{12}$ as an improper fraction.
   (b) Write the sum $\frac{1}{3} + \frac{2}{5} + \frac{7}{12}$ as a mixed number.
8. Perform the indicated operation. Simplify if possible.
   (a) $\frac{2}{7} + \frac{4}{7}$ (b) $\frac{5}{8} - \frac{3}{8}$
   (c) $\frac{2}{5} \times \frac{3}{4}$ (d) $\frac{4}{5} \div \frac{2}{3}$
   (e) $2\frac{1}{4} + 3\frac{2}{5}$ (f) $7\frac{1}{4} - 4\frac{2}{3}$
   (g) $6\frac{1}{2} \times 3\frac{1}{3}$ (h) $2\frac{2}{3} \div 5\frac{1}{3}$
9. Perform the indicated operation. Simplify if possible.
   (a) $\frac{5}{9} + \frac{1}{4}$ (b) $\frac{7}{8} - \frac{2}{3}$
   (c) $\frac{15}{8} \times \frac{4}{5}$ (d) $\frac{9}{8} \div \frac{3}{16}$
   (e) $2\frac{3}{8} + 5\frac{1}{2} + 4\frac{1}{3}$ (f) $17\frac{1}{8} - 9\frac{3}{4}$
   (g) $13\frac{1}{2} \times 2\frac{1}{6}$ (h) $7\frac{1}{3} \div 2\frac{5}{6}$
10. A heating unit produces 1 575 watts of power with an applied voltage of 115 V. The current in amperes flowing through this unit is $\frac{1575}{115}$. Write the current in simplest terms as (a) an improper fraction and (b) a mixed number.

Added end-of-chapter questions reinforce critical concepts and relate to worked-out examples in the chapter.

MATHEMATICS – The science of numbers and their operations, and the relationships between them, including space configurations, structure, and measurement. The systematic treatment of magnitude, relationships between figures and forms, and relations between quantities expressed symbolically. Simply put, it is the "how" and the "why" of computation with numbers.

Without mathematics, modern life would be next to impossible. In fact, it is hard to think of a part of our lives where mathematics is not used in one form or another. Therefore, a working knowledge and understanding of mathematical concepts is absolutely essential to function productively in today's society. Make no mistake: foundational math skills are a very necessary component of everyone's educational base. In many cases, personal achievement can be linked, to some extent, to mathematical ability.

This textbook has been developed by the National Joint Apprenticeship and Training Committee for the Electrical Industry (NJATC) with the aforementioned in mind. *Building a Foundation in Mathematics, Second Edition,* is a text intended to provide the fundamental principles that are involved with all of the mathematical operations presented. Every attempt has been made to present the material in a down-to-earth format and to provide practical applications of the various concepts in everyday situations. As you will see, the text begins with very basic concepts and works its way up, all the way to an introduction to Boolean algebra. It offers a way to build a confident working knowledge of each subject before progressing to the next step. This building-block approach will systematically help you build a solid foundation in mathematics and will provide the reference information that is needed for a lifetime.

As the electronic calculator has become a part of our lives, it has also become a part of mathematical instruction. For this reason, you will find numerous calculator computation examples throughout this text. Here again, every effort has been made to give step-by-step instructions on how to use the calculator properly, including what each keystroke (entry) should be and exactly what should appear in the display. All of these examples are based on the Texas Instruments TI-30X IIS scientific calculator. By following along with these examples, you will not only apply the mathematical principle(s) being presented, but will also become proficient in the use of the calculator. However, please do not make the mistake of thinking that there is no need to study mathematics, adopting the philosophy, "I'll just use my calculator." If you don't understand the mathematical principles that are involved and how to solve the problem "longhand," you will end up hopelessly lost and confused, instead of confident in your abilities. The calculator is a tool, and often a very useful tool, but it is not a replacement for your mind! Allow your calculator to *expand* your mind, not stifle it.

As the title suggests, *Building a Foundation in Mathematics* provides the means to develop a solid foundation in mathematics. From there, you can build your mathematical capabilities as high as you wish. Remember, the higher you go, the further you can see. Have fun with it, and enjoy the climb!

Should you decide on a career in the electrical industry, training provided by the International Brotherhood of Electrical Workers and the National Electrical Contractors Association (IBEW-NECA) is the most comprehensive the industry has to offer. If you are accepted into one of their local apprenticeship programs, you'll be trained for one of four career specialties: journeyman lineman, residential wireman, journeyman wireman, or telecommunications installer/technician. Most importantly, you'll be paid while you learn. To learn more visit http://www.njatc.org.

The efforts for continuous enhancement have produced the product you see before you: this technically precise, academically superior edition of *Building a Foundation in Mathematics*. Essential terms are presented, used, and thoroughly explained, and their relationship to the electrical field is emphasized. The use of mathematics in the electrical field is essential; here are the building blocks to prepare each student for the mathematical concepts necessary in their future electrical careers.

Building a Foundation in Mathematics

Writing and Rounding Whole Numbers

Adding Whole Numbers

Subtracting Whole Numbers

Multiplying Whole Numbers

Dividing Whole Numbers

Overview

All electricians need to know how to use mathematics. The basic operations of mathematics are addition, subtraction, multiplication, and division. One area in which an electrician would need to use addition, subtraction, multiplication, and division would be in ordering and inventorying supplies and materials for a job. These operations are based on the decimal system. This means that you need to understand the structure of the decimal system before doing the basic operations.

The development of the decimal system can be tracked back many centuries and is based on the fact that people have 10 fingers. Small numbers were counted by comparing the number of objects with the number of fingers. To count numbers larger than 10, an additional object, such as a pebble or stick, was used. As the number of pebbles or sticks increased, they were placed in groups of 10. Our present number system, the decimal system, is based on this practice of grouping by 10.

Whole Numbers

Objectives

After completing this chapter, you will be able to:

- Express the digit place value of whole numbers
- Write whole numbers in expanded form
- Round whole numbers to a designated place value
- Estimate answers
- Order, add, subtract, multiply, and divide whole numbers
- Solve problems using a combination of addition, subtraction, multiplication, and division of whole numbers
- Solve practical problems using addition, subtraction, multiplication, and division of whole numbers
- Turn your calculator on and off, enter and delete whole numbers in the calculator, and use a calculator to add, subtract, multiply, and divide whole numbers

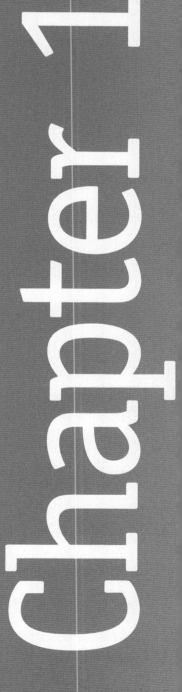

Chapter 1

WRITING AND ROUNDING WHOLE NUMBERS

Whole Numbers

The numbers 0, 1, 2, 3, 4, 5, . . ., make up the set of **whole numbers.** The whole numbers, along with the operations of addition, subtraction, multiplication, and division, form a mathematical system. This system is called the **decimal system,** from the Latin word, *decem,* meaning 10. The decimal system has only 10 symbols, the **digits** 0, 1, 2, 3, 4, 5, 6, 7, 8, and 9. The decimal system is said to have a base of 10.

To understand what a base-10 system is, consider how the odometer of a car functions.

The odometer on a new, undriven car reads: 0 0 0 0 0 0

After the first mile, the odometer reads: 0 0 0 0 0 1

After 9 miles, the odometer reads: 0 0 0 0 0 9

After 10 miles, the odometer reads: 0 0 0 0 1 0

At the tenth mile, the tens place has a one in it, and the ones place resets to 0.

Language Box

When regrouping, the word *carry* is sometimes used. The digit that is carried is written in the next column to the left.

The odometer illustration shows how we can use just 10 symbols to express any whole number. The symbols 0 through 9 are written in the ones place to express the single digit numbers 0 through 9. When we get to the whole number 10, the ones digit resets to 0 and the tens place to the left contains a 1.

We refer to this operation of reset and carry as *regrouping.* It is the basis of the decimal system.

EXAMPLE A

Show 1 mile more than 99 miles on an odometer.

After 99 miles, the odometer reads: 0 0 0 0 9 9

After 100 miles, the odometer reads: 0 0 0 1 0 0

To show 100, both the ones and the tens reset to zero, and 1 appears in the hundreds place.

Place Value

The value of any digit in a particular number is determined by the position it occupies in the number. Each position represents a power of 10. The place value chart below shows the value of each digit of the number 11,453,688. For example, the value of the digit 4 is four hundred thousand and the value of the digit 6 is six hundred.

Millions			Thousands			Ones		
hundred millions	ten millions	millions	hundred thousands	ten thousands	thousands	hundreds	tens	ones
100,000,000	10,000,000	1,000,000	100,000	10,000	1,000	100	10	1
	1	1	4	5	3	6	8	8

In **standard form,** we write four hundred thousand as 400,000 and six hundred as 600.

We can also express a number in **expanded form** to show the value of each digit of the number. In expanded form, 688 is written as $(6 \times 100) + (8 \times 10) + (8 \times 1)$.

In **word form,** 11,453,688 is written: *eleven million, four hundred fifty-three thousand, six hundred eighty-eight.* Insert commas after the words *million* and *thousand.* Use hyphens between all tens and ones.

EXAMPLE B

Write 47,902 in expanded form and word form.

Thousands			Ones		
hundred thousands	ten thousands	thousands	hundreds	tens	ones
100,000	10,000	1,000	100	10	1
	4	7	9	0	2

Expanded form: $(4 \times 10,000) + (7 \times 1,000) + (9 \times 100) + (2 \times 1)$

Notice how the expanded form shows each nonzero digit times its place value.

Word form: forty-seven thousand, nine hundred two

Notice the comma after *thousand* and the hyphen between *forty* and *seven.*

EXAMPLE C

Convert the number at the right from expanded form to both standard form and word form.

$(5 \times 100,000) + (7 \times 10,000) + (1 \times 10) + (4 \times 1)$

Helpful Hint

When you align numbers, be sure the ones digit of each number to be added is in the rightmost column.

STEP 1

Find the sum of the values.

```
   500,000
    70,000
        10
 +       4
   570,014
```

STEP 2

Write the number.

Standard form: 570,014

Word form: five hundred seventy thousand, fourteen

$(5 \times 100,000) + (7 \times 10,000) + (1 \times 10) + (4 \times 1) = 570,014.$
In word form, the number is five hundred seventy thousand, fourteen.

You may already be familiar with the decimal system and the various ways to represent numbers. Or, you may have studied this material years ago and just need a review to sharpen your skills for the studies that lie ahead. One cannot succeed without good basic math skills. The skills you develop will be used over and over throughout your career.

Rounding

The rounding of numbers depends on the accuracy that is desired. For example, to round 53 to the nearest ten, notice that 53 is closer to 50 than it is to 60. So 53 rounded to the nearest ten is 50.

What about a case where the number to be rounded lies exactly halfway between two numbers? Suppose 650 is to be rounded to the nearest hundred. Here, 650 is exactly halfway between 600 and 700. In such cases, round up. So, 650 rounded to the nearest hundred is 700.

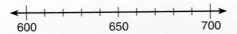

Different circumstances call for different levels of accuracy in rounding. For example, a weekly salary may be rounded to the nearest dollar (ones place), and a yearly salary to the nearest thousand dollars (thousands place).

    Mr. Smith's weekly salary: about $825

    Mr. Smith's yearly salary: about $43,000

The rules below summarize how to round a whole number.

Rules for Rounding		
Step 1	**Step 2**	**Step 3**
Identify the digit in the place you want to round to.	Look at the digit immediately to the right. If it is 5 or greater, increase the digit in the place being rounded by one (round up). Otherwise, keep that digit the same (round down).	Write zeros in the places to the right of the digit that was rounded.

EXAMPLE D

Round 2,485 to the nearest hundred using the Rules for Rounding.

STEP 1

The digit in the hundreds place is 4.

2 , 4 8 5

STEP 2

The digit immediately to the right of 4 is 8. Because 8 is greater than 5, we will round up. The 4 in the hundreds place will become 5.

2 , 4 8 5

↓

_ , 5 _ _

STEP 3

The digits in the ones and tens places will become zeros.

2 , 5 0 0

On the number line, 2,485 lies closer to 2,500 than to 2,400.

2,400   2,485   2,500

2,485 rounded to the nearest hundred is 2,500.

EXERCISES 1-1

Write the following numbers in expanded form and word form.

  1. 51,031   **2.** 1,707,925   **3.** 416,501

Write the following numbers in standard form and word form.

  4. $(7 \times 10{,}000) + (2 \times 100) + (8 \times 10) + (2 \times 1)$

  5. $(3 \times 100{,}000) + (6 \times 1{,}000) + (1 \times 10) + (8 \times 1)$

  6. $(4 \times 1{,}000{,}000) + (1 \times 10{,}000) + (5 \times 100) + (2 \times 10)$

Round each number to the indicated level of accuracy.

  7. 49,219 to the nearest ten

  8. 467,083 to the nearest thousand

  9. 51,250 to the nearest hundred

Answer the following.

  10. All the raffle tickets sold at a dance had the same digit in the thousands place. What digit is in the thousands place on the ticket shown?

COMMON ERROR

ALERT

To write a number in standard form, be sure to use the powers of 10 to write each digit in the correct place. Use zeros to "hold places" as needed.

11. A certain wiring job needs 17,843 feet of cable. How much cable is this to the nearest 100 feet?

12. After Ellie drives her car one more mile, the odometer will read 40,000 miles. What is the reading now?

13. Mike used the following amounts of cable during the last three months of the year: October: 9,467 feet; November: 8,749 feet; and December: 6,950 feet. Round the amount for each month to the nearest:

    (a) 10 feet

    (b) 100 feet

    (c) 1,000 feet

14. **Challenge:** The digits in a certain six-digit number decrease in value by 1 from left to right. The digit in the hundreds place is 6. What is the number?

ADDING WHOLE NUMBERS

The plus "+" symbol is the sign used for addition. The numbers being added are the **addends;** the result of addition is the **sum.**

Carla spent $3 for a can of tennis balls and $5 for a water bottle.

Find the total amount Carla spent.

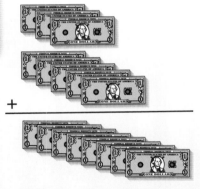

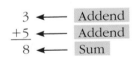

$$
\begin{array}{r}
3 \leftarrow \text{Addend} \\
+5 \leftarrow \text{Addend} \\
\hline
8 \leftarrow \text{Sum}
\end{array}
$$

It is possible to find the total amount Carla spent by counting dollars, but solving addition problems by counting is much too slow when the numbers are larger. Knowing addition facts such as 3 + 5 = 8 is essential.

Appendix A contains frequently used symbols and phrases in mathematics. Appendix B contains tables of basic addition facts. If you do not know the basic addition facts, you should commit them to memory as soon as possible.

Addition facts come in pairs. For example:

$$3 + 5 = 8 \qquad 5 + 3 = 8$$

When you know the answer for an addition fact, you also know the answer for the other addition fact in the pair.

To add one-digit numbers, place the numbers in a single column. For example, find the sum of 6, 3, 2, 7, and 9.

$$
\begin{array}{r}
6 \\
3 \\
2 \\
7 \\
+9 \\
\hline
27
\end{array}
$$

When one or more of the numbers being added has more than one digit, you *must* align the numbers so that all the ones digits are in a column, all the tens digits are in a column, all the hundreds digits are in a column, and so on.

Suppose we want to add 104 and 5. If we mistakenly place the 5 in the tens column, we arrive at an incorrect sum.

$$
\begin{array}{r}
104 \\
+\ 5\ \ \\
\hline
154
\end{array}
$$
← Incorrect placement
← Incorrect sum

$$
\begin{array}{r}
104 \\
+\ \ 5 \\
\hline
109
\end{array}
$$
← Correct placement
← Correct sum

The concept of regrouping was introduced earlier. Suppose we want to add 256 and 16. Place the digits in the correct columns. Then add, beginning in the ones column.

We cannot write a two-digit number in a column, so we regroup by writing the "carried" digit in the next column.

EXAMPLE A

Find the sum of 56 and 16.

STEP 1

Align the places, starting with the ones at the right. Add the ones.

$$
\begin{array}{r}
56 \\
+\ 16 \\
\hline
2
\end{array}
$$
— 6 + 6 = 12; write the 2.

STEP 2

Regroup 12 by writing 1 at the top of the tens column.

$$
\begin{array}{r}
1\ \ \\
56 \\
+\ 16 \\
\hline
2
\end{array}
$$

STEP 3

Add the tens, including the 1 that was carried.

$$
\begin{array}{r}
1\ \ \\
56 \\
+\ 16 \\
\hline
72
\end{array}
$$
← Add the 1 to other digits in the tens column.
← Sum

The sum of 56 and 16 is 72.

EXAMPLE B

Find the sum of 285 and 1,053.

Align the places and add. Regroup where necessary.

$$
\begin{array}{r}
{\scriptstyle 1} \\
285 \\
+\,1{,}053 \\
\hline
1{,}338
\end{array}
$$

The sum in the tens column is 13.

The sum of 285 and 1,053 is 1,338.

Introduction to the TI-30X IIS

It is to your advantage to practice using a calculator to check each example as you study this text. There are calculator icons inserted throughout the text that provide the steps necessary to solve the problem in the examples given. All steps are shown on the calculator icon along with the answer. The keystrokes necessary to begin using the TI-30X IIS are given after the next paragraph. Additional ways to use the calculator will be given throughout the text.

The calculator is a powerful tool, and you should learn to use it properly and efficiently. However, it is not intended to replace your thought processes—that is why it is so important that you understand and memorize mathematical processes. After all, you may or may not have a calculator with you at a given time. It is important to be able to compute either with or without a calculator.

Turning the Calculator On and Off

To turn the calculator on, press the [ON] key. The [ON] key is in the lower left corner of the calculator. You may see a black box, called the cursor, blinking on and off in the upper left corner of the screen or display. If you see something other than the cursor, press the [CLEAR] key.

To turn the calculator off, press the [2nd] key and then the [ON] key. The word *OFF* is printed on the calculator directly above the [ON] key.

Most keys have two different uses. You have already seen that the [ON] key can also be used as an "OFF" key by pressing the [2nd] [ON] keys. From now on, we will indicate a second key by showing it as a key with the keystrokes in brackets immediately after. Using this method, the "OFF" key would be written as OFF [[2nd] [ON]].

If you do not press any of the calculator keys for about 5 minutes, the calculator will turn off. To turn it back on, press [ON].

Basic Arithmetic on Your Calculator

The basic operations of addition, subtraction, multiplication, and division are performed using the following arithmetic keys:

and the digit keys

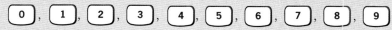

EXAMPLE 1

Use your calculator to add 47 + 139.

Solution

Press the following keys in this order:

You should see the following in the display:

This is a two-line display. The top line shows the problem you entered, and the bottom line shows the answer. So, from this display, you can see that 47 + 139 = 186.

From now on, we will not show the digits as a calculator key. Asking you to press 47 means that you should press the 4 and 7 keys in that order.

EXAMPLE 2

Use your calculator to calculate 357 − 54 + 28.

Solution

Press the following keys in this order. Notice that as soon as you press the 3, the old display disappears and the 3 is in the upper left corner of the display.

357 − 54 + 28 ENTER

You should see the following in the display:

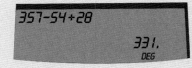

The result of 357 − 54 + 28 is 331.

Notice that you used the − key for subtraction. In the bottom row of the calculator is the (-) key. This key looks as if it could also be used for subtraction, but it cannot. We will use the (-) key in Chapter 4, when we study integers.

What happens if you press 357 [(-)] 54 instead of 357 [—] 54? Try it and see. After you press [ENTER =], you should see

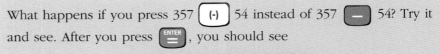

This indicates that you made some kind of error when you entered the problem into the calculator. Press the [CLEAR] key and your original entry will appear with the cursor blinking on the "-". This indicates that you pressed the [(-)] instead of [—]. Press [—] [ENTER =] and the answer will appear.

EXAMPLE 3

Use your calculator to multiply 268 × 75.

Solution

Press the following keys in this order.

268  75 [ENTER =]

You should see the following in the display:

Thus, 268 × 75 = 20,100.

Look at the top line of the display. Between the 268 and 75 is an *, which is the multiplication sign used on most calculators and computers.

Oops! I Made a Mistake

What happens if you make a mistake? You have three options: (1) erase everything and start over, (2) correct part of the problem, or (3) insert a missing number or operation.

Erase Everything   To erase everything in the display, press [CLEAR].

Correct Part of the Problem   Suppose that in Example 3, you pressed 268 [×] 74 [ENTER =] instead of 268 [×] 75 [ENTER =]. Rather than rekeying the entire problem, press the [◄] key. This will erase the second line of the display and put the cursor just to the right of the 4. Press [◄] again to move the cursor over the 4. Now, press 5 and [ENTER =]. The correct answer will be displayed.

What if you entered 268 [×] 95 in Example 3? To change the 9 to a 7, keep pressing the [◄] key until the cursor is over the 9. To change the 9 to a 7, just press 7 and then [ENTER =]. You do not have to move the cursor to the end of the line.

If your error was near the beginning of the problem, use the ▶ key until the cursor is over the error, press the correction, and ENTER ꞊ to get the correct answer.

What if you put too many digits in a number? For example, what if you keyed in 57742 instead of 5742? Use the ◀ to move the cursor over one of the 7s and press the DEL key. The DEL key will delete the character at the cursor.

Insert a Missing Number or Operation   If you forgot to enter part of the number, use the INS (for insert) key. It is over the DEL key, so to get INS press 2nd DEL . If you keyed in 3765 instead of 37 × 65, use the ◀ to move the cursor to the place where you want the missing character, in this problem at the 6, press INS [2nd DEL] then the correct character, × .

Reusing Answers

If you want to reuse the last answer, press ANS [2nd (-)].

Keyed	Display	Result
5 × 17 ENTER ꞊	5*17	85
ANS × 46 ENTER ꞊	Ans * 46	3910

The last answer, 3,910, is the product of 85 × 46.

As new topics are introduced in this book, we will look at other ways to use your calculator.

EXAMPLE C

Find the sum of 619, 297, and 3,142.

In this example, it is necessary to regroup three times. Follow the steps below to see how the sum in each column is recorded and regrouped.

STEP 1	STEP 2	STEP 3	STEP 4
1	1 1	1 1 1	1 1 1
619	619	619	619
297	297	297	297
+3,142	+3,142	+3,142	+3,142
8	58	058	4,058

STEP 1	STEP 2	STEP 3	STEP 4
The sum in the ones column is 18. Regroup 18 by writing 1 at the top of the tens column.	The sum in the tens column is 15. Regroup 15 by writing 1 at the top of the hundreds column.	The sum in the hundreds column is 10. Regroup 10 by writing 1 at the top of the thousands column.	The sum in the thousands column is 4.

The sum of 619, 297, and 3,142 is 4,058.

EXAMPLE D

Find the sum of 384, 291, 157, and 632.

In this example, you will see that the number that is "carried" can be greater than 1.

STEP 1	STEP 2	STEP 3	STEP 4

  1
384
291
157
+632
‾‾‾‾
   4
	
	  2 1
	384
	291
	157
	+632
	‾‾‾‾
	  64
		
		1 2 1
		384
		291
		157
		+ 632
		‾‾‾‾
		  464
			
			1 2 1
			384
			291
			157
			+ 632
			‾‾‾‾
			1,464

The sum in the ones column is 14; regroup 14 and carry 1.	The sum in the tens column is 26; regroup 26 and carry 2.	The sum in the hundreds column is 14; regroup 14 and carry 1.	The sum in the thousands column is 1.

The sum of 384, 291, 157, and 632 is 1,464.

EXAMPLE E

An electrician uses switch outlet boxes on seven different jobs. The number of boxes needed for each of the jobs were 12, 17, 22, 15, 27, 18, and 16. How many boxes were needed for all the jobs?

Align the numbers and add. Regroup as necessary.

STEP 1	STEP 2

  3
12
17
22
15
27
18
+ 16
‾‾‾
  7
	
	13
	12
	17
	22
	15
	27
	18
	+ 16
	‾‾‾
	127

The sum in the ones columns is 37. Regroup 37 and carry the 3.	The sum in the tens column is 12. Regroup 12 and carry the 1.

The seven jobs required 127 switch outlet boxes.

EXERCISES 1-2

Find each sum.

1. 28 + 7 =

2. 216 + 5 =

3. 95 + 19 =

4.   62
     +483

5.   1,067
      +3,503

6.   28,046
      +34,173

7.   52
        147
       +839

8.   1,013
          216
       +26,974

9.   65,250
         375,150
        +925,775

10. The following electrical items were ordered for a new house: 8 ceiling light fixtures, 27 wall outlets, 9 single-pole switches, 3 three-way wall switches, and one 220-volt wall outlet. If each item was packaged separately and delivered to the site, how many packages should the electrician find?

11. Mike is an electrician. He used the following amounts of cable during the first three months of the year: January: 8,320 feet; February: 7,650 feet; and March: 4,972 feet.

(a) Round the amount for each month to the nearest hundred feet and add to find the approximate amount of cable used during these three months.

(b) Find the actual total amount of cable used during this three-month period.

12. Mike used the following amounts of cable during the second quarter of the year: April: 9,630 feet; May: 10,286 feet; and June: 12,577 feet.

(a) Round the amount for each month to the nearest hundred feet and add to find the approximate amount of cable used during this quarter.

(b) Find the total amount of cable used during this quarter.

13. Based on the previous two exercises, what was the total amount of cable used by Mike during this first half of the year?

SUBTRACTING WHOLE NUMBERS

Subtraction is an operation that takes one number away from another number. Subtraction can answer three types of questions. They are:

- How many are left?
- What is the difference?
- How many more are needed?

A contractor had 12 posts. He used 8 of the posts for a project. How many posts does he have left?

Subtract 8 from 12 by counting.

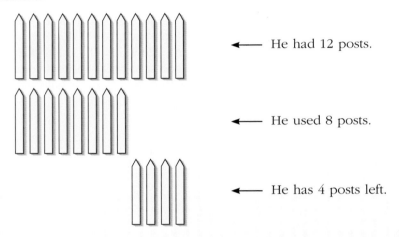

←—— He had 12 posts.

←—— He used 8 posts.

←—— He has 4 posts left.

Sometimes, subtraction can be performed by counting. In the figure above, the row of 12 represents the number of posts the contractor had to start with. The row of 8 shows how many posts he used. You can count the number of posts that are left by counting the posts in the last row.

The same diagram of fence posts can be used to answer the question *"What is the difference between the number of posts the contractor started with and the number he used?"* Also, the diagram can be used to answer the question *"If a contractor needed 12 posts and had 8 posts, how many more were needed?"*

Language Box

The number being subtracted is called the subtrahend. The number from which you are subtracting is called the minuend. The result is called the difference.

But subtraction, like addition, is slow when performed by counting. Subtraction facts should be learned so that recall is immediate. See Appendix B for a table of subtraction facts. Subtraction is the inverse operation of addition, and it can actually be thought of in terms of addition.

12 minus 8 = ?   *What number added to 8 equals 12?*

Obviously, the answer is 4.

The symbol for subtraction is the minus "−" sign.

$$
\begin{array}{r}
17 \\
-\ 7 \\
\hline
10
\end{array}
$$

17 ←—— Minuend
− 7 ←—— Subtrahend
10 ←—— Difference

Recall that we used regrouping in addition. We use another form of regrouping in subtraction. Within a column, if the digit in the subtrahend is greater than the digit in the minuend, you must regroup by borrowing from the next higher place in the minuend. This form of regrouping reduces that place value by 1 and adds ten to the column from which you are subtracting.

EXAMPLE A

Find the difference of 534 and 269.

STEP 1	STEP 2	STEP 3	STEP 4
$\begin{array}{r}53\vert4\\-269\\\hline\end{array}$	$\begin{array}{r}2\ 14\\5\cancel{3}4\\-269\\\hline 5\end{array}$	$\begin{array}{r}4\ 12\ 14\\\cancel{5}\cancel{3}4\\-269\\\hline 5\end{array}$	$\begin{array}{r}4\ 12\ 14\\\cancel{5}\cancel{3}4\\-269\\\hline 265\end{array}$
Notice that the 9 in the subtrahend is larger than the 4 in the minuend.	Regroup by borrowing 1 from the 3 in the tens place. Replace the 3 with a 2. Add 10 to the 4 in the ones place to make 14. Subtract in the ones column.	Regroup by borrowing 1 from 5 in the hundreds place. Replace 5 with a 4, add 10 to the 2 in the tens place.	Subtract in the tens column. Then, subtract in the hundreds column.

Notice that each time we borrow from the next higher place value, 10 must be added to the lower place value. This is because each place value is 10 times greater than the one to its right.

To check subtraction, add the difference to the subtrahend. If the sum is equal to the minuend, the subtraction is correct.

The difference of 534 and 269 is 265.

EXAMPLE B

Subtract 1,928 from 2,843. Check your answer.

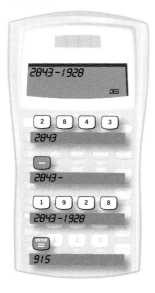

Solve.

$$\begin{array}{r}2,843 \quad\longleftarrow\ \text{Minuend}\\-1,928 \quad\longleftarrow\ \text{Subtrahend}\\\hline 915 \quad\longleftarrow\ \text{Difference}\end{array}$$

Check.

$$\begin{array}{r}\overset{1\quad\ 1}{1,928} \quad\longleftarrow\ \text{Subtrahend}\\+\ \ 915 \quad\longleftarrow\ \text{Difference}\\\hline 2,843 \quad\longleftarrow\ \text{Sum = Minuend}\end{array}$$

The difference between 2,843 and 1,928 is 915.

Special care must be taken when there are zeros in the minuend. In Example C that follows, 1 must be borrowed from the thousands place. To make each digit in the minuend greater than the digit below it, we regroup 1,000. This is done by thinking of 1,000 as 900 + 90 + 10. We can apply this concept in Example C by rewriting 6,008 as 5,000 + 900 + 90 + 18.

EXAMPLE C

An electrical supply store had 6,008 switch boxes in stock at the beginning of the week. During that week, 429 boxes were sold. How many were left in stock?

To solve this, you need to subtract 429 from 6,008. Check your answer.

Check.

$$\begin{array}{r} 5\ 9\,9\,18 \\ \cancel{6,008} \\ -\quad 429 \\ \hline \end{array}$$

$$\begin{array}{r} 5\ 9\,9\,18 \\ \cancel{6,008} \\ -\quad 429 \\ \hline 5,579 \end{array}$$

$$\begin{array}{r} 1\ 1\,1 \\ 5,579 \\ +\quad 429 \\ \hline 6,008 \end{array}$$

Regrouping is necessary because 9 in the subtrahend is greater than 8 in the minuend. You must go to the thousands place to borrow. Now the minuend is 5,000 + 900 + 90 + 18 and subtraction is possible.

Subtract in each column, using the new form of the minuend.

Add the difference and the subtrahend. The sum is equal to the minuend. The subtraction is correct.

The difference between 6,008 and 429 is 5,579.

EXERCISES 1-3

Subtract each of the following.

1. 54 − 28 =

2. 108 − 74 =

3. 200 − 125 =

4.
$$\begin{array}{r} 572 \\ -268 \\ \hline \end{array}$$

5.
$$\begin{array}{r} 963 \\ -781 \\ \hline \end{array}$$

6.
$$\begin{array}{r} 5,605 \\ -\quad 728 \\ \hline \end{array}$$

7.
$$\begin{array}{r} 4,832 \\ -1,320 \\ \hline \end{array}$$

8.
$$\begin{array}{r} 10,000 \\ -\ 5,037 \\ \hline \end{array}$$

9.
$$\begin{array}{r} 36,005 \\ -29,241 \\ \hline \end{array}$$

10. For a particular job, 3,500 feet of NM wire are purchased. On the first day, 783 feet were used.

    (a) Round off to the nearest hundred feet the wire used on the first day and subtract to determine the approximate number of feet of wire left.

    (b) Determine the actual number of feet of wire left.

11. At the beginning of the week, an electrical supply house has 853 solenoids in stock. During the week the following number of solenoids were sold: Monday: 57; Tuesday: 73; Wednesday: 64; Thursday: 49; and Friday: 62.

(a) Round off the numbers of solenoids sold to the nearest ten and add to determine the approximate the number of solenoids sold during the week.

(b) Round off the number of solenoids in stock at the beginning of the week to the nearest ten. Use this number and your answer to (a) to approximate the number of solenoids in stock at the end of the week.

(c) Determine the actual number of solenoids in stock at the end of the week.

12. A contractor is paid $1,147 for a certain job. Expenses for this job were $865. If the profit is the difference between the amount paid and the expenses, what was the profit for this job?

13. On the particular job, the expenses were $794 for material, $537 in electricians' labor, and $486 for taxes and insurance. The contractor is paid $1,974.

(a) What was the total of the expenses?

(b) What was the profit?

MULTIPLYING WHOLE NUMBERS

Multiplication is a way to express repeated addition.

To find the number of days in 5 weeks, you could add:

$7 + 7 + 7 + 7 + 7 = 35$

This addition statement can be written as the multiplication fact $5 \times 7 = 35$. See Appendix B for a table of multiplication facts.

Find the number of days in 5 weeks.

$5 \times 7 = 35$

Various symbols are used to indicate multiplication.

$4 \times 8 = 32$	$4 \cdot 8 = 32$	$(4)(8) = 32$
		$4(8) = 32$
The times "$\times$" sign between numbers indicates multiplication.	A raised dot between numbers indicates multiplication.	Enclosing one or several numbers in parentheses indicates multiplication.

$$
\begin{array}{l}
7 \leftarrow \text{Multiplicand} \leftarrow \text{Factor} \\
\underline{\times 5} \leftarrow \text{Multiplier} \leftarrow \text{Factor} \\
35 \leftarrow \text{Product}
\end{array}
$$

The number that is being multiplied is known as the **multiplicand.** The number by which you are multiplying is the **multiplier.** The multiplicand and the multiplier are both **factors.** The answer in multiplication is the **product.**

EXAMPLE A

Find the product of 276 and 4.

STEP 1

Multiply the ones digit by the multiplier.

$$\begin{array}{r} {\scriptstyle 2} \\ 276 \\ \times\ \ 4 \\ \hline 4 \end{array}$$

The product is 24. Regroup 24 and write the 2 above the tens digit, 7.

STEP 2

Multiply the tens digit by the multiplier.

$$\begin{array}{r} {\scriptstyle 3\,2} \\ 276 \\ \times\ \ 4 \\ \hline 04 \end{array}$$

The product is 28. Add the 2 tens that were carried. Regroup 30 and write the 3 above the hundreds digit, 2.

STEP 3

Multiply the hundreds digit by the multiplier.

$$\begin{array}{r} {\scriptstyle 1\,3\,2} \\ 276 \\ \times\ \ 4 \\ \hline 104 \end{array}$$

The product is 8. Add the 3 hundreds that were carried. Regroup 11 and write 1 above the thousands column.

STEP 4

There is no thousands digit.

$$\begin{array}{r} {\scriptstyle 1\,3\,2} \\ 276 \\ \times\ \ 4 \\ \hline 1{,}104 \end{array}$$

Write the 1 thousand that was carried in the thousands place in the product.

The product of 276 and 4 is 1,104.

Before exploring the multiplication process further, there are some special cases (properties) to look at involving multiplication by zero or by one.

 KEY IDEA

The product of 1 and any number is that number.

$1 \times 8 = 8 \qquad 8 \times 1 = 8 \qquad 1 \times n = n \qquad n \times 1 = n$

The product of zero and any number is zero.

$0 \times 5 = 0 \qquad 5 \times 0 = 0 \qquad 0 \times 0 = 0 \qquad 0 \times n = 0 \qquad n \times 0 = 0$

Here we have used an n to represent any number.

When the multiplier has two digits, such as in Example B that follows, the product of each of its digits and the multiplicand is a **partial product.**

EXAMPLE B

Find the product of 67 and 805.

Helpful Hint

Align each partial product under the digit you are multiplying by.

STEP 1

Multiply 805 by 7 in the multiplier to get the first partial product.

$$\begin{array}{r} {\scriptstyle 3} \\ 805 \\ \times\ \ 67 \\ \hline 5635 \end{array}$$

STEP 2

Multiply 805 by 6 in the multiplier to get the second partial product.

$$\begin{array}{r} {\scriptstyle 3} \\ {\scriptstyle 3} \\ 805 \\ \times\ \ 67 \\ \hline 5635 \\ 4830\ \ \end{array}$$

STEP 3

Add the partial products.

$$\begin{array}{r} {\scriptstyle 3} \\ {\scriptstyle 3} \\ 805 \\ \times\ \ 67 \\ \hline 5635 \\ +\,4830\ \ \\ \hline 53{,}935 \end{array}$$

The product of 67 and 805 is 53,935.

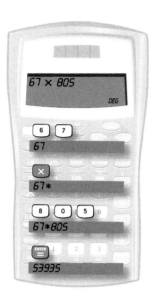

KEY IDEA

Remember to regroup whenever the product is greater than 9.

In Example C below, you must regroup twice for each digit of the multiplier. Notice how the digits being carried are written above the multiplicand in the work.

EXAMPLE C

An electrician needs 563 feet of conduit for each floor of a 49-story building. What is the total length of conduit needed?

STEP 1	STEP 2	STEP 3
Multiply 563 by 9 in the multiplier. Regroup whenever the product is greater than 9.	Multiply 563 by 4 in the multiplier. Align the partial product under the 4.	Add the partial products.

STEP 1:

$$\begin{array}{r} 5\,2 \\ 563 \\ \times\ \ 49 \\ \hline 5067 \end{array}$$

STEP 2:

$$\begin{array}{r} 2\,1 \\ 5\,2 \\ 563 \\ \times\ \ 49 \\ \hline 5067 \\ 2252\ \ \\ \hline \end{array}$$

STEP 3:

$$\begin{array}{r} 2\,1 \\ 5\,2 \\ 563 \\ \times\ \ 49 \\ \hline 5067 \\ +2252\ \ \\ \hline 27{,}587 \end{array}$$

The total length will be the product of 563 and 49. The product of 49 and 563 is 27,587. So, 27,587 feet of conduit will be needed for this building.

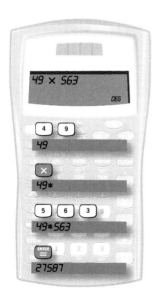

EXERCISES 1-4

Multiply each of the following.

1. 50 · 3 =   **2.** 18 · 8 =   **3.** 200 · 13 =

4. (7)(36) =   **5.** (50)(42) =   **6.** (47)(9) =

7.   45   **8.**   87   **9.**   936
        ×24   ×26   × 32

10.   453   **11.**   764   **12.**   4,285
         ×209   ×128   ×   16

13. A very small electromagnet is wound with 84 layers. Each layer requires 237 turns.

    (a) Round each number to the nearest ten and approximate the total number of turns on the electromagnet.

    (b) What are the actual number of turns on the electromagnet?

14. In a new subdivision, 1,547 feet of NM wire are needed for each house. The subdivision will have 43 houses.

    (a) Round each number to the nearest ten and approximate the total length of wire needed for the subivision.

    (b) Determine the actual number of feet of wire needed.

15. Each of the 43 homes in a new subdivision needs 8 outlets per room for 6 of the rooms, 2 outlets for each of the 2 bathrooms, and 1 outlet for the half-bath. How many outlets are needed for the entire subdivision?

16. An electrician is paid $14 an hour. How much is earned in a 37-hour week?

DIVIDING WHOLE NUMBERS

Division is a process used to separate a group into equal parts. For example, suppose there are eight quarters to be divided equally between two friends. The group of 8 quarters needs to be divided into equal parts.

This could be done by removing one quarter at a time from the group, forming two equal groups, and counting the number in each equal group. However, this problem can be answered with a math fact. See Appendix B for a table of division facts.

Divide 8 by 2.

8 ÷ 2 = 4

Division is the inverse operation of multiplication, and it can actually be thought of in terms of multiplication.

21 ÷ 3 = ?   *What number times 3 equals 21?*

Because 7 × 3 = 21, 21 ÷ 3 = 7.

Language Box

The number being divided is called the **dividend.** The number by which you are dividing is called the **divisor.** The answer is called the **quotient.** The result of the last subtraction is called the **remainder.**

Various symbols are used to indicate division. All of the symbols below indicate that 21 is divided into 3 groups of 7 each.

$21 \div 3 = 7$

The "÷" symbol is used as a sign for division.

$3\overline{)21}$ with 7 on top

The symbol $\overline{)}$ is used for division.

$\dfrac{21}{3} = 7$

A fraction bar indicates division.

$21/3 = 7$

A slash can also indicate division.

$$
\begin{array}{r}
8\ \text{R1} \quad \leftarrow \text{Quotient} \\
\text{Divisor} \rightarrow 4\overline{)33}\quad \leftarrow \text{Dividend} \\
\underline{-\ 32} \\
1 \quad \leftarrow \text{Remainder}
\end{array}
$$

Remember that the long division process includes the steps: divide, multiply, subtract, and bring down. Notice how this cycle of steps is carried out in Example A below.

EXAMPLE A

Find the quotient of 93 and 4.

STEP 1

$$
\begin{array}{r}
2 \\
4\overline{)93} \\
\underline{-8} \\
1
\end{array}
$$

There are 2 groups of 4 in 9. Write 2 in the quotient, multiply 2 × 4, and subtract.

STEP 2

$$
\begin{array}{r}
23 \\
4\overline{)93} \\
\underline{-8}\downarrow \\
13 \\
\underline{-12} \\
1
\end{array}
$$

Bring down the 3. There are 3 groups of 4 in 13. Write 3 in the quotient, multiply 3 × 4, and subtract. The remainder is 1.

The quotient of 93 divided by 4 is 23 R1.

Before exploring the division process further, there are some special cases (properties) to look at involving division by zero or by one.

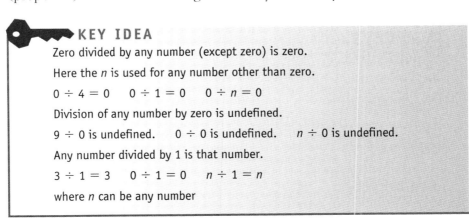

KEY IDEA

Zero divided by any number (except zero) is zero.

Here the *n* is used for any number other than zero.

$0 \div 4 = 0 \qquad 0 \div 1 = 0 \qquad 0 \div n = 0$

Division of any number by zero is undefined.

$9 \div 0$ is undefined. $0 \div 0$ is undefined. $n \div 0$ is undefined.

Any number divided by 1 is that number.

$3 \div 1 = 3 \qquad 0 \div 1 = 0 \qquad n \div 1 = n$

where *n* can be any number

You may want to try some of the problems involving division by 0 or 1 on your calculator. Remember, division by zero is not defined. If you try to divide by zero on your calculator, it will display "DIVIDE BY 0 Error" every time.

Example B shows the importance of placing the digits in the quotient in the proper place. Notice how zero in the dividend is brought down just as any other digit would be.

EXAMPLE B

Find the quotient of 960 and 32.

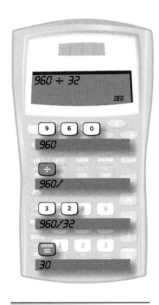

STEP 1

$$\begin{array}{r} 3 \\ 32\overline{)960} \\ -96 \\ \hline 0 \end{array}$$

There are 3 groups of 32 in 96. Write 3 in the quotient over the 6, multiply 3×32, and subtract.

STEP 2

$$\begin{array}{r} 30 \\ 32\overline{)960} \\ -96\downarrow \\ \hline 00 \\ -00 \\ \hline 0 \end{array}$$

Bring down the 0. There are no groups of 32 in 00. Write 0 in the quotient, multiply 0×32, and subtract.

The quotient of 960 and 32 is 30.

Every time you subtract in a long division problem, the remainder must be less than the divisor. If it is not, the digit placed in the quotient was too small. Note the final remainder in Example C below.

EXAMPLE C

Find the quotient of 127 and 8, including any remainder.

STEP 1

$$\begin{array}{r} 1 \\ 8\overline{)127} \\ -8 \\ \hline 4 \end{array}$$

There is 1 group of 8 in 12. Write 1 in the quotient above the 2, multiply 1×8, and subtract.

STEP 2

$$\begin{array}{r} 15 \\ 8\overline{)127} \\ -8\downarrow \\ \hline 47 \\ -40 \\ \hline 7 \end{array}$$

Bring down the 7. There are 5 groups of 8 in 47. Write 5 in the quotient, multiply 5×8, and subtract.

The remainder is 7. Write the quotient as 15 R7 or $15\frac{7}{8}$. The decimal equivalent of $\frac{7}{8}$ shown on a calculator is 0.875, so the quotient can be written 15.875.

The quotient of 127 and 8 is 15 R7.

Correct placement of the first digit in the quotient is critical. In Example D there are no groups of 63 in 44, but there are 7 groups of 63 in 442 (7 × 63 = 441). Note that the placement of the first digit 7 in the quotient is over 2 in the hundreds place.

EXAMPLE D

Find the quotient of 44,238 and 63.

STEP 1

$$\begin{array}{r} 7 \\ 63\overline{)44238} \\ -441 \\ \hline 1 \end{array}$$

Write 7 in the quotient above the 2, multiply 7 × 63, and subtract.

STEP 2

$$\begin{array}{r} 70 \\ 63\overline{)44238} \\ -441 \\ \hline 13 \\ -\ 0 \\ \hline 13 \end{array}$$

There are 0 groups of 63 in 13. Be sure to write 0 above the 3 in the quotient to "hold the place." Multiply 0 × 63, and subtract.

STEP 3

$$\begin{array}{r} 702\ \text{R12} \\ 63\overline{)44238} \\ -441 \\ \hline 13 \\ -\ 0 \\ \hline 138 \\ -126 \\ \hline 12 \end{array}$$

Bring down the 8. There are 2 groups of 63 in 138. Multiply 2 × 63, and subtract. The remainder is 12.

The quotient of 44,238 and 63 is 702 R12.

EXAMPLE E

A wiring job requires 37 outlets. They are to be spaced equally over 432 feet, with the center of one outlet at the beginning of the 432 feet and the center of another at the very end. Determine the center-to-center distance between the outlets.

One way to solve a hard problem is to first look at an easier problem.

Suppose that the distance was 72 feet and there were only five outlets, with one at the beginning of the 72 foot distance and one at the end. The figure below shows this situation. Notice that the number of spaces between outlets is 4, one less than the number of outlets.

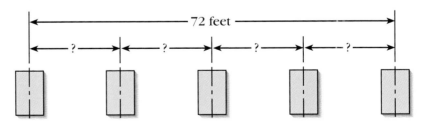

Now consider the original problem. There are 37 outlets but only 36 spaces, so the center-to-center distance between outlets is 432 feet ÷ 36 outlets. Dividing, we get

$$\begin{array}{r} 12 \\ 36\overline{)432} \\ 36 \\ \hline 72 \\ 72 \\ \hline 0 \end{array}$$

The center-to-center distance between two outlets is 12 feet.

EXERCISES 1-5

Find each of the following quotients.

1. $84 \div 7 =$ **2.** $99 \div 11 =$ **3.** $78 \div 5 =$

4. $21\overline{)420}$ **5.** $15\overline{)525}$ **6.** $25\overline{)750}$

7. $614 \div 18 =$ **8.** $947 \div 23 =$ **9.** $2{,}890 \div 24 =$

10. $36\overline{)10260}$ **11.** $29\overline{)101239}$ **12.** $637\overline{)108290}$

13. A total load of 170,478 watts is distributed evenly among 123 circuits. You need to determine the load, in watts, per circuit.

    (a) What is the divisor?

    (b) What is the dividend?

    (c) Round each number to the nearest hundred and approximate the load per circuit in watts.

    (d) What is the actual load per circuit in watts?

14. During a two-week period, an electrician works 76 hours and has a gross pay of $1,216. You need to determine the electrician's hourly pay.

    (a) What is the divisor?

    (b) What is the dividend?

    (c) What was the electrician's actual hourly pay?

15. An electrician's gross pay for a 38-hour week is $1,178. Deductions for taxes, Social Security, Medicare, and insurance amount to $456. You need to determine the electrician's hourly take-home pay.

    (a) What is the electrician's net pay (gross pay minus deductions) for that week?

    (b) What was the electrician's actual hourly take-home pay?

16. An 18-story apartment building requires 12,312 outlets.

    (a) If each floor requires the same number of outlets, how many outlets are needed for each floor?

    (b) Each floor has 12 identical apartments. How many outlets are needed for each apartment?

CHAPTER 1 REVIEW EXERCISES

1. Write 72,356 in both expanded form and word form.

2. Round 942,745 to the nearest thousand.

In Exercises 3–10, perform the indicated operation.

3. $47 + 58 =$
4.
$$
\begin{array}{r}
21{,}362 \\
791 \\
356{,}925 \\
+ \quad 5{,}471 \\
\hline
\end{array}
$$
     5. $183 - 75 =$
6.
$$
\begin{array}{r}
23{,}745 \\
- \quad 6{,}943 \\
\hline
\end{array}
$$

7. $14 \times 37 =$
8.
$$
\begin{array}{r}
7{,}903 \\
\times \quad 57 \\
\hline
\end{array}
$$
     9. $98 \div 7 =$ **10.** $34\overline{)21768}$

11. Resistance in an electric circuit is measured in ohms (Ω). The total resistance in a series circuit is the sum of all the resistances in that circuit. What is the total resistance in a series circuit that has resistances of 20 Ω, 24 Ω, and 35 Ω?

12. The total resistance in a series circuit with two resistors is 63 Ω. If one of the resistors has a resistance of 37 Ω, what is the resistance in the other resistor?

13. A coil of wire is wound in 16 layers with 17 turns per layer. What is the total number of turns on the coil?

14. A 192-foot length of Romex cable needs to be stapled. The staples are evenly spaced, with a staple at the beginning and another at the end of the cable. If 49 staples are used, how far apart should the staples be placed?

15. A certain wiring job requires 17,842 feet of Romex cable.

(a) Round this amount to the nearest thousand feet.

(b) Romex cable is packaged in 250-foot rolls. How many rolls should be ordered for this job?

16. When a new job is begun, there are 35 rolls of Romex cable in stock. Each roll contains 250 feet of cable. The first day, the electrician uses 1,347 feet of cable. How much unused cable is left in stock?

17. During a one-week period, an electrician works 37 hours and has a gross pay of $555. You need to determine the electrician's hourly pay.

(a) What is the divisor?

(b) What is the dividend?

(c) What was the electrician's actual hourly pay?

18. An electrical contractor uses the following amounts of cable during the third quarter of the year: July: 12,789 feet; August: 11,951 feet; and September: 7,472 feet.

(a) Round the amount for each month to the nearest hundred feet and add to find the approximate amount of cable used during this quarter.

(b) Find the total amount of cable used during this quarter.

19. An electrical contractor employs 23 people. Six of them earn $14 per hour, 12 people earn $15 per hour, and the rest earn $18 per hour.

(a) What is the total hourly wage for all 23 people?

(b) If each employee works 38 hours a week, what are the total wages in one week?

(c) The contractor must pay an additional $1,029 to the federal government to cover Social Security and Medicare for these employees. How much did the employer have to pay this week to the employees and government?

20. Current is measured in amperes (A). The total current in a parallel circuit containing several resistors is the sum of the current in each of the resistors.

(a) A parallel circuit has four resistors with currents of 12 A, 29 A, 45 A, and 27 A. What is the total current in this circuit?

(b) A parallel circuit has a total current of 37 A. If the circuit has three resistors and the current in two of them is 11 A and 17 A, what is the current in the third resistor?

Building a Foundation in Mathematics

Writing and Ordering Fractions

Converting between Improper
Fractions and Mixed Numbers

Adding Fractions

Subtracting Fractions

Multiplying Fractions and Mixed
Numbers

Dividing Fractions and Mixed
Numbers

Reading a Customary Rule

Converting between Customary
Units of Length

Overview

Most measurements and calculations done by an electrician are not limited to whole numbers. Electricians need to use arithmetic operations that use fractions—fractions of an inch, fractions of a foot, fractions of ohms, fractions of volts, and so on. They must also be able to measure accurately and convert the measurement to different units. For example, electricians use fractions when measuring and installing conduit.

Fractions

Objectives

After completing this chapter, you will be able to:

- Determine greatest common factors
- Determine lowest common multiples and denominators
- Express fractions as equivalent fractions
- Express fractions in lowest terms
- Express mixed numbers as improper fractions and improper fractions as mixed numbers
- Add, subtract, multiply, and divide fractions, mixed numbers, and whole numbers
- Solve problems using a combination of addition, subtraction, multiplication, and division of fractions, mixed numbers, and whole numbers
- Measure objects using a customary ruler or tape measure
- Convert feet and inches to yards, feet, and inches
- Convert a given customary length to inches

Chapter 2

WRITING AND ORDERING FRACTIONS

Equivalent Fractions

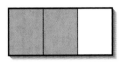

A fraction is often used to name a part of a whole. It is possible to represent a fraction with a diagram. In the diagram shown, 2 of 3 rectangles are shaded. The fraction $\frac{2}{3}$ has a **numerator** of 2 and a **denominator** of 3. The numerator and denominator can be referred to as the terms of a fraction.

$$\text{numerator} \longrightarrow \frac{2}{3} \longleftarrow$$
$$\text{denominator} \longrightarrow 3$$

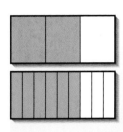

The fractions $\frac{2}{3}$ and $\frac{6}{9}$ are **equivalent fractions** because they represent the same part of a whole. To create an equivalent fraction in higher terms, multiply both the numerator and denominator by the same whole number greater than 1. To create an equivalent fraction in lower terms, divide both the numerator and the denominator by the same whole number greater than 1.

$$\frac{2}{3} = \frac{2 \times 3}{3 \times 3} = \frac{6}{9}$$

$$\frac{6}{9} = \frac{6 \div 3}{9 \div 3} = \frac{2}{3}$$

So, $\frac{2}{3} = \frac{6}{9}$.

EXAMPLE A

Change $\frac{6}{8}$ to an equivalent fraction in higher terms with a denominator of 16.

$$\frac{6}{8} = \frac{?}{16}$$

Multiply the numerator and denominator by 2.

$$\frac{6}{8} = \frac{6 \times 2}{8 \times 2} = \frac{12}{16}$$

$$\frac{6}{8} = \frac{12}{16}$$

EXAMPLE B

Change $\frac{6}{8}$ to an equivalent fraction in lower terms with a denominator of 4.

$$\frac{6}{8} = \frac{?}{4}$$

Divide the numerator and denominator by 2.

$$\frac{6}{8} = \frac{6 \div 2}{8 \div 2} = \frac{3}{4}$$

$$\frac{6}{8} = \frac{3}{4}$$

EXAMPLE C

Write two fractions that are equivalent to $\frac{7}{14}$. Write one in higher terms and one in lower terms.

Higher terms:

You can use any whole number greater than 1 to multiply the numerator and the denominator. For example, use 10.

$$\frac{7}{14} = \frac{7 \times 10}{14 \times 10} = \frac{70}{140}$$

Both $\frac{70}{140}$ and $\frac{1}{2}$ are equivalent to $\frac{7}{14}$.

Lower terms:

Divide both numerator and denominator by the same number. Only 7 and 1 divide both of them evenly. Use 7 since it is greater than 1.

$$\frac{7}{14} = \frac{7 \div 7}{14 \div 7} = \frac{1}{2}$$

Lowest Terms

Free Throw Record

Free Throws Attempted	Free Throws Made
35	21

To write a fraction in lowest terms, also known as **simplest form**, divide the numerator and denominator by their **greatest common factor (GCF)**. A **factor** divides evenly into a number, and there is no remainder. To find the GCF, list the factors of the numerator and the denominator. Choose the greatest factor that appears in both lists. Writing fractions in simplest form is sometimes called reducing to lowest terms or simplifying.

A basketball player attempted 35 free throws and made 21 of them. The player made $\frac{21}{35}$ of the free throws attempted.

Write $\frac{21}{35}$ in simplest form.

Helpful Hint

If the greatest common factor of the numerator and the denominator is 1, the fraction is already in simplest form. For example, $\frac{2}{9}$, $\frac{4}{7}$, and $\frac{1}{6}$ are all in simplest form.

STEP 1

Find the GCF of the numerator and denominator.

Factors of 21: 1, 3, ⑦ 21

Factors of 35: 1, 5, ⑦ 35

The GCF of 21 and 35 is 7.

STEP 2

Divide the numerator and denominator by the GCF.

$$\frac{21}{35} = \frac{21 \div 7}{35 \div 7} = \frac{3}{5}$$

In simplest form, $\frac{21}{35}$ is $\frac{3}{5}$.

EXAMPLE D

In a certain circuit the resistance is 24 Ω (ohms) and the voltage is 30 V (volts). According to Ohm's law, the current, in amperes (A), is $\frac{24}{30}$. Simplify this fraction.

STEP 1

Find the GCF of 24 and 30.

Factors of 24: ①, ②, ③, 4, ⑥, 8, 12, 24

Factors of 30: ①, ②, ③, 5, ⑥, 10, 15, 30

The common factors of 24 and 30 are 1, 2, 3, and 6. Since 6 is the greatest of these, 6 is the GCF of 24 and 30.

STEP 2

Divide the numerator and denominator by the GCF.

$$\frac{24}{30} = \frac{24 \div 6}{30 \div 6} = \frac{4}{5}$$

The simplest form of $\frac{24}{30}$ is $\frac{4}{5}$. Thus, the current, when written in simplest form, is $\frac{4}{5}$ A.

Ordering Fractions

To order a set of numbers, write them in a list from *least to greatest* or *greatest to least*. To order fractions, start by writing equivalent fractions with the same denominator.

Use the **least common multiple (LCM)** of the original denominators as the denominator of the equivalent fractions. The LCM of two numbers is the least whole number that is a multiple of both numbers. Multiply each number by 1, 2, 3, 4, . . . , to obtain a list of multiples for each number.

The LCM of the original denominators is also called the *least common denominator* (LCD).

During a football game, a quarterback completed 3 of 4 passes. His passing record for the game can be expressed as the fraction $\frac{3}{4}$. The backup quarterback completed 4 of 5 passes, indicated by the fraction $\frac{4}{5}$. Which player had the better passing record for that game?

Is $\frac{3}{4}$ greater than, less than, or equal to $\frac{4}{5}$?

STEP 1

Find the LCM of the denominators.

Multiples of 4: 4, 8, 12, 16, ⑳, 24, . . .

Multiples of 5: 5, 10, 15, ⑳, 25, . . .

The LCM of 4 and 5 is 20.

Language Box

Inequality Symbols
$<$ means *is less than*.
$>$ means *is greater than*.

STEP 2

Write equivalent fractions with denominators equal to the LCM.

$$\frac{3}{4} = \frac{3 \times 5}{4 \times 5} = \frac{15}{20}$$

$$\frac{4}{5} = \frac{4 \times 4}{5 \times 4} = \frac{16}{20}$$

Compare $\frac{15}{20}$ and $\frac{16}{20}$. Relate these fractions to the original fractions.

$$\frac{15}{20} < \frac{16}{20}$$
$$\downarrow \qquad \downarrow$$
$$\frac{3}{4} < \frac{4}{5}$$

The backup quarterback, who completed 4 of 5 passes, had the better passing record for that game because $\frac{4}{5} > \frac{3}{4}$.

EXAMPLE E

Order $\frac{3}{5}$, $\frac{4}{9}$, and $\frac{8}{15}$ from least to greatest.

STEP 1

Find the LCM of the denominators.

Multiples of 5: 5, 10, 15, 20, 25, 30, 35, 40, ㊺, . . .

Multiples of 9: 9, 18, 27, 36, ㊺, . . .

Multiples of 15: 15, 30, ㊺, . . .

The LCM of 5, 9, and 15 is 45.

STEP 2

Write equivalent fractions with denominators equal to the LCM.

$$\frac{3}{5} = \frac{27}{45} \qquad \frac{4}{9} = \frac{20}{45} \qquad \frac{8}{15} = \frac{24}{45}$$

Use the numerators of the equivalent fractions to compare and order the fractions.

$$\frac{20}{45} < \frac{24}{45} < \frac{27}{45}$$
$$\downarrow \qquad \downarrow \qquad \downarrow$$
$$\frac{4}{9} < \frac{8}{15} < \frac{3}{5}$$

The fractions in order from *least to greatest* are $\frac{4}{9}$, $\frac{8}{15}$, $\frac{3}{5}$.

To write the fractions in order from *greatest* to *least,* write $\frac{3}{5}$, $\frac{8}{15}$, $\frac{4}{9}$.

EXERCISES 2-1

Find an equivalent fraction with the given denominator.

1. $\frac{2}{6} = \frac{}{18}$

2. $\frac{3}{7} = \frac{}{56}$

3. $\frac{17}{34} = \frac{}{2}$

4. $\frac{10}{11} = \frac{}{33}$

5. $\frac{12}{48} = \frac{}{24}$

6. $\frac{5}{45} = \frac{}{9}$

Simplify if possible. If not possible, write *already in simplest form*.

7. $\frac{8}{10} =$

8. $\frac{9}{16} =$

9. $\frac{40}{88} =$

10. $\frac{100}{150} =$

11. $\frac{25}{65} =$

12. $\frac{8}{21} =$

Solve.

13. When a doorbell rings, it uses a voltage of 6 V and has a resistance of 28 Ω. If the current, in amperes (A), is $\frac{6}{28}$, what is the current in lowest terms?

14. Carolyn and Damon are reading the same book. Carolyn has read $\frac{3}{8}$ of the book. Damon has read $\frac{2}{5}$ of the book. Who has read more of the book?

15. Trent works 8 hours of a 12-hour shift. José works $\frac{3}{4}$ of the shift, and Lisa works $\frac{1}{2}$ of the shift. List the workers in order from the least to the greatest fraction of a shift worked.

16. A certain circuit has a resistance of 64 Ω and a voltage of 20 V. The current is $\frac{20}{64}$ A. What is this in lowest terms?

17. What's the Error? Your coworker says that $\frac{5}{16}$ is greater than $\frac{5}{8}$. Draw and shade a diagram to show why this statement is incorrect.

CONVERTING BETWEEN IMPROPER FRACTIONS AND MIXED NUMBERS

Improper Fractions to Mixed Numbers

A **proper fraction** is a fraction in which the numerator is less than the denominator, such as $\frac{2}{7}$, $\frac{1}{3}$, or $\frac{8}{10}$. An **improper fraction** is a fraction in which the numerator is greater than or equal to the denominator, such as $\frac{15}{7}$, $\frac{5}{1}$, or $\frac{3}{3}$. An improper fraction has a value greater than or equal to 1. A **mixed number,** such as $12\frac{3}{4}$, is a whole number together with a fraction.

To convert an improper fraction to a mixed number, divide the numerator by the denominator. The whole number in the quotient becomes the whole number part in the mixed number. Any remainder becomes the numerator of the fraction part of the mixed number. Use the original denominator in the fraction.

During his workout, Joe runs 11 quarter-mile sections of the track. That distance can be expressed by the improper fraction $\frac{11}{4}$. How many miles does Joe run?

Helpful Hint

If the numerator equals the denominator, the fraction is equal to 1.

$\frac{3}{3} = 3 \div 3 = 1$

If the denominator equals 1, the fraction is equal to the numerator.

$\frac{5}{1} = 5 \div 1 = 5$

Convert $\frac{11}{4}$ to a mixed number.

Divide the numerator by the denominator.

$$\text{Denominator of fraction} \longrightarrow 4\overline{)11} \begin{array}{l} 2 \leftarrow \text{Whole number} \\ \leftarrow \text{Numerator of fraction} \\ \underline{-\ 8} \\ 3 \leftarrow \text{Remainder} \\ (\text{new numerator}) \end{array}$$

$$\frac{11}{4} = 2 + \frac{3}{4} = 2\frac{3}{4}$$

Joe runs $2\frac{3}{4}$ miles.

EXAMPLE A

Convert $\frac{15}{7}$ to a mixed number.

Divide the numerator by the denominator.

$$\text{Denominator of fraction} \longrightarrow 7\overline{)15} \begin{array}{l} 2 \leftarrow \text{Whole number} \\ \leftarrow \text{Numerator of fraction} \\ \underline{-\ 14} \\ 1 \leftarrow \text{Remainder} \\ (\text{new numerator}) \end{array}$$

$$\frac{15}{7} = 2\frac{1}{7}$$

EXAMPLE B

The total resistance, in ohms (Ω), of a 6 Ω and an 8 Ω resistor connected in parallel is $\frac{6 \times 8}{6 + 8}$. Find the total resistance of this circuit and write the answer as a mixed number in simplest form.

The total resistance is $\frac{6 \times 8}{6 + 8} = \frac{48}{14}$ Ω.

To write this improper fraction as a mixed number, divide the numerator by the denominator.

$$\begin{array}{r} 3 \\ 14\overline{)48} \\ \underline{42} \\ 6 \end{array}$$

$$\frac{48}{14} = 3\frac{6}{14}$$

The fraction $\frac{6}{14}$ can be simplified to $\frac{3}{7}$.

The total resistance is $3\frac{3}{7}$ Ω.

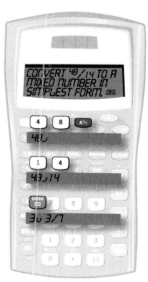

Mixed Numbers to Improper Fractions

To convert a mixed number to an improper fraction, multiply the whole number by the denominator of the fraction part of the mixed number. Add the product to the numerator of the fraction part of the mixed number. This equals the numerator of the improper fraction. Retain the denominator from the fraction part of the mixed number.

Jerry is making dinner for some friends. He has $5\frac{1}{2}$ pounds of hamburger. How many half-pound hamburgers can he make?

Convert $5\frac{1}{2}$ to an improper fraction.

Find the numerator. $\longrightarrow$ $2 \times 5 + 1 = 11$

Write the improper fraction. $\longrightarrow$ $5\frac{1}{2} = \frac{11}{2}$

The mixed number $5\frac{1}{2}$ is equal to 11 halves.

Jerry can make 11 half-pound hamburgers.

Using a Calculator for Fractions

The ⒜ᵇ∕꜀ key is used to enter a fraction into your calculator.

To enter the mixed number $5\frac{2}{7}$, press 5 ⒜ᵇ∕꜀ 2 ⒜ᵇ∕꜀ 7 ⒠ⁿᵗᵉʳ. You should see

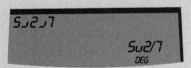

The *u* in the display is used to separate the whole number from the fraction.

To enter $\frac{5}{12}$, press 5 ⒜ᵇ∕꜀ 12 ⒠ⁿᵗᵉʳ; you should see 5/12 on the second line of the display.

An improper fraction is entered in the calculator in the same way. Enter the fraction $\frac{9}{6}$ by pressing 9 ⒜ᵇ∕꜀ 6 ⒠ⁿᵗᵉʳ. The display is 1 *u* 1/2. Notice that the answer is shown as a mixed number in lowest terms.

Any fraction entered in the calculator will be given in lowest terms. For example, enter $\frac{8}{36}$ in your calculator. After you press ⒠ⁿᵗᵉʳ the displayed result is 2/9.

EXAMPLE C

Convert $4\frac{3}{7}$ to an improper fraction.

Find the numerator. ⟶ $7 \times 4 + 3 = 31$

Write the improper fraction. ⟶ $4\frac{3}{7} = \frac{31}{7}$

$4\frac{3}{7} = \frac{31}{7}$

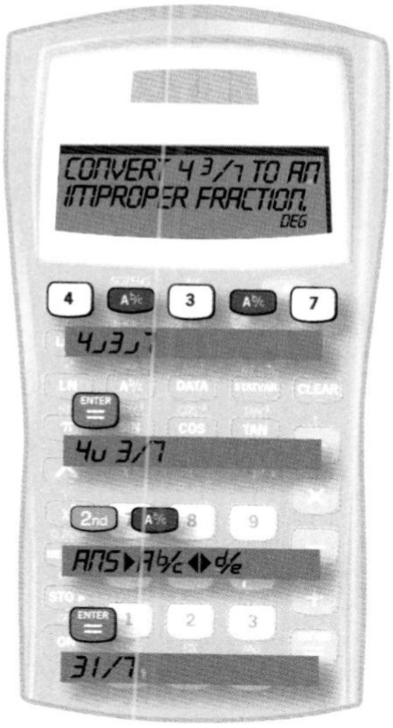

EXAMPLE D

Convert $3\frac{4}{8}$ to an improper fraction in simplest form.

First, write the fraction part of the mixed number in simplest form.

$3\frac{4}{8} = 3\frac{1}{2}$

Then convert the mixed number $3\frac{1}{2}$ to an improper fraction.

Find the numerator. ⟶ $2 \times 3 + 1 = 7$

Write the improper fraction. ⟶ $\frac{7}{2}$

$3\frac{4}{8} = \frac{7}{2}$

Using a Calculator to Change Mixed Fractions to Improper Fractions

To use your calculator to convert mixed numbers to an improper fraction, you use the [A b/c ◂▸ d/e] [[2nd] [A b/c]] key. (Remember, this notation means that to get the [A b/c ◂▸ d/e] key, you press the [2nd] [A b/c] keys.) For example, to convert $7\frac{5}{6}$ to an improper fraction, press

7 [A b/c] 5 [A b/c] 6 [A b/c ◂▸ d/e] [[2nd] [A b/c]] [ENTER =]

The result should be $\frac{47}{6}$.

EXERCISES 2-2

Convert the improper fractions to mixed numbers. Simplify if possible.

1. $\frac{6}{5} =$ **2.** $\frac{23}{23} =$ **3.** $\frac{33}{6} =$ **4.** $\frac{100}{10} =$

5. $\frac{15}{12} =$ **6.** $\frac{7}{4} =$ **7.** $\frac{51}{25} =$ **8.** $\frac{21}{8} =$

Convert the mixed numbers to improper fractions in simplest form.

9. $7\frac{1}{4} =$ **10.** $1\frac{2}{9} =$ **11.** $5\frac{2}{6} =$ **12.** $4\frac{7}{10} =$

13. $2\frac{12}{16} =$ **14.** $5\frac{1}{5} =$ **15.** $50\frac{1}{2} =$ **16.** $2\frac{6}{9} =$

Solve.

17. Maria is running on a $\frac{1}{4}$-mile oval track and decided to stop after running $7\frac{3}{4}$ miles. How many times did Maria run around the track?

18. The total resistance when an 18 Ω and a 24 Ω resistor are connected in parallel is $\frac{18 \times 24}{18 + 24}$ Ω.

(a) Find the total resistance of this circuit.
(b) Write the answer as a mixed number in simplest form.

19. The total inductance in henrys (H) when a 6 H and a 10 H inductor are connected in parallel is $\frac{6 \times 10}{6 + 10}$ H.

(a) Find the total inductance of this circuit.
(b) Write the answer as a mixed number in simplest form.

20. The total resistance when three resistors of 4 Ω, 6 Ω, and 14 Ω are connected in parallel is $\frac{4 \times 6 \times 14}{4 \times 6 + 4 \times 14 + 6 \times 14}$ Ω.

(a) Find the total resistance of this circuit.
(b) Write the answer as a mixed number in simplest form.

21. Challenge: Jeb put $14\frac{1}{2}$ gallons of fuel in a tank. Is that amount more or less than $\frac{57}{4}$ gallons? Explain.

ADDING FRACTIONS

Fractions with Like Denominators

To add fractions that have the same denominator, add the numerators only. The denominator will remain the same. Reduce the answer to lowest terms.

Mike and Joey ordered pizza. Mike ate $\frac{1}{8}$ of the pizza and Joey ate $\frac{5}{8}$ of the pizza. How much of the pizza did they eat altogether?

Add $\frac{1}{8}$ and $\frac{5}{8}$. Simplify if possible.

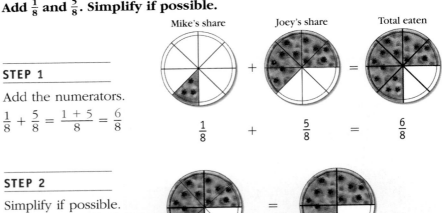

STEP 1

Add the numerators.

$\frac{1}{8} + \frac{5}{8} = \frac{1+5}{8} = \frac{6}{8}$

STEP 2

Simplify if possible.

$\frac{6 \div 2}{8 \div 2} = \frac{3}{4}$

Mike and Joey ate $\frac{3}{4}$ of the pizza altogether.

EXAMPLE A

Add $\frac{3}{5}$ and $\frac{1}{5}$.
Simplify if
possible.

STEP 1

Add the numerators.

$$\frac{3}{5} + \frac{1}{5} = \frac{3+1}{5} = \frac{4}{5}$$

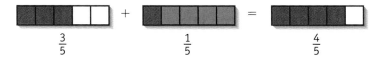

$\frac{3}{5}$ $\frac{1}{5}$ $\frac{4}{5}$

STEP 2

Simplify if possible.
$\frac{4}{5}$ is in lowest terms already.

$$\frac{3}{5} + \frac{1}{5} = \frac{4}{5}$$

EXAMPLE B

Add $\frac{9}{16}$ and $\frac{11}{16}$.
Simplify if
possible.

STEP 1

Add the numerators.

$$\frac{9}{16} + \frac{11}{16} = \frac{9+11}{16} = \frac{20}{16}$$

STEP 2

Simplify if possible.

$$\frac{20 \div 4}{16 \div 4} = \frac{5}{4} = 1\frac{1}{4}$$

$$\frac{9}{16} + \frac{11}{16} = 1\frac{1}{4}$$

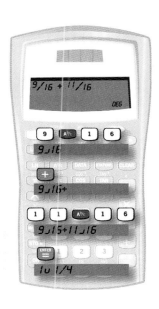

Unlike Denominators

To add fractions that have unlike denominators, start by writing equivalent fractions with the same denominator. This is also referred to as the least common denominator (LCD). Remember, the LCD is the least common multiple (LCM) of the denominators.

To get an equivalent fraction, multiply the original denominator by the number that gives the LCM. Multiply the numerator and denominator by the same number so that the value of the fraction is not changed.

When the fractions have the same denominator, add them together as you did in previous examples.

Two motors will be installed in a workstation. They are rated at $\frac{5}{8}$ horsepower and $\frac{1}{2}$ horsepower. What is the combined horsepower for the two motors?

Add $\frac{5}{8}$ and $\frac{1}{2}$. Simplify if possible.

STEP 1

Find the least common denominator.

Multiples of 8: 8, 16, 24, . . .

Multiples of 2: 2, 4, 6, 8, 10, . . .

8 is the least common denominator.

STEP 2

Rewrite the fractions so that they have a common denominator.

$\frac{5}{8}$ already has the common denominator 8.

$\frac{1}{2} \times \frac{}{?} = \frac{}{8}$ ⟵ Decide what 2 must be multiplied by to get 8.

$\frac{1 \times 4}{2 \times 4} = \frac{4}{8}$ ⟵ Multiply the numerator and denominator by the same number.

STEP 3

Add the fractions. Simplify if possible.

$\frac{5}{8} + \frac{4}{8} = \frac{5 + 4}{8} = \frac{9}{8}$

$\frac{9}{8} = 1\frac{1}{8}$

The combined length is $1\frac{1}{8}$ feet.

EXAMPLE C

The total resistance, in ohms, in a series circuit is the sum of all the resistances in that circuit. Find the total resistance of a series circuit with the two resistances $\frac{1}{8}$ Ω and $\frac{5}{6}$ Ω. Simplify if possible.

STEP 1

Find the least common denominator.

Multiples of 8: 8, 16, ⟨24⟩, 32, . . .

Multiples of 6: 6, 12, 18, ⟨24⟩, 30, . . .

24 is the least common denominator.

STEP 2

Rewrite the fractions so that they have a common denominator.

$\frac{1 \times 3}{8 \times 3} = \frac{3}{24}$

$\frac{5 \times 4}{6 \times 4} = \frac{20}{24}$

The least common
multiple (LCM) of
a set of denominators
is called the least
common denominator
(LCD).

STEP 3

Add the fractions. Simplify if possible.

$$\frac{3}{24} + \frac{20}{24} = \frac{3 + 20}{24} = \frac{23}{24}$$

$\frac{23}{24}$ is in lowest terms already.

$$\frac{1}{8} + \frac{5}{6} = \frac{23}{24}$$

The total resistance is $\frac{23}{24}$ Ω.

Mixed Numbers

To add mixed numbers, the fractional parts
of the mixed numbers must have the same
denominator.

For example, to find the cost of a rectangular
picture frame, you need to calculate the United
Inches. The United Inches measure for a frame
is the sum of its length and width. What is the
United Inches measure for a frame that measures $22\frac{3}{8}$ inches by $33\frac{1}{4}$ inches?

Add $22\frac{3}{8}$ and $33\frac{1}{4}$. Simplify if possible.

STEP 1

Find a common denominator.

8 is a common denominator of 8 and 4.

STEP 2

Rewrite the fractional parts of the mixed numbers so that they have a common denominator.

$\frac{3}{8}$ already has the common denominator 8.

$$\frac{1 \times 2}{4 \times 2} = \frac{2}{8} \longleftarrow \quad \text{Rewrite } \frac{1}{4} \text{ to have a denominator of 8.}$$

STEP 3

Add the mixed numbers. Simplify if possible.

$$
\begin{aligned}
22\frac{3}{8} &= 22\frac{3}{8} \\
+\ 33\frac{1}{4} &= 33\frac{2}{8} \\
\hline
&\ \ 55\frac{5}{8}
\end{aligned}
$$

⟵ Add the fractions in a column.
Add the whole numbers in a column.

⟵ $55\frac{5}{8}$ is in lowest terms already.

The United Inches measure for the frame is $55\frac{5}{8}$ inches.

EXAMPLE D

Add $1\frac{3}{4}$ and $5\frac{1}{3}$.

STEP 1

Find the least common denominator.

Multiples of 4: 4, 8, ⑫, 16, . . .

Multiples of 3: 3, 6, 9, ⑫, 15, . . .

12 is the least common denominator.

STEP 2

Rewrite the fractional parts so that they have a common denominator.

$$\frac{3 \times 3}{4 \times 3} = \frac{9}{12}$$

$$\frac{1 \times 4}{3 \times 4} = \frac{4}{12}$$

STEP 3

Add the mixed numbers. Simplify if possible.

$$
\begin{aligned}
1\frac{3}{4} &= 1\frac{9}{12} \\
+\ 5\frac{1}{3} &= 5\frac{4}{12} \\
\hline
&\ 6\frac{13}{12} = 6 + 1\frac{1}{12} = 7\frac{1}{12}
\end{aligned}
$$

$$1\frac{3}{4} + 5\frac{1}{3} = 7\frac{1}{12}$$

EXAMPLE E

Add $4\frac{2}{3}$, 2, and $\frac{5}{6}$.

$$4\frac{2}{3} = 4\frac{4}{6}$$

$$2 \ \ = 2$$

$$+\ \frac{5}{6} = \ \frac{5}{6}$$

$$6\frac{9}{6} = 6\frac{3}{2} = 6 + 1\frac{1}{2} = 7\frac{1}{2}$$

Rewrite the fractional parts so that they have a common denominator. Place fractions and whole numbers in separate columns.

Simplify.

$$4\frac{2}{3} + 2 + \frac{5}{6} = 7\frac{1}{2}$$

COMMON ERROR

ALERT

Students sometimes forget to multiply the numerator by the same number as the denominator when rewriting fractions. Be careful to rewrite fractions correctly. For example:
$\frac{3}{4} \times \frac{3}{3} = \frac{9}{12}$.

EXERCISES 2-3

Add. Simplify if possible.

1. $\frac{1}{5} + \frac{2}{5}$
2. $\frac{3}{10} + \frac{1}{10}$
3. $\frac{11}{15} + \frac{7}{15}$
4. $\frac{1}{8} + \frac{3}{4}$

5. $\frac{3}{4} + \frac{5}{6}$
6. $\frac{2}{3} + \frac{4}{9}$
7. $4\frac{4}{9} + 7\frac{2}{9}$
8. $3\frac{1}{3} + 3\frac{8}{15}$

9. $\frac{3}{4} + \frac{1}{12}$
10. $12\frac{4}{5} + 7\frac{1}{2}$
11. $\frac{7}{12} + \frac{3}{12} + \frac{5}{12}$
12. $1\frac{1}{5} + 2\frac{7}{15} + \frac{1}{3}$

Solve.

13. A wiring job calls for four pieces of $\frac{1}{2}$-inch conduit of the following lengths: $8\frac{3}{4}$ feet, $5\frac{1}{2}$ feet, $\frac{7}{8}$ foot, and $6\frac{1}{4}$ feet. What is the total length of the conduit needed for this job?

14. Janice needs six strips of fiber in the following lengths: $9\frac{1}{2}$ inches, $7\frac{3}{8}$ inches, $5\frac{7}{16}$ inches, $3\frac{1}{4}$ inches, $8\frac{13}{16}$ inches, and $4\frac{7}{8}$ inches.

    (a) Round the length of each strip to the nearest inch and add to find the approximate total length of strips that are needed.

    (b) Determine the actual total length needed.

    (c) What is the shortest strip of fiber that can be used if $\frac{1}{8}$ inch is allowed for each saw cut?

15. Rick mixed $1\frac{1}{2}$ pints of yellow paint with $\frac{1}{4}$ pint of red paint to make orange paint. How much orange paint did Rick make?

16. *Amperage* refers to the strength of the current of electricity expressed in amps. Based on a 120-volt system, a 19-inch color TV uses $\frac{3}{5}$ amp of electricity and a Blu-ray Disc (BD) uses $\frac{1}{3}$ amp of electricity. How many amps do the TV and BD use altogether?

17. **What's the Error?** Several students in the class said the sum of $\frac{2}{5}$ and $\frac{1}{5}$ is $\frac{3}{10}$. Explain what the students did wrong and give the correct answer.

18. **Challenge:** Find the missing fraction in the equation $\frac{1}{4} + \frac{\square}{\square} = \frac{3}{8}$.

SUBTRACTING FRACTIONS

Proper Fractions

To subtract two fractions that have the same denominator, subtract the numerators only. The denominator will remain the same.

Usually, Kevin walks his dog for $\frac{3}{4}$ mile. Yesterday he had less time than usual and had to omit the last $\frac{1}{4}$ mile. How far did Kevin walk his dog yesterday?

Subtract $\frac{1}{4}$ from $\frac{3}{4}$. Simplify if possible.

STEP 1

Subtract the numerators.

$$\frac{3}{4} - \frac{1}{4} = \frac{3-1}{4} = \frac{2}{4}$$

STEP 2

Simplify if possible.

$$\frac{2 \div 2}{4 \div 2} = \frac{1}{2}$$

Kevin walked his dog $\frac{1}{2}$ mile yesterday.

EXAMPLE A

Subtract $\frac{3}{8}$ from $\frac{5}{8}$. Simplify if possible.

STEP 1

Subtract the numerators.

$$\frac{5}{8} - \frac{3}{8} = \frac{5-3}{8} = \frac{2}{8}$$

STEP 2

Simplify if possible.

$$\frac{2 \div 2}{8 \div 2} = \frac{1}{4}$$

$$\frac{5}{8} - \frac{3}{8} = \frac{1}{4}$$

To subtract two fractions that have unlike denominators, start by writing equivalent fractions with the same denominator. Use the LCM of the original denominators as the denominator of the equivalent fractions.

EXAMPLE B

Subtract $\frac{5}{9}$ from $\frac{5}{6}$.
Simplify if possible.

STEP 1

Find the least common denominator.

Multiples of 6: 6, 12, ⑱ 24, 30, . . .

Multiples of 9: 9, ⑱ 27, 36, . . .

The LCM of 6 and 9 is 18.

Use 18 as the least common denominator.

STEP 2

Rewrite the fractions so that they have a common denominator.

$\frac{5}{6} = \frac{5 \times 3}{6 \times 3} = \frac{15}{18}$

$\frac{5}{9} = \frac{5 \times 2}{9 \times 2} = \frac{10}{18}$

STEP 3

Subtract the fractions. Simplify if possible.

$\frac{5}{6} - \frac{5}{9} = \frac{15}{18} - \frac{10}{18} = \frac{15 - 10}{18} = \frac{5}{18}$

$\frac{5}{18}$ is in lowest terms already.

$\frac{5}{6} - \frac{5}{9} = \frac{5}{18}$

EXAMPLE C

Subtract $\frac{1}{6}$ from $\frac{2}{3}$.
Simplify if possible.

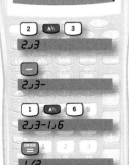

STEP 1

Find a common denominator.

6 is the least common denominator of 6 and 3.

STEP 2

Rewrite the fractions so that they have a common denominator.

$\frac{2}{3} = \frac{2 \times 2}{3 \times 2} = \frac{4}{6}$

$\frac{1}{6}$ already has a denominator of 6.

STEP 3

Subtract the fractions. Simplify if possible.

$$\frac{2}{3} - \frac{1}{6} = \frac{4}{6} - \frac{1}{6} = \frac{4-1}{6} = \frac{3}{6}$$

$$\frac{3 \div 3}{6 \div 3} = \frac{1}{2}$$

$$\frac{2}{3} - \frac{1}{6} = \frac{1}{2}$$

Fractions from Whole Numbers

To subtract a fraction from a whole number, start by rewriting the whole number as a mixed number. The fraction part of the mixed number should have the same denominator as the original fraction.

Dan is starting a 5-day vacation. After the first $\frac{1}{2}$ day is over, how much of his vacation does Dan have left?

Subtract $\frac{1}{2}$ from 5. Simplify if possible.

Helpful Hint

It is possible to write the number 1 as a fraction with any denominator, except 0. When 1 is written as a fraction, the numerator and the denominator are the same. Examples:

$$1 = \frac{2}{2} = \frac{7}{7} = \frac{24}{24}$$

STEP 1

Rewrite the whole number as a mixed number, using 2 as both the numerator and denominator of the fraction part of the number.

$$5 = 4 + 1$$

$$5 = 4 + \frac{2}{2} \longleftarrow \text{ because } 1 = \frac{2}{2}$$

$$5 = 4\frac{2}{2}$$

STEP 2

Subtract. Simplify if possible.

$$\begin{array}{r} 4\frac{2}{2} \\ - \frac{1}{2} \\ \hline 4\frac{1}{2} \end{array} \longleftarrow 4\frac{1}{2} \text{ is in lowest terms already.}$$

After $\frac{1}{2}$ day, Dan has $4\frac{1}{2}$ days of vacation left.

EXAMPLE D

Subtract $\frac{9}{15}$ from 2.
Simplify if possible.

$$2 - \frac{9}{15} = 1\frac{2}{5}$$

Mixed Numbers, Whole Numbers, and Fractions

If a subtraction problem contains whole numbers or mixed numbers, subtract the fraction parts and whole number parts in separate columns. Be sure the fractions have a common denominator.

Manny orders $3\frac{1}{2}$ cubic yards of concrete for a patio and a sidewalk. After he pours $1\frac{1}{4}$ cubic yards for the sidewalk, how many cubic yards does Manny have left?

Subtract $1\frac{1}{4}$ from $3\frac{1}{2}$. Simplify if possible.

STEP 1

Find the least common denominator.

The LCM of 4 and 2 is 4.

STEP 2

Subtract.

$$
\begin{array}{c}
3\frac{1}{2} \\
-1\frac{1}{4} \\
\hline
\end{array}
\longrightarrow
\begin{array}{c}
3\frac{2}{4} \\
-1\frac{1}{4} \\
\hline
2\frac{1}{4}
\end{array}
$$

Manny has $2\frac{1}{4}$ cubic yards of concrete left for the patio.

EXAMPLE E

Subtract $\frac{1}{6}$ from $9\frac{4}{5}$. Simplify if possible.

STEP 1

Find the least common denominator.

The LCM of 6 and 5 is 30.

STEP 2

Subtract.

$$
\begin{array}{r}
9\frac{4}{5} \\
-\ \frac{1}{6} \\
\end{array}
\longrightarrow
\begin{array}{r}
9\frac{24}{30} \\
-\ \frac{5}{30} \\
\hline
9\frac{19}{30}
\end{array}
$$

$$9\frac{4}{5} - \frac{1}{6} = 9\frac{19}{30}$$

EXAMPLE F

An electrician needs $8\frac{3}{4}$ feet of cable for a job and has $9\frac{1}{2}$ feet in stock. How much cable will be left in stock once the job is finished? Simplify if possible.

Helpful Hint

If the fraction in the mixed number you are subtracting is larger than the fraction in the mixed number you are subtracting from, just "borrow" 1 from the whole number and write it as a fraction.

To solve this problem, you need to subtract $8\frac{3}{4}$ from $9\frac{1}{2}$.

STEP 1

Find the least common denominator.

The LCM of 4 and 2 is 4.

STEP 2	STEP 3	STEP 4	STEP 5
Use 4 as the common denominator.	"Borrow" 1 from the 9; write the 1 as $\frac{4}{4}$.	Add $\frac{4}{4}$ and $\frac{2}{4}$.	Subtract.

$$
\begin{array}{r}
9\frac{1}{2} = 9\frac{2}{4} \\
-8\frac{3}{4} = -8\frac{3}{4} \\
\hline
\end{array}
$$

$$
\begin{array}{r}
8 + \frac{4}{4} + \frac{2}{4} \\
-8 \qquad \frac{3}{4} \\
\hline
\end{array}
$$

$$
\begin{array}{r}
8\frac{6}{4} \\
-8\frac{3}{4} \\
\hline
\end{array}
$$

$$
\begin{array}{r}
8\frac{6}{4} \\
-8\frac{3}{4} \\
\hline
\frac{3}{4}
\end{array}
$$

$$9\frac{1}{2} - 8\frac{3}{4} = \frac{3}{4}$$

There will be $\frac{3}{4}$ feet of cable left in stock.

EXERCISES 2-4

Subtract. Simplify if possible.

1. $\frac{3}{4} - \frac{3}{8} =$

2. $5\frac{4}{9} - 3\frac{1}{3} =$

3. $\frac{11}{12} - \frac{5}{12} =$

4. $\frac{6}{11} - \frac{2}{11} =$

5. $\frac{5}{8} - \frac{5}{16} =$

6. $10 - \frac{2}{7} =$

7. $7\frac{3}{8} - 1\frac{1}{16} =$

8. $7 - 3\frac{2}{5} =$

9. $9 - \frac{4}{16} =$

10. $\frac{3}{5} - \frac{2}{7} =$

11. $12\frac{1}{3} - \frac{3}{4} =$

12. $\frac{2}{3} - \frac{3}{8} =$

13. $5 - \frac{3}{8} =$

14. $4\frac{5}{9} - \frac{2}{3} =$

15. $8\frac{5}{12} - \frac{1}{4} =$

Solve. Simplify fractions if possible.

16. Sean spends $\frac{2}{5}$ of his budget on rent and $\frac{1}{6}$ of his budget on groceries. What is the difference between the fraction of his budget Sean spends on rent and the fraction he spends on groceries?

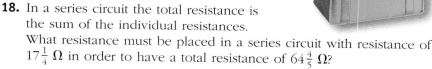

17. A work crew plans to lay 12 miles of cable. After $2\frac{2}{5}$ miles are completed, how many miles remain to be done?

18. In a series circuit the total resistance is the sum of the individual resistances. What resistance must be placed in a series circuit with resistance of $17\frac{1}{4}$ Ω in order to have a total resistance of $64\frac{4}{5}$ Ω?

19. The total current in a parallel circuit is equal to the sum of all the branches of the circuit. A given parallel circuit has two branches, and the total current in the circuit is $7\frac{1}{4}$ A. If the current in one branch is $4\frac{5}{8}$ A, what is the current in the other branch?

20. What's the Error? Where does the error occur in the subtraction problem shown below?

$$\begin{array}{r} 10 \\ \hline -\,7\frac{1}{3} \end{array} \longrightarrow \begin{array}{r} 10\frac{3}{3} \\ \hline -\,7\frac{1}{3} \end{array} \longrightarrow \begin{array}{r} 10\frac{3}{3} \\ -\,7\frac{1}{3} \\ \hline 3\frac{2}{3} \end{array}$$

MULTIPLYING FRACTIONS AND MIXED NUMBERS

Proper Fractions

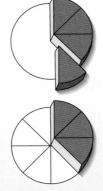

To multiply fractions, first multiply the numerators to find the numerator of the product. Then multiply the denominators to find the denominator of the product.

Jeff brought out $\frac{1}{2}$ of a cake for dessert. He and his friends ate $\frac{3}{4}$ of the cake he served. What part of the whole cake did they eat?

Multiply $\frac{3}{4}$ and $\frac{1}{2}$. Simplify if possible.

$$\frac{3}{4} \times \frac{1}{2} = \frac{3 \times 1}{4 \times 2} = \frac{3}{8}$$

$\frac{3}{8}$ is in lowest terms already.

Jeff and his friends ate $\frac{3}{8}$ of the whole cake.

EXAMPLE A

Multiply $\frac{2}{3}$ and $\frac{5}{8}$.
Simplify if possible.

$$\frac{2}{3} \times \frac{5}{8} = \frac{2 \times 5}{3 \times 8} = \frac{10}{24} = \frac{5}{12}$$

$$\frac{2}{3} \times \frac{5}{8} = \frac{5}{12}$$

Example B shows that you can simplify either after multiplying or before multiplying. The result will be the same.

EXAMPLE B

Multiply $\frac{5}{9}$ and $\frac{3}{10}$.
Simplify if possible.

One Way: Multiply, then simplify.

$$\frac{5}{9} \times \frac{3}{10} = \frac{5 \times 3}{9 \times 10} = \frac{15}{90} = \frac{15 \div 15}{90 \div 15} = \frac{1}{6}$$

Another Way: Look for a numerator-denominator pair with a common factor. You can "cancel," or simplify, these common factors before you multiply.

$$\frac{5}{9} \times \frac{3}{10} = \frac{\overset{1}{\cancel{5}} \times \overset{1}{\cancel{3}}}{9 \times \underset{2}{\cancel{10}}} = \frac{1 \times 1}{3 \times 2} = \frac{1}{6}$$

$$\frac{5}{9} \times \frac{3}{10} = \frac{1}{6}$$

Mixed Numbers, Whole Numbers, and Fractions

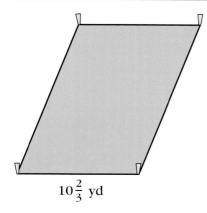

$10\frac{2}{3}$ yd

If a multiplication problem contains a fraction or a mixed number, express all factors in fraction form. Convert whole numbers and mixed numbers to improper fractions.

A garden has a width of $10\frac{2}{3}$ yards. The length of the garden is $1\frac{1}{2}$ times the width. What is the length of the garden?

Multiply $1\frac{1}{2}$ and $10\frac{2}{3}$.

STEP 1

Convert the mixed numbers to improper fractions.

$$1\frac{1}{2} = \frac{2 \times 1 + 1}{2} = \frac{3}{2}$$

$$10\frac{2}{3} = \frac{3 \times 10 + 2}{3} = \frac{32}{3}$$

STEP 2

Multiply the improper fractions, cancelling common factors.

$$\frac{3}{2} \times \frac{32}{3} = \frac{\overset{1}{\cancel{3}} \times \overset{16}{\cancel{32}}}{\underset{1}{\cancel{2}} \times \underset{1}{\cancel{3}}} = \frac{16}{1} = 16$$

The length of the garden is 16 yards.

Order of Operations

What is the answer to $3 + 5 \times 4$? It should be the same as the answer to $4 \times 5 + 3$. The order in which operations are performed is based on some basic agreements made by mathematicians so that both problems give the same answer. These agreements are in the form of a series of steps to be followed. The first steps are presented below. If a step does not apply to an expression, go to the next step.

Order of Operations

- Complete all calculations within grouping symbols such as parentheses.
- Perform all multiplications and divisions in order from left to right.
- Perform all additions and subtractions in order from left to right.

Evaluate 3 + 5 × 4.

$$3 + 5 \times 4 =$$

$$3 + 20 = \quad \longleftarrow \text{Perform multiplication before}$$
addition and subtraction.

$$23 = \quad \longleftarrow \text{Perform addition.}$$

When a fraction has some operations in the numerator or denominator, it is implied that the numerator and denominator are placed in parentheses. It is very important that you include these parentheses when you use your calculator. The above problem where you converted $1\frac{1}{2}$ to $\frac{2 \times 1 + 1}{2}$ should be thought of as $\frac{(2 \times 1 + 1)}{2}$.

EXAMPLE C

Write $\frac{5 + 12}{6}$ as a mixed number.

$$\frac{(5 + 12)}{6} = \quad \longleftarrow \text{According to the order of operations, think of the numera-} $$
tor as being in parentheses.

$$\frac{17}{6} = \quad \longleftarrow \text{Add the numbers in parentheses.}$$

$$\frac{17}{6} = 2\frac{5}{6} \quad \longleftarrow \text{Convert to a mixed number.}$$

As yet, you cannot use your calculator to convert $\frac{5 + 12}{6}$ to a mixed number. For example, if you keyed in

5 ⊞ 12 ÷ 6 ENTER=

the result would be 7, a wrong answer.

If instead, you key in

(5 ⊞ 12) A^b/c 6 ENTER=

you will get a "SYNTAX Error" message. In the next chapter you will see how to use your calculator to convert $\frac{5 + 12}{6}$ to a mixed number.

Multiplying Fractions with a Calculator

The calculator does most of the work for you. For example, to solve $5\frac{2}{3} \times 7\frac{4}{5}$, press the following keys:

5 A^b/c 2 A^b/c 3 ⊠ 7 A^b/c 4 A^b/c 5 ENTER=

The display should show

```
5⌐2⌐3 * 7⌐4⌐5

              44⌐1/5
                  DEG
```

The second line of the display shows that the product is $44\frac{1}{5}$.

EXAMPLE D

Multiply 7 by $\frac{5}{9}$. Simplify if possible.

Rewrite the whole number as a fraction.

$$7 \times \frac{5}{9} = \frac{7}{1} \times \frac{5}{9} = \frac{7 \times 5}{1 \times 9} = \frac{35}{9}$$

Convert the improper fraction to a mixed number.

$$\frac{35}{9} = 3\frac{8}{9}$$

$$7 \times \frac{5}{9} = 3\frac{8}{9}$$

EXAMPLE E

Multiply $9\frac{1}{2}$ by $5\frac{3}{4}$. Simplify if possible.

$$9\frac{1}{2} \times 5\frac{3}{4} = \frac{19}{2} \times \frac{23}{4} = \frac{19 \times 23}{2 \times 4} = \frac{437}{8} = 54\frac{5}{8}$$

$$9\frac{1}{2} \times 5\frac{3}{4} = 54\frac{5}{8}$$

EXAMPLE F

A neon sign uses $3\frac{1}{2}$ W of power per foot of tubing. How many watts are used for a sign with 7 feet of tubing? Simplify if possible.

$$7 \times 3\frac{1}{2} = \frac{7}{1} \times \frac{7}{2} = \frac{7 \times 7}{1 \times 2} = \frac{49}{2} = 24\frac{1}{2}$$

$$7 \times 3\frac{1}{2} = 24\frac{1}{2}$$

The sign used $24\frac{1}{2}$ W of power.

EXAMPLE G

Multiply $\frac{1}{8}$ by $6\frac{4}{5}$.
Simplify if possible.

$$6\frac{4}{5} \times \frac{1}{8} = \frac{34}{5} \times \frac{1}{8} = \frac{34 \times 1}{5 \times 8} = \frac{34}{40} = \frac{17}{20}$$

$$6\frac{4}{5} \times \frac{1}{8} = \frac{17}{20}$$

EXERCISES 2-5

Multiply. Simplify if possible.

1. $\frac{3}{5} \times \frac{2}{5}$
2. $\frac{3}{11} \times 10$
3. $\frac{6}{7} \times \frac{35}{36}$
4. $1\frac{1}{4} \times 3\frac{3}{7}$
5. $3 \times \frac{5}{9}$
6. $\frac{1}{7} \times \frac{4}{9}$
7. $2\frac{3}{4} \times 5\frac{1}{4}$
8. $\frac{1}{8} \times \frac{2}{3}$
9. $9\frac{2}{3} \times 3\frac{3}{4}$
10. $\frac{5}{12} \times \frac{4}{15}$
11. $\frac{1}{7} \times \frac{1}{7}$
12. $8\frac{2}{5} \times 2\frac{5}{7}$

Solve.

13. Tom spends $\frac{5}{8}$ of his time at work supervising a crew. How many hours of a 40-hour workweek does he spend supervising?

14. If $7\frac{1}{2}$ feet of conduit are needed between electrical outlets, how many feet are required for six of these outlets if they are equally spaced?

15. A contractor pays a helper $16 per hour for the first 40 hours of work in a week, and he pays the helper $1\frac{1}{2}$ times that rate for overtime. How much did he pay the helper for working 44 hours in a week?

16. Electricians get paid time and one half for overtime. One week, Juanita worked $6\frac{1}{4}$ hours of overtime. How many hours of regular time would she need to work to earn the same amount of money?

17. **What's the Error?** A student says that the product of $3\frac{3}{5}$ and $1\frac{1}{8}$ is $3\frac{3}{40}$. Explain what the student did wrong and give the correct answer.

DIVIDING FRACTIONS AND MIXED NUMBERS

Proper Fractions

To divide by a fraction, multiply by its **reciprocal.** To get the reciprocal of a fraction, interchange the numerator and the denominator.

Examples of reciprocals:

$\frac{3}{4}$ and $\frac{4}{3}$, $\frac{1}{7}$ and $\frac{7}{1}$, $\frac{10}{9}$ and $\frac{9}{10}$

To find the reciprocal of a mixed number, first write the number as an improper fraction.

For example, to find the reciprocal of $5\frac{2}{3}$, first change it to an improper fraction: $5\frac{2}{3} = \frac{3 \times 5 + 2}{3} = \frac{17}{3}$. The reciprocal of $\frac{17}{3}$ is $\frac{3}{17}$, so the reciprocal of $5\frac{2}{3}$ is $\frac{3}{17}$.

A baker needs $\frac{1}{8}$ cup of cinnamon for each batch of cookies. He has $\frac{3}{4}$ cup of cinnamon. How many batches can he make?

How many $\frac{1}{8}$s are in $\frac{3}{4}$?

Divide $\frac{3}{4}$ by $\frac{1}{8}$. Simplify if possible.

To divide by $\frac{1}{8}$, multiply by its reciprocal.

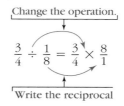

$$\underset{\text{Write the reciprocal}}{\underbrace{\frac{3}{4} \div \frac{1}{8} = \frac{3}{4} \times \frac{8}{1}}}$$

$$\frac{3}{4} \div \frac{1}{8} = \frac{3}{4} \times \frac{8}{1} = \frac{3 \times \overset{2}{\cancel{8}}}{\underset{1}{\cancel{4}} \times 1} = \frac{6}{1} = 6$$

There is enough cinnamon for 6 batches of cookies.

Language Box

Finding the reciprocal of a fraction is sometimes called **inverting the fraction** because the numerator and the denominator switch places.

EXAMPLE A

Divide $\frac{3}{5}$ by $\frac{7}{8}$. Simplify if possible.

$$\frac{3}{5} \div \frac{7}{8} = \frac{3}{5} \times \frac{8}{7} = \frac{3 \times 8}{5 \times 7} = \frac{24}{35}$$

$$\frac{3}{5} \div \frac{7}{8} = \frac{24}{35}$$

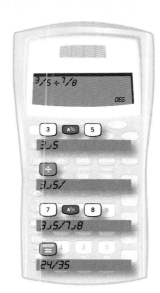

EXAMPLE B

A parallel circuit has four resistors of 16 Ω, 12 Ω, 10 Ω, and 8 Ω. Determine the total resistance, R, in ohms, if $\frac{1}{R} = \frac{1}{16} + \frac{1}{12} + \frac{1}{10} + \frac{1}{8}$.

Solution:

First, add the fractions to the right of the equal sign.

$$\frac{1}{16} + \frac{1}{12} + \frac{1}{10} + \frac{1}{8} = \frac{89}{240}$$

So, $\frac{1}{R} = \frac{89}{240}$.

Since R is the reciprocal of $\frac{1}{R}$, it is also the reciprocal of $\frac{89}{240}$.

$$R = \frac{240}{89} = 2\frac{62}{89}.$$

The total resistance in this circuit is $2\frac{62}{89}$ Ω.

Finding Reciprocals with a Calculator

The [x^{-1}] key is used to find the reciprocal of a number. For example, to find the reciprocal of $\frac{89}{240}$, press 89 [A$\frac{b}{c}$] 240 [x^{-1}] [ENTER =]. The result is $2\frac{62}{89}$, the reciprocal as a mixed number. Pressing [2nd] [A$\frac{b}{c}$] [ENTER =] will give the reciprocal as the improper fraction $\frac{240}{89}$. Notice how the [x^{-1}] key was used in the last line of the calculator in the previous example.

Mixed Numbers, Whole Numbers, and Fractions

If a division problem contains a fraction or a mixed number, express both dividend and divisor in fraction form. Convert whole numbers and mixed numbers to improper fractions.

Rich has $4\frac{1}{2}$ rolls of tape and uses $1\frac{1}{2}$ rolls of tape to wrap a hockey stick. How many sticks can he tape?

Divide $4\frac{1}{2}$ by $1\frac{1}{2}$. Simplify if possible.

$$4\frac{1}{2} \div 1\frac{1}{2} = \frac{9}{2} \div \frac{3}{2} = \frac{9}{2} \times \frac{2}{3} = \frac{\overset{3}{\cancel{9}} \times \overset{1}{\cancel{2}}}{\underset{1}{\cancel{2}} \times \underset{1}{\cancel{3}}} = \frac{3}{1} = 3$$

Express both dividend and divisor in fraction form.

There is enough tape for 3 hockey sticks.

EXAMPLE C

Divide $\frac{2}{5}$ by 10.
Simplify if possible.

$$\frac{2}{5} \div 10 = \frac{2}{5} \div \frac{10}{1} = \frac{2}{5} \times \frac{1}{10} = \frac{\cancel{2} \times 1}{5 \times \underset{5}{\cancel{10}}} = \frac{1}{25}$$

$$\frac{2}{5} \div 10 = \frac{1}{25}$$

EXAMPLE D

Divide 15 by $\frac{6}{7}$.
Simplify if possible.

$$15 \div \frac{6}{7} = \frac{15}{1} \div \frac{6}{7} = \frac{15}{1} \times \frac{7}{6} = \frac{\overset{5}{\cancel{15}} \times 7}{1 \times \underset{2}{\cancel{6}}} = \frac{35}{2} = 17\frac{1}{2}$$

$$15 \div \frac{6}{7} = 17\frac{1}{2}$$

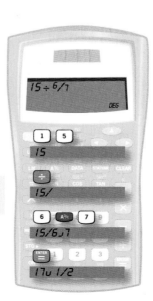

EXAMPLE E

Divide $5\frac{5}{8}$ by $2\frac{1}{2}$.
Simplify if possible.

$$5\frac{5}{8} \div 2\frac{1}{2} = \frac{45}{8} \div \frac{5}{2} = \frac{45}{8} \times \frac{2}{5} = \frac{\overset{9}{\cancel{45}} \times \overset{1}{\cancel{2}}}{\underset{4}{\cancel{8}} \times \underset{1}{\cancel{5}}} = \frac{9}{4} = 2\frac{1}{4}$$

$$5\frac{5}{8} \div 2\frac{1}{2} = 2\frac{1}{4}$$

COMMON ERROR

ALERT

A common error when dividing fractions is to forget to use the **reciprocal** of the divisor.

EXERCISES 2-6

Divide. Simplify if possible.

1. $\frac{5}{6} \div \frac{5}{8} =$

2. $\frac{4}{6} \div \frac{1}{10} =$

3. $10 \div \frac{1}{10} =$

4. $\frac{2}{5} \div 2\frac{1}{2} =$

5. $\frac{1}{3} \div \frac{1}{3} =$

6. $\frac{1}{6} \div 2 =$

7. $\frac{3}{7} \div \frac{3}{14} =$

8. $\frac{4}{7} \div 2 =$

9. $1\frac{1}{10} \div 2\frac{1}{4} =$

10. $2\frac{4}{5} \div \frac{7}{9} =$

11. $8\frac{1}{4} \div 4\frac{1}{2} =$

12. $\frac{9}{16} \div 9 =$

Solve.

13. What is the total resistance in a parallel circuit with resistances of $2\ \Omega$, $3\ \Omega$, and $5\ \Omega$?

14. If $15\frac{3}{8}$ watts is distributed equally over each of nine resistors, what is the average number of watts per resistor?

15. A $27\frac{1}{2}$-inch length of cable is cut into $3\frac{5}{8}$-inch lengths. Assume there is no waste in cutting.

(a) Round each length to the nearest inch and approximate the number of $3\frac{5}{8}$-inch pieces that will be made.

(b) How many $3\frac{5}{8}$-inch pieces can be made?

(c) What is the length of cable that is left over?

16. The current, in amperes (A), in a circuit with a voltage of $8\frac{1}{2}$ volts and a resistance of $24\ \Omega$ is $\frac{8\frac{1}{2}}{24}$. What is the current in this circuit?

17. Challenge: In the problem $\frac{1}{4} \div 4$, is the quotient less than or greater than $\frac{1}{4}$? Explain the steps you took to answer the question.

READING A CUSTOMARY RULE

Most lengths in the customary or English system of measurement are given in inches (in.), feet (ft), and yards (yd). One yard is equal to 3 feet, or 36 inches. One foot is equal to 12 inches. A length of 1 yd is shown in the illustration below (smaller than actual size)

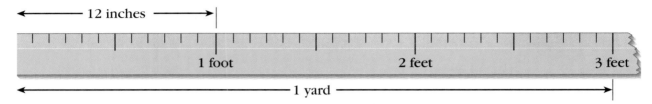

12 inches

1 foot   2 feet   3 feet

1 yard

(smaller than actual size)

Language Box

The equal sections between *tick marks* on a ruler are often called *graduations*. If 1 inch is divided into 16 sections on a ruler, each graduation is $\frac{1}{16}$ inch.

A section of a rule for measuring inches is shown below in actual size. Every inch on this ruler is divided into 16 equal parts, or 16ths, by tick marks. The distance between each pair of tick marks is $\frac{1}{16}$ inch.

inches   1   2   3

(actual size)

To understand how to read a rule and give a measure in simplest form, use the illustration below (larger than actual size). Figure A shows an inch divided into 16ths. Figure B is the same as Figure A, but with all measures in lowest terms. The length represented by the line segment above each figure is $\frac{4}{16}$ inch, or $\frac{1}{4}$ inch.

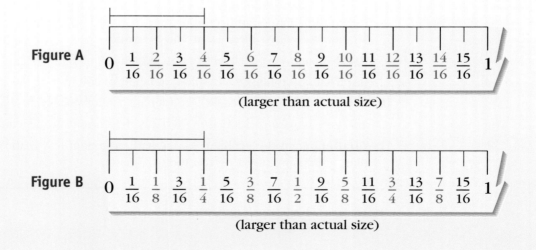

Figure A 0 $\frac{1}{16}$ $\frac{2}{16}$ $\frac{3}{16}$ $\frac{4}{16}$ $\frac{5}{16}$ $\frac{6}{16}$ $\frac{7}{16}$ $\frac{8}{16}$ $\frac{9}{16}$ $\frac{10}{16}$ $\frac{11}{16}$ $\frac{12}{16}$ $\frac{13}{16}$ $\frac{14}{16}$ $\frac{15}{16}$ 1

(larger than actual size)

Figure B 0 $\frac{1}{16}$ $\frac{1}{8}$ $\frac{3}{16}$ $\frac{1}{4}$ $\frac{5}{16}$ $\frac{3}{8}$ $\frac{7}{16}$ $\frac{1}{2}$ $\frac{9}{16}$ $\frac{5}{8}$ $\frac{11}{16}$ $\frac{3}{4}$ $\frac{13}{16}$ $\frac{7}{8}$ $\frac{15}{16}$ 1

(larger than actual size)

Cal measures a nail with the ruler shown below. What is the length of the nail in lowest terms?

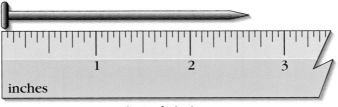

(actual size)

The nail is longer than 2 inches and shorter than 3 inches. Count the number of sixteenths to the right of the 2-inch mark. There are 10 sixteenths, so the length of the nail is $2\frac{10}{16}$ inches, or $2\frac{5}{8}$ inches in lowest terms. Another way to determine the length is by using the pattern of different lengths of tick marks on the rule. The longest tick mark on the rule between 2 and 3 represents $2\frac{1}{2}$ inches. The tip of the nail extends $\frac{1}{8}$ inch beyond $2\frac{1}{2}$ inches. Add to find the length of the nail:

$$2\frac{1}{2} + \frac{1}{8} = 2\frac{4}{8} + \frac{1}{8} = 2\frac{5}{8}.$$

The nail is $2\frac{5}{8}$ inches long.

EXAMPLE A

What is the length of the tubing in lowest terms?

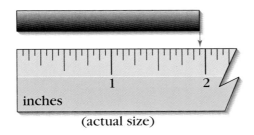

(actual size)

The tubing is longer than 1 inch and shorter than 2 inches. It measures 15 sixteenths to the right of the 1-inch mark, so the length of the tubing is $1\frac{15}{16}$ inches. Another way to determine the length is to subtract. The end of the tube is $\frac{1}{16}$ to the left of the 2-inch mark. Subtract to find the length:

$$2 - \frac{1}{16} = 1\frac{16}{16} - \frac{1}{16} = 1\frac{15}{16}.$$

The tubing is $1\frac{15}{16}$ inches long.

CONVERTING BETWEEN CUSTOMARY UNITS OF LENGTH

Common customary units of length are shown in the table below.

Customary Units of Linear Measure

1 yard (yd) = 3 feet (ft)
1 yard (yd) = 36 inches (in.)
1 foot (ft) = 12 inches (in.)

To convert from one unit to another, multiply by a fraction made from one of the lines in the above table. The denominator of the fraction will have the given unit and the numerator will have the unit to which you are converting.

EXAMPLE B

Express $3\frac{1}{4}$ ft in inches.

You are converting feet to inches, so use the third line of the table and the fraction $\frac{12 \text{ in.}}{1 \text{ ft}}$.

$$3\frac{1}{4} \text{ ft} \times \frac{12 \text{ in.}}{1 \text{ ft}} = 3\frac{1}{4} \times 12 \text{ in.}$$

Notice how the units "cancel." If you can cancel the given unit, then you are using the right fraction.

$3\frac{1}{4}$ ft = 39 in.

EXAMPLE C

What is the length of the board in lowest terms? Give the answer in inches, in feet and inches, and in yards, feet, and inches.

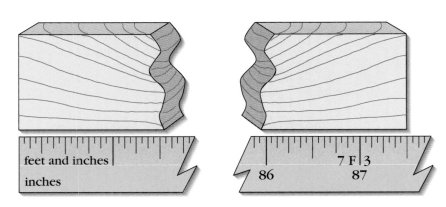

The "F" on the tape measure stands for feet.

The length of the board is $87\frac{1}{2}$ inches.

To convert the length from inches to feet and inches, multiply by the fraction $\frac{1 \text{ ft}}{12 \text{ in.}}$. To begin, convert the 87 inches to feet.

$$87 \text{ in.} \times \frac{1 \text{ ft}}{12 \text{ in.}} = \frac{87}{12} \text{ ft}$$

$$= 7\frac{1}{4} \text{ ft or } 7 \text{ ft} + \frac{1}{4} \text{ ft}$$

$\frac{1}{4}$ ft = $\frac{3}{12}$ ft. Since 1 inch is $\frac{1}{12}$ ft, $\frac{3}{12}$ ft would be 3 inches.

So, $87\frac{1}{2}$ inches is the same as 7 feet $3\frac{1}{2}$ inches.

$87\frac{1}{2}$ inches = 7 feet $3\frac{1}{2}$ inches.

To convert 7 feet $3\frac{1}{2}$ inches to yards, feet, and inches, you need to convert the 7 feet to yards, feet, and inches and add the $3\frac{1}{2}$ inches to that answer.

Begin by converting the 7 feet to yards and feet. Multiply by the fraction $\frac{1 \text{ yd}}{3 \text{ ft}}$.

$$7 \text{ ft} \times \frac{1 \text{ yd}}{3 \text{ ft}} = 2\frac{1}{3} \text{ yd or } 2 \text{ yd} + \frac{1}{3} \text{ yd}$$

$\frac{1}{3}$ yd = 1 ft since there are 3 feet in 1 yard.

So, 7 feet $3\frac{1}{2}$ inches is 2 yards 1 foot $3\frac{1}{2}$ inches.

The length of the board is $87\frac{1}{2}$ inches, or 7 feet $3\frac{1}{2}$ inches, or 2 yards 1 foot $3\frac{1}{2}$ inches.

EXERCISES 2-7

What is the measurement in lowest terms at each point labeled on the rule? Give the answer in inches, and in feet and inches for Exercises 7–9.

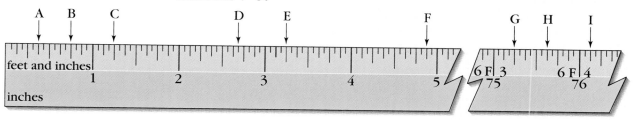

1. Point A

2. Point B

3. Point C

4. Point D

5. Point E

6. Point F

7. Point G

8. Point H

9. Point I

10. A cable is $23\frac{1}{2}$ feet long. How many inches is this?

11. A cable is 90 inches long.

(a) What is the length of the cable in feet and inches?

(b) What is the length of the cable in feet?

(c) What is the length of the cable in yards?

12. A cable is 32 feet $10\frac{1}{2}$ inches long and is to be cut into $8\frac{1}{2}$-inch lengths. Allow $\frac{1}{16}$ inch for each saw cut.

(a) What is the length of the cable in inches?

(b) Round each length to the nearest inch and approximate the number of $8\frac{1}{2}$-inch pieces that can be made.

(c) How many $8\frac{1}{2}$-inch pieces can be made from this cable?

(d) What is the length of cable left over?

13. Challenge: A measurement is given as $3\frac{3}{8}$ inches. How many sixteenths of an inch is that?

CHAPTER 2 REVIEW EXERCISES

1. (a) What is the GCF of 45 and 33?

(b) Write $\frac{33}{45}$ in simplest form.

2. (a) What is the LCM of 3, 5, and 12?

(b) Write equivalent fractions for $\frac{1}{3}$, $\frac{2}{5}$, and $\frac{7}{12}$ using the LCM you found in (a).

3. Convert the mixed number $4\frac{2}{3}$ to an improper fraction.

4. Convert $\frac{27}{12}$ to a mixed number in simplest form.

5. What is the reciprocal of $\frac{15}{17}$?

6. What is the reciprocal of $3\frac{5}{8}$?

7. (a) Write the sum $\frac{1}{3} + \frac{2}{5} + \frac{7}{12}$ as an improper fraction.

(b) Write the sum $\frac{1}{3} + \frac{2}{5} + \frac{7}{12}$ as a mixed number.

8. Perform the indicated operation. Simplify if possible.

(a) $\frac{2}{7} + \frac{4}{7}$ (b) $\frac{5}{8} - \frac{3}{8}$

(c) $\frac{2}{3} \times \frac{5}{4}$ (d) $\frac{4}{5} \div \frac{2}{7}$

(e) $2\frac{1}{5} + 3\frac{2}{5}$ (f) $7\frac{1}{3} - 4\frac{2}{3}$

(g) $6\frac{1}{2} \times 3\frac{1}{4}$ (h) $2\frac{2}{3} \div 5\frac{1}{3}$

9. Perform the indicated operation. Simplify if possible.

(a) $\frac{5}{3} + \frac{1}{4}$ (b) $\frac{7}{8} - \frac{2}{3}$

(c) $\frac{15}{8} \times \frac{4}{5}$ (d) $\frac{9}{8} \div \frac{3}{16}$

(e) $2\frac{3}{4} + 5\frac{1}{2} + 4\frac{1}{3}$ (f) $17\frac{5}{8} - 9\frac{3}{4}$

(g) $13\frac{1}{2} \times 2\frac{1}{6}$ (h) $7\frac{1}{3} \div 2\frac{5}{6}$

10. A heating unit produces 1575 watts of power with an applied voltage of 115 V. The current in amperes flowing through this unit is $\frac{1575}{115}$. Write the current in simplest terms as (a) an improper fraction and (b) a mixed number.

11. It take 14 850 watts of power to light a certain building. If the applied voltage is 120 V, the current in amperes is $\frac{14\,850}{120}$.

   (a) What is the current for the building?

   (b) There are 350 lights in this building. What is the average current through each light?

12. In budgeting the costs for a certain job, Ben figured that it would take three electricians who each work $5\frac{1}{4}$ hours, five other electricians who each work $7\frac{1}{5}$ hours, and six others who each work $6\frac{1}{3}$ hours. How many hours would this job take?

13. Resistance in an electric circuit is measured in ohms (Ω). The total resistance in a parallel circuit is the reciprocal of the sum of the reciprocals of all the resistances in that circuit. So, a parallel circuit that has resistances of 20 Ω, 24 Ω, and 35 Ω has a total resistance R, where $\frac{1}{R} = \frac{1}{20} + \frac{1}{24} + \frac{1}{35}$. What is the total resistance of this circuit?

14. The total resistance when a 2 Ω and a 32 Ω resistor are connected in parallel is $\frac{24 \times 32}{24 + 32}$ Ω.

   (a) Find the total resistance of this circuit.

   (b) Write the answer as a mixed number in simplest form.

15. José needs five pieces of cable for a wiring job. The pieces have the following lengths: $9\frac{5}{8}$ feet, $6\frac{3}{4}$ feet, $1\frac{5}{6}$ feet, $\frac{1}{3}$ foot, and $5\frac{1}{4}$ feet.

   (a) Round each length to the nearest foot and add to find the approximate total length that José needs for this job.

   (b) Determine the actual length of cable needed in feet and inches.

   (c) What is the actual length of cable needed in yards, feet, and inches?

16. In a series circuit, the total resistance is the sum of the individual resistances. If a four-resistor series has the resistances of $3\frac{1}{4}$ MΩ (megaohms), $2\frac{1}{2}$ MΩ, $\frac{7}{10}$ MΩ, and $4\frac{2}{5}$ MΩ, what is the total resistance?

17. Luigi needs five sections of cable that are each $7\frac{3}{4}$ feet long.

   (a) Determine the actual length of cable needed in feet and inches.

   (b) What is the total length of cable needed in yards, feet, and inches?

18. In a series circuit, the total resistance is the sum of the individual resistances. What resistance must be placed in a series circuit with a $15\frac{2}{5}$ Ω resistance in order to have a total resistance of $21\frac{1}{3}$ Ω?

19. A parallel circuit has four resistors of 3 Ω, 5 Ω, 6 Ω, and 8 Ω. Determine the total resistance, R, in ohms, if $\frac{1}{R} = \frac{1}{3} + \frac{1}{5} + \frac{1}{6} + \frac{1}{8}$. Express the answer as an improper fraction and a mixed number.

20. A cable that is 5 yards 2 feet $3\frac{1}{4}$ inches long is to be cut into $7\frac{5}{8}$-inch lengths. Allow $\frac{1}{16}$-inch for each saw cut.

   (a) Round each length to the nearest inch and approximate the number of $7\frac{5}{8}$-inch pieces that will be made.

   (b) How many $7\frac{5}{8}$-inch pieces can actually be made?

   (c) What is the length of cable that is left over?

Building a Foundation in Mathematics

Writing and Rounding Decimals

Adding Decimals

Subtracting Decimals

Multiplying Decimals

Dividing Decimals

Converting Decimals and Fractions

Reading a Metric Rule

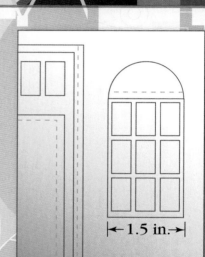

Overview

Most of the time we use a set of numbers called decimal fractions. Decimals form the basis of much of what we do. Our money is expressed in decimals, with the dollars representing whole numbers and the cents showing the decimal fractions. Much of the reason decimal fractions are used is because performing calculations using decimals is often faster and easier than fraction computation.

Decimals

Objectives

After completing this chapter, you will be able to:

- Write decimal numbers in word form
- Write numbers expressed in word form as decimal fractions
- Express fractions as decimal numbers
- Express decimals as fractions
- Round decimal numbers to a designated place value
- Add, subtract, multiply, and divide decimal numbers
- Solve problems using addition, subtraction, multiplication, and division of decimal numbers
- Solve problems using a combination of addition, subtraction, multiplication, and division of decimal numbers
- Measure objects using a metric ruler or tape measure
- Convert between millimeters, centimeters, and meters

Chapter 3

WRITING AND ROUNDING DECIMALS

Place Value

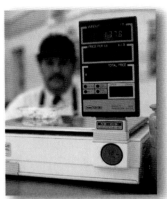

A **decimal** is a number that contains a decimal point. The **decimal point** separates the whole number from the part of the number less than a whole. The places to the right of the decimal point are called **decimal places,** and the digits in those places form the **decimal part** of the number. The place value table below shows some whole number and decimal places.

Whole Number Places			Decimal Point	Decimal Places		
hundreds	tens	ones	.	tenths	hundredths	thousandths
100	10	1	.	$\frac{1}{10}$	$\frac{1}{100}$	$\frac{1}{1,000}$
2	3	6	.	5	4	1

In **standard form** the number in the table is 236.541. In **word form** it is two hundred thirty-six and five hundred forty-one thousandths.

The table does not show all the whole number places, nor does it show all the decimal places. The next decimal place to the right is ten thousandths.

A scale gives a weight of 1.378 pounds. **Express that value in word form and expanded form.**

$$1 \quad . \quad 3 \quad\quad 7 \quad\quad 8$$
one **and** three hundred seventy-eight thousandths

Use the decimal name *thousandths* because the last digit is in the thousandths place.

Expanded form:
In *expanded form,* 1.378 is written as the sum of the values of the places.

$$1.378 = 1 + \left(\frac{3}{10}\right) + \left(\frac{7}{100}\right) + \left(\frac{8}{1,000}\right)$$
$$= 1 + 0.3 + 0.07 + 0.008$$

Language Box

The whole number part and the decimal part of a number are separated by *and* in the word form of a number. Do not use *and* anywhere else when reading or writing a number.

EXAMPLE A

Write 900.026 in word form and expanded form.

$$9 \; 0 \; 0 \quad . \quad 0 \; 2 \; 6$$
Word form: **nine hundred and twenty-six thousandths**

Expanded form:
$$900.026 = (9 \times 100) + \left(\frac{2}{100}\right) + \left(\frac{6}{1,000}\right)$$
$$= 900 + 0.02 + 0.006$$

Word form: 900.026 ⟶ *nine hundred and twenty-six thousandths*
Expanded form: 900.026 = 900 + 0.02 + 0.006

Example A shows the use of the word *and* in the word form of a number. Compare Example A and Example B, which follows. Note that *and* is not needed in Example B.

EXAMPLE B

Write 0.926 in word form and expanded form.

0 . 9 2 6

Word form: nine hundred twenty-six thousandths

Expanded form:

$$0.926 = \left(\frac{9}{10}\right) + \left(\frac{2}{100}\right) + \left(\frac{6}{1,000}\right)$$
$$= 0.9 + 0.02 + 0.006$$

Word form: 0.926 $\longrightarrow$ *nine hundred twenty-six thousandths.*
Expanded form: 0.926 = 0.9 + 0.02 + 0.006

EXAMPLE C

Write 12.05 in word form and expanded form.

12 . 05

Word form: twelve and five hundredths

Expanded form:

$$12.05 = (1 \times 10) + (2 \times 1) + \left(\frac{5}{100}\right)$$
$$= 10 + 2 + 0.05$$

Word form: 12.05 $\longrightarrow$ *twelve and five hundredths*
Expanded form: 12.05 = 12 + 0.05

The expanded table below shows decimal places to millionths. Below the table are examples of decimal numbers with 4, 5, or 6 decimal places. To read the decimal part of a number correctly, read the digits in the decimal places as a whole number and then add the name of the last decimal place. For example: Read 120.0305 as *one hundred twenty and three hundred five ten-thousandths*.

Whole Number Places			Decimal Point	Decimal Places					
hundreds	tens	ones	.	tenths	hundredths	thousandths	ten-thousandths	hundred-thousandths	millionths
100	10	1	.	$\frac{1}{10}$	$\frac{1}{100}$	$\frac{1}{1,000}$	$\frac{1}{10,000}$	$\frac{1}{100,000}$	$\frac{1}{1,000,000}$
1	2	0	.	0	3	0	5		

Examples:

0.0004 is read *four ten-thousandths*
0.0109 is read *one hundred nine ten-thousandths*

0.00007 is read *seven hundred-thousandths*
0.00451 is read *four hundred fifty-one hundred-thousandths*

0.000066 is read *sixty-six millionths*
0.003002 is read *three thousand two millionths*

Equivalent Decimals

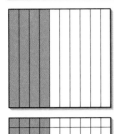

$$0.4 = \frac{4}{10} = \frac{4 \times 10}{10 \times 10} = \frac{40}{100} = 0.40$$

Decimals that represent the same part of a whole are **equivalent decimals.** Placing additional zeros to the right of the last digit of a decimal does not change the value of the decimal.

Of the 10 columns in a square, 4 are shaded, so 0.4 of the square is shaded.

Write equivalent decimals for 0.4.

Recall that 0.4 can be written as $\frac{4}{10}$. Multiply $\frac{4}{10}$ by $\frac{10}{10}$ and multiply $\frac{4}{10}$ by $\frac{100}{100}$ to write equivalent decimals for 0.4.

$$0.4 = \frac{4}{10} = \frac{4 \times 10}{10 \times 10} = \frac{40}{100} = 0.40$$

$$0.4 = \frac{4}{10} = \frac{4 \times 100}{10 \times 100} = \frac{400}{1,000} = 0.400$$

EXAMPLE D

Write two equivalent decimals for 0.20.

Express 0.20 as tenths:
$$0.20 = \frac{20}{100} = \frac{20 \div 10}{100 \div 10} = \frac{2}{10} = 0.2$$

Express 0.20 as thousandths:
$$0.20 = \frac{20}{100} = \frac{20 \times 10}{100 \times 10} = \frac{200}{1,000} = 0.200$$

Both 0.2 and 0.200 are equivalent to 0.20.

Comparing Decimals

In 2002, Chris Tomlinson broke the old long jump record of 8.23 m with a jump of 8.27 m—exceeding the old record by 0.04 m.

To compare two decimals, write them so that they each have the same number of decimal places. For example, to compare 0.05 and 0.505, write 0.05 with three decimal places. When both numbers are written with the same number of decimal places, it is easier to see which is less and which is greater.

Compare 0.05 and 0.505.

$$0.05 = 0.050 \longleftarrow \text{fifty \textbf{thousandths}}$$
$$0.505 = 0.505 \longleftarrow \text{five hundred five \textbf{thousandths}}$$

Fifty **thousandths** is less than five hundred five **thousandths**.

$$0.050 < 0.505$$
$$\downarrow \qquad\qquad \downarrow$$
$$0.05 < 0.505$$

EXAMPLE E

Compare 2.4 and 2.064.

$$2.4 = 2.400 \; \longleftarrow \; \text{two and four hundred thousandths}$$
$$2.064 = 2.064 \; \longleftarrow \; \text{two and sixty-four thousandths}$$

Two and four hundred thousandths is greater than two and sixty-four thousandths.

$$2.400 \; > \; 2.064$$
$$\downarrow \qquad\qquad \downarrow$$
$$2.4 \; > \; 2.064$$

Rounding Decimals

The rules for rounding a number to a decimal place are similar to the rules for rounding a whole number. Steps 1 and 2 are the same, but Step 3 is different. Instead of inserting placeholders, you will be eliminating some digits to the right of the decimal point.

A giant pumpkin weighed 1,206.392 lb.

Round 1,206.392 to the nearest tenth.

The digit in the tenths place is 3. The digit 9 to the right indicates that 3 should be increased to 4. Eliminate the digits in the hundredths and thousandths places.

$$1,206.392 \longrightarrow 1,206.4$$

Rules for Rounding to a Decimal Place		
Step 1	**Step 2**	**Step 3**
Identify the digit in the place you want to round to.	Look at the digit to the right. If it is 5 or greater, increase the digit in the place being rounded by one (round up). Otherwise, keep that digit the same (round down).	Eliminate all digits to the right of the decimal place to which the number was rounded.

EXAMPLE F

Round 14.3617 to the nearest hundredth.

The digit in the hundredths place is 6. The digit to the right is 1, which is less than 5; this indicates that 6 should remain the same. Eliminate the digits to the right of 6 — 1 in the thousandths place and 7 in the ten-thousandths place.

$$14.3617 \longrightarrow 14.36$$

To the nearest hundredth, 14.3617 is 14.36.

EXERCISES 3-1

Write the following numbers in word form and expanded form.

1. 0.063　　　　**2.** 5.9　　　　**3.** 20.07

4. 0.175　　　　**5.** 3.0004　　　　**6.** 8.35

If the following decimals are equivalent, write *yes*. If the decimals are not equivalent, write *no*.

7. 0.06, 0.60　　　　**8.** 3.7, 3.70　　　　**9.** 0.014, 0.104

10. 9.03, 9.030　　　　**11.** 2.55, 2.555　　　　**12.** 0.147, 0.15

Compare the following pairs of numbers. Use < or > .

13. 2.16  2.61　　**14.** 0.925  0.952　　**15.** 0.01  0.5

16. 64.7  6.47　　**17.** 8.405  8.504　　**18.** 0.34  0.304

Round the following numbers to the nearest tenth.

19. 6.55　　　　**20.** 0.07　　　　**21.** 75.74

Round the following numbers to the nearest hundredth.

22. 6.007　　　　**23.** 25.339　　　　**24.** 100.702

25. Dan's job requires 1785.475 meters of cable for a wiring job. How much cable is this to the nearest meter?

26. Emily used the following amounts of cable during the last three months of the year: October—14 726.57 meters; November—12 562.51 meters; and December—9754.486 meters. Round the amount for each month to the nearest

    (a) ten meters.

    (b) meter.

    (c) tenth meter.

ADDING DECIMALS

Receipt
$14.95 TX
5.49 TX
24.99 TX
$45.43 Subtotal
2.27 Tax
$47.70 Total
Thank you for
shopping with us

The key to adding whole numbers is to align the numbers so that ones are added to ones, tens are added to tens, and so on. The same is true for adding *decimals*. We want to add tenths to tenths, and hundredths to hundredths. If all the decimal points are written in a vertical column, this will happen automatically.

When all the decimals are properly aligned, regroup as needed by the same method used for regrouping whole numbers.

Examples:

$$\begin{array}{r} \overset{1}{1}7 \\ +48 \\ \hline 65 \end{array} \qquad \begin{array}{r} \overset{1}{5}.9 \\ +0.4 \\ \hline 6.3 \end{array} \qquad \begin{array}{r} \overset{1}{2}.05 \\ +13.48 \\ \hline 15.53 \end{array}$$

When numbers to be added do not have the same number of decimal places, write equivalent decimals so that all the numbers have the same number of decimal places.

Language Box

The zeros that are written to make equivalent decimals are sometimes called *placeholders*.

Find the sum of 4.5, 23.037, and 0.78.

STEP 1	STEP 2	STEP 3
Write the numbers with the decimal points in a column.	Write zeros as needed so that all the numbers have the same number of decimal places.	Add. Write the decimal point in the sum directly below the other decimal points.
$$\begin{array}{r} 4.5 \\ 23.037 \\ +\ 0.78 \\ \hline \end{array}$$	$$\begin{array}{r} 4.500 \\ 23.037 \\ +\ 0.780 \\ \hline \end{array}$$	$$\begin{array}{r} 4.500 \\ 23.037 \\ +\ 0.780 \\ \hline 28.317 \end{array}$$

The sum is 28.317.

EXAMPLE A

Find the sum of 8.98, 7, and 16.402.

STEP 1	STEP 2	STEP 3
Write the numbers with the decimal points in a column.	Write zeros as needed so that all the numbers have the same number of decimal places.	Add. Write the decimal point in the sum directly below the other decimal points.
$$\begin{array}{r} 8.98 \\ 7. \\ +16.402 \\ \hline \end{array}$$	$$\begin{array}{r} 8.980 \\ 7.000 \\ +16.402 \\ \hline \end{array}$$	$$\begin{array}{r} 8.980 \\ 7.000 \\ +16.402 \\ \hline 32.382 \end{array}$$

$8.98 + 7 + 16.402 = 32.382$

EXAMPLE B

The total resistance in a series circuit is the sum of the individual resistances. What is the total resistance in a circuit with resistances of 0.15 Ω and 0.908 Ω?

STEP 1	**STEP 2**	**STEP 3**
Write the numbers with the decimal points in a column.	Write zeros as needed so that all the numbers have the same number of decimal places.	Add. Write the decimal point in the sum directly below the other decimal points.

STEP 1:
$$0.15$$
$$+0.908$$

STEP 2:
$$0.150$$
$$+0.908$$

STEP 3:
$$0.150$$
$$+0.908$$
$$1.058$$

$$0.15 + 0.908 = 1.058$$

The total resistance in this circuit is 1.058 Ω.

Helpful Hint

A calculator may delete rightmost zeros in a result. For example, it may show 6.7 instead of 6.70 as the result for 0.53 + 6.17.

EXERCISES 3-2

Add.

1. 3.46 + 2.1

2. 98.01 + 0.453

3. 60.1 + 7.9

4. 3.0005 + 1.078

5. 0.53 + 6.17

6. 0.45 + 0.8

7. 6.7 + 1.23 + 6.008

8. 3.29 + 5.71 + 83

Solve.

9. A chemist combined 10.6 grams of a compound with 3.5 grams of another compound. What is the total number of grams she combined?

10. Would a $20 bill be enough to pay for items that cost $15.25, $2.89, and $1.59?

11. Bill connected the following resistors together in a series circuit: 0.64 Ω, 1.007 Ω, 1.3 Ω, 0.286 Ω, and 1.35 Ω. What was the total resistance in this circuit?

12. Emma purchased some electrical equipment and paid the following amount for each: digital voltmeter, $44.62; phase rotation meter, $99.95; AC digital ammeter, $269.28; AC millivoltmeter, $120; digital insulation tester, $427.51; digital capacitor tester, $119.50.

(a) Round each price to the nearest $10 and add to get the approximate total cost for all of these items, before tax.

(b) What was the total purchase price before tax?

13. What's the Error? The statement 0.3 + 0.3 = 0.6 is true. However, the statement 0.8 + 0.8 = 0.16 is false. What is the error in 0.8 + 0.8 = 0.16?

SUBTRACTING DECIMALS

The key to subtracting decimals is to align the numbers so that the decimal points are written in a vertical column. Just as in addition, zeros are written as needed so that both numbers have the same number of decimal places.

When regrouping is necessary in a decimal subtraction problem, it is done the same way that it is done with whole numbers.

A water station is placed at the 7.25-km mark in a 10-km race. How far from the finish line is the water station?

Subtract 7.25 from 10.

STEP 1	STEP 2	STEP 3
Write the numbers with the decimal points in a column.	Write zeros as needed so that both numbers have the same number of decimal places.	Subtract. Write the decimal point in the difference directly below the other decimal points.

$$\begin{array}{r} 10. \\ -\ 7.25 \\ \hline \end{array}$$

$$\begin{array}{r} 10.00 \\ -\ 7.25 \\ \hline \end{array}$$

$$\begin{array}{r} 10.00 \\ -\ 7.25 \\ \hline 2.75 \end{array}$$

The water station is 2.75 km from the finish line.

EXAMPLE A

Subtract 0.24 from 5.3.

STEP 1	STEP 2	STEP 3
Write the numbers with the decimal points in a column.	Write zeros as needed so that both numbers have the same number of decimal places.	Subtract. Write the decimal point in the difference directly below the other decimal points.

$$\begin{array}{r} 5.3 \\ -0.24 \\ \hline \end{array}$$

$$\begin{array}{r} 5.30 \\ -0.24 \\ \hline \end{array}$$

$$\begin{array}{r} 5.30 \\ -0.24 \\ \hline 5.06 \end{array}$$

$$5.3 - 0.24 = 5.06$$

EXAMPLE B

A locating pin is supposed to be 9.5 mm in diameter. Its actual diameter is 9.875 mm. How much is it oversized?

The answer will be the solution to 9.875 − 9.5.

STEP 1	STEP 2	STEP 3
Write the numbers with the decimal points in a column.	Write zeros as needed so that both numbers have the same number of decimal places.	Subtract. Write the decimal point in the difference directly below the other decimal points.

$$\begin{array}{r} 9.875 \\ -9.5 \\ \hline \end{array}$$

$$\begin{array}{r} 9.875 \\ -9.500 \\ \hline \end{array}$$

$$\begin{array}{r} 9.875 \\ -9.500 \\ \hline 0.375 \end{array}$$

9.875 − 9.5 = 0.375

The pin is 0.375 mm too large.

EXERCISES 3-3

Subtract.

1. 3.24 − 2.1   **2.** 8.01 − 0.923   **3.** 200 − 7.18

4. 5.37 − 4.73   **5.** 6.121 − 1.4   **6.** 41.44 − 23.008

7. 0.3 − 0.054   **8.** 15.7 − 4.005   **9.** 29 − 5.1

Solve.

10. A 60-watt incandescent lightbulb produces about 13.52 lumens/watt. A 60-watt Compact Fluorescent Light (CFL) gives about 59.33 lumens/watt. How many more lumens per watt does a CFL bulb produce than a comparable incandescent bulb?

11. The voltage drop in a transmission line is 4.8 V. If the source voltage is 240V, what is the voltage at the end of the transmission line?

12. What is the difference between the length and the width of the rectangle shown?

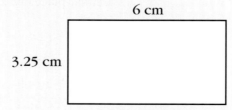

6 cm

3.25 cm

13. Challenge: The width of a certain machine part is 2.3 centimeters with a tolerance of ±0.008 centimeter. Find the maximum width by adding 0.008 to the measurement; find the minimum width by subtracting 0.008 from the measurement.

Helpful Hint

When the difference between two numbers is less than 1, write a zero in the ones place in the difference. Example:

$$\begin{array}{r} 1.6 \\ -1.2 \\ \hline 0.4 \end{array}$$

MULTIPLYING DECIMALS

Multiplying decimals is similar to multiplying whole numbers. The difference is in the placement of the decimal point in the product.

To multiply decimals, follow these steps:

STEP 1

Multiply as you would multiply whole numbers.

STEP 2

Find the sum of the decimal places in the factors.

STEP 3

Insert a decimal point so that the number of decimal places in the product is the sum you found in Step 2. If the decimal is less than 1, it is preferred to write a zero in the ones place.

Find the cost of 2.2 pounds of grapes at $1.59 per pound.

Multiply $1.59 by 2.2.

STEP 1	STEP 2	STEP 3
Multiply as you would multiply whole numbers.	Find the sum of the decimal places in the factors.	Insert a decimal point in the product.

STEP 1

$$\begin{array}{r} \$1.59 \\ \times\ \ 2.2 \\ \hline 318 \\ 318\ \ \\ \hline 3498 \end{array}$$

STEP 2

←——— 2 places
←——— 1 place

The product will have 2 + 1 = 3 places.

STEP 3

$$\begin{array}{r} \$1.59 \\ \times\ \ 2.2 \\ \hline 318 \\ 318\ \ \\ \hline \$3.498 \end{array}$$ ←— 3 places

Round the product $3.498 to $3.50.

The cost of 2.2 pounds of grapes at $1.59 per pound is $3.50.

EXAMPLE A

Find the product of 0.6 and 0.8.

STEP 1

Multiply as you would multiply whole numbers.

$$\begin{array}{r} 0.8 \\ \times 0.6 \\ \hline 48 \end{array}$$

STEP 2

Find the sum of the decimal places in the factors.

←——— 1 place
←——— 1 place

The product will have 1 + 1 = 2 places.

STEP 3

Insert a decimal point in the product.

$$\begin{array}{r} 0.8 \\ \times 0.6 \\ \hline .48 \end{array}$$ ←— 2 places

$0.6 \times 0.8 = 0.48.$

In Example B below, it is necessary to place an additional zero at the left of the product before inserting the decimal point. Notice also that the zero in the hundred-thousandths place must be kept until after the decimal point is inserted.

EXAMPLE B

Dylan needs 0.098 meter (about $3\frac{7}{8}$ inches) of wire that costs $0.75 a meter. How much will Dylan need to pay for this wire?

To solve this problem, you need to find the product of 0.098×0.75.

STEP 1	STEP 2	STEP 3
Multiply as you would multiply whole numbers.	Find the sum of the decimal places in the factors.	Insert a decimal point in the product.

STEP 1

$$
\begin{array}{r}
0.098 \\
\times\ 0.75 \\
\hline
490 \\
+\ 686 \\
\hline
7\ 350
\end{array}
$$

STEP 2

←——— 3 places
←——— 2 places

The product will have $2 + 3 = 5$ places.

STEP 3

$$
\begin{array}{r}
0.098 \\
\times\ 0.75 \\
\hline
490 \\
+\ 686 \\
\hline
0.07350
\end{array}
$$
←— 5 places

$0.098 \times 0.75 = 0.0735$

Since this answer is in dollars, it is rounded off to the nearest hundredth.

The price of this wire is $0.07 = 7¢.

EXERCISES 3-4

Multiply.

1. 1.28×3.1

2. $2.5 \times \$8.49$

3. 0.06×0.1

4. 5.03×4.7

5. 90×0.004

6. $1.25 \times \$6.75$

7. 0.029×6.6

8. 0.625×8

9. 100×0.0007

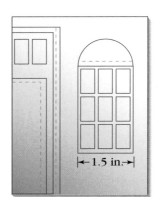

←1.5 in.→

Solve.

10. A window shown on a blueprint is 1.5 inches wide. If the actual window is 15 times that wide, what is its actual width?

11. The mass of one penny is 2.67 grams. What is the mass of 250 pennies?

12. How many liters are there in 12 bottles of water if each bottle contains 0.71 liter?

13. Number 14 copper wire has a resistance of 0.0002525 Ω/ft.

(a) What is the resistance of 8.75 feet of #14 wire?

(b) Round your answer in (a) to the nearest thousandth ohm.

14. Chloé worked 12.75 hours on a wiring job. Her salary is $16.90/hour. How much did she earn for this job?

15. Challenge: $0.8 \times 0.08 \times 0.008 = ?$

DIVIDING DECIMALS

When the divisor in a division problem is a whole number and the dividend is a decimal, place the decimal point in the quotient directly above the decimal point in the dividend. Then divide as you would with whole numbers. This is shown in the division below.

$$4\overline{)8.4} = 2.1$$

When the divisor contains one or more decimal places, follow these steps:

STEP 1

Move the decimal point in the divisor to the right until the divisor becomes a whole number.

STEP 2

Move the decimal point in the dividend to the right the same number of places you moved the decimal point in the divisor. Place the decimal point in the quotient directly above the newly established decimal point in the dividend.

STEP 3

Divide.

How many dimes are there in $23.70?

Divide $23.70 by $0.10.

STEP 1

Move the decimal point in the divisor until the divisor is a whole number.

$$0.10\overline{)23.70}$$

STEP 2

Move the decimal point in the dividend the same number of places. Place the decimal point in the quotient.

$$0.10\overline{)23.70.}$$

STEP 3

Divide.

$$
\begin{array}{r}
237. \\
010.\overline{)2370.} \\
-20 \\
\hline
37 \\
-30 \\
\hline
70 \\
-70 \\
\hline
0
\end{array}
$$

How many dimes are there in $23.70?
There are 237 dimes in $23.70

EXAMPLE A

Divide 38.626 by 3.1.

STEP 1	STEP 2	STEP 3
Move the decimal point in the divisor until the divisor is a whole number.	Move the decimal point in the dividend the same number of places. Place the decimal point in the quotient.	Divide.

$3.1.\overline{)38.626}$

$3.1.\overline{)38.6.26}$

$$\begin{array}{r} 12.46 \\ 31.\overline{)386.26} \\ -31 \\ \hline 76 \\ -62 \\ \hline 142 \\ -124 \\ \hline 186 \\ -186 \\ \hline 0 \end{array}$$

$38.626 \div 3.1 = 12.46$

In the examples that follow, "short division" is used. The digits are shown in the quotient, but the "multiply," "subtract," and "bring down" steps are not shown. Arrows indicate the movement of the decimal points, and color is used for emphasis.

EXAMPLE B

Divide 14.4 by 0.12.

$0.12.\overline{)14.40.}^{120.}$

Annex a 0 in the hundredths place of the dividend to be able to move the decimal point the required number of places.

Language Box

We *annex* a zero to 14.4 to get the equivalent decimal 14.40.

$14.4 \div 0.12 = 120$

EXAMPLE C

Divide 64 by 0.008.

$0.008.\overline{)64.000.}^{8,000.}$

Write a decimal point and annex three zeros to the dividend to be able to move the decimal point the required number of places.

$64 \div 0.008 = 8,000$

EXAMPLE D

Divide 4.22 by 0.4.

$0.4.\overline{)4.2.20}^{10.55}$

Annex a 0 in the dividend to be able to finish dividing.

$4.22 \div 0.4 = 10.55$

EXAMPLE E

Divide 1.212 by 4.

$$\begin{array}{r} 0.303 \\ 4\overline{)1.212} \end{array}$$

Write a 0 in the ones place in the quotient that would otherwise be empty. This is not required, but it is preferred.

$$1.212 \div 4 = 0.303$$

EXAMPLE F

A generator delivers 31.4 kW of electric power. The generator has an efficiency of 0.9. If its power input is $\frac{31.4\text{kW}}{0.9}$, determine its power input to the nearest tenth kilowatt.

To find the answer, divide 31.4 by 0.9.

$$\begin{array}{r} 34.88 \\ 0.9\overline{)31.4.00} \\ \underline{-27} \\ 4\ 4 \\ \underline{-3\ 6} \\ 8\ 0 \\ \underline{-7\ 2} \\ 8\ 0 \\ \underline{-7\ 2} \\ 8 \end{array}$$

The quotient must be rounded to the nearest tenth, so divide until the quotient has a digit in the hundredths place.

The digit in the hundredths place is greater than 5. Increase the 8 in the tenths place to 9.

$$34.88 \longrightarrow 34.9$$

To the nearest tenth, $31.4 \div 0.9 = 34.9$. Thus, to the nearest tenth kilowatt, this generator's power input is 34.9 kW.

Helpful Hint

Your calculator can be set to display a predetermined number of decimal places. Use the keystrokes [2nd] [FIX] to select the number of decimal places for an answer.

EXERCISES 3-5

Divide.

1. $5.25 \div 0.25$ **2.** $13.31 \div 1.1$ **3.** $81 \div 0.09$

4. $49.49 \div 7$ **5.** $17.68 \div 2.6$ **6.** $54.3 \div 0.5$

7. $66 \div 3.3$ **8.** $62 \div 6.2$ **9.** $33.5 \div 0.04$

Divide. Round the quotient to the nearest tenth.

10. $20 \div 0.3$ **11.** $125.5 \div 6$ **12.** $0.21 \div 0.08$

Solve.

13. It took a race car driver 0.7 hour to drive a distance of 74.83 miles. What was the average speed during that time?

14. A current of 0.0478 Ω flows when the applied voltage is 45.41 V. If the resistance, in ohms, is $\frac{45.41}{0.0478}$, what is the resistance in this circuit?

15. A series circuit has nine equal resistors. If the total voltage for the circuit is 20.025 V, what is the voltage across each resistor?

16. What's the Error? Explain the error in the work shown below.

$$\begin{array}{r} 0.1 \\ 2.4\overline{)0.024} \end{array}$$

CONVERTING DECIMALS AND FRACTIONS

Decimal to a Fraction or a Mixed Number

Thickness 0.03 in.

2.12 in.

3.38 in.

A decimal number less than 1 can be converted to a proper fraction. To convert, use the word form of the decimal number. The first part of the word form is the numerator of the fraction, and the second part is the denominator. The decimal 0.03 is converted below.

$$0.03 \longrightarrow \textit{three hundredths} \longrightarrow \frac{3}{100}$$

A decimal number greater than 1 can be converted to a mixed number. The whole number part remains the same, and the decimal part is converted to a fraction.

Write 3.38 as a mixed number in simplest form.

$$3.38 \longrightarrow \textit{three and thirty-eight hundredths} \longrightarrow 3\frac{38}{100}$$

The fraction $\frac{38}{100}$ can be simplified.

$$\frac{38 \div 2}{100 \div 2} = \frac{19}{50}$$

So, $3.38 = 3\frac{19}{50}$.

EXAMPLE A

Convert 0.215 to a fraction in simplest form.

$$0.215 \longrightarrow \textit{two hundred fifteen thousandths} \longrightarrow \frac{215}{1,000}$$

The fraction $\frac{215}{1,000}$ can be simplified.

$$\frac{215 \div 5}{1,000 \div 5} = \frac{43}{200}$$

$$0.215 = \frac{43}{200}$$

EXAMPLE B

Convert 4.071 to a mixed number in simplest form.

$$4.071 \longrightarrow \textit{four and seventy-one thousandths} \longrightarrow 4\frac{71}{1,000}$$

The fraction $\frac{71}{1,000}$ is already in simplest form.

$$4.071 = 4\frac{71}{1,000}$$

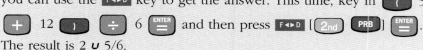

Your calculator can be used to convert a decimal to a fraction by using the keystroke F◄►D [2nd   PRB]. To convert 4.071 to a mixed number in simplest form, press 4.071 F◄►D [2nd   PRB] ENTER/= . The result is 4 u 71/1000.

Remember back in Chapter 2 when you tried to use your calculator to convert $\frac{5 + 12}{6}$ to a mixed number? The calculator gave a "SYNTAX Error" message. However, if you first convert the number to a decimal, you can use the F◄►D key to get the answer. This time, key in (5 + 12) ÷ 6 ENTER/= and then press F◄►D [2nd   PRB] ENTER/= . The result is 2 u 5/6.

Fraction or Mixed Number to a Decimal

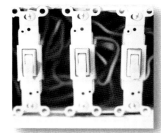

Using division, any proper fraction can be converted to a decimal. Divide the numerator by the denominator to convert a fraction to a decimal. In some cases, the division will result in a remainder of zero. In other cases, no matter how many zeros are annexed, there will always be a remainder.

Of 25 switches, 9 were defective.

Convert $\frac{9}{25}$ to a decimal.

Divide 9 by 25.

$$\begin{array}{r} 0.36 \\ 25\overline{)9.00} \\ -75 \\ \hline 150 \\ -150 \\ \hline 0 \end{array}$$

Annex zeros in the dividend as needed to carry out the division.

$\frac{9}{25} = 0.36$

EXAMPLE C

Convert $\frac{5}{16}$ to a decimal.

Divide 5 by 16.

$$\begin{array}{r} 0.3125 \\ 16\overline{)5.0000} \\ -4\,8 \\ \hline 20 \\ -16 \\ \hline 40 \\ -32 \\ \hline 80 \\ -80 \\ \hline 0 \end{array}$$

Annex zeros in the dividend as needed to carry out the division.

When a remainder of 0 is obtained, the decimal quotient is equal to the fraction.

$\frac{5}{16} = 0.3125$

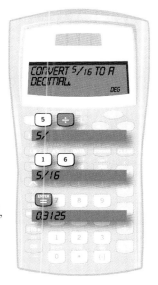

EXAMPLE D

Convert $\frac{1}{3}$ to a decimal.

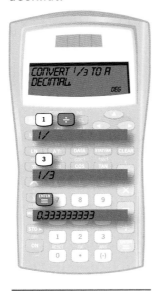

Divide 1 by 3.

$$\begin{array}{r} 0.333 \\ 3\overline{)1.000} \\ \underline{-9} \\ 10 \\ \underline{-9} \\ 10 \\ \underline{-9} \\ 1 \end{array}$$

The result of every subtraction step in this problem is 1. The remainder will never be zero. This quotient is called a repeating decimal. The quotient can be rounded to give an approximation for the fraction or can be written with a repeating decimal symbol. A bar written over a digit (or digits) indicates that it repeats.

$\frac{1}{3} = 0.\overline{3}$ The bar over the 3 indicates a repeating decimal that repeats 3.

EXAMPLE E

A transistor has $\frac{7}{22}$ mA flowing through it. Express this current as a decimal.

Divide 7 by 22.

$$\begin{array}{r} 0.3181818 \\ 22\overline{)7.0000000} \end{array}$$

The remainder in this problem will never be zero. The digits 1 and 8 repeat in the quotient. To show the repeating digits in your answer, place a bar over the digits 1 and 8.

$\frac{7}{22} = 0.3\overline{18}$

EXAMPLE F

Convert $\frac{3}{11}$ to a decimal. Round to the nearest hundredth.

Language Box

The symbol $\approx$ means "is approximately equal to."

$$\begin{array}{r} 0.272 \\ 11\overline{)3.000} \\ \underline{-22} \\ 80 \\ \underline{-77} \\ 30 \\ \underline{-22} \\ 8 \end{array}$$

Carry out the division to the thousandths place in the quotient.

The digit in the hundredths place is 7. The digit 2 to the right indicates that 7 should remain the same.

0.272 $\longrightarrow$ 0.27

$\frac{3}{11} \approx 0.27$

To convert a mixed number to a decimal, first identify the whole number. The whole number will be the same in the decimal number. Then convert the fraction part of the mixed number to a decimal.

EXAMPLE G

Write $6\frac{1}{2}$ as a decimal.

The whole number is 6. Convert $\frac{1}{2}$ to a decimal. Combine the whole number and the decimal part to form the decimal number.

$$1\overline{)1.0}^{\,0.5}$$

$$6\frac{1}{2} = 6 + 0.5 = 6.5$$

$$6\frac{1}{2} = 6.5$$

It is sometimes possible to convert a fraction to a decimal by first writing the fraction as an equivalent fraction with a denominator of 10, 100, or another power of 10.

EXAMPLE H

Convert $\frac{7}{20}$ to a decimal by means of equivalent fractions.

$$\frac{7}{20} = \frac{?}{100}$$

$$\frac{7}{20} = \frac{7 \times 5}{20 \times 5} = \frac{35}{100}$$

The fraction $\frac{7}{20}$ can be written as an equivalent fraction with a denominator of 100 because $20 \times 5 = 100$. Multiply both the numerator and denominator by 5 to obtain the equivalent fraction.

The fraction $\frac{35}{100}$ can be written 0.35.

$$\frac{7}{20} = 0.35$$

EXERCISES 3-6

Convert each decimal to a fraction or a mixed number in simplest form.

1. 0.64   **2.** 5.3   **3.** 11.07

4. 0.19   **5.** 3.08   **6.** 5.95

Convert each fraction or mixed number to a decimal.

7. $1\frac{3}{4}$ **8.** $\frac{7}{8}$ **9.** $8\frac{7}{100}$

10. $\frac{25}{40}$ **11.** $10\frac{1}{10}$ **12.** $\frac{40}{50}$

Convert each fraction or mixed number to a decimal. Round your answer to the nearest hundredth.

13. $\frac{4}{9}$ **14.** $5\frac{2}{7}$ **15.** $\frac{5}{6}$

Use equivalent fractions to convert each fraction or mixed number to a decimal.

16. $\frac{21}{25}$ **17.** $4\frac{3}{50}$ **18.** $\frac{13}{20}$

Solve.

19. A lamp requires 0.75 A. If the total current in a circuit is 60.0 A, what fraction of the total current is used by the lamp?

20. Ryan's wages were raised from $15.84/hour to $16.83/hour.

(a) How much money was this increase in his hourly wage?

(b) Express the increase as a fraction of his wages before the raise.

(c) Convert your answer in (b) to a decimal fraction.

21. A package weighs 3.08 kilograms. Express the weight of the package as a mixed number in simplest form.

22. In one polling district, $\frac{5}{8}$ of the registered voters voted in the last election. What is the decimal form of $\frac{5}{8}$?

23. Challenge: Jerry's black lab pup weighed 35 pounds 4 ounces at his last visit to the vet. Convert the pup's weight to pounds. (16 oz = 1 lb)

READING A METRIC RULE

Many lengths in the metric system of measurement are given in millimeters (mm), centimeters (cm), and meters (m). One meter is equal to 100 centimeters, or 1000 millimeters. One centimeter is equal to 10 millimeters. A length of 1 meter is shown in the illustration below.

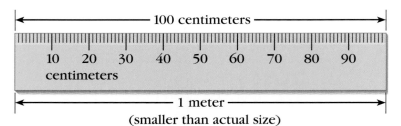

(smaller than actual size)

A part of a metric rule is shown below in actual size. Every centimeter on this ruler is divided into 10 equal parts by tick marks. The distance between each pair of tick marks is $\frac{1}{10}$ centimeter, or 1 millimeter.

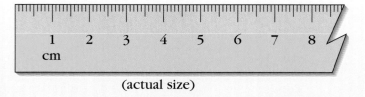

(actual size)

To understand how to read a metric rule and give a metric measure, use the diagram below.

What is the width of the strip of film?

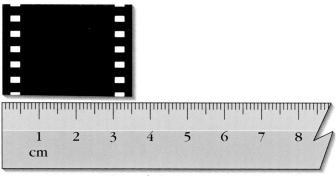

(actual size)

The width of the film is 35 millimeters. To express 35 millimeters as centimeters, divide 35 by 10.

35 ÷ 10 = 3.5

The width of the film is 35 mm, or 3.5 cm.

EXAMPLE A

What is the size of the opening of the wrench?

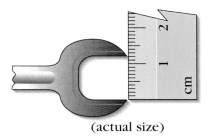

(actual size)

The size of the opening of the wrench is 12 mm.

EXAMPLE B

One end of a dowel is placed at 0 on a metric rule. The other end is shown at right. What is the length of the dowel? Give the answer in millimeters, in centimeters, and in meters.

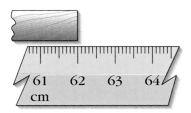

The end of the dowel is shown at 62 cm. Multiply the length in centimeters by 10 to express it in millimeters. Divide the length in centimeters by 100 to express it in meters.

$$620 \text{ mm} \xleftarrow{\times 10} 62 \text{ cm} \xrightarrow{\div 100} 0.62 \text{ m}$$

The length of the dowel is 620 mm, 62 cm, or 0.62 m.

EXERCISES 3-7

What measurement is indicated at each point labeled on the rule? Give each measurement in both millimeters and centimeters.

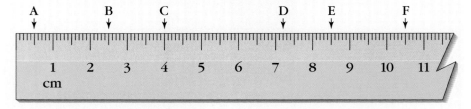

1. Point A   **2.** Point B   **3.** Point C

4. Point D   **5.** Point E   **6.** Point F

7. A cable is 90 meters long. What is the length of the cable in centimeters?

8. A cable is 9.27 meters long.

(a) What is the length of the cable in centimeters?

(b) What is the length of the cable in millimeters?

9. A cable is 7.045 m long and is to be cut into pieces that are each 23.4 cm long. Allow 1.5 mm for each saw cut.

(a) What is the length of the cable in millimeters?

(b) What is the length of each piece in millimeters?

(c) Round each length to the nearest centimeter and approximate the number of 23.4-cm pieces that can be made.

(d) How many 23.4-cm pieces can be made from this cable?

(e) What is the length of cable left over?

10. What's the Error? Your coworker describes a 4-millimeter drill bit as being 40 centimeters in diameter. Describe the error and give the correct measurement in centimeters.

CHAPTER 3 REVIEW EXERCISES

1. Write the number 27.053 in word form and expanded form.

2. Round 142.0547

(a) to the nearest ten.

(b) to the nearest tenth.

(c) to the nearest hundredth.

3. Perform the indicated operation.

   (a) 25.24 + 3.27

   (b) 6.57 − 4.281

   (c) 4.2 × 2.5

   (d) 11.2 ÷ 1.75

   (e) 107.35 + 29.561 + 1,039.4

4. Perform the indicated operation.

   (a)   4.278   (b)   12.057   (c)   46.527
       + 37.146   3.0685   − 18.23
                                 143.
                               + 5.07

   (d)   102.035   (e) 0.54)‾43.268‾
         × 2.8

5. Convert each decimal to a fraction or mixed number in simplest form.

   (a) 0.875

   (b) 17.25

6. Convert each fraction or mixed number to a decimal.

   (a) $\frac{3}{5}$

   (b) $9\frac{13}{25}$

7. Convert each fraction or mixed number to a decimal. Round your answer to the nearest hundredth.

   (a) $\frac{13}{16}$

   (b) $12\frac{17}{22}$

8. Emily used the following amounts of cable during the first three months of the year: January—7608.85 meters; February—9006.028 meters; and March—11 157.203 meters. Round the amount for each month to the nearest

   (a) ten meters.

   (b) meter.

   (c) tenth meter.

9. In a series circuit, the total resistance is the sum of the individual resistances. The following resistors are in a series circuit: 1.64 Ω, 0.88 Ω, 1.46 Ω, 0.654 Ω, 1.12 Ω, and 1.35 Ω. What was the total resistance in this circuit?

10. One week Sam earned $758.35. Her income tax of $106.40 was deducted from her paycheck. She also had $47.02 withheld for Social Security. Assuming that nothing else was withheld, what was her take-home pay?

11. A certain dry cell battery with a terminal voltage of 1.2 V has a wire of 0.25 Ω connected to it. According to Ohm's law, the current, in amperes (A), is $\frac{1.2}{0.25}$. Determine the current in the wire.

12. Nick buys a new television set for $598.32. He will not have to pay any interest if he can pay for it in one year. He decides to pay one half of the amount when he purchases the set and the rest in 12 equal payments. How much is each of the 12 payments?

13. A parallel circuit that has resistances of 6.25 Ω and 7.5 Ω has a total resistance of R, where $\frac{1}{R} = \frac{1}{6.25} + \frac{1}{7.5}$. What is the total resistance of this circuit?

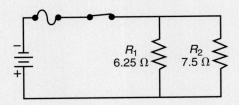

14. A four-resistor series circuit has resistances of 4.75 MΩ (megaohms), 2.6 MΩ, 0.95 MΩ, and 3.15 MΩ.

(a) Round each number to the nearest whole number and find the approximate total resistance.

(b) What is the actual total resistance?

15. What resistance must be placed in a series circuit with a resistance of 15.75 Ω in order to have a total resistance or 22.42 Ω?

16. A parallel circuit has four resistors of 12.5 Ω, 7.25 Ω, 6.25 Ω, and 8 Ω. The total resistance, R, in ohms, is given by $\frac{1}{R} = \frac{1}{12.5} + \frac{1}{7.25} + \frac{1}{6.25} + \frac{1}{8}$.

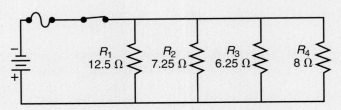

(a) Convert each of the fractions to a decimal. Round each decimal to the nearest thousandth.

(b) What is the sum of the decimals in (a)?

(c) What is the total resistance of this circuit? Write the answer as a decimal to the nearest thousandth.

(d) Write the answer as a mixed number in simplest form.

17. A cable that is 4.35 m long is to be cut into 18.5-cm lengths. Allow 1.5 mm for each saw cut.

(a) What is the length of the cable in millimeters?

(b) What is the length of each piece in millimeters?

(c) Round each length to the nearest centimeter and approximate the number of 18.5-cm pieces that can be made.

(d) How many 18.5-cm pieces can actually be made from this cable?

(e) What is the length of cable left over?

18. The total resistance when three resistors of 4.5 Ω, 6.2 Ω, and 14.7 Ω are connected in parallel is

$$R = \frac{4.5 \times 6.2 \times 14.7}{4.5 \times 6.2 + 4.5 \times 14.7 + 6.2 \times 14.7} \; \Omega$$

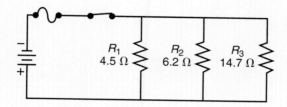

(a) What is the numerator of this fraction?

(b) Rewrite the denominator by finding each of the products. Do not do the addition.

(c) Add your answers in (b) to find the value of the denominator.

(d) Find the total resistance of this circuit. Round your answer to the nearest hundredth.

(e) Write the answer as a mixed number in simplest form.

Building a Foundation in Mathematics

Introducing Integers

Adding Integers

Subtracting Integers

Multiplying Integers

Dividing Integers

Mon	Tue	Wed	Thu
☁	❄	❄	☀
−2°	1°	5°	10°

Overview

Integers are common in everyday use and in many electrical applications. For example, it is not unusual in the winter to see temperatures below zero written as negative numbers. Integers are also used to indicate direction and distance from a reference point.

Opposites, such as up and down, left and right, and north and south, may be written using positive and negative signs. In this chapter you will learn about integers and how to perform the basic arithmetic operations with them.

Integers

Objectives

After completing this chapter, you will be able to:

- Write integers using a number scale
- Determine the absolute value of an integer
- Add, subtract, multiply, and divide integers
- Solve problems using addition, subtraction, multiplication, and division of integers
- Use your calculator to work with integers

INTRODUCING INTEGERS

Opposites

Badwater in Death Valley State Park is at −282 feet.

Every number except zero has an opposite. **Opposites** are a pair of numbers that are the same distance from zero on the number line.

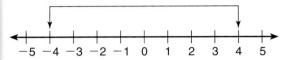

A pair of opposites is shown on the number line above. All of the following state the same fact:

• 4 and −4 are a pair of opposites.

• −4 is the opposite of 4.

• 4 is the opposite of −4.

The set of **integers** is composed of the counting numbers (1, 2, 3, 4, . . .) and their opposites, together with zero.

Negative integers are less than zero.	Zero is an integer that is neither positive nor negative.	Positive integers are greater than zero.

Integers are sometimes called *signed numbers*, because a "−" sign is used to indicate a negative number, and a "+" sign is sometimes used to indicate a positive number. In this book, positive numbers are not usually shown with a "+" sign.

Positive and negative integers can be used to show that a measure is greater than zero or less than zero. For example, 5°F (Fahrenheit) indicates a temperature 5 degrees greater than zero, and −5°F indicates a temperature 5 degrees less than zero.

Another important use for positive and negative integers is to show increases and decreases. For example, a $40 deposit to a bank account could be indicated by +40, and a $40 withdrawal could be indicated by −40.

Use an integer to show a payment of $200 from a checking account.

The integer −200 could be used to describe a payment from a checking account.

Helpful Hint

You can write a positive number *with* or *without* the "+" sign. Positive five can be written +5 or 5.

EXAMPLE A

Use an integer to represent each situation.

The hikers descended into a canyon until they were 150 ft *below* sea level.

The price of a stock *decreased* $2.

The length of a fence is *increased* 15 ft.

−150   **−2**   **+15**

EXAMPLE B

Graph −7 and its opposite on the number line.

The opposite of −7 is 7. Be careful to read −7 as *negative seven,* not *minus seven.* You may read 7 as *positive 7,* but not as *plus 7.*

The numbers –7 and 7 are opposite integers; they are the same distance from zero on the number line.

Language Box

The opposite of negative 7 can be written −(−7).

Absolute Values

The **absolute value** of a number is its distance from zero on the number line. Because absolute value is defined as a distance, it is always nonnegative.

Opposite integers have the same absolute value. For example, 7 and –7 have the same absolute value because they are the same distance from zero on the number line. In symbols, we write:

$|-7| = 7$ and $|7| = 7$

The statement $|-7| = 7$ is read *the absolute value of negative seven is seven.*

The statement $|7| = 7$ is read *the absolute value of 7 is 7.*

Name two numbers that have an absolute value of 18.

$|-18| = 18$ and $|18| = 18$

So, two numbers that have an absolute value of 18 are −18 and 18.

EXAMPLE C

Name two integers that have an absolute value of 3.

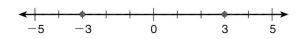

The distance from 0 to -3 is 3 units.

The distance from 0 to 3 is 3 units.

$|-3| = |3| = 3$

Two integers that have an absolute value of 3 are -3 and 3.

Think of absolute value signs as grouping symbols just as you do parentheses.

EXAMPLE D

Evaluate $-|-7|$.

$-|-7| =$

$-(-7) =$ ⟵——— The absolute value of -7 is 7.

$-7 =$ ⟵——— The opposite of 7 is -7.

Thus, $-|-7| = -7$

Comparing and Ordering

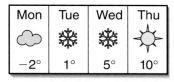

Mon	Tue	Wed	Thu
☁	❄	❄	☀
$-2°$	$1°$	$5°$	$10°$

Comparing the expected temperatures in the 4-day forecast shown at the left, the temperature expected each day will be greater than the temperature the day before. It is possible to compare two or more integers using each integer's position on the number line as a guide. A lesser number will lie to the left of another number.

Compare -2 and 1.

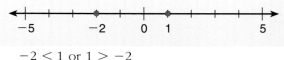

$-2 < 1$ or $1 > -2$

EXAMPLE E

Compare -3 and -5.

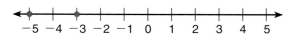

On the number line, -5 lies to the left of -3. Therefore, -5 is the lesser number and -3 is the greater number.

$-3 > -5$ or $-5 < -3$

EXAMPLE F

Complete the
statement
$|-5|$ ⬭ $|-2|$
using > or <.

Find the absolute value of each integer, and then compare the absolute
values.

$|-5|$ ⬭ $|-2|$

5 ⬭ 2

5 > 2

So, $|-5| > |-2|$.

To order three or more integers, use their positions on the number line.

EXAMPLE G

Order 8, −4, 0,
and −9 from least to
greatest.

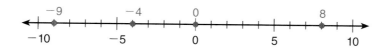

To order the numbers from least to greatest, read them from left to right on
the number line.

The order from least to greatest is −9, −4, 0, 8.

This can also be written −9 < −4 < 0 < 8.

EXAMPLE H

Order −2, 0,
and −3 from least
to greatest.

Read the numbers from left to right on the number line.

The order from least to greatest is −3, −2, 0.

EXERCISES 4-1

Write the number represented by each point on the number line,
and give the opposite of that number.

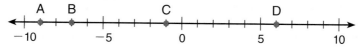

1. Point A   **2.** Point B   **3.** Point C   **4.** Point D

Find the absolute value of each number.

5. $|-15| =$ **6.** $|20| =$ **7.** $|-19| =$

8. $|-1| =$ **9.** $|-30| =$ **10.** $|1| =$

11. $-|9| =$ **12.** $-|-12| =$ **13.** $-|25 - 17| =$

Complete each number sentence. Use <, >, or =.

14. -4 ⬤ -5 **15.** $|-12|$ ⬤ $|12|$ **16.** 28 ⬤ -30

17. 0 ⬤ -21 **18.** -35 ⬤ 35 **19.** $|-7|$ ⬤ $|3|$

Arrange the numbers in order from least to greatest.

20. $16, -10, 8$ **21.** $0, 45, -26$ **22.** $30, -20, -15$

Solve.

23. An automobile ammeter can show whether a battery is charging $(+)$ or discharging $(-)$. Write an integer to indicate that an automobile ammeter is discharging 7 amperes.

24. In an LC circuit, the inductive reactance (X_L) increases as frequency increases, whereas capacitive reactance (X_C) decreases with an increase in frequency. We can think of X_L as positive and X_C as negative. Write an integer to represent:

(a) An inductive reactance of 15 Ω

(b) A capacitive reactance of 21 Ω

(c) A capacitive reactance of 13 Ω

25. What's the Error? Explain the error in this statement: *The absolute value of 16 is negative 16 because the absolute value of a positive number is a negative number.*

ADDING INTEGERS

Same Signs

To add two or more numbers that have the same sign, add their absolute values. The sum has the same sign as the addends.

In the first two rounds of a tournament, a golfer shot a 4 under par and a 5 under par. What was her score after two rounds?

Add -4 and -5.

$|-4| = 4, |-5| = 5$ ⟵ Find the absolute value of each addend.

$4 + 5 = 9$ ⟵ Add the absolute values.

$-4 + (-5) = -9$ ⟵ Use the same sign as the addends for the sum.

It is also helpful to use a number line to add integers. Start at zero. To add a *negative* number, move *left* on the number line.

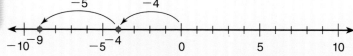

$-4 + (-5) = -9$

After two rounds, her score is 9 under par.

EXAMPLE A

Add −1 and −7.

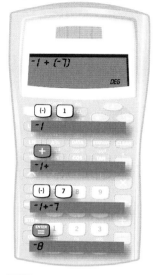

$|-1| = 1, |-7| = 7$ ◀─── Find the absolute value of each addend.

$1 + 7 = 8$ ◀─── Add the absolute values.

$-1 + (-7) = -8$ ◀─── Use the same sign as the addends for the sum.

$-1 + (-7) = -8$

On your calculator, the (-) key is used for a negative sign and the ⊟ is used for subtraction. If you make a mistake and use the (-) key for subtraction, you will see a **SYNTAX ERROR** message in the calculator display.

EXAMPLE B

Add 3 and 4.

$|3| = 3, |4| = 4$ ◀─── Find the absolute value of each addend.

$3 + 4 = 7$ ◀─── Add the absolute values.

$3 + 4 = 7$ ◀─── Use the same sign as the addends for the sum.

To use a number line, start at zero. To add a *positive* number on the number line, move *right*.

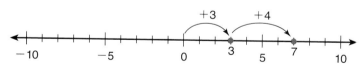

$3 + 4 = 7$

EXAMPLE C

Add −1, −2, and −3.

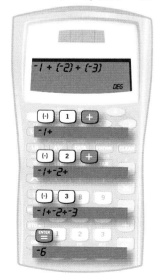

$|-1| = 1, |-2| = 2,$ and $|-3| = 3$ ◀─── Find the absolute value of each addend.

$1 + 2 + 3 = 6$ ◀─── Add the absolute values.

$-1 + (-2) + (-3) = -6$ ◀─── Use the same sign as the addends for the sum.

$-1 + (-2) + (-3) = -6$

Different Signs

To add two numbers that have different signs, subtract the lesser absolute value from the greater absolute value. The sum has the sign of the number with the greater absolute value.

Arturo spent $25 for a book at the beginning of a semester. When the course was over, he sold the used book for $10. What was Arturo's overall cost for the book?

Add −25 and 10.

$|-25| = 25, |10| = 10$ ⟵ Find the absolute value of each addend.

$25 - 10 = 15$ ⟵ Subtract the absolute values.

$-25 + 10 = -15$ ⟵ Use the sign of the addend with the greater absolute value for the sum.

$-25 + 10 = -15$

Arturo's overall cost for the book was $15.

EXAMPLE D

Add −4 and 7.

$|-4| = 4, |7| = 7$ ⟵ Find the absolute value of each addend.

$7 - 4 = 3$ ⟵ Subtract the absolute values.

$-4 + 7 = 3$ ⟵ Use the sign of the addend with the greater absolute value for the sum.

To use a number line, start at zero. To add a *negative* number, move *left*. To add a *positive* number, move *right*.

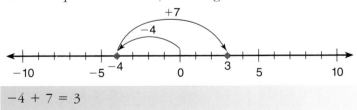

$-4 + 7 = 3$

EXAMPLE E

Add 8 and −12.

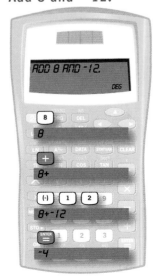

$|8| = 8, |-12| = 12$ ⟵ Find the absolute value of each addend.

$12 - 8 = 4$ ⟵ Subtract the absolute values.

$8 + (-12) = -4$ ⟵ Use the sign of the addend with the greater absolute value for the sum.

To use a number line, start at zero. To add a *positive* number, move *right*. To add a *negative* number, move *left*.

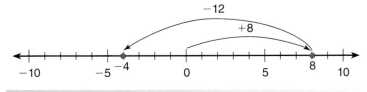

$8 + (-12) = -4$

EXAMPLE F

Add 22, −8, −13, and 3.

To add more than two numbers with different signs, add the positive numbers, and add the negative numbers. Then add the two sums using the rules for addition of integers.

$22 + 3 = 25$ ◀—— Add the positive numbers.

$-8 + (-13) = -21$ ◀—— Add the negative numbers.

$|25| = 25, |-21| = 21$ ◀—— Find the absolute value of each addend.

$25 - 21 = 4$ ◀—— Subtract the absolute values.

$25 + (-21) = 4$ ◀—— Use the sign of the addend with the greater absolute value for the sum.

$22 + (-8) + (-13) + 3 = 4$

Rules for Addition

To add numbers that have the same sign, add the absolute values. The sum has the same sign as the addends.

To add two numbers that have different signs, subtract the absolute values. The sum has the sign of the addend with the greater absolute value.

Order of Operations

The absolute value symbols act as a set of grouping symbols just as parentheses do.

EXAMPLE G

Solve $-15 + |8 + -12|$.

$-15 + |8 + -12| =$ ◀—— According to the order of operations, treat the absolute value just as you would parentheses.

$-15 + |-4| =$ ◀—— Add the numbers in absolute value.

$-15 + 4 =$ ◀—— Determine the absolute value.

-11 ◀—— Add integers.

$-15 + |8 + -12| = -11$

EXAMPLE H

Evaluate $|-9 + 15| + |4 + (-12)|$.

$|-9| = 9, |15| = 15$ ◄——— Begin with the first set of absolute value signs. Find the absolute value of each addend.

$15 - 9 = 6$ ◄——— Subtract the absolute values.

$-9 + 15 = 6$ ◄——— Use the sign of the addend with the greater absolute value for the sum.

$|4| = 4, |-12| = 12$ ◄——— Now work with the second set of absolute value signs. Find the absolute value of each addend.

$12 - 4 = 8$ ◄——— Subtract the absolute values.

$4 + (-12) = -8$ ◄——— Use the sign of the addend with the greater absolute value for the sum.

$|6| + |-8|$ ◄——— Rewrite with the sums in absolute values.

$|6| = 6, |-8| = 8$ ◄——— Find the absolute value of each addend.

$6 + 8 = 14$ ◄——— Add the absolute values.

$|-9 + 15| + |4 + (-12)| = 14$

EXERCISES 4-2

Add.

1. $-4 + (-2)$ 2. $6 + (-5)$ 3. $-13 + 9$

4. $-3 + 6$ 5. $-7 + (-1)$ 6. $4 + (-12)$

7. $-7 + 35$ 8. $6 + 28$ 9. $-15 + 3$

10. $-14 + (-4) + 15$ 11. $-2 + 16 + (-11)$ 12. $-5 + |-9 + 6|$

13. $|12 + (-7)| + |-15 + 18|$

14. $|19 + (-25)| + 16$

15. $|-15 + (-23)| + |-42 + 13|$

Solve.

16. Materials worth $198 are purchased for a project. At the end of the project, the contractor receives a $38 credit for unused materials but is charged a $5 restocking fee. What is the net cost of the materials?

17. Going clockwise around a circuit, the voltages are −5.3 V, −5.4 V, −5.5 V, and −5.8 V and the voltage increases are 7.2 V, 7.3 V, and 7.5 V. What is the total voltage change in this circuit?

18. Two AC voltages are out of phase. Their instantaneous values are 35 V and −112 V. What is the total instantaneous voltage?

19. In an LC circuit, the voltage across the inductive reactance (V_L) can be considered positive and the voltage across the capacitive reactance (V_C) can be considered negative. Find the total reactive voltage for each of the following.

(a) $V_L = 65$ V, $V_C = 37$ V

(b) $V_L = 42$ V, $V_C = 68$ V

(c) $V_L = 38$ V, $V_C = 38$ V

20. Challenge: When will the sum of two or more numbers be zero?

SUBTRACTING INTEGERS

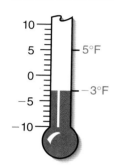

To subtract a number, add its opposite.

How much higher is 5°F than −3°F?

Subtract −3 from 5.

$5 - (-3) =$

$5 + \quad 3 \quad = 8$

To show subtraction of a *negative* number on a number line, move *right*.

$5 - (-3) = 5 + 3 = 8$

So, 5°F is 8 degrees higher than −3°F.

EXAMPLE A

Subtract 1 from −4.

$-4 - \quad 1 \quad =$

$-4 + \quad (-1) = -5$

To show subtraction of a *positive* number on the number line, move *left*.

$-4 - 1 = -4 + (-1) = -5$

$-4 - 1 = -5$

EXAMPLE B

Subtract -4
from -3.

$-3 - (-4) = -3 + 4 = 1$

$-3 - (-4) = 1$

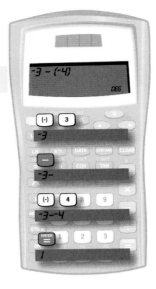

EXAMPLE C

Subtract 19 from 10.

$10 - 19 = 10 + (-19) = -9$

$10 - 19 = -9$

EXAMPLE D

One point in a circuit
is 45 V with respect
to ground. Another
point is -21 V with
respect to ground.
What is the potential
difference between
these two points?

To find the answer, subtract -21 from 45.

$45 - (-21) = 45 + 21 = 66$

$45 - (-21) = 66$

The potential difference between these
two points is 66 V.

Rule for Subtraction

To subtract a number, add its opposite.

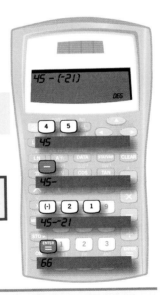

Helpful Hint

Do not change the sign
of the first number in
a subtraction problem.
Change subtraction to
addition, and change
the sign of the second
number.

$-5 - \quad 2 =$

$\downarrow \quad \downarrow$

$-5 + (-2) = -7$

EXERCISES 4-3

Subtract.

1. $-3 - (-9)$ **2.** $4 - 11$ **3.** $-8 - 6$

4. $5 - (-7)$ **5.** $-19 - (-23)$ **6.** $-43 - (-15)$

7. $0 - (-4)$ **8.** $29 - 74$ **9.** $25 - 0$

10. $64 - (-2)$ **11.** $10 - 26$ **12.** $20 - 20$

13. $|-5 - 17|$ **14.** $|-12| - |17|$ **15.** $|-13 + 7| - |18 - (-9)|$

Solve.

16. The voltage in a circuit with respect to ground is 13 V. When the ground connection is changed, the voltage becomes −21 V. What was the amount of change?

17. Mercury freezes at approximately −40°C and boils at approximately 357°C. How much higher is the boiling point than the freezing point?

18. What's the Error? Ray says that −7 − (−7) = −14. Explain the error, and then find the correct answer.

MULTIPLYING INTEGERS

Multiplication of integers can be expressed as repeated addition. Consider the following problem.

The water level of Lake March dropped 4 feet each month during June, July, and August. What was the total change in the water level during this three-month period?

Multiply 3 and −4.

You can find the product of 3 and −4 by adding.

(−4) + (−4) + (−4) = −12

3 × (−4) = −12

The total change in the water level during June, July, and August was −12 feet.

EXAMPLE A

Multiply 5 and −1.

5 × (−1) = (−1) + (−1) + (−1) + (−1) + (−1) = −5

5 × (−1) = −5

Instead of using repeated addition, you can just use the following rules:

If the signs of the factors are the *same*, the product is *positive*.
If the signs of the factors are *different*, the product is *negative*.

EXAMPLE B

Multiply 7 and 3.

The signs of the factors are the same, so the product is positive.

7 × 3 = 21

EXAMPLE C

Multiply 8 and −5.

The signs of the factors are different, so the product is negative.

8 × (−5) = −40

EXAMPLE D

Multiply −6 and −9.

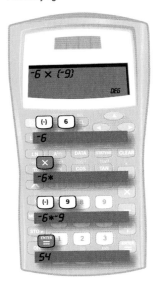

The signs of the factors are the same, so the product is positive.

$$-6 \times (-9) = 54$$

Rules for Multiplication

To multiply two numbers that have the *same* sign, multiply the absolute values. The product is positive.

To multiply two numbers that have *different* signs, multiply the absolute values. The product is negative.

The product of *any* number and zero is zero.

COMMON ERROR

ALERT

Beware: A common error is to multiply two negative numbers and make the product negative. This is not correct. The product of two negative numbers is positive.

EXERCISES 4-4

Multiply.

1. -1×6 **2.** $3 \times (-8)$ **3.** 1×12

4. $-20 \times (-9)$ **5.** 11×7 **6.** -2×13

7. $-7 \times (-30)$ **8.** -3×0 **9.** $6 \times (-18)$

10. $-9 \times (-1)$ **11.** -4×11 **12.** 8×5

Solve.

13. Sigrid, the company bookkeeper, writes expenses with negative numbers and receipts with positive numbers. Indicate how she would record

  (a) the purchase of 5 decorator white 4-way switches at $11 each.

  (b) the purchase of 7 bat-handle toggle switches at $6 each.

  (c) the sale of 3 Mediterranean-style wall lanterns at $47 each.

  (d) the purchase of 13 decorator white 20 A duplex switches at $8 each.

14. How would Sigrid record

  (a) the sale of 4 Kidde Hardwired CO/Smoke Voice Combo Alarms at $43 each?

  (b) the purchase of 5 halogen quartz lightbulbs at $15 each?

  (c) the sale of 3 Progress 52-inch ceiling fans at $128 each?

  (d) the purchase of 75 two-packs of 23-watt soft white compact fluorescent lights at $7 each?

15. Challenge: Is the product $12 \times (-3) \times (5) \times (-6) \times 10$ positive or negative? Explain how you found your answer.

DIVIDING INTEGERS

Value of Property

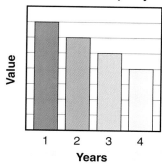

The rules for the sign of a quotient in division are similar to the rules for the sign of a product in multiplication. If two numbers have the *same* sign, the quotient is *positive;* if they have *different* signs, the quotient is *negative.*

The value of a building declined steadily over four years. If the total change was −$3,600, what was the average change per year?

Divide −3,600 by 4.

The numbers have different signs, so the quotient is negative.

$$-3,600 \div 4 = -900$$

The average change in the value of the building was --$900 per year.

EXAMPLE A

Divide 48 by −6.

The signs of the numbers are different, so the quotient is negative.

$$48 \div (-6) = -8$$

EXAMPLE B

Divide 30 by 6.

The signs of the numbers are the same, so the quotient is positive.

$$30 \div 6 = 5$$

EXAMPLE C

Divide −28 by −7.

The signs of the numbers are the same, so the quotient is positive.

$$-28 \div (-7) = 4$$

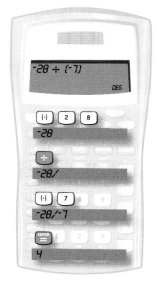

Rules for Division

To divide numbers that have the *same* sign, divide the absolute values. The quotient is positive.

To divide numbers that have *different* signs, divide the absolute values. The quotient is negative.

Zero divided by any nonzero number is zero; division by zero is undefined.

The signs of the numbers are different, so the quotient is negative.

Divide −12 by 4.

$$-12 \div 4 = -3$$

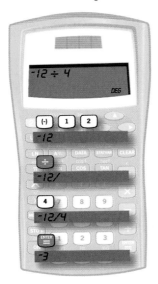

4-5 EXERCISES

Divide.

1. $-10 \div (-2)$ **2.** $18 \div (-2)$ **3.** $-14 \div 7$

4. $150 \div 50$ **5.** $-14 \div 1$ **6.** $2{,}100 \div (-3)$

7. $30 \div 15$ **8.** $-11 \div 11$ **9.** $-24 \div (-8)$

10. $0 \div (-4)$ **11.** $-26 \div (-13)$ **12.** $13 \div (-1)$

Solve.

13. A company lost $3,786 during a three-month period. Use positive or negative integers to indicate its average loss for each month.

14. During the first three months of the year, a company recorded the following profit (+) or loss (−): January, $−27,539: February, $−13,890; March, $19,742.

 (a) What was the total profit or loss for the three months?

 (b) Was this a profit or loss?

 (c) What was the average profit (or loss) for these three months?

15. An airplane descends 10,000 feet in 10 minutes. What is the airplane's average change in altitude per minute?

16. Challenge: What is the value of $\frac{-24 \div 2}{8 \div (-4)}$? Show your work.

CHAPTER 4 REVIEW EXERCISES

1. Find the absolute value of each number.

 (a) $|-137|$

 (b) $|43|$

 (c) $|12 - 18|$

2. Arrange the numbers −7, 12, −4, |−6|, 9, and −11 in order from least to greatest.

3. Perform the indicated operation.

   (a) −5 + (−9)

   (b) 12 − (−8)

   (c) −4 × 7

   (d) 28 ÷ −4

   (e) |15 − 32| − 16

   (f) −17 − 23

   (g) −152 ÷ −8

   (h) −53 + 82 − (−19)

   (i) −14 + (−4) − 15

4. In a certain electrostatic field, the voltage at one point is 1,575 V and the voltage at another point is −975 V. Find the potential difference between the two points by subtracting the second voltage from the first.

5. In an LC circuit, think of the inductive reactance (X_L) as positive and the capacitive reactance (X_C) as negative. Write an integer to represent

   (a) a capacitive reactance of 37 Ω.

   (b) an inductive reactance of 45 Ω.

   (c) a capacitive reactance of 51 Ω.

6. In an LC circuit, consider the voltage across the inductive reactance (V_L) as positive and the voltage across the capacitive reactance (V_C) as negative. Find the total reactive voltage for each of the following.

   (a) $V_L = 37$ V, $V_C = 58$ V

   (b) $V_L = 24$ V, $V_C = 19$ V

   (c) $V_L = 36$ V, $V_C = 43$ V

7. Sigrid, the company bookkeeper, writes expenses with negative numbers and receipts with positive numbers. Indicate how she would record

   (a) the purchase of eight 13-watt 120-volt white spiral CFL bulbs at $9 each.

   (b) the sale of three wall-mount outdoor light fixtures at $52 each.

   (c) the purchase of twelve 15 W CFL dimmable lightbulbs at $13 each.

8. During the second three months of the year, a company recorded the following profit (+) or loss (−): April, $156,837: May, $−78,932: June, $−41,359.

   (a) What was the total profit or loss for the three months?

   (b) Was this a profit or loss?

   (c) What was the average profit (or loss) for these three months?

Building a Foundation in Mathematics

- Adding Rational Numbers
- Subtracting Rational Numbers
- Multiplying Rational Numbers
- Dividing Rational Numbers

Overview

In the last chapter we studied integers, and in this chapter we look at rational numbers. As with fractions, rational numbers give us the ability to represent points between the integers. In this chapter you will see that the rules for working with rational numbers combine the rules used with integers and those used with fractions and decimals.

People who drive recreational vehicles (RVs) are cautioned to check the voltage whenever they enter a different RV park. Many voltmeters measure the voltage as a rational number to the nearest tenth of a volt.

Rational Numbers

Objectives

After completing this chapter, you will be able to:

- Write rational numbers in the $\frac{p}{q}$ form
- Convert decimal fractions to rational numbers
- Determine the absolute value of a rational number
- Add, subtract, multiply, and divide rational numbers
- Solve problems using addition, subtraction, multiplication, and division of rational numbers
- Use your calculator to work with rational numbers

ADDING RATIONAL NUMBERS

Rational numbers are all the numbers that can be expressed as the ratio of two integers. Rational numbers include fractions, decimals, and integers. The form $\frac{p}{q}$, where p and q are integers, but q is not zero, is used to describe rational numbers. Because p and q are integers, rational numbers may be positive, negative, or zero. Some examples of rational numbers are shown below.

Rational Number	In $\frac{p}{q}$ Form
$-\frac{4}{9}$	$\frac{-4}{9}, \frac{4}{-9}$
$1\frac{3}{5}$	$\frac{8}{5}$
-7	$\frac{-7}{1}, \frac{7}{-1}$
0.9	$\frac{9}{10}$
-2.1	$\frac{-21}{10}, \frac{21}{-10}$

The rules for adding rational numbers are like the rules for adding integers.

- To add two or more numbers that have the same sign, add their absolute values. The sum has the same sign as the addends.

- To add two numbers that have different signs, subtract their absolute values. The sum has the sign of the number with the greater absolute value.

If $3\frac{1}{4}$ gallons of water flow into a tank and $4\frac{3}{4}$ gallons of water leak out of the tank at the same time, what is the overall change in the amount of water in the tank?

Add $3\frac{1}{4}$ and $-4\frac{3}{4}$.

$$\left|3\frac{1}{4}\right| = 3\frac{1}{4}, \left|-4\frac{3}{4}\right| = 4\frac{3}{4} \qquad \longleftarrow \text{Find the absolute value of each addend.}$$

$$4\frac{3}{4} - 3\frac{1}{4} = 1\frac{2}{4} = 1\frac{1}{2} \qquad \longleftarrow \text{Subtract the absolute values.}$$

$$3\frac{1}{4} + \left(-4\frac{3}{4}\right) = -1\frac{1}{2} \qquad \longleftarrow \text{Use the sign of the number with the greater absolute value for the sum.}$$

$$3\frac{1}{4} + \left(-4\frac{3}{4}\right) = -1\frac{1}{2}$$

The overall change in the amount of water in the tank is a decrease of $1\frac{1}{2}$ gallons.

EXAMPLE A

Add $-6\frac{1}{2}$ and $\frac{1}{2}$.

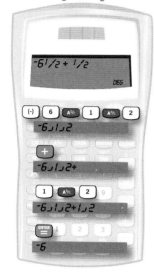

$\left|-6\frac{1}{2}\right| = 6\frac{1}{2}, \left|\frac{1}{2}\right| = \frac{1}{2}$ ◄—— Find the absolute value of each addend.

$6\frac{1}{2} - \frac{1}{2} = 6$ ◄—— Subtract the absolute values.

$-6\frac{1}{2} + \frac{1}{2} = -6$ ◄—— Use the sign of the number with the greater absolute value for the sum.

$-6\frac{1}{2} + \frac{1}{2} = -6$

EXAMPLE B

Add -6.7 and -5.4.

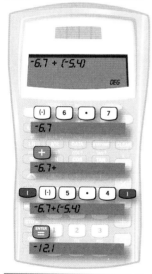

$|-6.7| = 6.7, |-5.4| = 5.4$ ◄—— Find the absolute value of each addend.

$6.7 + 5.4 = 12.1$ ◄—— Add the absolute values.

$-6.7 + (-5.4) = -12.1$ ◄—— Use the same sign as the addends for the sum.

$-6.7 + (-5.4) = -12.1$

EXAMPLE C

Two AC voltages are out of phase. Their instantaneous values are $-\frac{1}{2}$ V and 3 V. What is the total instantaneous voltage?

To answer this question, we add $-\frac{1}{2}$ and 3.

$\left|-\frac{1}{2}\right| = \frac{1}{2}, |3| = 3$ ◄—— Find the absolute value of each addend.

$3 - \frac{1}{2} = 2\frac{1}{2}$ ◄—— Subtract the absolute values.

$-\frac{1}{2} + 3 = 2\frac{1}{2}$ ◄—— Use the sign of the number with the greater absolute value for the sum.

The total instantaneous voltage is $2\frac{1}{2}$ V.

When a fraction and a decimal are given in a problem, one number must be converted to the other form before adding. In Example D, $-\frac{1}{4}$ is converted to a decimal before adding.

EXAMPLE D

Add -1.3 and $-\frac{1}{4}$.

Convert $-\frac{1}{4}$ to a decimal. $\frac{1}{4} = 0.25$, so $-\frac{1}{4} = -0.25$.

$$-\frac{1}{4} = -0.25 \quad \longleftarrow \text{Write } -\frac{1}{4} \text{ as a decimal.}$$

$$\left|-1.3\right| = 1.3, \; \left|-\frac{1}{4}\right| = \left|-0.25\right| = 0.25 \quad \longleftarrow \text{Find the absolute value of each addend.}$$

$$1.3 + 0.25 = 1.55 \quad \longleftarrow \text{Add the absolute values.}$$

$$-1.3 + \left(-\frac{1}{4}\right) = -1.55 \quad \longleftarrow \text{Use the same sign as the addends for the sum.}$$

$$-1.3 + \left(-\frac{1}{4}\right) = -1.55$$

Helpful Hint

On the TI-30X IIS calculator it is possible to enter one addend as a decimal and one addend as a fraction. The sum will be given as a decimal.

Rules for Addition

To add rational numbers that have the *same* sign, add the absolute values. The sum has the same sign as the addends.

To add two rational numbers that have *different* signs, subtract the absolute values. The sum has the sign of the addend with the greater absolute value.

EXERCISES 5-1

Add.

1. $-\frac{1}{3} + \left(-\frac{1}{3}\right)$

2. $7.5 + (-6.8)$

3. $-4 + \frac{2}{5}$

4. $0.875 + \frac{1}{2}$

5. $5\frac{1}{2} + \left(-4\frac{3}{8}\right)$

6. $-3.35 + (-4.82)$

7. $12.2 + (-1.5)$

8. $-1.75 + \frac{3}{4}$

9. $8\frac{3}{5} + (-0.3)$

10. $-100 + (-6.5)$

11. $-4\frac{5}{12} + \left(-2\frac{1}{6}\right)$

12. $24 + (-2.19)$

Solve.

13. Greg's bank account is overdrawn by $20.00. If he deposits $35.00, what is his new account balance?

14. Two AC voltages are out of phase. Their instantaneous values are -42.5 V and $13\frac{3}{4}$ V. What is the total instantaneous voltage?

15. In an LC circuit, the voltage across the inductive reactance (V_L) can be considered positive and the voltage across the capacitive reactance (V_C), as negative. Find the total reactive voltage for each of the following:

    (a) $V_L = 48.35$ V, $V_C = 58.125$ V

    (b) $V_L = 37\frac{5}{8}$ V, $V_C = 12\frac{2}{5}$ V

    (c) $V_L = 63.8$ V, $V_C = 75.55$ V

16. Challenge: Which is greater? $-5\frac{1}{2} + \left(-5\frac{1}{2}\right)$ or $-5 + \left(\frac{1}{2}\right) + (-5) + \left(\frac{1}{2}\right)$? Show your work to explain your answer.

SUBTRACTING RATIONAL NUMBERS

To subtract a number, you can add its opposite.

The bottom of a mineshaft is 150 feet below the surface. A crew extends the shaft downward another 30.5 feet. How far below the surface is the bottom of the shaft then?

Subtract 30.5 from –150.

$$-150 - 30.5 =$$

$$-150 + (-30.5) = -180.5 \qquad \longleftarrow \text{Add the opposite of } 30.5.$$

$$-150 - 30.5 = -180.5$$

The bottom of the shaft is 180.5 feet below the surface after the crew extends the shaft.

EXAMPLE A

Subtract −0.45 from 0.39.

$$0.39 - (-0.45) =$$

$$0.39 + 0.45 = 0.84 \qquad \longleftarrow \text{Add the opposite of } -0.45.$$

$$0.39 - (-0.45) = 0.84$$

EXAMPLE B

Subtract $\frac{4}{5}$ from $\frac{1}{5}$.

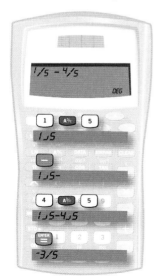

When subtraction is rewritten as addition in this example, the addends have different signs. Use the rules for adding two rational numbers with different signs to complete the problem.

$$\frac{1}{5} - \frac{4}{5} = \frac{1}{5} + \left(-\frac{4}{5}\right)$$

$$\left|\frac{1}{5}\right| = \frac{1}{5}, \left|-\frac{4}{5}\right| = \frac{4}{5} \qquad \longleftarrow \text{Find the absolute value of each addend.}$$

$$\frac{4}{5} - \frac{1}{5} = \frac{3}{5} \qquad \longleftarrow \text{Subtract the absolute values.}$$

$$\frac{1}{5} + \left(-\frac{4}{5}\right) = -\frac{3}{5} \qquad \longleftarrow \text{Use the sign of the addend with the greater absolute value for the sum.}$$

$$\frac{1}{5} - \frac{4}{5} = -\frac{3}{5}$$

EXAMPLE C

Subtract $3\frac{6}{7}$ from $1\frac{5}{7}$.

When subtraction is rewritten as addition in this example, the addends have different signs. Use the rules for adding two rational numbers with different signs to complete the problem.

$$1\frac{5}{7} - 3\frac{6}{7} = 1\frac{5}{7} + \left(-3\frac{6}{7}\right)$$

$$\left|1\frac{5}{7}\right| = 1\frac{5}{7}, \left|-3\frac{6}{7}\right| = 3\frac{6}{7} \quad \longleftarrow \text{Find the absolute value of each addend.}$$

$$3\frac{6}{7} - 1\frac{5}{7} = 2\frac{1}{7} \quad \longleftarrow \text{Subtract the absolute values.}$$

$$1\frac{5}{7} + \left(-3\frac{6}{7}\right) = -2\frac{1}{7} \quad \longleftarrow \text{Use the sign of the addend with the greater absolute value for the sum.}$$

$$1\frac{5}{7} - 3\frac{6}{7} = -2\frac{1}{7}$$

EXAMPLE D

One point in a circuit is $3\frac{1}{5}$ V with respect to ground. Another point is 8.625 V with respect to ground. What is the potential difference between these two points?

In Example D below, the mixed number $3\frac{1}{5}$ is written in decimal form as 3.2 to perform the calculation.

To find the answer, subtract $3\frac{1}{5}$ from 8.625.

Rewrite $3\frac{1}{5}$ in decimal form as 3.2.

$$8.625 - 3\frac{1}{5} = 8.625 - 3.2$$

The problem can be completed using the rules for addition and absolute values, or it can be set up as a regular subtraction problem as shown below.

$$\begin{array}{r} 8.625 \\ -3.200 \\ \hline 5.425 \end{array}$$

$$8.625 - 3\frac{1}{5} = 5.425$$

The potential difference between these two points is 5.425 V.

EXAMPLE E

Subtract 0.3 from $-9\frac{3}{4}$.

The mixed number $-9\frac{3}{4}$ is written in decimal form as −9.75.

$$-9\frac{3}{4} - 0.3 = -9.75 - 0.3 = -9.75 + (-0.3)$$

$$|-9.75| = 9.75, |-0.3| = 0.3 \quad \longleftarrow \text{Find the absolute value of each addend.}$$

$$9.75 + 0.3 = 10.05 \quad \longleftarrow \text{Add the absolute values.}$$

$$-9\frac{3}{4} + (-0.3) = -10.05 \quad \longleftarrow \text{Use the same sign as the addends for the sum.}$$

$$-9\frac{3}{4} - 0.3 = -10.05$$

EXAMPLE F

Subtract $101\frac{1}{4}$ from $10\frac{4}{5}$.

$$10\frac{4}{5} - 101\frac{1}{4} = 10\frac{4}{5} + \left(-101\frac{1}{4}\right)$$

$$\left|10\frac{4}{5}\right| = 10\frac{4}{5}, \ \left|-101\frac{1}{4}\right| = 101\frac{1}{4}$$ ←— Find the absolute value of each addend.

$$\begin{array}{ccc} 101\frac{1}{4} & 101\frac{5}{20} & 100\frac{25}{20} \\ \longrightarrow & \longrightarrow & \\ -10\frac{4}{5} & -10\frac{16}{20} & -10\frac{16}{20} \\ \hline & & 90\frac{9}{20} \end{array}$$ ←— Borrow 1 from 101 and add it to $\frac{5}{20}$ to make subtraction possible. Then subtract.

$$10\frac{4}{5} + \left(-101\frac{1}{4}\right) = -90\frac{9}{20}$$ ←— Use the sign of the addend with the greater absolute value.

$$10\frac{4}{5} - 101\frac{1}{4} = -90\frac{9}{20}$$

Rule for Subtraction

> To subtract a rational number, add its opposite.

EXERCISES 5-2

Subtract.

1. $7.5 - (-3.1)$

2. $-\frac{4}{11} - \left(-\frac{6}{11}\right)$

3. $\frac{15}{16} - 3$

4. $2\frac{1}{5} - 0.7$

5. $-3\frac{1}{2} - 0.35$

6. $-0.3 - (-8)$

7. $12 - \left(-2\frac{1}{4}\right)$

8. $-\frac{2}{7} - \frac{3}{4}$

9. $-9.8 - 8.9$

10. $-15 - 7\frac{2}{5}$

11. $0.04 - 2.6$

12. $3\frac{3}{5} - \left(-4\frac{3}{5}\right)$

13. $6.8 - 4\frac{3}{8}$

14. $-22\frac{2}{5} - 0.9$

15. $2\frac{5}{8} - 7\frac{1}{3}$

Solve.

16. The voltage in a circuit with respect to ground is $12\frac{1}{4}$ V. When the ground connection is changed, the voltage becomes $-5\frac{1}{2}$ V. What was the amount of change?

17. In a certain electrostatic field, the voltage at one point is 675.36 V and the voltage at another point is -953.97 V. Find the potential difference between the two points by subtracting the second voltage from the first.

18. In another electrostatic field, the voltage at one point is $450\frac{4}{5}$ V and the voltage at another point is $-354\frac{3}{4}$ V. Find the potential difference between the two points by subtracting the second voltage from the first.

19. **What's the Error?** Explain the error in the statement:
$$6\frac{2}{3} - \left(\frac{-1}{-3}\right) = 7.$$

MULTIPLYING RATIONAL NUMBERS

To multiply rational numbers, multiply as you did with positive numbers. But remember to follow the rules for multiplying signed numbers.

For example, $\frac{1}{2} \times \frac{3}{5} = \frac{3}{10}$, and $\frac{1}{2} \times \left(-\frac{3}{5}\right) = -\frac{3}{10}$.

Sarah borrowed $40 for some equipment. With her first payment, she repaid $\frac{3}{5}$ of the amount. How much did she still owe?

She repaid $\frac{3}{5}$ of what she owed, so she still owed $1 - \frac{3}{5} = \frac{2}{5}$ of what she borrowed. To find out how much she owed, find $\frac{2}{5}$ of -40.

Multiply $\frac{2}{5}$ and -40.

The signs of the factors are different, so the product is negative.

$$\frac{2}{5} \times -40 = \frac{2}{5} \times \frac{-40}{1} = \frac{2}{\overset{1}{\cancel{5}}} \times \frac{\overset{-8}{\cancel{-40}}}{1} = \frac{-16}{1} = -16$$

$$\frac{2}{5} \times -40 = -16$$

Sarah still owed $16.

EXAMPLE A

Multiply $-\frac{1}{12}$ and $-\frac{8}{15}$.   The signs of the factors are the same, so the product is positive.

> **Helpful Hint**
>
> A negative fraction such as $-\frac{1}{12}$ can be written three ways:
> $-\frac{1}{12} = \frac{-1}{12} = \frac{1}{-12}$.
> For calculations, it is often best to use the form $\frac{-1}{12}$.

$$-\frac{1}{12} \times \left(-\frac{8}{15}\right) = \frac{-1}{12} \times \frac{-8}{15} = \frac{-1}{\underset{3}{\cancel{12}}} \times \frac{\overset{-2}{\cancel{-8}}}{15} = \frac{2}{45}$$

$$-\frac{1}{12} \times \left(-\frac{8}{15}\right) = \frac{2}{45}$$

EXAMPLE B

Multiply $-1\frac{1}{3}$ and 4.5.

In Example B that follows, both factors are converted to improper fractions before multiplication is performed.

Convert $-1\frac{1}{3}$ to an improper fraction: $-1\frac{1}{3} = \frac{-4}{3}$

Convert 4.5 to an improper fraction: $4.5 = 4\frac{5}{10} = 4\frac{1}{2} = \frac{9}{2}$

The signs of the factors are different, so the product is negative.

$-1\frac{1}{3} \times 4.5 = \frac{-4}{3} \times \frac{9}{2} = \frac{-36}{6} = -6$

$-1\frac{1}{3} \times 4.5 = -6$

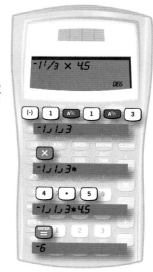

EXAMPLE C

Multiply and round the product of $\frac{4}{7} \times$ 3.31 to the nearest hundredth.

The signs of the factors are the same, so the product is positive.

$\frac{4}{7} \times 3.31 = \frac{4}{7} \times \frac{3.31}{1} = \frac{13.24}{7}$

$\begin{array}{r} 1.891 \\ 7\overline{)13.240} \end{array}$ ◀——Divide until the quotient has a digit in the thousandths place.

To the nearest hundredth, 1.891 is 1.89.

$\frac{4}{7} \times 3.31 \approx 1.89$

EXAMPLE D

Multiply -2.4 and 1.9.

The signs of the factors are different, so the product is negative.

$\begin{array}{r} 2.4 \\ \times 1.9 \\ \hline 216 \\ 24 \\ \hline 4.56 \end{array}$ ◀——Multiply the absolute values.

$-2.4 \times 1.9 = -4.56$

EXAMPLE E

Multiply −0.57 and −0.4.

The signs of the factors are the same, so the product is positive.

$$\begin{array}{r} 0.57 \\ \times\ 0.4 \\ \hline 0.228 \end{array}$$ ←—Multiply the absolute values.

$$-0.57 \times -0.4 = 0.228$$

Rules for Multiplication

> To multiply two rational numbers that have the *same* sign, multiply the absolute values. The product is positive.
>
> To multiply two rational numbers that have *different* signs, multiply the absolute values. The product is negative.
>
> The product of *any* number and zero is zero.

EXERCISES 5-3

Multiply.

1. $27 \times -3\frac{1}{3}$

2. -7.8×9.35

3. $-\frac{14}{15} \times \left(-\frac{5}{28}\right)$

4. $5 \times (-0.2)$

5. $6\frac{2}{5} \times (-8)$

6. $-0.08 \times \left(-6\frac{1}{4}\right)$

7. $\left(-\frac{1}{5}\right) \times \left(-\frac{1}{5}\right)$

8. $0.625 \times (-0.5)$

9. $\frac{2}{50} \times (0.9)$

Multiply and round the product to the nearest hundredth.

10. $1.4 \times \left(-\frac{3}{11}\right)$

11. $-\frac{2}{9} \times 1\frac{1}{5}$

12. $-0.77 \times (-1.6)$

Solve.

13. When the multiplication is finished for the expression $-\frac{7}{8} \times 6\frac{3}{4} \times 1 \times \left(-\frac{1}{3}\right)$, will the product be positive or negative? Explain.

14. Sigrid, the company bookkeeper, writes expenses with negative numbers and receipts with positive numbers. Indicate how she would record each of (a) through (d). Then answer (e).

(a) The purchase of five decorator white four-way switches at $10.95 each

(b) The purchase of seven bat-handle toggle switches at $3.34 each

(c) The sale of three Sturbridge-style wall lanterns at $317.85 each

(d) The purchase of 18 white three-way 15 A duplex switches at $6.81 each

(e) If this represents all the sales and purchases for one day, how much did the company make (or lose)?

15. Referring back to question 14, how would Sigrid record the following sales and purchases?

    (a) The sale of six Kidde Hardwired CO/Smoke Voice Combo Alarms at $42.26 each

    (b) The purchase of eight 14/19/32 watt CFL lightbulbs at $13.21 each

    (c) The sale of five flush ceiling lights at $101.16 each

    (d) The purchase of six 19 watt Compact Fluorescent Reflector flood lights at $9.76 each

16. When the expression $-2.25 \times (-7.5) \times (-6.4)$ is multiplied, will the product be positive or negative? Explain.

17. **Challenge:** A formula for converting Fahrenheit temperatures to Celsius temperatures is $C = \frac{5}{9}(F - 32)$, where F is a temperature in degrees Fahrenheit and C is the temperature in degrees Celsius. If a temperature is $-4°F$, what is the temperature in degrees Celsius?

DIVIDING RATIONAL NUMBERS

To divide rational numbers, divide as you did with positive numbers, but remember to follow the rules for dividing signed numbers.

For example, $10 \div 2.5 = 4$ and $10 \div (-2.5) = -4$.

A drop in barometric pressure suggests that rain or snow is likely. If the barometric pressure drops 0.6 inch in 4 hours, what is the average rate of change in pressure per hour?

Divide –0.6 by 4.

The numbers have different signs, so the quotient is negative.

$$\begin{array}{r} 0.15 \\ 4\overline{)0.60} \end{array}$$ ◄——Divide the absolute values.

$-0.6 \div 4 = -0.15$

The average rate of change in pressure is a drop of 0.15 inch per hour.

EXAMPLE A

Divide −0.6 by −0.125.

The signs of the numbers are the same, so the quotient is positive.

$$\begin{array}{r} 4.8 \\ 0.125\overline{)0.6000} \end{array}$$ ◄——Divide the absolute values.

$-0.6 \div (-0.125) = 4.8$

EXAMPLE B

Divide $-\frac{3}{4}$ by $\frac{2}{3}$.

The signs of the numbers are different, so the quotient is negative.

$$-\frac{3}{4} \div \frac{2}{3} = \frac{-3}{4} \times \frac{3}{2} = \frac{-9}{8} = -1\frac{1}{8}$$

$$-\frac{3}{4} \div \frac{2}{3} = -1\frac{1}{8}$$

EXAMPLE C

Divide $-\frac{2}{3}$ by -12.

The signs of the numbers are the same, so the quotient is positive.

$$-\frac{2}{3} \div (-12) = \frac{-2}{3} \div \frac{-12}{1} = \frac{\overset{1}{\cancel{-2}}}{3} \times \frac{1}{\underset{6}{\cancel{-12}}} = \frac{1}{18}$$

$$-\frac{2}{3} \div (-12) = \frac{1}{18}$$

Examples D and E illustrate different ways to divide when decimals and fractions are involved.

EXAMPLE D

Divide $-8\frac{1}{3}$ by 12.5.

The signs of the numbers are different, so the quotient is negative.

$$-8\frac{1}{3} \div 12.5 = -8\frac{1}{3} \div 12\frac{1}{2} = \frac{-25}{3} \div \frac{25}{2} = \frac{\overset{-1}{\cancel{-25}}}{3} \times \frac{2}{\underset{1}{\cancel{25}}} = \frac{-2}{3}$$

$$-8\frac{1}{3} \div 12.5 = -\frac{2}{3}$$

Rules for Division

To divide rational numbers that have the *same* sign, divide the absolute values. The quotient is positive.

To divide rational numbers that have *different* signs, divide the absolute values. The quotient is negative.

Zero divided by any nonzero number is zero; division by zero is undefined.

EXAMPLE E

The current, in amperes (A), in a circuit with a voltage of 1.03 V and a resistance of $1\frac{1}{4}$ Ω is $\frac{1.03}{1\frac{1}{4}}$. What is the current in this circuit? Round your answer to the nearest tenth.

The answer is the quotient of $1.03 \div 1\frac{1}{4}$.

The signs of the numbers are the same, so the quotient is positive.

$$1.03 \div 1\frac{1}{4} = 1.03 \div 1.25$$

$$1.25\overline{)1.0300}\quad\overset{0.82}{}$$ ←— Divide until the quotient has a digit in the hundredths place.

To the nearest tenth, 0.82 is 0.8.

$$1.03 \div 1\frac{1}{4} \approx 0.8$$

The current to the nearest tenth is 0.8 A.

EXERCISES 5-4

Divide.

1. $\frac{3}{4} \div \left(-\frac{3}{5}\right)$

2. $-\frac{9}{10} \div 36$

3. $0 \div -0.01$

4. $7.2 \div (-90)$

5. $6.75 \div \left(-\frac{3}{4}\right)$

6. $-\frac{4}{9} \div \left(-3\frac{3}{5}\right)$

7. $2\frac{2}{3} \div 0.1$

8. $-\frac{5}{32} \div \left(-\frac{1}{8}\right)$

9. $20 \div \left(-2\frac{2}{7}\right)$

10. $0.256 \div 1.6$

11. $-3.2 \div 1\frac{5}{11}$

12. $-\frac{4}{21} \div \frac{8}{35}$

Divide and round the quotient to the nearest tenth.

13. $-8.2 \div (-2.3)$

14. $1.5 \div 1\frac{4}{5}$

15. $2 \div \left(-\frac{6}{7}\right)$

Solve.

16. Going clockwise around a closed path in a circuit, the voltage drops are -9.3 V, $-5\frac{2}{3}$ V, and $-7\frac{1}{12}$ V. What is the average voltage drop?

17. Rick wants to pay a $426.75 bill in three equal installments. If he makes three equal withdrawals from his checking account to make the payments, how much is each withdrawal?

18. During the third quarter of the year, a company recorded the following profit (+) or loss (−): July, $−36,785.39; August, $15,046.31; September, $12,842.67.

    (a) What was the total profit or loss for the three months?

    (b) Was this a profit or loss?

    (c) What was the average profit (or loss) for these three months?

19. **Challenge:** Evaluate: $\left[-\frac{1}{2} \div \left(-\frac{1}{2}\right)\right] \div \left(-\frac{1}{2}\right)$. Show your work.

CHAPTER 5 REVIEW EXERCISES

1. Find the absolute value of each number.

   (a) $\left|-15\frac{1}{2}\right|$

   (b) $|97.325|$

   (c) $\left|5\frac{1}{2} - 7.35\right|$

2. Arrange the numbers $-8\frac{1}{2}$, 3.47, -4.74, -8.4, 5.48, $-4\frac{3}{4}$ in order from least to greatest.

3. Perform the indicated operation.

   (a) $-2.5 \times 3\frac{3}{4}$

   (b) $-4.93 + (-7.35)$

   (c) $5.672 - (-12\frac{3}{16})$

   (d) $-4.72 \div -5$

   (e) $18.36 + (-35.217)$

   (f) $18\frac{5}{11} \div -1.20$

   (g) $-53 + (-3.5) \times 2.5$

   (h) $7\frac{1}{2} - \left|8\frac{3}{4} - (-3\frac{1}{6})\right| \times (-2.1)$

   (i) $-127.35 - (-84.35) - 29.13$

4. In an LC circuit, consider the voltage across the inductive reactance (V_L) as positive and the voltage across the capacitive reactance (V_C) as negative. Find the total reactive voltage for each of the following.

   (a) $V_L = 119\frac{3}{5}$ V, $V_C = 95.26$ V

   (b) $V_L = 27\frac{3}{4}$ V, $V_C = 56\frac{5}{6}$ V

   (c) $V_L = 64.825$ V, $V_C = 97.397$ V

5. Going clockwise around a closed path in a circuit, the voltage drops are -2.45 V, $-4\frac{1}{6}$ V, and $-8\frac{5}{12}$ V.

   (a) What is the total voltage drop?

   (b) What is the average voltage drop? Round the answer to the nearest thousandth.

6. The current, in amperes (A), in a circuit with a voltage of 15.55 V and a resistance of $5\frac{1}{3}$ Ω is $\frac{15.55}{5\frac{1}{3}}$. What is the current in this circuit? Round your answer to the nearest hundredth.

7. Sigrid, the company bookkeeper, writes expenses with negative numbers and receipts with positive numbers. Indicate how she would record each of the following.

   (a) The purchase of nine 20 watt white spiral CFL bulbs at $6.13 each

   (b) The purchase of six 19 watt CFL reflector lightbulbs at $10.25 each

   (c) The sale of five low-voltage non-IC mini-recessed light housings at $61.49 each

8. During the last quarter of the year, a company recorded the following profit (+) or loss (−): October, 52,951.37; November, $−62,483.95; December, $14,796.32.

(a) What was the total profit or loss for the three months?

(b) Was this a profit or loss?

(c) What was the average profit (or loss) for these three months?

Building a Foundation in Mathematics

Introducing Exponents

Applying Laws of Exponents

Using Scientific Notation

Simplifying Expressions with Powers of Ten

Using Engineering Notation

Introducing Roots and Fractional Exponents

2^3

2^4

$A = lw$

Overview

Until now the numbers we have used have not been very large or very small. Both large and small numbers can be hard to read because of the many place values that are involved. In this chapter we look at methods we can use to quickly write large numbers. This is especially useful in electronics because powers of ten are used to express units and measurements. All electrical units are based on the metric system, and powers often are used to convert from one metric unit to a larger or smaller version of that unit.

The inverse of raising a number to a power is a finding the root of a number. Some electrical formulas use square roots or cube roots. Just as we used powers of ten, we will also use roots of ten. Calculators are very helpful for finding powers and roots of numbers. For example, if we know the resistance of and wattage of a heater, we use square roots to determine its rated voltage.

Two ways to express very large or small numbers involve scientific notation and engineering notation. Both of these use very similar techniques, and your calculator can be used to help with the calculation.

Exponents

Objectives

After completing this chapter, you will be able to:

- Compute powers and roots of signed numbers
- Simplify numbers with radical signs
- Express decimal numbers in scientific or engineering notation
- Express numbers written in scientific or engineering notation as decimal numbers
- Compute expressions using scientific or engineering notation

Chapter 6

INTRODUCING EXPONENTS

Positive Exponents

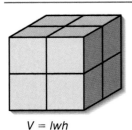

$V = lwh$

The expression $2 \times 2 \times 2$ can be written using a **base** and an **exponent**.

$2 \times 2 \times 2 = 2^3$ ← **exponent**

└── **base**

$2 \times 2 \times 2 = 2^3$

The factor 2 is the base. The exponent is 3 because the factor appears 3 times.

An expression such as 2^3 is called a **power.** Read the expression 2^3 as *2 to the third power* or *2 raised to the third power.*

An exponent may be positive, negative, or zero.

- A positive exponent indicates how many times the base is to be used as a factor.

- A negative exponent indicates how many times the reciprocal of the base is to be used as a factor.

- A zero exponent with a nonzero base indicates that the value of the power is 1.

- The expression 0^0 is undefined.

In the diagram at the top left, the edge of the cube is 2 cm. What is the volume of the entire figure?

Evaluate 2^3.

$2^3 = 2 \times 2 \times 2 = 8$ ← Use 2 as a factor 3 times.

The volume of the figure is 8 cubic centimeters, or 8 cm³.

Language Box

When no exponent is written in an expression, the exponent is understood to be 1.
Examples:
$4 = 4^1$; $-7 = (-7)^1$;
$\frac{2}{3} = \left(\frac{2}{3}\right)^1$

There are two keys on your calculator that can be used for powers. The x^2 key can be used only for numbers with an exponent of 2—that is, for finding the second power of a number. The $\wedge$ key can be used with any exponent. You can find the value of 7^2 by either pressing 7 x^2 or by using 7 $\wedge$ 2. Both methods will give an answer of 49. However, to find 2^3 on a calculator, you must press 2 $\wedge$ 3. Notice that after you use the $\wedge$ key, you have to key in the exponent.

EXAMPLE A

Evaluate 3^4.

$3^4 = 3 \times 3 \times 3 \times 3 = 81$ ← Use 3 as a factor 4 times.

$3^4 = 81$

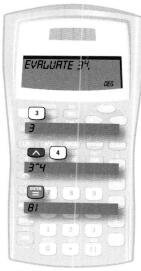

EXAMPLE B

Evaluate $\left(\frac{1}{2}\right)^3$.

In Example B, the base is a fraction.

Parentheses are used to indicate that the exponent applies to the entire fraction.

$\left(\frac{1}{2}\right)^3 = \frac{1}{2} \times \frac{1}{2} \times \frac{1}{2} = \frac{1}{8}$ ← Use $\frac{1}{2}$ as a factor 3 times.

$\left(\frac{1}{2}\right)^3 = \frac{1}{8}$

EXAMPLE C

Evaluate $(-2)^3$.

In this example parentheses are used to indicate that a negative number is a base.

$(-2)^3 = (-2) \times (-2) \times (-2) = -8$ ← Use -2 as a factor 3 times.

$(-2)^3 = -8$

EXAMPLE D

Evaluate $(-4.1)^2$.

Parentheses can be used to show multiplication.

$(-4.1)^2 = (-4.1)(-4.1) = 16.81$ ← Use -4.1 as a factor 2 times.

$(-4.1)^2 = 16.81$

EXAMPLE E

Evaluate $(-2)^4$.

$(-2)^4 = (-2)(-2)(-2)(-2) = 16$ ← Use -2 as a factor 4 times.

$(-2)^4 = 16$

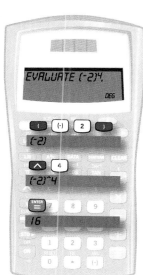

EXAMPLE F

Evaluate -2^4.

In an expression such as -2^4, the negative sign is not part of the base because it is not enclosed in parentheses. The negative sign represents the factor -1.

$$-2^4 = -1 \times 2 \times 2 \times 2 \times 2 = -1 \times 16 = -16$$

$$-2^4 = -16$$

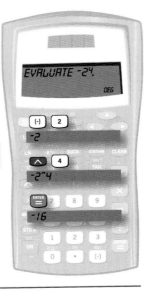

Order of Operations

What is the answer to a problem like $3 + 5 \times 6^2$? The order of operations that was introduced in Chapter 2 has to be expanded to take care of exponents. The new step is listed below and is highlighted in red.

Order of Operations

- Complete all calculations within grouping symbols such as parentheses.
- Evaluate all numbers with exponents in order from left to right.
- Perform all multiplications and divisions in order from left to right.
- Perform all additions and subtractions in order from left to right.

EXAMPLE G

Evaluate $3 + 5 \times 6^2$.

$3 + 5 \times 6^2 =$ ← There are no parentheses, so according to the order of operations, evaluate 6^2 first.

$3 + 5 \times 36 =$ ← Next, perform the multiplication.

$3 + 180 = 183$ ← Add.

$$3 + 5 \times 6^2 = 183$$

EXAMPLE H

The power, in watts, dissipated by a 5.6 A current flowing through a 7.25 Ω resistor is given by 7.25×5.6^2. How many watts are dissipated in this example?

$7.25 \times 5.6^2 = 7.25 \times 31.36$

$7.25 \times 31.36 = 227.36$

The power is 227.36 W.

Negative Exponents

EXAMPLE I

Evaluate 5^{-2}.

To evaluate an expression that has a negative exponent, replace the base with its reciprocal and replace the exponent with its opposite.

$5^{-2} = \left(\frac{1}{5}\right)^2$ ← Use the reciprocal of the base and the opposite of the exponent.

$5^{-2} = \left(\frac{1}{5}\right)^2 = \frac{1}{5} \times \frac{1}{5} = \frac{1}{25}$ ← Use $\frac{1}{5}$ as a factor 2 times.

$5^{-2} = \frac{1}{25}$

EXAMPLE J

Evaluate $\left(-\frac{3}{4}\right)^{-2}$.

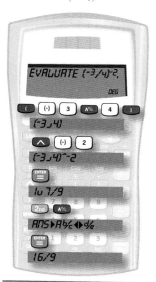

$\left(-\frac{3}{4}\right)^{-2} = \left(-\frac{4}{3}\right)^2$ ← Use the reciprocal of the base and the opposite of the exponent.

$\left(-\frac{3}{4}\right)^{-2} = \left(-\frac{4}{3}\right)^2 = \left(-\frac{4}{3}\right)\left(-\frac{4}{3}\right) = \frac{16}{9}$ ← Use $-\frac{4}{3}$ as a factor 2 times.

$\left(-\frac{3}{4}\right)^{-2} = \frac{16}{9}$

EXAMPLE K

Evaluate $\left(-\frac{3}{4}\right)^{-1}$.

$\left(-\frac{3}{4}\right)^{-1} = \left(-\frac{4}{3}\right)^1$ ← Use the reciprocal of the base and the opposite of the exponent.

$\left(-\frac{3}{4}\right)^{-1} = \left(-\frac{4}{3}\right)^1 = -\frac{4}{3}$

$\left(-\frac{3}{4}\right)^{-1} = -\frac{4}{3}$

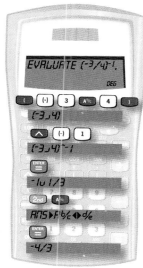

EXAMPLE L

Evaluate 8^0.

Any nonzero base to the zero power is 1.

$8^0 = 1$

EXAMPLE M

Evaluate $\left(\frac{-2}{3}\right)^0$.

$\left(\frac{-2}{3}\right)^0 = 1$

EXERCISES 6-1

COMMON ERROR

ALERT

It is a mistake to think that a negative exponent will automatically produce a negative result. For example, $5^{-2} = \frac{1}{25}$, which is positive.

Evaluate the following.

1. 2^5
2. 6^{-3}
3. $\left(\frac{2}{3}\right)^3$

4. $(-1)^2$
5. $\left(-\frac{1}{3}\right)^4$
6. 3.2^2

7. $(-2)^{-5}$
8. $(-2.5)^2$
9. $\left(\frac{1}{4}\right)^0$

10. $\left(-\frac{2}{5}\right)^{-2}$
11. 7^{-2}
12. $(-0.9)^3$

13. 16^0
14. -3^4
15. $\left(\frac{1}{3}\right)^{-1}$

16. $5 + 6^2 - 7$
17. $(1.2 + 3.8)^{-2} + 4 \times 3.6$
18. $\frac{2 \times 7.5}{3^2 - 1} + \frac{13}{2^3}$

19. Evaluate the powers of ten.

   a. $-10^{-3} =$ _____ $(-10)^{-3} =$ _____

   b. $-10^{-2} =$ _____ $(-10)^{-2} =$ _____

   c. $-10^{-1} =$ _____ $(-10)^{-1} =$ _____

   d. $-10^0 =$ _____ $(-10)^0 =$ _____

   e. $-10^1 =$ _____ $(-10)^1 =$ _____

   f. $-10^2 =$ _____ $(-10)^2 =$ _____

   g. $-10^3 =$ _____ $(-10)^3 =$ _____

20. The power, in watts, dissipated by a 9.4 A current flowing through a 8.75 Ω resistor is given by $9.4^2 \times 8.75$. How many watts are dissipated in this circuit?

21. The power, in watts, dissipated when 80.25 V are applied to a 225 Ω resistor is given by $\frac{80.25^2}{225}$. How many watts are dissipated in this circuit?

22. The total resistance when an 8 Ω and a 10 Ω conductor are connected in parallel is $\frac{1}{8^{-1} + 10^{-1}}$.

   (a) Write the expression with positive exponents.

   (b) Determine the total resistance.

23. The total resistance, in ohms, in a certain series-parallel circuit is $\left(\frac{1}{0.8} + \frac{1}{0.45}\right)^{-1} + 0.76$. What is the total resistance in this circuit?

Answer the following.

24. Is the following statement true or false? *Any power of a negative number is negative.* Justify your answer.

25. Evaluate $(-5)^2$ and -5^2. Explain why the two expressions do not have the same value.

26. Challenge: Express the number 64 as three different powers. Use 2, 3, and 6 as the three exponents.

APPLYING LAWS OF EXPONENTS

Multiplication

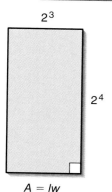

2^3

2^4

$A = lw$

To find the area of the rectangle at the left, find the product $2^4 \times 2^3$.

$2^4 \times 2^3 = (2 \times 2 \times 2 \times 2) \times (2 \times 2 \times 2) = 2^7$

To multiply powers that have the same base, add the exponents and use the same base.

$2^4 \times 2^3 = 2^{4+3} = 2^7$

Product of Powers

> For any number a except $a = 0$, and all numbers m and n:
> $a^m \times a^n = a^{m+n}$

Simplify 4×4^2.

Recall that when an exponent is not written, it is understood to be 1.

$4 \times 4^2 = 4^1 \times 4^2 = 4^{1+2} = 4^3$ ← The base is the same so you add the exponents.

$$= 64$$

$$4 \times 4^2 = 64$$

EXAMPLE A

Simplify $1.5^2 \times 1.5^4$.

$1.5^2 \times 1.5^4 = 1.5^{2+4} = 1.5^6$ ← The base is the same, so you add the exponents. The expression 1.5^6 is not easily evaluated without a calculator. Leave it in exponential form.

$1.5^2 \times 1.5^4 = 1.5^6$

EXAMPLE B

Simplify $\left(\frac{1}{3}\right)^2 \times \left(\frac{1}{3}\right)^2$.

$\left(\frac{1}{3}\right)^2 \times \left(\frac{1}{3}\right)^2 = \left(\frac{1}{3}\right)^{2+2} = \left(\frac{1}{3}\right)^4 = \frac{1}{81}$ ← The base is the same, so you add the exponents.

$\left(\frac{1}{3}\right)^2 \times \left(\frac{1}{3}\right)^2 = \frac{1}{81}$

The expression in Example C is evaluated in two ways. One of the ways involves a negative exponent.

EXAMPLE C

Simplify $\frac{1}{6^3} \times 6^2$.

It is possible to evaluate this expression in two ways.

$\frac{1}{6^3} \times 6^2 = \frac{1}{6^3} \times \frac{6^2}{1} = \frac{1 \times \cancel{6} \times \cancel{6}}{\cancel{6} \times \cancel{6} \times 6 \times 1} = \frac{1}{6}$ ← Write the second factor as a fraction, and then multiply.

or

$\frac{1}{6^3} \times 6^2 = 6^{-3} \times 6^2 = 6^{-3+2} = 6^{-1} = \frac{1}{6}$ ← Write $\frac{1}{6^3}$ as 6^{-3} and apply the rule for the product of powers.

$\frac{1}{6^3} \times 6^2 = \frac{1}{6}$

EXAMPLE D

Simplify $8^3 \times 8^{-7}$.

$8^3 \times 8^{-7} = 8^{3+(-7)}$ ← Apply the rule for the product of powers.

$= 8^{-4}$ ← Add the exponents.

$= \frac{1}{8^4}$ ← Write the result using a positive exponent.

$8^3 \times 8^{-7} = \frac{1}{8^4}$

EXAMPLE E

Simplify $10 \cdot 10^2 \cdot 10^6$.

$10 \cdot 10^2 \cdot 10^6 = 10^{1+2+6} = 10^9$ ← The base is the same, so you add the exponents.

$10 \cdot 10^2 \cdot 10^6 = 10^9$

Division

2^3

Area = 2^7 square units

$l = \dfrac{A}{w}$

To find the length of the rectangle at the left, find the quotient $\dfrac{2^7}{2^3}$.

$$\frac{2^7}{2^3} = \frac{\cancel{2} \times \cancel{2} \times \cancel{2} \times 2 \times 2 \times 2 \times 2}{\cancel{2} \times \cancel{2} \times \cancel{2}} = \frac{2^4}{1} = 2^4$$

To divide powers that have the same base, subtract the exponents and use the same base. Always subtract the exponent in the denominator from the exponent in the numerator.

$$\frac{2^7}{2^3} = 2^{7-3} = 2^4$$

Quotient of Powers

For any number a except $a = 0$, and all numbers m and n:

$$\frac{a^m}{a^n} = a^{m-n}$$

Simplify $\dfrac{6^3}{6^8}$.

$$\frac{6^3}{6^8} = 6^{3-8} = 6^{-5} = \frac{1}{6^5}$$

$$\frac{6^3}{6^8} = \frac{1}{6^5}$$

← The base is the same, so you subtract the exponents. It is customary to write the answer using a positive exponent.

EXAMPLE F

Simplify $\dfrac{9^2}{9^{-4}}$.

It is possible to evaluate this expression in two ways.

$$\frac{9^2}{9^{-4}} = 9^{2-(-4)} = 9^{2+4} = 9^6$$

← The base is the same, so you subtract the exponents.

or

$$\frac{9^2}{9^{-4}} = \frac{9^2 \times 9^4}{1} = \frac{9^{2+4}}{1} = \frac{9^6}{1} = 9^6$$

← Write $\dfrac{1}{9^{-4}}$ as 9^4; add the exponents.

$$\frac{9^2}{9^{-4}} = 9^6$$

EXAMPLE G

Simplify $\dfrac{(-4)^5}{(-4)^2}$.

$$\frac{(-4)^5}{(-4)^2} = (-4)^{5-2} = (-4)^3 = -64$$

← The power $(-4)^3$ can be easily evaluated.

$$\frac{(-4)^5}{(-4)^2} = -64$$

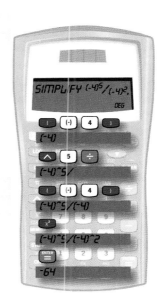

EXAMPLE H

Simplify $\frac{8^3}{8^3}$.

Recall that a base raised to the zero power is 1. The expression in Example H is evaluated in two ways to show why this is true.

It is possible to evaluate this expression in two ways.

$$\frac{8^3}{8^3} = \frac{8 \times 8 \times 8}{8 \times 8 \times 8} = 1$$

or

$$\frac{8^3}{8^3} = 8^{3-3} = 8^0$$

Notice that $\frac{8^3}{8^3}$ is equal to both 1 and 8^0. Therefore, $8^0 = 1$.

$$\frac{8^3}{8^3} = 1$$

Power of a Quotient

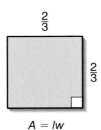

$A = lw$

A power of a quotient (fraction) can be evaluated by methods shown previously or by using the following law.

Power of a Quotient

For any nonzero numbers a and b, and all numbers n:

$$\left(\frac{a}{b}\right)^n = \frac{a^n}{b^n}$$

To find the area of the square at the left, you may use either method shown below.

$$\left(\frac{2}{3}\right)^2 = \frac{2}{3} \times \frac{2}{3} = \frac{4}{9} \quad \longleftarrow \text{ Use } \frac{2}{3} \text{ as a factor 2 times.}$$

$$\left(\frac{2}{3}\right)^2 = \frac{2^2}{3^2} = \frac{4}{9} \quad \longleftarrow \text{ Use the rule for power of a quotient.}$$

Simplify $\left(\frac{3}{4}\right)^3$.

$$\left(\frac{3}{4}\right)^3 = \frac{3^3}{4^3} = \frac{27}{64}$$

EXAMPLE I

Simplify $\left(\frac{1}{2}\right)^4$.

$$\left(\frac{1}{2}\right)^4 = \frac{1^4}{2^4} = \frac{1}{2^4} = \frac{1}{16}$$

$$\left(\frac{1}{2}\right)^4 = \frac{1}{16}$$

The expression in Example J is evaluated in two ways. Notice that the result is the reciprocal of the previous result.

EXAMPLE J

Simplify $\left(\frac{1}{2}\right)^{-4}$.

$\left(\frac{1}{2}\right)^{-4} = \frac{1^{-4}}{2^{-4}} = \frac{2^4}{1^4} = \frac{16}{1} = 16$ ← Use the law for power of a quotient.

or

$\left(\frac{1}{2}\right)^{-4} = \left(\frac{2}{1}\right)^4 = 2^4 = 16$ ← Use the reciprocal of the base and the opposite of the exponent.

$\left(\frac{1}{2}\right)^{-4} = 16$

Power of a Power

2^3

$A = s^2$

An expression such as $(2^3)^2$ is known as a power of a power.

A power of a power can be evaluated by methods shown previously or by using the following law.

To simplify a power of a power, multiply the exponents and use the same base.

Power of a Power

For any nonzero number a, and all numbers m and n:

$(a^m)^n = a^{mn}$

Simplify $(5^4)^7$.

$(5^4)^7 = 5^{4 \times 7} = 5^{28}$

EXAMPLE K

Simplify $(10^2)^{-5}$.

$(10^2)^{-5} = 10^{2 \times (-5)} = 10^{-10} = \left(\frac{1}{10}\right)^{10} = \frac{1}{10^{10}}$ ← Multiply the exponents.

$(10^2)^{-5} = \frac{1}{10^{10}}$

EXAMPLE L

Simplify $\left[(-8)^3\right]^3$.

$\left[(-8)^3\right]^3 = (-8)^{3 \times 3} = (-8)^9$ ← Multiply the exponents.

$\left[(-8)^3\right]^3 = (-8)^9$

The last example used both parentheses and brackets, but there are no brackets on your calculator, so how can you use a calculator to evaluate an expression like $[(-8)^3]^3$? Brackets are another grouping symbol just like parentheses. They are mainly used to make a problem easier to read. Because there are no brackets on your calculator, whenever you see a bracket, substitute a parenthesis. To evaluate $[(-8)^3]^3$ with a calculator, key in

$$(\quad (\quad (\text{-}) \; 8 \;) \quad \wedge \; 3 \;) \quad \wedge \; 3 \; \boxed{\text{ENTER}}$$

The result should be $-134{,}217{,}728$, which is $(-8)^9$.

EXAMPLE M

Simplify $(1.2^3)^5$.

$(1.2^3)^5 = 1.2^{3\times5} = 1.2^{15}$ ← Multiply the exponents.

$(1.2^3)^5 = 1.2^{15}$

EXERCISES 6-2

Simplify.

1. $0.4^2 \times 0.4$
2. $\left[\left(\frac{2}{5}\right)^{-2}\right]^{-2}$
3. $(3.6^2)^5$
4. $\frac{5^3}{5^{-1}}$
5. $\frac{(-8)^2}{(-8)^4}$
6. $0.2^4 \times 0.2$
7. $\left(\frac{4}{9}\right)^2$
8. $\left(\frac{1}{2}\right) \times \left(\frac{1}{2}\right)^2$
9. $4^{-2} \times 4$
10. $(2^2)^{-3}$
11. $5.1^2 \times 5.1^0$
12. $[(-4)^5]^3$
13. $\left(\frac{1}{10}\right)^{-3}$
14. $\frac{1}{7^3} \times 7^5$
15. $\frac{9^{-2}}{9^{-2}}$
16. $(10^3)^4$
17. $(-3)^2(-3)^4(-3)^5$
18. $\frac{10^2}{10^3}$
19. $\left[\frac{3^2}{(4^3 \times 4^2)}\right]^{-2}$
20. $\left[\left(\frac{2}{3}\right)^3 \times \left(\frac{2}{3}\right)^{-5}\right]^4$

Answer the following.

21. True or False? $2^{-3} = \frac{1}{2^3}$. Explain your answer.
22. Which is greater, $5^3 \times 5^6$ or $(5^5)^2$? Justify your answer.
23. **What's the Error?** A student says that 6^6 divided by 6^3 is 6^2. Explain what the student did wrong and give the correct answer.

USING SCIENTIFIC NOTATION

Standard Form to Scientific Notation

The Golden Gate Bridge cost $35,000,000 to complete.

Scientific notation is a useful way to represent very large or very small quantities.

A number in **scientific notation** is a product of the form:

(a number that is at least 1 and less than 10) $\times$ (a power of 10)

To write a number in scientific notation, first move the decimal point to the immediate right of the first nonzero digit. Then count the number of places you moved the decimal point. Use that number as the exponent of 10. If the decimal point was moved to the left, the exponent is positive.

Write 35,000,000 in scientific notation.

35,000,000.   ← The decimal point is understood to be at the end of a whole number.

3.5,000,000.   ← Move the decimal point 7 places to form a number that is at least 1 and less than 10.

$35,000,000 = 3.5 \times 10^7$ ← Use 7 as the exponent. Use positive 7 because the decimal point was moved to the left.

$35,000,000 = 3.5 \times 10^7$

EXAMPLE A

Write 0.00065 in scientific notation.

0.0006.5   ← Move the decimal point 4 places.

$0.00065 = 6.5 \times 10^{-4}$ ← Use −4 as the exponent. Use negative 4 because the decimal point was moved to the right.

$0.00065 = 6.5 \times 10^{-4}$

EXAMPLE B

Write each number given in standard form in scientific notation.

Standard Form	Scientific Notation
30,410	3.041×10^4
0.029	2.9×10^{-2}
100.3	1.003×10^2
0.000081	8.1×10^{-5}
66,000,500	6.60005×10^7
0.25	2.5×10^{-1}
5.24	5.24×10^0

When some calculators, such as the TI-30X IIS, are placed in scientific notation mode, they will display a number or a result of a calculation in scientific notation.

To put a TI-30X IIS calculator in scientific notation mode, press SCI/ENG [2nd DRG] ▶. You might notice that when you pressed the ▶ key, the word "SCI" was underlined. Now, press ENTER and you should see the word "SCI" at the very bottom center of the display. The word "SCI" will be displayed as long as your calculator is in scientific notation. If your calculator is in this mode, any number entered into the calculator will be displayed in scientific notation.

To get the calculator out of scientific notation mode and back to "normal" mode, press SCI/ENG [2nd DRG] ◀. The underlining should move from "SCI" to "FLO". Press ENTER. The word "SCI" is no longer displayed. "FLO" is short for FLOating point, which tells the calculator to "float" the decimal point to the correct position.

EXAMPLE C

Use your calculator to write each number in scientific notation.

Standard Form	Scientific Notation
30,410	3.041×10^4
0.029	2.9×10^{-2}

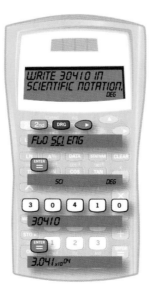
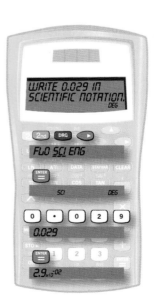

Scientific Notation to Standard Form

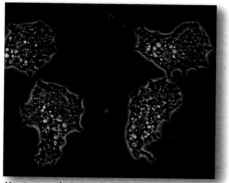

Most amoebas are less than 2.0×10^{-5} meters across.

A number in scientific notation can be written in standard form by moving the decimal point and filling in zeros as necessary. Use the exponent of 10 as the number of places to move the decimal point. If the exponent is positive, move the decimal point right; if the exponent is negative, move the decimal point left.

Write 2.0×10^{-5} in standard form.

$2.0 \times 10^{-5} = 0.00002.$ ← Move the decimal point 5 places. Move the decimal point to the left because the exponent is negative.

$2.0 \times 10^{-5} = 0.00002$

EXAMPLE D

The total resistance of a certain series circuit is 4.09×10^4 Ω. Write 4.09×10^4 in standard form.

$4.09 \times 10^4 = 40,900.$ ← Move the decimal point 4 places, filling in with zeros. Move the decimal point to the right because the exponent is positive.

$4.09 \times 10^4 = 40,900$

EXAMPLE E

Write each number given in scientific notation in standard form.

Scientific Notation	Standard Form
3.0×10^{-2}	0.03
6.0003×10^3	6000.3
9.875×10^{-3}	0.009875
2.41×10^6	2,410,000
4.4×10^{-1}	0.44
1.0×10^{-5}	0.00001

You can use the Enter Exponent (EE) key on the TI-30X IIS calculator to enter a number in scientific notation and convert it to standard form. First, be sure the calculator is set to floating notation (FLO) by pressing [SCI/ENG] [2nd] [DRG], and choosing _FLO_. For example, to convert 3.0×10^{-2} to standard form, press [3] [.] [0] [EE] [2nd] [x^{-1}] [(-)] [2] [ENTER =].
The calculator will display the number in standard form, 0.03.

COMMON ERROR

ALERT

It is a mistake to think that the exponent in scientific notation indicates how many zeros the standard form has. For example, $3.5 \times 10^2 = 350$.

EXERCISES 6-3

Write each number in scientific notation.

1. 219
2. 0.00601
3. 50.5001
4. 0.0012
5. 71.6
6. 100,000
7. 1,835
8. 0.308
9. 3,650,000

Write each number in standard form.

10. 5.006×10^6
11. 2.2×10^{-5}
12. 8.0×10^4
13. 7.07×10^{-2}
14. 6.111×10^3
15. 4.5×10^{-1}
16. 3.0005×10^2
17. 9.2×10^0
18. 1.9406×10^3

Answer the following.

19. One watt is equivalent to 0.000 2389 kilocalories/second (kcal/s). Write this number in scientific notation.

20. The total resistance of a certain series circuit is 6.605×10^9 Ω. Write this number in standard form.

21. **Challenge:** Which is greater: 4.9×10^{-2} or 9.4×10^{-3}? Explain your answer.

SIMPLIFYING EXPRESSIONS WITH POWERS OF TEN

Multiplication

To multiply two numbers that are powers of ten, add the exponents and use 10 as the base. This follows the rules for multiplication with exponents.

Example:

$$10^3 \times 10^4 = 10^{3+4} = 10^7$$

To multiply numbers in scientific notation, multiply the numbers that are between 1 and 10. Add the exponents of the powers of 10.

Example:

$(2.4 \times 10^4) \times (3.5 \times 10^3) =$

$(2.4 \times 3.5) \times (10^4 \times 10^3) =$

$8.4 \times 10^{4+3} =$

8.4×10^7

Helpful Hint

If the TI-30X IIS calculator is set in scientific notation mode, the result of the multiplication will be given in scientific notation. The product at the right will appear as

Multiply (6.7×10^{-4}) and (3.0×10^5). Express the answer in scientific notation.

$(6.7 \times 10^{-4}) \times (3.0 \times 10^5) =$

$(6.7 \times 3.0) \times (10^{-4} \times 10^5) =$

$20.1 \times 10^{-4+5} =$

$20.1 \times 10^1 =$ ⟵ The number is not written in scientific notation because $20.1 > 10$.

2.01×10^2 ⟵ The number is now written in scientific notation.

EXAMPLE A

Multiply (4.7×10^{-2}) and (5.001×10^{-1}). Express the answer in scientific notation.

$(4.7 \times 10^{-2}) \times (5.001 \times 10^{-1}) =$

$(4.7 \times 5.001) \times (10^{-2} \times 10^{-1}) =$

$23.5047 \times 10^{-2+(-1)} =$

$23.5047 \times 10^{-3} =$ ⟵ The number is not written in scientific notation because $23.5 > 10$.

2.35047×10^{-2} ⟵ The number is now written in scientific notation.

$(4.7 \times 10^{-2}) \times (5.001 \times 10^{-1}) = 2.35047 \times 10^{-2}$

EXAMPLE B

Multiply 46,000 and 512,000,000. Express the answer in scientific notation.

To multiply very large or very small numbers, first restate the numbers in scientific notation. Then multiply.

$46,000 = 4.6 \times 10^4$ and $512,000,000 = 5.12 \times 10^8$

$46,000 \times 512,000,000 =$

$(4.6 \times 10^4) \times (5.12 \times 10^8) =$

$(4.6 \times 5.12) \times (10^4 \times 10^8) =$

$23.552 \times 10^{12} =$ ⟵ The product is correct, but it is not written in scientific notation.

2.3552×10^{13} ⟵ The product is written in scientific notation.

$46,000 \times 512,000,000 = 2.3552 \times 10^{13}$

Division

To divide two numbers that are powers of ten, subtract the exponents and use 10 as the base. This follows the rules for division with exponents.

Example:

$$\frac{10^6}{10^{-2}} = 10^{6-(-2)} = 10^{6+2} = 10^8$$

Helpful Hint

To perform the division on the TI-30X IIS calculator, enclose the entire denominator in parentheses as you enter the number.

To divide numbers in scientific notation, divide the numbers that are between 1 and 10. Subtract the exponents of the powers of 10.

Example:

$$\frac{8.4 \times 10^6}{2.1 \times 10^{-2}} = \frac{8.4}{2.1} \times \frac{10^6}{10^{-2}} = 4.0 \times 10^{6-(-2)} = 4.0 \times 10^8$$

Divide 8.8×10^5 by 1.6×10^2. Express the answer in scientific notation.

$$\frac{8.8 \times 10^5}{1.6 \times 10^2} = \frac{8.8}{1.6} \times \frac{10^5}{10^2} = 5.5 \times 10^{5-2} = 5.5 \times 10^3$$

EXAMPLE C

In a certain circuit, the current, in amps, is $\frac{4.5 \times 10^{-2}}{2.25 \times 10^1}$. Evaluate this fraction and express the answer in scientific notation.

$$\frac{4.5 \times 10^{-2}}{2.25 \times 10^1} = \frac{4.5}{2.25} \times \frac{10^{-2}}{10^1} = 2.0 \times 10^{-2-1} = 2.0 \times 10^{-3}$$

$$\frac{4.5 \times 10^{-2}}{2.25 \times 10^1} = 2.0 \times 10^{-3}$$

To divide very large or very small numbers, first restate the numbers in scientific notation. Then divide. State the result in scientific notation.

EXAMPLE D

Divide 0.0039 by 0.000013. Express the answer in scientific notation.

$0.0039 = 3.9 \times 10^{-3}$ and $0.000013 = 1.3 \times 10^{-5}$

$$\frac{0.0039}{0.000013} = \frac{3.9 \times 10^{-3}}{1.3 \times 10^{-5}} = \frac{3.9}{1.3} \times \frac{10^{-3}}{10^{-5}} = 3.0 \times 10^{-3-(-5)} = 3.0 \times 10^2$$

$$0.0039 \div 0.000013 = 3.0 \times 10^2$$

Helpful Hint

If a product such as 32.4×10^5 must be written in scientific notation, move the decimal point one place to the left in 32.4. Then increase the exponent in the power by 1.

$32.4 \times 10^5 = 3.24 \times 10^6$

EXERCISES 6-4

Multiply or divide as indicated. Express the result in scientific notation.

1. $(7.2 \times 10^3) \times (4.5 \times 10^2)$

2. $(6.0 \times 10^{-4}) \times (1.02 \times 10^{-2})$

3. $\frac{2.0 \times 10^5}{5.0 \times 10^1}$

4. $\frac{4.6 \times 10^{-7}}{2.3 \times 10^{-8}}$

5. $4{,}200 \times 8{,}700{,}000$

6. $100{,}000 \times 0.003$

7. $0.005 \div 0.025$

8. $0.36 \div 18{,}000$

Solve.

9. The current in a certain circuit is $\dfrac{4.95 \times 10^{-3}}{3.30 \times 10^{3}}$ A. What is this in scientific notation?

10. The inductive reactance in a certain circuit is $6.2832 \times 15{,}000{,}000 \times 0.0025 \ \Omega$. What is this in scientific notation?

11. **Challenge:** How many times heavier is a seed with a mass of 2.0×10^{2} grams than a seed with a mass of 4.0×10^{-2} grams?

USING ENGINEERING NOTATION

Engineering notation is similar to scientific notation, but it requires that the exponent in the power of 10 be divisible by 3. The multiplier in front of the power of 10 must be at least 1 but less than 1,000.

To write a number in engineering notation, write it as a product in the following form:

(a number that is at least 1 and less than 1,000) $\times$ (a power of 10 whose exponent is divisible by 3)

Write 0.0351 in engineering notation.

$0.0351 = 35.1 \times 10^{-3}$ ◄── Move the decimal point 3 places to the right to form a number that is at least 1 and less than 1,000 multiplied by a power of 10 whose exponent is divisible by 3.

$0.0351 = 35.1 \times 10^{-3}$

Electronic multimeters, used to measure voltage, current, and resistance, are designed for use with engineering notation. A measurement of 4.05×10^{-5} amps would have no corresponding setting on an electronic multimeter, but a setting for the equivalent value 40.5×10^{-6} amp could easily be found.

EXAMPLE A

Write 64,500 in engineering notation.

The number 64,500 is greater than 1,000.

$64{,}500 = 64.5 \times 10^{3}$ ◄── Move the decimal point 3 places to the left.

$64{,}500 = 64.5 \times 10^{3}$

EXAMPLE B

Write 5,019,000 in engineering notation.

Moving the decimal point 3 places is not enough in this case. Therefore, you must move the decimal point 6 places.

$5{,}019{,}000 = 5.019 \times 10^{6}$ ◄── Move the decimal point 6 places to the left.

$5{,}019{,}000 = 5.019 \times 10^{6}$

EXAMPLE C

Write 2.84 × 10⁴ in engineering notation.

Notice that the number is given in scientific notation but the exponent is not a multiple of 3.

Move the decimal point 1 place to the right and decrease the exponent by 1.

$2.84 \times 10^4 = 28.4 \times 10^3$ ← Rewrite the expression so that the exponent is divisible by 3.

$$2.84 \times 10^4 = 28.4 \times 10^3$$

The prefixes associated with the powers of ten are shown in the table below.

Number Value	Prefix	Symbol	Power of 10
1,000,000,000,000	tera	T	10^{12}
1,000,000,000	giga	G	10^9
1,000,000	mega	M	10^6
1,000	kilo	k	10^3
1	units		10^0
0.001	milli	m	10^{-3}
0.000,001	micro	μ	10^{-6}
0.000,000,001	nano	n	10^{-9}
0.000,000,000,001	pico	p	10^{-12}

The names of the prefixes and the powers of ten should be committed to memory. They are used frequently in electronic measurements. For example, when a number is given in engineering notation with the exponent −6, it is described with the prefix *micro* or the symbol μ.

Examples:

40.5×10^{-6} amp = 40.5 microamps or 40.5 μamps

100×10^{-9} second = 100 nanoseconds or 100 nsec

EXAMPLE D

Express 4,500,000 ohms in engineering notation. Then write the measurement using the correct symbol.

The symbol for ohms is Ω.

$$4,500,000 \text{ ohms} = 4.5 \times 10^6 \ \Omega = 4.5 \ M\Omega$$

EXERCISES 6-5

Write in engineering notation.

1. 0.078

2. 30,000,000

3. 16

4. 0.00042

5. 3,750

6. 0.0008

Write each measurement in engineering notation. Then write the measurement using the correct prefix and symbol.

7. 4,700 meters

8. 0.002 amp

9. 0.018 watt

Solve.

10. The charge of one proton is 0.000 000 000 000 000 000 160 22 coulomb (C). Write this in engineering notation.

11. The voltage in a certain series circuit is $5 \times (1.25 \times 10^2) \times (7.5 \times 10^5)$ V. Evaluate this number and write the answer in engineering notation.

12. The power in a circuit with a current of 38.0×10^{-6} A and resistance of 290.0×10^3 Ω is given by $(38.0 \times 10^{-6})^2 \times (290.0 \times 10^3)$ W.

   (a) Determine the power in this circuit, and write the answer in engineering notation.

   (b) Write the answer in (a) using the correct symbol.

13. **What's the Error?** Find and correct the error in the statement:
5×10^3 meters = 5 millimeters.

INTRODUCING ROOTS AND FRACTIONAL EXPONENTS

Radicals

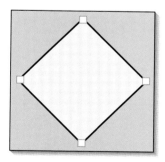

The square of a number is the second power of that number. The square of 6 is 36 because $6^2 = 6 \times 6 = 36$.

A **square root** of a number is a number that can be squared to get the given number. The square roots of 36 are 6 and -6 because $6^2 = (6)(6) = 36$ and $(-6)^2 = (-6)(-6) = 36$.

Every positive number has a positive square root and a negative square root. We are normally concerned with the positive square root of a number, so we call it the **principal square root.** The principal square root of a number is indicated by the radical sign, $\sqrt{}$.

$\sqrt{36} = 6$ ← $\sqrt{36}$ indicates the principal square root of 36.

$-\sqrt{36} = -6$ ← $-\sqrt{36}$ indicates the opposite of the principal square root of 36, which is -6.

From now on, "square root" will mean "principal square root" in this book.

Helpful Hint

On the TI-30X IIS cal-culator, use the keys

to find the square root of 8,100. The calcula-tor gives the principal square root of a number.

The area inside the bases on a regulation baseball field is 8,100 square feet. What is the distance between bases?

Evaluate $\sqrt{8,100}$.

$90^2 = 8,100$. So, the square root of 8,100 is 90.

$\sqrt{8,100} = 90$

The distance between bases is 90 feet.

There is an algorithm, or set of steps, that can be used to find or approximate the square root of any positive number.

EXAMPLE A

Evaluate $\sqrt{121}$.

$11^2 = 121$

$\sqrt{121} = 11$

EXAMPLE B

Evaluate $-\sqrt{64}$.

$8^2 = 64$

$\sqrt{64} = 8$ ← Find the square root first.

$-\sqrt{64} = -8$ ← Find the opposite of the square root.

$-\sqrt{64} = -8$

EXAMPLE C

Evaluate $\sqrt{\frac{4}{25}}$.

$\left(\frac{2}{5}\right)^2 = \frac{4}{25}$

$\sqrt{\frac{4}{25}} = \frac{2}{5}$

To take a square root using a TI-30X IIS calculator, use the $\boxed{\sqrt{}}$ [$\boxed{2nd}$ $\boxed{x^2}$] keys. For example, to find $\sqrt{25}$, press $\boxed{\sqrt{}}$ [$\boxed{2nd}$ $\boxed{x^2}$] 25 $\boxed{)}$ $\boxed{\text{ENTER} =}$. Notice that when you press the $\boxed{\sqrt{}}$ [$\boxed{2nd}$ $\boxed{x^2}$] keys, the calculator dis-plays $\sqrt{}($. You have to enter the right parenthesis.

The radical sign is another grouping symbol in the order of operations. Think of anything under a radical sign as being in parentheses.

EXAMPLE D

Evaluate $\sqrt{1.44}$.

$(1.2)^2 = 1.44$

$\sqrt{1.44} = 1.2$

EXAMPLE E

Evaluate $\sqrt{1.69} + 12$.

Notice that the bar of the radical sign covers the 1.69 and not the +12, so on your calculator you should put a right parenthesis after 1.69.

$1.3^2 = 1.69$

$\sqrt{1.69} = 1.3$

$\sqrt{1.69} + 12 = 1.3 + 12 = 13.3$

$\sqrt{1.69} + 12 = 13.3$

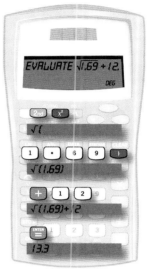

Language Box

A number that is not rational is **irrational.** The numbers $\sqrt{8}$, $\sqrt{5}$, and $\sqrt{2}$ are examples of irrational numbers because there is no rational number that can be squared to get 8, 5, or 2.

All of the square roots discussed to this point are rational numbers. However, not every square root is a rational number. For example, $\sqrt{8}$ is not rational, because there is no rational number that can be used to complete the statement $(\ ?\ )^2 = 8$.

An irrational square root can be approximated. $\sqrt{8} \approx 2.83$. An irrational square root can also be simplified using the rule below.

Product Property of Square Roots

For any numbers a and b, where $a \geq 0$ and $b \geq 0$,
$\sqrt{ab} = \sqrt{a} \cdot \sqrt{b}$

To simplify $\sqrt{8}$, write 8 as $4 \cdot 2$. A raised dot is used instead of a "times sign" to show multiplication.

$\sqrt{8} = \sqrt{4 \cdot 2} = \sqrt{4} \cdot \sqrt{2} = 2\sqrt{2}$

The expression $2\sqrt{2}$ is read *two times the square root of 2.*

In Example F, 45 could be written as either $15 \cdot 3$ or $9 \cdot 5$. Use $9 \cdot 5$ because $\sqrt{9}$ is rational.

EXAMPLE F

Simplify $\sqrt{45}$.

Write 45 as $9 \cdot 5$.

$\sqrt{45} = \sqrt{9 \cdot 5} = \sqrt{9} \cdot \sqrt{5} = 3\sqrt{5}$

$\sqrt{45} = 3\sqrt{5}$

EXAMPLE G

Simplify $-\sqrt{200}$.

Write 200 as $100 \cdot 2$. The negative sign is not used in the simplifying process, but it will be part of the answer.

$$\sqrt{200} = \sqrt{100 \cdot 2} = \sqrt{100} \cdot \sqrt{2} = 10\sqrt{2}$$

$$-\sqrt{200} = -10\sqrt{2}$$

Fractional Exponents

The square root of a number can be expressed using the fraction $\frac{1}{2}$ as the exponent. To understand why this is true, compare the following products:

$5^{\frac{1}{2}} \cdot 5^{\frac{1}{2}} = 5^{\frac{1}{2}+\frac{1}{2}} = 5^1 = 5$ ⟵ The base is the same, so you add the exponents.

$\sqrt{5} \cdot \sqrt{5} = 5$ ⟵ Apply the definition of square root.

The example above shows that $5^{\frac{1}{2}}$ and $\sqrt{5}$ both represent the square root of 5. So, $5^{\frac{1}{2}} = \sqrt{5}$.

The time t in seconds that it takes a pebble to fall s feet is given by the equation $t = 0.25 \cdot s^{\frac{1}{2}}$. How long does it take for a pebble to reach the water if it is dropped from a height of 64 feet?

Evaluate $0.25 \cdot 64^{\frac{1}{2}}$.

$$0.25 \cdot 64^{\frac{1}{2}} = 0.25\sqrt{64} = (0.25)(8) = 2$$

It takes 2 seconds for a pebble dropped from a height of 64 feet to reach the water (neglecting any effects of air resistance).

EXAMPLE H

Evaluate $144^{\frac{1}{2}}$.

$$144^{\frac{1}{2}} = \sqrt{144} = 12$$

$$144^{\frac{1}{2}} = 12$$

EXAMPLE I

Evaluate $-\left(225^{\frac{1}{2}}\right)$.

$$-\left(225^{\frac{1}{2}}\right) = -\sqrt{225} = -15$$

$$-\left(225^{\frac{1}{2}}\right) = -15$$

EXAMPLE J

Evaluate $\left(\dfrac{121}{49}\right)^{\frac{1}{2}}$.

$$\left(\frac{121}{49}\right)^{\frac{1}{2}} = \sqrt{\frac{121}{49}} = \frac{\sqrt{121}}{\sqrt{49}} = \frac{11}{7} = 1\frac{4}{7}$$

$$\left(\frac{121}{49}\right)^{\frac{1}{2}} = 1\frac{4}{7}$$

In addition to square roots, there are other roots. There are cube roots, fourth roots, fifth roots, and so on. Every root can be expressed with either a radical symbol or a fractional exponent. The relationships between powers and roots are summarized in the table below, along with examples.

Language Box

The number 3 in the symbol $\sqrt[3]{}$ (cube root) is called the *index*. The index 2 is understood in the symbol $\sqrt{}$ (square root).

Powers		Roots
$6^2 = 36$ 6 squared equals 36. 6 to the second power equals 36.	$36^{\frac{1}{2}} = 6$ 36 to the $\frac{1}{2}$ power equals 6.	$\sqrt{36} = 6$ The square root of 36 equals 6. The second root of 36 equals 6.
$10^3 = 1,000$ 10 cubed equals 1,000. 10 to the third power equals 1,000.	$1,000^{\frac{1}{3}} = 10$ 1,000 to the $\frac{1}{3}$ power equals 10.	$\sqrt[3]{1,000} = 10$ The cube root of 1,000 equals 10. The third root of 1,000 equals 10.
$2^4 = 16$ 2 to the fourth power equals 16.	$16^{\frac{1}{4}} = 2$ 16 to the $\frac{1}{4}$ power equals 2.	$\sqrt[4]{16} = 2$ The fourth root of 16 equals 2.
$3^5 = 243$ 3 to the fifth power equals 243.	$243^{\frac{1}{5}} = 3$ 243 to the $\frac{1}{5}$ power equals 3.	$\sqrt[5]{243} = 3$ The fifth root of 243 equals 3.

There are two ways to use a TI-30X IIS calculator to take any root. Although both of these can be used to find a square root, it is usually easier to use the [√] [[2nd] [x²]] keys for square roots.

(a) Use fractional exponents. Since $\sqrt[5]{243} = 243^{1/5}$, you can use the following keys; 243 [∧] [(] 1 [÷] 5 [)] [ENTER =]. The result should be 3.

(b) Use the [ˣ√] key. This key can be used to find any root. To find $\sqrt[5]{243}$ using this key, press 5 [ˣ√] [[2nd] [∧]] 243 [ENTER]. Again, the result should be 3. Notice that the index is entered in the calculator before pressing [ˣ√] [[2nd] [∧]].

EXAMPLE K

Find the cube root of 125.

Find a number that gives a product of 125 when used as a factor three times.

$5 \times 5 \times 5 = 125$

$\sqrt[3]{125} = 5$

The cube root of 125 is 5.

EXAMPLE L

Evaluate $\sqrt[4]{10,000}$.

Find a number that gives a product of 10,000 when used as a factor four times.

$$10 \times 10 \times 10 \times 10 = 10^4 = 10,000$$

$$\sqrt[4]{10,000} = 10$$

EXERCISES 6-6

Evaluate.

1. $\sqrt{49}$

2. $-\sqrt{169}$

3. $\sqrt[3]{\dfrac{8}{27}}$

4. $\sqrt[4]{625}$

5. $\sqrt{\dfrac{81}{4}}$

6. $\sqrt[3]{1,000}$

7. $\sqrt{\dfrac{1}{9}}$

8. $-\sqrt{9}$

9. $\sqrt[3]{1}$

Simplify.

10. $\sqrt{12}$

11. $-\sqrt{90}$

12. $\sqrt{44}$

13. $\sqrt{18}$

14. $-\sqrt{28}$

15. $\sqrt{250}$

Evaluate.

16. $64^{\frac{1}{3}}$

17. $8^{\frac{1}{3}}$

18. $-\left(144^{\frac{1}{2}}\right)$

19. $\left(\dfrac{9}{16}\right)^{\frac{1}{2}}$

20. $-\left(81^{\frac{1}{4}}\right)$

21. $49^{\frac{1}{2}}$

22. $-\left(4^{\frac{1}{2}}\right)$

23. $\left(\dfrac{25}{4}\right)^{\frac{1}{2}}$

24. $\left(\dfrac{49}{36}\right)^{\frac{1}{2}}$

25. $1^{\frac{1}{4}}$

26. $\left(\dfrac{1}{1,000}\right)^{\frac{1}{3}}$

27. $32^{\frac{1}{5}}$

Solve.

28. The applied voltage of a circuit with a resistance of 1444 Ω that dissipates 0.4225 W of power is $\sqrt{1444 \times 0.4225}$. Evaluate this expression.

29. If its power rating is not to be exceeded, the maximum current in an 80 Ω, 16.2 W resistor is $\sqrt{\dfrac{16.2}{80}}$ A. Evaluate this expression.

30. **Challenge:** $\left(\dfrac{32}{100,000}\right)^{\frac{1}{5}} = ?$ Express your answer as a fraction.

CHAPTER 6 REVIEW EXERCISES

Evaluate the following.

1. -7^2

2. 5^{-3}

3. $\left(-\frac{9}{11}\right)^2$

4. $(-1.2)^2$

5. 1.87^0

6. $\sqrt{196}$

7. $\sqrt{2.25}$

8. $-\sqrt[3]{8}$

9. $\sqrt[4]{\frac{16}{81}}$

10. $0.125^{\frac{1}{3}}$

Simplify.

11. $0.5^2 \times 0.5^6$

12. $\left[(-3)^4\right]^5$

13. $4.2^{-3} \times 4.2^3$

14. $2.5^{-4} \times \left(\frac{1}{2.5}\right)^3$

15. $\sqrt{75}$

16. $\sqrt{\frac{27}{50}}$

17. Write the number 12,960,000,000,000 in scientific notation.

18. Write the number 0.000000047 in scientific notation.

19. Write the number 37,500,000,000 in engineering notation.

20. Write the number 0.000000261 in engineering notation.

21. The power, in watts, dissipated when 80 V are applied to a circuit with a resistance of 250 Ω is given by $\frac{80^2}{250}$. How much power is dissipated in this circuit?

22. The rated voltage of a 4000 watt electric wall heater with a resistance of 14.4 Ω is $\sqrt{4000 \times 14.4}$. What is the rated voltage of this heater?

23. One ampere-hour (Ah) is 3.6×10^3 coulombs. What is 1 Ah in standard form?

24. The current in a certain circuit is $\frac{3.60 \times 10^{-3}}{3.20 \times 10^3}$ A. Evaluate this fraction and express the answer using scientific notation.

25. The power in a circuit with a current of 1.25×10^{-6} and resistance of 4.8×10^5 Ω is given by $(1.25 \times 10^{-6})^2 \times (4.8 \times 10^5)$ W.

   (a) Determine the power in this circuit and write the answer in scientific notation.

   (b) Write the answer in engineering notation.

   (c) Write the answer using the correct symbol.

Building a Foundation in Mathematics

Finding Logarithms

Properties of Logarithms

Using Logarithms

Natural Logarithms

Applying Logarithms

Overview

Before the invention of calculators and computers, logarithms were used to make it much easier to compute with multiplication, division, powers, and roots. Not only was it faster to use logarithms for computing, but the answers were often more accurate. Now, with the availability of calculators and computers, it is not necessary to make use of logarithms for computations.

However, electricians need to know about logarithms because the analyses of circuits often introduce calculations involving logarithms. In this chapter you will learn the relationship between exponents and logarithms. You will also learn the rules, or properties, for working with logarithms, and how they can be used to solve problems. Finally, you will see how logarithms can be used to determine such things as the power gain when an antenna is installed.

Logarithms

Objectives

After completing this chapter, you will be able to:

- Convert expressions between exponential and logarithmic form
- Find common and natural logarithms
- Find the antilogarithm of a number
- Use logarithms to re-express data
- Use the three basic rules of logarithms and antilogarithms
- Evaluate, manipulate, and simplify logarithmic expressions
- Solve exponential and logarithmic equations
- Use the decibel equations to determine gain or loss
- Use logarithms to solve voltage, current, and resistance problems

chapter 7

FINDING LOGARITHMS

Exponential Form and Logarithmic Form

The intensity of earthquakes is measured on the Richter scale, which is based on logarithms.

You have learned how to convert numbers between standard notation and scientific notation or engineering notation. In this lesson you will learn how to convert between equations in exponential form and equations in logarithmic form.

A **logarithm** is an exponent. When a base and a power of that base are known, the logarithm is the exponent needed to achieve that power. A fact such as $3^2 = 9$ involves a base, an exponent, and a power. This same fact can be expressed in logarithmic form.

Exponential Form

exponent
↓
$$3^2 = 9$$
↑   ↑
base power

Read: *3 squared equals 9.*

Logarithmic Form

logarithm
↓
$$\log_3 9 = 2$$
↑ ↑
base power

Read: *The logarithm of 9, base 3, equals 2.*

Notice that in the exponential form, 2 is called the exponent. In the logarithmic form, 2 is called the logarithm. This is because *a logarithm is an exponent*. Both equations convey the same information. The exponential form answers the question: *What is 3 to the second power?* The logarithmic form answers the question: *What exponent must be used with base 3 to get a result of 9?*

The answer is 2. We say that the logarithm of 9, base 3, is 2.

Write $2^5 = 32$ in logarithmic form.

STEP 1

Identify the base, the exponent, and the power.

exponent
↓
$$2^5 = 32$$
↑   ↑
base power

STEP 2

Write the equation in logarithmic form.

logarithm
↓
$$\log_2 32 = 5$$
↑ ↑
base power

The logarithmic form of the equation indicates that an exponent of 5 is used with a base of 2 to get a result of 32.

$$\log_2 32 = 5$$

Language Box

Frequently, the word *logarithm* is shortened to "log" when an equation is read aloud. For example, read $\log_2 32 = 5$ as *the log of 32, base 2, is 5.*

EXAMPLE A

Write $5^4 = 625$ in logarithmic form.

STEP 1

Identify the base, exponent, and power.

exponent
↓
$5^4 = 625$
↑   ↑
base power

STEP 2

Write the equation in logarithmic form.

logarithm
↓
$\log_5 625 = 4$
↑   ↑
base power

$\log_5 625 = 4$

EXAMPLE B

Write $\log_4 \left(\frac{1}{64}\right) = -3$ in exponential form.

STEP 1

Identify the base, logarithm, and power.

logarithm
↓
$\log_4 \left(\frac{1}{64}\right) = -3$ ← In this case, the power is a fraction.
↑   ↑
base power

STEP 2

Write the equation in exponential form.

$4^{-3} = \frac{1}{64}$ ← Recall that 4^{-3} is equal to $\frac{1}{4^3}$ and is simplified to $\frac{1}{64}$.

$4^{-3} = \frac{1}{64}$

The exponential form of an equation can be used to find the logarithm of a number. This is shown in Example C.

EXAMPLE C

Find $\log_3 81$.

$\log_3 81 = x$ ← Let x represent $\log_3 81$.

$3^x = 81$ ← Write the equation in exponential form.

$3^x = 3^4$ ← Rewrite 81, using base 3.

$x = 4$ ← Solve for x. (If the bases are the same, the exponents must be the same.)

$\log_3 81 = 4$

The relationship between the base, the exponent, and the power are summarized in the definition of logarithm below.

Definition of Logarithm

For all positive numbers b, except $b = 1$, for all positive numbers N, and for all numbers p:

If $b^p = N$, then $\log_b N = p$ and if $\log_b N = p$, then $b^p = N$.

There are two special cases of equations in exponential form and logarithmic form to examine. Consider the following equations:

Exponential Form	Logarithmic Form
$6^1 = 6$	$\log_6 6 = 1$
$10^1 = 10$	$\log_{10} 10 = 1$
$25^1 = 25$	$\log_{25} 25 = 1$

When the base b and the number N are the same, $\log_b N = 1$.

Recall that any number to the zero power is 1. Consider the equations below:

Exponential Form	Logarithmic Form
$6^0 = 1$	$\log_6 1 = 0$
$10^0 = 1$	$\log_{10} 1 = 0$
$25^0 = 1$	$\log_{25} 1 = 0$

No matter what the base is, the logarithm of 1 is zero.

EXAMPLE D

Write $\log_{12} x = 1$ in exponential form. Solve for x.

Write the equation in exponential form.

$12^1 = x$

Solve for x by evaluating the expression 12^1.

$x = 12$

$12^1 = x;\ x = 12$

EXAMPLE E

Write $8^x = 1$ in
logarithmic form.
Solve for x.

$\log_8 1 = x$ ← Write the logarithmic form of the equation.

No matter what the base is, the logarithm of 1 is zero.

$\log_8 1 = x;\ x = 0$

Common Logarithms

Exponential Form	Logarithmic Form
$10^0 = 1$	$\log 1 = 0$
$10^1 = 10$	$\log 10 = 1$
$10^2 = 100$	$\log 100 = 2$
$10^3 = 1{,}000$	$\log 1{,}000 = 3$
$10^4 = 10{,}000$	$\log 10{,}000 = 4$

A logarithm with base 10 is called a **common logarithm.** If a logarithm is written without a base, it is understood to be a common logarithm, and the base is 10. You can use the LOG key on a calculator to find common logarithms.

Although tables of common logarithms exist, modern calculators are often used to evaluate common logarithms. Take time to verify the logarithms given in the table at the left. Use the LOG key on the TI-30X IIS calculator.

Find the following common logarithm, using a calculator:
log 8.21, log 82.1, log 821, and log 8,210.

$\log 8.21 \ = \log (8.21 \times 10^0) = 0.9143$
$\log 82.1 \ = \log (8.21 \times 10^1) = 1.9143$
$\log 821 \ \ = \log (8.21 \times 10^2) = 2.9143$
$\log 8{,}210 = \log (8.21 \times 10^3) = 3.9143$

The logarithms listed above illustrate that if a number is multiplied by a power of 10, the logarithm of the result can be found by simply adding the exponent of 10 to the logarithm of the original number.

Language Box

In 0.9143, the whole number 0 is called the **characteristic**, and the decimal part of the number, 9143, is called the **mantissa.**

EXAMPLE F

Find log 50 using
a calculator.

_1.6990_

A calculator will display the logarithm of a number using all available decimal places on the screen. In this book, we will use four decimal places for logarithms.

$\log 50 = 1.6990$

EXAMPLE G

Find log 5,000
without using
a calculator.

$\log 50 = 1.6990$ ← From above

$\log 5{,}000 = 3.6990$ ← Add 2 to log 50 (1.6990 + 2) because
$5{,}000 = 50 \times 10^2$.

$\log 5{,}000 = 3.6990$

Antilogarithms

A logarithm is an exponent, and an antilogarithm is a power. If the logarithm of N is p, then the **antilogarithm** of p is N. This is stated below for common logarithms, using abbreviations.

> If log $N = p$, then antilog $p = N$.

The two examples below show how logarithms and antilogarithms are related.

$10^2 = 100$, so log $100 = 2$ and antilog $2 = 100$

$10^{2.4} \approx 251$, so log $251 \approx 2.4$ and antilog $2.4 \approx 251$

Find antilog 3.6. Round your answer to the nearest hundredth.

Using the TI-30X IIS calculator, the keystrokes are

[10ˣ] [2nd] [LOG] [3] [.] [6] [)] [ENTER =].

antilog $3.6 \approx 3{,}981.07$

When you use the keystrokes above to find the antilogarithm of 3.6, the screen will display: 10^(3.6) 3981.071706. This is because the common antilogarithm of 3.6 is the power $10^{3.6}$.

EXAMPLE H

Find antilog 5.

Using the TI-30X IIS calculator, the keystrokes are

[10ˣ] [2nd] [LOG] [5] [)] [ENTER =].

antilog $5 = 100{,}000$

EXAMPLE I

Find antilog 2.5. Round your answer to the nearest tenth.

Using the TI-30X IIS calculator, the keystrokes are

[10ˣ] [2nd] [LOG] [2] [.] [5] [)] [ENTER =].

antilog $2.5 \approx 316.2$

EXERCISES 7-1

Write each equation in logarithmic form.

1. $5^3 = 125$

2. $10^6 = 1{,}000{,}000$

3. $4^2 = 16$

4. $6^{-2} = \dfrac{1}{36}$

5. $7^4 = 2{,}401$

6. $10^{-1} = 0.1$

Write each equation in exponential form.

7. $\log_8 64 = 2$

8. $\log 1{,}000 = 3$

9. $\log_3 243 = 5$

10. $\log 10{,}000 = 4$

11. $\log_9 \left(\dfrac{1}{81}\right) = -2$

12. $\log 10 = 1$

Write each equation in exponential form. Solve for x.

13. $\log_6 1 = x$

14. $\log_2 x = 2$

15. $\log_{11} 121 = x$

16. $\log 10{,}000 = x$

17. $\log_3 9 = x$

18. $\log_8 x = 3$

19. $\log_2 8 = x$

20. $\log_2 \left(\frac{1}{8}\right) = x$

21. $\log_5 \left(\frac{1}{25}\right) = x$

Use a calculator to find each of the following. Round to the nearest hundredth.

22. antilog 0.25

23. antilog (-2)

24. antilog 1.1

Solve.

25. A circuit has a current of 7500 A. The common logarithm of 7.5 is 0.8751. How can you determine the common logarithm of 7,500 without a calculator or table of values?

26. The resistance in a circuit is $2.85 \times 10^5\ \Omega$. If the common logarithm of 2.85 is 0.4548, determine the common logarithm of 2.85×10^5 without using a calculator.

27. The number 16 can be expressed as 2^4 and 4^2. Write two logarithmic equations using these facts.

28. **Challenge:** Solve for x. $\log \sqrt{10} = x$

PROPERTIES OF LOGARITHMS

The pH scale, based on a logarithm, is used to measure the *acidity* or *alkalinity* of a substance. Soil pH is important to farmers because it affects crop production.

Because logarithms are exponents, the properties of logarithms follow directly from the properties of exponents.

Consider the following common logarithms:

$\log 100 = 2$ ← because $10^2 = 100$

$\log 1{,}000 = 3$ ← because $10^3 = 1{,}000$

$\log 100{,}000 = 5$ ← because $10^5 = 10{,}000$

We can use logarithms to rewrite the equation $5 = 2 + 3$.

5	=	2	+	3
$\log 100{,}000$	=	$\log 100$	+	$\log 1{,}000$
$\log (100 \cdot 1{,}000)$	=	$\log 100$	+	$\log 1{,}000$

The discussion above illustrates our first property of logarithms.

Product Property of Logarithms

If b, M, and N are positive numbers and $b \neq 1$,
$\log_b MN = \log_b M + \log_b N$.

Use the Product Property to rewrite the expression log $(17 \cdot 28)$.

$\log (17 \cdot 28) = \log 17 + \log 28$

EXAMPLE A

Use the Product Property to rewrite the expression log (6 · 11).

$$\log (6 \cdot 11) = \log 6 + \log 11$$

EXAMPLE B

Rewrite log 25 + log 4 as a single logarithm, and then evaluate without a calculator.

$$\log 25 + \log 4$$
$$= \log (25 \cdot 4) \quad \longleftarrow \text{ Apply the Product Property.}$$
$$= \log 100 \quad \longleftarrow \text{ Simplify the product.}$$
$$= 2 \quad \longleftarrow \text{ Because } 10^2 = 100, \log 100 = 2.$$

$$\log 25 + \log 4 = 2$$

EXAMPLE C

Rewrite $\log_5 10$ + $\log_5 12.5$ as a single logarithm, and then evaluate without a calculator.

$$\log_5 10 + \log_5 12.5$$
$$= \log_5 (10 \times 12.5) \quad \longleftarrow \text{ Apply the Product Property.}$$
$$= \log_5 125 \quad \longleftarrow \text{ Simplify the product.}$$
$$= 3 \quad \longleftarrow \text{ Because } 5^3 = 125, \log_5 125 = 3.$$

$$\log_5 10 + \log_5 12.5 = 3$$

COMMON ERROR

ALERT

You cannot use the Quotient Property to evaluate an expression such as $\dfrac{\log 1{,}000}{\log 10}$. To evaluate an expression containing the log of a number divided by the log of another number, such as $\dfrac{\log 1{,}000}{\log 10}$, find each logarithm first and then divide.

$$\frac{\log 1{,}000}{\log 10} = \frac{3}{1} = 3$$

We can illustrate the next property of logarithms by rewriting the equation $2 = 6 - 4$.

2	=	6	−	4
log 100	=	log 1,000,000	−	log 10,000
$\log \left(\dfrac{1{,}000{,}000}{10{,}000}\right)$	=	log 1,000,000	−	log 10,000

Quotient Property of Logarithms

If b, M, and N are positive numbers and $b \neq 1$,
$$\log_b \frac{M}{N} = \log_b M - \log_b N.$$

Use the Quotient Property to rewrite the expression $\log \dfrac{18}{5}$.

$$\log \frac{18}{5} = \log 18 - \log 5$$

EXAMPLE D

Use the Quotient Property to rewrite the expression $\log \frac{1,000}{10}$, and then evaluate the expression.

$\log \frac{1,000}{10}$

$= \log 1,000 - \log 10$ ← Apply the Quotient Property.

$= 3 - 1$ ← Evaluate each logarithm.

$= 2$

$\log \frac{1,000}{10} = 2$

EXAMPLE E

Rewrite log 7,000 − log 70 as a single logarithm, and then evaluate without a calculator.

$\log 7,000 - \log 70$

$= \log \frac{7,000}{70}$ ← Apply the Quotient Property.

$= \log 100$ ← Simplify the quotient.

$= 2$

$\log 7,000 - \log 70 = 2$

EXAMPLE F

Rewrite $\log_2 48 - \log_2 1.5$ as a single logarithm, and then evaluate without a calculator.

$\log_2 48 - \log_2 1.5$

$= \log_2 \frac{48}{1.5}$ ← Apply the Quotient Property.

$= \log_2 32$ ← Simplify the quotient.

$= 5$ ← Because $2^5 = 32$, $\log_2 32 = 5$.

$\log_2 48 - \log_2 1.5 = 5$

To illustrate our third property of logarithms, we can evaluate $\log 100^3$ and $3 \log 100$.

$\log 100^3 = \log 1,000,000 = 6$

$3 \log 100 = 3 \cdot 2 = 6$

We see that $\log 100^3 = 3 \log 100$.

Power Property of Logarithms

If p is any real number, M and b are positive numbers, and $b \neq 1$,

$\log_b M^p = p \log_b M$.

Use the Power Property to rewrite the expression $\log_3 9^5$, and then evaluate without a calculator.

$\log_3 9^5$

$= 5 \log_3 9$ ← Apply the Power Property.

$= 5 \cdot 2$ ← Evaluate the logarithm.

$= 10$ ← Multiply.

So, $\log_3 9^5 = 10$.

EXAMPLE G

Use the Power Property to rewrite the expression log 10,000,000³, and then evaluate without a calculator.

$\log 10{,}000{,}000^3$

$= 3 \log 10{,}000{,}000$ ← Apply the Power Property.

$= 3 \cdot 7$ ← Evaluate the logarithm.

$= 21$ ← Multiply.

$\log 10{,}000{,}000^3 = 21$

EXAMPLE H

Use the Power Property to rewrite the expression $\log_4 16^{\frac{1}{2}}$, and then evaluate without a calculator.

$\log_4 16^{\frac{1}{2}}$

$= \frac{1}{2} \log_4 16$ ← Apply the Power Property.

$= \frac{1}{2} \cdot 2$ ← Evaluate the logarithm.

$= 1$ ← Multiply.

$\log_4 16^{\frac{1}{2}} = 1$

To rewrite some expressions, you may need to apply more than one property of logarithms.

EXAMPLE I

Rewrite the expression $\log_5 \left(\frac{25^2}{5}\right)$, and then evaluate without a calculator.

$\log_5 \left(\frac{25^2}{2}\right)$

$= \log_5 25^2 - \log_5 5$ ← Use the Quotient Property.

$= 2 \log_5 25 - \log_5 5$ ← Use the Power Property.

$= 2 \cdot 2 - 1$ ← Evaluate the logarithms.

$= 4 - 1$ ← Multiply.

$= 3$

$\log_5 \left(\frac{25^2}{2}\right) = 3$

EXERCISES 7-2

Rewrite each expression using one or more properties of logarithms. Do not evaluate the expression.

1. $\log \frac{3}{8}$

2. $\log (2.1 \times 5.8)$

3. $\log 5^3$

4. $\log \frac{2^3}{7}$

5. $\log (7.5^2 \times 9.4)$

6. $\log (4 \times 3.5 \times 90)$

7. $\log_2 (4 \times 7.5)$

8. $\log_5 \left(\frac{42}{5^3}\right)$

Rewrite each expression using one or more properties of logarithms. Then, evaluate each expression without a calculator.

9. $\log 2 + \log 5{,}000$

10. $\log_2 4 - \log_2 32$

11. $\log_5 50 - \log_5 2$

12. $\log 16 + \log 6.25$

13. $\log_4 \left(\frac{16^2}{4}\right)$

14. $\log_3 6 - \log_3 54$

15. Use the properties of logarithms to evaluate $\log\left(\frac{1}{10^{-3}}\right)$. Show your work.

16. The decibel (dB) gain in an amplifier is given as $10(\log P_1 - \log P_2)$. Rewrite this using one or more properties of logarithms.

17. The decibel gain in a certain amplifier is given as $10(\log 500 - \log 5)$.
(a) Rewrite this using one or more properties of logarithms.
(b) Evaluate this without using a calculator.

18. What's the Error? Two evaluations of the expression $\log 8$ are shown below. Which one is correct? Identify the error in the incorrect evaluation.

$$\begin{aligned} \log 8 &= \log 3 + \log 5 \\ &= 0.4771 + 0.6990 \\ &= 1.1761 \end{aligned} \qquad \begin{aligned} \log 8 &= \log 2 + \log 4 \\ &= 0.3010 + 0.602 \\ &= 0.9031 \end{aligned}$$

USING LOGARITHMS

All of these tools have been used over the years to perform calculations.

You can use common logarithms and antilogarithms to multiply, divide, and find powers. Use the formula given below to find a product.

To find the product of *x* and *y*, follow the rule:
$$x \times y = \text{antilog} (\log x + \log y)$$

Multiply 5 and 2 using logarithms and antilogarithms.

STEP 1

Find $\log 5$ and $\log 2$.

$\log 5 = 0.6990$
$\log 2 = 0.3010$

STEP 2

Find the sum of $\log 5$ and $\log 2$.

$$\begin{aligned} \log 5 &= 0.6990 \\ + \log 2 &= 0.3010 \\ \hline &1.0000 \end{aligned}$$

STEP 3

Find the antilogarithm of the sum of $\log 5$ and $\log 2$.

$5 \times 2 = 10$

$\text{antilog} (\log 5 + \log 2)$
$= \text{antilog} (1)$
$= 10$

EXAMPLE A

Multiply 4.5 × 0.03 using logarithms and antilogarithms.

Find antilog (log 4.5 + log 0.03).

STEP 1

Find log 4.5 and log 0.03.

log 4.5 = 0.6532
log 0.03 = −1.5229

STEP 2

Find the sum of log 4.5 and log 0.03.

$$\begin{array}{r} \log 4.5 = 0.6532 \\ + \log 0.03 = -1.5229 \\ \hline -0.8697 \end{array}$$

STEP 3

Find the antilogarithm of the sum of log 4.5 and log 0.03.

antilog (log 4.5 + log 0.03)
= antilog (−0.8697)
= 0.135

4.5 × 0.03 = 0.135

To find the quotient of *x* and *y*, follow the rule:
$$x \div y = \text{antilog } (\log x - \log y)$$

EXAMPLE B

Divide 31.6 ÷ 7.91 using logarithms and antilogarithms.

Find antilog (log 31.6 − log 7.91).

STEP 1

Find log 31.6 and log 7.91.

log 31.6 = 1.4997
log 7.91 = 0.8982

STEP 2

Find the difference of log 31.6 and log 7.91.

$$\begin{array}{r} \log 31.6 = 1.4997 \\ - \log 7.91 = 0.8982 \\ \hline 0.6015 \end{array}$$

STEP 3

Find the antilogarithm of the difference of log 31.6 and log 7.91.

antilog (log 31.6 − log 7.91)
= antilog (0.6015)
≈ 3.995

31.6 ÷ 7.91 ≈ 3.995

To find *x* raised to the *y* power, follow the rule:
$$x^y = \text{antilog } (y \log x)$$

EXAMPLE C

Evaluate 2.8^7.

Find antilog (7 log 2.8).

STEP 1

Find log 2.8.

$\log 2.8 = 0.4472$

STEP 2

Multiply log 2.8 and 7.

$7 \log 2.8$
$= 7 \times 0.4472$
$= 3.1304$

STEP 3

Find the antilogarithm of the product of 7 and log 2.8.

antilog (7 log 2.8)
$=$ antilog (3.1304)
$\approx 1,350$

$2.8^7 \approx 1,350$

EXERCISES 7-3

Evaluate each of the following using logarithms and antilogarithms. Round your final answer to the nearest whole number. Show your work.

1. 4.5×0.03
2. 56^4
3. $\frac{21}{18}$
4. $\frac{6^3}{7}$
5. $1,000 \times 1,000$
6. 0.5×236

7. The power, in watts, dissipated when 80.25 V is applied to a 225 Ω resistor is given by $\frac{80.25^2}{225}$.
   (a) Rewrite this using one or more properties of logarithms.
   (b) Evaluate your answer in (a) using logarithms. Round your answer to four decimal places.
   (c) Evaluate the given expression by using antilogarithms of your answer in (b). Round your answer to one decimal place.

8. If its power rating is not to be exceeded, the maximum current in an 80 Ω, 16.2 W resistor is $\sqrt{\frac{16.2}{80}}$ A.
   (a) Rewrite this using one or more properties of logarithms.
   (b) Evaluate your answer in (a) using logarithms. Round your answer to four decimal places.
   (c) Evaluate the given expression by using antilogarithms of your answer in (b). Round your answer to two decimal places.

9. The total resistance when a 28 Ω and a 32 Ω resistor are connected in parallel is $\frac{28 \times 32}{28 + 32}$ Ω.
   (a) Rewrite this expression using the properties of logarithms.
   (b) Evaluate your answer in (a) using logarithms. Round your answer to four decimal places.
   (c) Evaluate the given expression by using antilogarithms of your answer in (b). Round your answer to two decimal places.

NATURAL LOGARITHMS

The number e is named for the mathematician Leonard Euler, who first proposed it.

So far, we have worked with logarithms that have whole number bases. Many formulas in finance, chemistry, and physics use natural logarithms, which do not use a whole number as the base.

A **natural logarithm** is a logarithm whose base is the irrational number e. The number e can be defined as the sum of an infinite series. The first five terms of the series are shown in the definition below.

$$e = 2 + \frac{1}{2} + \frac{1}{2 \times 3} + \frac{1}{2 \times 3 \times 4} + \frac{1}{2 \times 3 \times 4 \times 5} + \cdots$$

The value of e is approximately 2.7183.

The natural logarithm is denoted by the symbol ln. The [LN] key on the TI-30X IIS calculator is used to find the natural logarithm of a number.

Compare the notations for common logarithms and natural logarithms.

Common Logarithm of N (base = 10)	Natural Logarithm of N (base = e)
log N	ln N

Recall that a logarithm is an exponent. The natural logarithm of a number is the exponent needed to achieve that number by using e as the base. In the statement below, p represents both a natural logarithm and an exponent.

> If ln $N = p$, then $e^p = N$.

For example, we can use a calculator to find that $e^3 \approx 20.09$. We can also use a calculator to find that ln $20.09 \approx 3$. These two facts represent the same relationship.

Exponential Form

exponent
↓
$e^3 \approx 20.09$
↑   ↑
base power

Logarithmic Form

logarithm
↓
ln $20.09 \approx 3$
↑   ↑
base e power

Find ln 45 rounded to the nearest hundredth.

Use the keystrokes [LN] [4] [5] [)] [ENTER] on the TI-30X IIS calculator.

The display is 3.80666249.
Round the number to the nearest hundredth.

ln $45 \approx 3.81$

EXAMPLE A

Find ln 1,000 rounded to the nearest tenth.

Use the keystrokes **LN** **1** **0** **0** **0** **)** **ENTER =** on the TI-30X IIS calculator.
The display is 6.907755279.
Round the number to the nearest tenth.

ln 1,000 ≈ 6.9

EXAMPLE B

Find ln $\left(\frac{5}{3}\right)$ rounded to the nearest thousandth.

Use the keystrokes **LN** **5** **÷** **3** **)** **ENTER =** on the TI-30X IIS calculator.
The display is 0.510825624.
Round the number to the nearest thousandth.

ln $\left(\frac{5}{3}\right)$ ≈ 0.511

All of the properties of logarithms apply to natural logarithms. These properties are summarized in the box below. You will use these properties to rewrite expressions involving natural logarithms.

Properties of Logarithms, Including Natural Logarithms

If p is any real number, b, M, and N are positive numbers, and $b \neq 1$:

$\log_b MN = \log_b M + \log_b N$ Product Property

$\log_b \frac{M}{N} = \log_b M - \log_b N$ Quotient Property

$\log_b M^p = p \log_b M$ Power Property

EXAMPLE C

Use the Product Property to rewrite the expression ln (50 × 2.6).

ln (50 × 2.6) = ln 50 + ln 2.6

EXAMPLE D

Use the Quotient Property to rewrite the expression ln $\frac{2}{9}$.

ln $\frac{2}{9}$ = ln 2 − ln 9

EXAMPLE E

Use the Power Property to rewrite the expression ln 3.94^5.

ln 3.94^5 = 5 ln 3.94

EXERCISES 7-4

Find each natural logarithm. Round to the nearest hundredth.

1. $\ln 38$

2. $\ln \left(\frac{2}{7}\right)$

3. $\ln 0.28$

4. $\ln \left(\frac{6}{5}\right)$

5. $\ln 45{,}306$

6. $\ln 4.75$

7. $\ln 80.6$

8. $\ln 461.2$

9. $\ln \left(\frac{8}{15}\right)$

Rewrite each expression using the properties of logarithms.

10. $\ln (76 \times 8.3)$

11. $\ln 0.3^5$

12. $\ln \frac{1}{8}$

13. $\ln \frac{5}{6}$

14. $\ln (10^4 \times 10^5)$

15. $\ln 6.7^3$

Solve.

16. Some investments pay *compound interest*. If interest is paid continuously, then the formula $t = \left(\frac{1}{r}\right) \ln \left(\frac{A}{A_0}\right)$ can be used, where t is the number of years the money is invested, r is the annual interest rate as a decimal. A_0 is the money invested at the start of the time period, and A is the amount of money after t years (original investment plus interest). Carlos invested \$200 in a savings account that pays an annual interest rate of 4.5% (0.045). compounded continuously.

(a) How many years will it take for Carlos's investment to reach \$250? (Round your answer to the nearest tenth year.)

(b) How many years will it take for Carlos's investment to double, that is, reach \$400? (Round your answer to the nearest tenth year.)

17. If interest is compounded annually (once a year), then the formula $t = \dfrac{\ln\left(\frac{A}{A_0}\right)}{\ln(1 + r)}$ can be used, where t is the number of years the money is invested, r is the annual interest rate as a decimal, A_0 is the money invested at the start of the time period, and A is the amount of money after t years (original investment plus interest). Brianna invested \$200 in a savings account that pays an annual interest rate of 0.045, compounded annually.

(a) How many years will it take for Brianna's investment to reach \$250? (Round your answer to the nearest tenth year.)

(b) How many years will it take for Brianna's investment to double, that is, reach \$400? (Round your answer to the nearest tenth year.)

18. What's the Error? Explain why the following statement is not correct:
$-1 \cdot \ln 5 = \ln (-5)$.

APPLYING LOGARITHMS

The Decibel

When a radio picks up a signal through its antenna, the signal is too weak to be heard by the human ear. The radio must amplify the weak input signal to get a stronger output signal that people can hear. This increase in signal power is called *gain*.

The common measure of gain for signal level changes is the **decibel** (dB). A decibel is based on a ratio of two values of power, the output power and the input power. The formula to calculate gain is:

$$dB = 10 \log \frac{P_1}{P_2} \quad \begin{array}{l} \leftarrow \text{Output power} \\ \leftarrow \text{Input power} \end{array}$$

where dB represents decibels, P_1 represents output power, and P_2 represents input power. Power is measured in watts (W), or milliwatts (mW).

An amplifier has an input power P_2 of 5 mW. The output power P_1 is 100 mW. What is the gain, rounded to the nearest decibel?

$$dB = 10 \log \frac{P_1}{P_2}$$

$$= 10 \log \frac{100}{5}$$

$$= 10 \log 20$$

$$= 10 \times 1.3010$$

$$= 13.010$$

$$\approx 13$$

The gain is approximately 13 dB.

Language Box

The basic unit of gain is the **bel**, named after inventor Alexander Graham Bell. A **decibel** is one tenth of a bel.

EXAMPLE A

A circuit has an input power of 4.5 mW. Its output power is 63 mW. What is the gain, rounded to the nearest decibel?

$$dB = 10 \log \frac{P_1}{P_2}$$

$$= 10 \log \frac{63}{4.5}$$

$$= 10 \times 1.1461$$

$$= 11.461$$

$$\approx 11$$

The gain is approximately 11 dB.

When the output power is less than the input power, there is a decibel *loss*. A loss of power is indicated by a negative decibel value. Consider the following example.

EXAMPLE B

A circuit has an input power of 6 mW. Its output power is 4 mW. What is the loss, rounded to the nearest decibel?

$$\text{dB} = 10 \log \frac{P_1}{P_2}$$

$$= 10 \log \frac{4}{6}$$

$$= 10 \times (-0.1761)$$

$$= -1.761$$

$$\approx -2$$

The loss is approximately 2 dB.

The same decibel formula can be used to find output power or input power.

EXAMPLE C

An amplifier has a gain of 12 dB. If the input power is 6 mW, what is the output power, rounded to the nearest whole number?

$$\text{dB} = 10 \log \frac{P_1}{P_2}$$

$$12 = 10 \log \frac{P_1}{6} \qquad \longleftarrow \text{ Substitute 12 for dB and 6 for } P_2.$$

$$1.2 = \log \frac{P_1}{6} \qquad \longleftarrow \text{ Divide both sides by 10.}$$

$$1.2 = \log P_1 - \log 6 \qquad \longleftarrow \text{ Apply the Quotient Property.}$$

$$1.2 = \log P_1 - 0.7782 \qquad \longleftarrow \text{ Use a calculator to find log 6.}$$

$$1.9782 = \log P_1 \qquad \longleftarrow \text{ Add 0.7782 to both sides.}$$

$$P_1 = 10^{1.9782} \qquad \longleftarrow \text{ Rewrite the equation in exponential form.}$$

$$P_1 = 95.1043 \qquad \longleftarrow \text{ Use a calculator to find } 10^{1.9782}.$$

$$P_1 \approx 95$$

The output power is approximately 95 mW.

Keeping in mind that the decibel measurement is based on the ratio of the output power to the input power, consider the case where the output power is *twice* the input power. Then the ratio $\frac{P_1}{P_2}$ is $\frac{2}{1}$, or 2.

EXAMPLE D

Find the gain whenever the output power is twice the input power.

$$\text{dB} = 10 \log \frac{P_1}{P_2}$$

$$= 10 \log 2$$

$$= 10 \times 0.3010$$

$$= 3.010$$

$$\approx 3$$

There is a gain of about 3 dB whenever the output power is twice the input power.

EXAMPLE E

Find the decibel loss when the output power is *half* the input power.

Language Box

The half-power point is 3 dB. For every 3 dB of gain, the output power is doubled. For every 3 dB of loss, the output power is cut in half.

In this case, the ratio $\frac{P_1}{P_2}$ equals $\frac{1}{2}$, or 0.5.

$$dB = 10 \log \frac{P_1}{P_2}$$
$$= 10 \log 0.5$$
$$= 10 \times (-0.3010)$$
$$= -3.010$$
$$\approx -3$$

There is a loss of about 3 dB whenever the output power is half the input power.

Voltage, Current, and Resistance

Voltage, current, resistance, and power in a circuit are related by these equations:

$$E \quad = \quad I \quad \times \quad R$$

voltage current resistance
(volts) (amps) (ohms)

$$P \quad = \quad E \quad \times \quad I$$

power voltage current
(watts) (volts) (amps)

If input power and output power are known, the formula $dB = 10 \log \frac{P_1}{P_2}$ can be used to find a gain or loss in decibels.

However, since power is the product of voltage and current, there are two other formulas that can be used to find a gain or a loss. If there is no change in resistance, we can use the formulas below.

$$dB = 20 \log \frac{E_1}{E_2} \qquad dB = 20 \log \frac{I_1}{I_2}$$

E_1 = output voltage $\qquad$ I_1 = output current
E_2 = input voltage $\qquad$ I_2 = input current

The following discussion will explain how the formula using voltage is derived from the basic formula. The formula using current is derived in a similar way.

If the formula $E = I \times R$ is solved for I, we have $I = \frac{E}{R}$. Then, for input power P_2 and output power P_1, we have:

$$P_1 = E_1 \times I_1 \qquad P_2 = E_2 \times I_2$$
$$= E_1 \times \frac{E_1}{R_1} \qquad = E_2 \times \frac{E_2}{R_2}$$
$$= \frac{E_1^2}{R_1} \qquad = \frac{E_2^2}{R_2}$$

If there is no change in resistance, then $R_1 = R_2$, and the basic formula for gain or loss can be written:

$$dB = 10 \log \frac{P_1}{P_2}$$

$$= 10 \log \frac{\dfrac{E_1^2}{R_1}}{\dfrac{E_2^2}{R_2}} \qquad \longleftarrow \text{ Substitute for } P_1 \text{ and } P_2.$$

$$= 10 \log \frac{E_1^2}{E_2^2} \qquad \longleftarrow \text{ Simplify, using } R_1 = R_2.$$

$$= 10 \log \left(\frac{E_1}{E_2}\right)^2 \qquad \longleftarrow \text{ Rewrite the fraction with a single exponent.}$$

$$= 10 \times 2 \log \frac{E_1}{E_2} \qquad \longleftarrow \text{ Apply the Power Property.}$$

$$= 20 \log \frac{E_1}{E_2} \qquad \longleftarrow \text{ Simplify.}$$

A constant tone is applied to a speaker. The voltage across the speaker is 3 volts. Then it is increased to 9 volts. What is the gain, rounded to the nearest tenth of a dB?

$$dB = 20 \log \frac{E_1}{E_2}$$

$$= 20 \log \frac{9}{3} \qquad \longleftarrow \text{ Substitute 9 for output voltage } E_1; \text{ substitute 3 for input voltage } E_2.$$

$$= 20 \log 3 \qquad \longleftarrow \text{ Divide 9 by 3.}$$

$$= 20 \times 0.4771 \qquad \longleftarrow \text{ Evaluate log 3.}$$

$$\approx 9.5$$

The gain is approximately 9.5 dB.

EXAMPLE F

The signal to a speaker causes 2.5 amps of current. The current is decreased to 0.5 amp. What is the loss, rounded to the nearest decibel?

Substitute 0.5 for I_1 and 2.5 for I_2 in the formula for decibel gain or loss of current.

$$dB = 20 \log \frac{I_1}{I_2}$$

$$= 20 \log \frac{0.5}{2.5} \qquad \longleftarrow \text{ Substitute 0.5 for output current } I_1; \text{ substitute 2.5 for input current } I_2.$$

$$= 20 \log 0.2 \qquad \longleftarrow \text{ Divide 0.5 by 2.5.}$$

$$= 20 \times (-0.6990) \qquad \longleftarrow \text{ Evaluate log 0.2.}$$

$$\approx -14$$

The loss is approximately 14 dB.

When the input and output resistances are *not* equal, use the formula Power = Voltage × Current to calculate the input and output power.

Then use the basic formula: $dB = 10 \log \frac{P_1}{P_2}$. This is shown in Example G.

EXAMPLE G

An amplifier has an input resistance of 200 ohms, with 0.25 volt applied across it. It has an output resistance of 3,200 ohms, with a voltage of 10 volts. What is the gain?

Input Power

$$P_2 = \frac{E_2{}^2}{R_2}$$

$$= \frac{0.25^2}{200}$$

$$= 0.0003125$$

Output Power

$$P_1 = \frac{E_1{}^2}{R_1}$$

$$= \frac{10^2}{3,200}$$

$$= 0.03125$$

Use these values for P_1 and P_2 for power in the basic formula.

$$dB = 10 \log \frac{P_1}{P_2}$$

$$= 10 \log \frac{0.03125}{0.0003125}$$

$$= 10 \log 100$$

$$= 10 \times 2$$

$$= 20$$

The gain is 20 dB.

EXERCISES 7-5

Solve.

1. An amplifier has a 50-watt output and a 5-watt input. Find the gain for this amplifier, rounded to the nearest decibel.

2. A CB antenna has a 4-dB gain. If the input power to the antenna is 3 microwatts, what is the power output of the antenna, rounded to the nearest hundredth?

3. Describe the half-power point in your own words.

4. A constant tone is applied to a speaker. The voltage across the speaker is 8 volts. It is decreased to 4 volts. What is the loss, rounded to the nearest decibel?

5. An electronic circuit has an input current of 5 amps and an output current of 2 amps. If the input resistance equals the output resistance, what is the loss, rounded to the nearest decibel?

6. **Challenge:** An amplifier has an input resistance of 180 ohms, with 0.35 volt applied across it. It has an output of 4,000 ohms with a voltage of 12 volts. What is the gain, rounded to the nearest decibel?

CHAPTER 7 REVIEW EXERCISES

Write each equation in logarithmic form.

1. (a) $12^2 = 144$
   (b) $5^{-3} = 0.008$

Write each equation in exponential form.

2. (a) $\log_3 729 = 6$
   (b) $\log_{1/2} 8 = -3$
   (c) $\log 100 = 2$
   (d) $\ln 5 \approx 1.6094$

3. Use your calculator to find the logarithm of each of the following numbers. Round the answers to four decimal places.
   (a) $\log 144$
   (b) $\log 1,440$
   (c) $\ln 225$
   (d) $\ln 2,250$

4. Use your calculator to find the antilog of each of the following numbers. Round the answers to two decimal places.
   (a) antilog 3
   (b) antilog 1.5
   (c) antilog 2.5
   (d) antilog 1.6990

5. Rewrite each expression using one or more properties of logarithms. Do not evaluate the expression.
   (a) $\log \frac{2}{3}$
   (b) $\log (5 \times 3^7)$
   (c) $\ln \frac{56}{7^3}$
   (d) $\ln \frac{125}{18 \times 4^3}$

6. The voltage in a certain series circuit is $5 \times (1.25 \times 10^{-2}) \times (7.5 \times 10^5)$ V.
   (a) Rewrite this using one or more properties of logarithms.
   (b) Evaluate your answer in (a) using logarithms. Round your answer to four decimal places.
   (c) Evaluate the given expression by using antilogarithms of your answer in (b). Round your answer to two decimal places.

7. A circuit has a current of 9.2×10^5 A.
   (a) Use your calculator to find log 9.2. Round the answer to four decimal points.
   (b) Use the answer from (a) to find $\log (9.2 \times 10^5)$.

8. The power in a circuit with a current of 7.25×10^{-6} A and resistance of 4.55×10^5 Ω is given by $(7.25 \times 10^{-6})^2 \times (4.55 \times 10^5)$ W. Rewrite $(7.25 \times 10^{-6})^2 \times (4.55 \times 10^5)$ using one or more properties of logarithms.

9. An amplifier has an output power of 12.5 W and an input level of 27.3 W. Find the gain of this amplifier to the nearest tenth decibel.

10. An HDTV antenna has a gain of 13.7 dB. If the power output of the antenna is 102 microwatts, what is the input power? Round your answer to the nearest hundredth.

11. An amplifier has an input resistance of 160 Ω when 0.4 V is applied across it. The amplifier has an output resistance of 3200 Ω, with a voltage of 20 V.

   (a) What is the input power of the amplifier?

   (b) What is the output power of the amplifier?

   (c) What is the gain?

Building a Foundation in Mathematics

Converting Units in the Customary System

Converting Units in the Metric System

Converting between Customary and Metric Units

Converting Measures of Temperature

Converting Measures of Area and Volume

Using the Mil

2.5 cm

2.2 cm

Overview

The United States uses both the American customary and the SI metric systems of weights and measures. If a measurement is to be understood or useful, its units must be carefully defined and standardized. Units of measurement have evolved over the centuries. The first units were based on a person's body measurement, and the names of some of our present units, such as foot and hand, reflect this. But, not everyone had the same-sized foot, so standardization was necessary.

During the French Revolution, the metric system was developed and was based on the Earth's measurements. When Napoleon conquered much of Europe, the French spread the metric system to every country they conquered. Since then, the metric system has been adopted by almost every country in the world, and most units in electricity are based on the metric system.

In this chapter, you will learn about the different units of measure in the American customary system and the SI metric system. Since there are times when you need to convert measurements from one system to another, you will learn how to do that in this chapter.

Units and Measurements

Objectives

After completing this chapter, you will be able to:

- Express given customary lengths in smaller or larger customary linear units
- State the basic units of the customary or metric (SI) system
- Express given metric lengths in smaller or larger metric linear units
- Convert between customary measures and metric measures
- Convert between mils and circular mils

CONVERTING UNITS IN THE CUSTOMARY SYSTEM

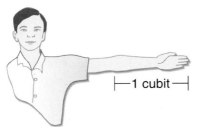

├─1 cubit─┤

The **customary system** is the system of measurement used most often in the United States. The customary system of measurement dates back to the time when many units of measure were based on a person's foot or arm. For example, the cubit was a unit of measure equivalent to the length of a person's arm from the elbow to the tip of the middle finger.

In the 13th century, King Edward I of England defined three units of length that we still use: the yard, the foot, and the inch. He ordered an iron measuring stick to serve as the standard *yard* for his entire kingdom. He then declared a *foot* to be $\frac{1}{3}$ of a yard, and an *inch* to be $\frac{1}{36}$ of a yard. Today, there are national standards for all units of measure. These standards have been established by the National Institute of Standards and Technology (NIST).

The most common units and equivalent measures in the customary system of measurement are listed in the table below.

Language Box

The customary system is still sometimes called the English System due to its origins in England.

Customary Table of Measures

Time	Length	Area
60 seconds (sec) = 1 minute (min)	12 inches (in.) = 1 foot (ft)	144 sq in. = 1 sq ft
60 minutes (min) = 1 hour (hr)	3 feet (ft) = 1 yard (yd)	9 sq ft = 1 sq yd
24 hours (hr) = 1 day	5.5 yards (yd) = 1 rod (rd)	30.25 sq yd = 1 sq rod
365 days = 1 year (yr)	320 rods (rd) = 1 mile (mi)	43,560 sq ft = 1 acre
366 days = 1 leap year	5,280 feet (ft) = 1 mile (mi)	640 acres = 1 sq mi

Volume (Liquid)	Volume (Dry)	Weights
8 fluid ounces (fl oz) = 1 cup (c)	2 pints (pt) = 1 quart (qt)	16 ounces (oz) = 1 pound (lb)
2 cups (c) = 1 pint (pt)	8 quarts (qt) = 1 peck (pk)	2,000 pounds (lb) = 1 ton (T)
2 pints (pt) = 1 quart (qt)	4 pecks (pk) = 1 bushel (bu)	
4 quarts (qt) = 1 gallon (gal)		

Architectural wiring plans often use ″ and ′ for inches and feet, respectively. For example, some drawings will have 8′-9″ for 8 ft-9 in.

You often need to convert measurements from one unit to another within the customary system. For example, you may need to convert miles to yards. To do this, a conversion factor is used. A **conversion factor** is a fraction that is equal to 1, with the numerator expressed in the units that you want to convert *to*, and the denominator expressed in the units that you want to convert *from*.

For example, to convert from quarts to gallons, use the conversion factor:

$\dfrac{1 \text{ gal}}{4 \text{ qt}}$ ← Convert to gallons.
      ← Convert from quarts.

To convert from gallons to quarts, use the conversion factor:

$\dfrac{4 \text{ qt}}{1 \text{ gal}}$ ← Convert to quarts.
      ← Convert from gallons.

At Bill's Garage, the mechanics use about 40 quarts of oil every day for oil changes. How many gallons of oil do they use every day for oil changes?

Convert 40 quarts to gallons.

$40 \text{ qt} = \dfrac{40 \text{ qt}}{1}$ ← Write the quantity to be converted as a fraction with denominator 1.

$40 \text{ qt} = \dfrac{40 \text{ qt}}{1} \times \dfrac{1 \text{ gal}}{4 \text{ qt}}$ ← Multiply by the appropriate conversion factor.

$40 \text{ qt} = \dfrac{40 \text{ qt}}{1} \times \dfrac{1 \text{ gal}}{4 \text{ qt}} = \dfrac{40 \text{ gal}}{4} = 10 \text{ gal}$ ← Cancel the unit that appears in both numerator and denominator, and simplify.

The mechanics use about 10 gallons of oil every day.

EXAMPLE A

Jasmine needs 14 yards of cable, but it is sold only in feet. Convert 14 yards to feet.

$14 \text{ yd} = \dfrac{14 \text{ yd}}{1}$ ← Write the quantity to be converted as a fraction with denominator 1.

$14 \text{ yd} = \dfrac{14 \text{ yd}}{1} \times \dfrac{3 \text{ ft}}{1 \text{ yd}}$ ← Multiply by the appropriate conversion factor.

$14 \text{ yd} = \dfrac{14 \text{ yd}}{1} \times \dfrac{3 \text{ ft}}{1 \text{ yd}} = \dfrac{42 \text{ ft}}{1} = 42 \text{ ft}$ ← Cancel the unit that appears in both numerator and denominator, and simplify.

14 yards = 42 feet

Jasmine needs 42 feet of cable.

For some conversions, you will need more than one conversion factor. This is shown in Example B.

EXAMPLE B

How many hours are there in 1.5 years?

Use the conversion factor $\dfrac{365 \text{ days}}{1 \text{ yr}}$ to convert years to days, and use the conversion factor $\dfrac{24 \text{ hr}}{1 \text{ day}}$ to convert days to hours.

$1.5 \text{ yr} = \dfrac{1.5 \text{ yr}}{1}$ ← Write the quantity as a fraction.

$1.5 \text{ yr} = \dfrac{1.5 \text{ yr}}{1} \times \dfrac{365 \text{ days}}{1 \text{ yr}} \times \dfrac{24 \text{ hr}}{1 \text{ day}}$ ← Multiply by the conversion factors.

$1.5 \text{ yr} = \dfrac{1.5 \text{ yr}}{1} \times \dfrac{365 \text{ days}}{1 \text{ yr}} \times \dfrac{24 \text{ hr}}{1 \text{ day}} = 13{,}140 \text{ hr}$ ← Cancel units.

There are 13,140 hours in 1.5 years.

EXAMPLE C

A map shows a plot of land that has an area of 140,000 square feet. About how many acres is this? Round your answer to the nearest tenth of an acre.

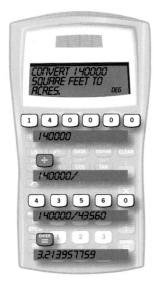

The Customary Table of Measures contains the fact 43,560 sq ft = 1 acre. Write the conversion factor $\frac{1 \text{ acre}}{43,560 \text{ sq ft}}$ and use it to convert square feet to acres.

$140,000 \text{ sq ft} = \dfrac{140,000 \text{ sq ft}}{1}$ ← Write the quantity as a fraction.

$140,000 \text{ sq ft} = \dfrac{140,000 \text{ sq ft}}{1} \times \dfrac{1 \text{ acre}}{43,560 \text{ sq ft}}$ ← Multiply by the conversion factor.

$140,000 \text{ sq ft} = \dfrac{140,000 \text{ sq ft}}{1} \times \dfrac{1 \text{ acre}}{43,560 \text{ sq ft}} \approx 3.2 \text{ acres}$ ← Cancel units.

An area of 140,000 square feet is approximately 3.2 acres.

EXERCISES 8-1

Write the conversion factor or factors needed for each conversion.

1. quarts to pints   **2.** square yards to square feet   **3.** ounces to pounds

4. days to minutes   **5.** pounds to tons   **6.** cups to gallons

Convert each measurement to the indicated unit.

7. 360 in. = ____ yd   **8.** 7,000 lb = ____ T

9. 1 gal = ____ pt   **10.** 13 bu = ____ qt

11. 72 hr = ____ days   **12.** 18,000 sec = ____ hr

13. 6 mi = ____ ft   **14.** 4 qt = ____ fl oz

Convert each measurement to the indicated unit. Round each answer to the nearest tenth, if rounding is necessary.

15. 1,000 hr = ____ days   **16.** 160 sq yd = ____ sq ft

17. 25.4 ft = ____ in.   **18.** 8.5 lb = ____ oz

Solve.

19. Nathan needs 73.25 yards of cable for a certain job. How long is this in feet and inches?

20. Number 14 copper wire weighs 12.68 lb/1,000 ft.
   (a) What is the weight, in pounds, of 75 feet of #14 wire?
   (b) Express the answer to (a) in ounces.

21. A box contains 1,200 bolts weighing 1 ounce each. How many pounds of bolts does this box contain?

22. **Challenge:** Ed wants to make a large amount of lemonade for a picnic. Each batch of the recipe calls for 4 cups of water, 1 cup of lemon juice, and 1 cup of sugar. If Ed has an unlimited amount of water and sugar, but only 256 fluid ounces of lemon juice, how many gallons of lemonade can he make? (Assume that the dissolved sugar does not add to the total amount of lemonade.)

Helpful Hint

The only abbreviation that is written with a period is in. (for inches). Do not use a period after any other abbreviation unless it occurs at the end of a sentence.

CONVERTING UNITS IN THE METRIC SYSTEM

The French implemented a system of measurement in the 1790s, called the **metric system.** Today, almost all countries use the updated system of metric measurements, called *SI Metric*. The initials SI stand for the French phrase *Systeme International*. We translate it in English as *System of International Units*. Although the metric system of measurement has been declared the preferred system of measurement by the U.S. Congress, conversion to the system in the United States has been largely voluntary. However, because so many measurements are given in the metric system, it is important to learn how to use it.

In the metric system, the meter is the base unit of length, the liter is the base unit of capacity, and the gram is the base unit of mass (weight). The table below gives some base units of the metric system, along with their symbols and uses.

Metric System of Measurement

Base Unit	SI Symbol	Used to Measure
meter	m	length
gram	g	mass
second	s	time
liter	L	volume
kelvin	K	temperature
ampere	A	electric current

> ### Helpful Hint
>
> In the metric system, the base units are the most basic (fundamental) units from which other units are formed.

The base units in the metric system are combined with prefixes to define smaller and larger units. For example, a kilometer is equal to 1000 meters, and a millimeter is $\frac{1}{1000}$ meter. The most common prefixes and their corresponding values are given in the next table.

Table of Prefixes and Values

Metric Prefix	Symbol	Power of Ten	Number Value
giga	G	10^9	1 000 000 000
mega	M	10^6	1 000 000
kilo	k	10^3	1000
hecto	h	10^2	100
deka	da	10^1	10
units	—	10^0	1
deci	d	10^{-1}	0.1
centi	c	10^{-2}	0.01
milli	m	10^{-3}	0.001
micro	μ	10^{-6}	0.000 001
nano	n	10^{-9}	0.000 000 001

Spaces are used instead of commas when writing numbers in the metric system. This is done because people in some countries use a comma just as we use a decimal point. Numbers with more than three digits to the right of the decimal point also include spaces to separate groups of three digits so that the numbers are easier to read.

Helpful Hint

Conversions in the metric system involve multiplying or dividing by powers of 10.

Some examples of the way base units are combined with prefixes are shown below.

1 centimeter = 0.01 meter
(The prefix *centi* has the value 0.01.)

A fingernail is about 1 centimeter in width.

1 kilogram = 1000 grams
(The prefix *kilo* has the value 1000.)

A textbook has a mass of about 1 kilogram.

1 milliamp = 0.001 amp
(The prefix *milli* has the value 0.001.)

A person can barely feel 1 milliamp of electric current.

Some of the most common metric units of linear measure and how they are related are shown in the following table.

Metric Units of Linear Measure

1 millimeter (mm) = 0.001 meter (m)	1000 millimeters (mm) = 1 meter (m)
1 centimeter (cm) = 0.01 meter (m)	100 centimeters (cm) = 1 meter (m)
1 meter (m) = 1 meter (m)	1 meter (m) = 1 meter (m)
1 kilometer (km) = 1000 meters (m)	0.001 kilometer (km) = 1 meter (m)

The Table of Prefixes and Values can be used to create the appropriate conversion factors in the metric system. For example, to convert from centimeters to meters, use the fact that a *centi*meter is 0.01 meter.

$$1 \text{ cm} = 0.01 \text{ m}$$

$$100 \times 1 \text{ cm} = 100 \times 0.01 \text{ m} \quad \longleftarrow \text{ Multiply both sides by 100.}$$

$$100 \text{ cm} = 1 \text{ m}$$

So, the conversion factor $\frac{1 \text{ m}}{100 \text{ cm}}$ can be used to convert from centimeters to meters.

Convert 5 kilometers to meters.

$$5 \text{ km} = \frac{5 \text{ km}}{1} \qquad\qquad \longleftarrow \text{ Write the quantity as a fraction.}$$

$$5 \text{ km} = \frac{5 \text{ km}}{1} \times \frac{1000 \text{ m}}{1 \text{ km}} \qquad \longleftarrow \text{ Multiply by the appropriate conversion factor.}$$

$$5 \text{ km} = \frac{5 \cancel{\text{ km}}}{1} \times \frac{1000 \text{ m}}{1 \cancel{\text{ km}}} = 5000 \text{ m} \qquad \longleftarrow \text{ Cancel units.}$$

5 kilometers = 5000 meters

EXAMPLE A

Convert 2500 milliliters (mL) to liters (L).

$$2500 \text{ mL} = \frac{2500 \text{ mL}}{1} \qquad\qquad \longleftarrow \text{ Write the quantity as a fraction.}$$

$$2500 \text{ mL} = \frac{2500 \text{ mL}}{1} \times \frac{1 \text{ L}}{1000 \text{ mL}} \qquad \longleftarrow \text{ Multiply by the conversion factor.}$$

$$2500 \text{ mL} = \frac{2500 \cancel{\text{ mL}}}{1} \times \frac{1 \text{ L}}{1000 \cancel{\text{ mL}}} = 2.5 \text{ L} \qquad \longleftarrow \text{ Cancel units and simplify.}$$

2500 milliliters = 2.5 liters

EXAMPLE B

Find the number of milliseconds in 8.5 seconds.

$$8.5 \text{ s} = \frac{8.5 \text{ s}}{1}$$ ← Write the quantity as a fraction.

$$8.5 \text{ s} = \frac{8.5 \text{ s}}{1} \times \frac{1,000 \text{ ms}}{1 \text{ s}}$$ ← Multiply by the conversion factor.

$$8.5 \text{ s} = \frac{8.5 \cancel{\text{ s}}}{1} \times \frac{1,000 \text{ ms}}{1 \cancel{\text{ s}}} = 8,500 \text{ ms}$$ ← Cancel units and simplify.

There are 8,500 milliseconds in 8.5 seconds.

In the metric system, the units of time are s, for second; m, for minute; and h, for hour. In the customary system they are sec, for second; min, for minute; and hr, for hour.

If the conversion problem does not mention a base unit, you can convert to the appropriate base unit, and then to the desired unit. This is shown in Example C.

EXAMPLE C

Convert 58 millimeters to centimeters.

Since 1 mm = 0.001 m, then 1000 mm = 1 m.

Use the conversion factor $\frac{1 \text{ m}}{1000 \text{ mm}}$ to convert millimeters to the base unit meters, and use the conversion factor $\frac{100 \text{ cm}}{1 \text{ m}}$ to convert meters to centimeters.

$$58 \text{ mm} = \frac{58 \text{ mm}}{1}$$ ← Write the quantity as a fraction.

$$58 \text{ mm} = \frac{58 \text{ mm}}{1} \times \frac{1 \text{ m}}{1000 \text{ mm}} \times \frac{100 \text{ cm}}{1 \text{ m}}$$ ← Multiply by the conversion factors.

$$58 \text{ mm} = \frac{58 \text{ mm}}{1} \times \frac{1 \text{ m}}{1000 \text{ mm}} \times \frac{100 \text{ cm}}{1 \text{ m}} = 5.8 \text{ cm}$$ ← Cancel units and simplify.

58 millimeters = 5.8 centimeters

EXERCISES 8-2

Write the conversion factor or factors needed for each conversion.

1. grams to kilograms
2. centimeters to meters
3. liters to milliliters
4. nanoseconds to seconds
5. millimeters to centimeters
6. A to mA

Convert each measurement to the indicated unit.

7. 20 mm = _____ cm
8. 7.75 L = _____ mL
9. 3000 mA = _____ A
10. 9200 g = _____ kg
11. 60 g = _____ mg
12. 4750 mL = _____ L
13. 2.75 cm = _____ mm
14. 45.7 kg = _____ g

Solve.

15. The diameter of a ceiling lamp is 97.155 cm. Convert this length to meters.

16. A digital television station broadcasts at the frequency range of 638 000 000 to 644 000 000 Hz (hertz). Convert these frequencies to (a) kilohertz (kHz) and (b) megahertz (MHz).

17. Electricity is transmitted at high voltages of 110 kV or more in order to reduce the energy lost in transmission. Convert this voltage to (a) volts and (b) megavolts (MV).

18. Challenge: Convert 10 nanoseconds to milliseconds.

CONVERTING BETWEEN CUSTOMARY AND METRIC UNITS

Converting a measurement between the customary and metric systems requires the same process used to convert within a system, that is, multiplying by a conversion factor. The conversion factor is formed using an appropriate fact found in a table of measures. Because the basic units in the customary and the metric system are different, conversions between systems do not result in exact answers. You will often have to round your answer when doing this type of conversion.

At one time the length of an inch in the United States was not the same as an inch in Great Britain. In 1958, it was decided to make them the same so that 1 in. = 2.54 cm. All other conversions between the two systems are approximations.

Table of Customary/Metric Equivalents
(All entries, except 1 in. = 2.54 cm, are approximations.)

Length	Capacity	Weight (Mass)
1 in. = 2.54 cm	1 fl oz = 29.57 mL	1 oz = 28.34 g
0.39 in. = 1 cm	0.03 fl oz = 1 mL	0.035 oz = 1 g
1 yd = 0.914 m	1 qt = 0.946 L	1 lb = 0.45 kg
1.094 yd = 1 m	1.057 qt = 1 L	2.22 lb = 1 kg
1 mi = 1.61 km		
0.621 mi = 1 km		

The fact 1 qt = 0.946 L is located in the center column of the table. It can be used to form the conversion factor $\frac{0.946 \text{ L}}{1 \text{ qt}}$ to convert quarts to liters.

Kevin is mixing up a pitcher of instant iced tea. The directions call for 3 quarts of water. His pitcher is labeled in liters. About how many liters of water does Kevin need?

Convert 3 quarts to liters. Round to the nearest tenth.

$3 \text{ qt} = \dfrac{3 \text{ qt}}{1}$ ← Write the quantity as a fraction.

$3 \text{ qt} = \dfrac{3 \text{ qt}}{1} \times \dfrac{0.946 \text{ L}}{1 \text{ qt}}$ ← Multiply by the conversion factor.

$3 \text{ qt} = \dfrac{3 \text{ q\!\!\!/t}}{1} \times \dfrac{0.946 \text{ L}}{1 \text{ q\!\!\!/t}} \approx 2.8 \text{ L}$ ← Cancel units and simplify.

Kevin needs about 2.8 liters of water.

The examples that follow involve length, capacity, and weight (mass). Conversions of temperature, area, and volume will be addressed later in this chapter.

EXAMPLE A

A conduit decreases in length by 0.6 inch due to a temperature change. How many centimeters is this change in length, to the nearest tenth?

$0.6 \text{ in.} = \dfrac{0.6 \text{ in.}}{1}$ ← Write the quantity as a fraction.

$0.6 \text{ in.} = \dfrac{0.6 \text{ in.}}{1} \times \dfrac{2.54 \text{ cm}}{1 \text{ in.}}$ ← Multiply by the conversion factor.

$0.6 \text{ in.} = \dfrac{0.6 \text{ i\!\!\!/n.}}{1} \times \dfrac{2.54 \text{ cm}}{1 \text{ i\!\!\!/n.}} \approx 1.5 \text{ cm}$ ← Cancel units.

The change in length is about 1.5 centimeters.

EXAMPLE B

Convert 120 miles to kilometers. Round to the nearest whole kilometer.

The Table of Customary/Metric Equivalents contains two facts about miles and kilometers. To convert miles to kilometers, use the fact that tells how many kilometers are in 1 mile, 1 mi = 1.61 km.

$120 \text{ mi} = \dfrac{120 \text{ mi}}{1}$ ← Write the quantity as a fraction.

$120 \text{ mi} = \dfrac{120 \text{ mi}}{1} \times \dfrac{1.61 \text{ km}}{1 \text{ mi}}$ ← Multiply by the conversion factor.

$120 \text{ mi} = \dfrac{120 \text{ m\!\!\!/i}}{1} \times \dfrac{1.61 \text{ km}}{1 \text{ m\!\!\!/i}} \approx 193 \text{ km}$ ← Cancel units.

A distance of 120 miles is about 193 kilometers.

For some conversions, you may need to use more than one table of measures. This is shown in Example C.

EXAMPLE C

A load of 250-ft spools of electric wire weighs 1200 kilograms. What is the equivalent weight in tons? Round to the nearest hundredth.

Language Box

A metric ton (or tonne) is 1000 kg. The symbol t is used for metric ton.

The Table of Customary/Metric Equivalents contains facts about kilograms (kg) and pounds (lb), so use the conversion factor $\frac{1\ lb}{0.45\ kg}$ to convert kilograms to pounds. The Customary Table of Measures contains a fact about pounds and tons, so use the conversion factor $\frac{1\ T}{2{,}000\ lb}$ to convert pounds to tons.

$$1200\ \text{kg} = \frac{1200\ \text{kg}}{1} \qquad \leftarrow \text{Write the quantity as a fraction.}$$

$$1200\ \text{kg} = \frac{1200\ \text{kg}}{1} \times \frac{1\ \text{lb}}{0.45\ \text{kg}} \times \frac{1\ \text{T}}{2{,}000\ \text{lb}} \qquad \leftarrow \text{Multiply by the conversion factors.}$$

$$1200\ \text{kg} = \frac{1200\ \cancel{\text{kg}}}{1} \times \frac{1\ \cancel{\text{lb}}}{0.45\ \cancel{\text{kg}}} \times \frac{1\ \text{T}}{2{,}000\ \cancel{\text{lb}}} = \frac{1{,}200\ \text{T}}{0.45 \times 2{,}000} \approx 1.33\ \text{T}$$

The load of bricks weighs about 1.33 tons.

8-3 EXERCISES

Convert each of the following to the indicated unit. Round each answer to the nearest tenth, if rounding is necessary.

1. 16 m = _____ yd
2. 78 km = _____ mi
3. 4.7 oz = _____ g
4. 800 fl oz = _____ L
5. 1000 mm = _____ ft
6. 10 L = _____ gal
7. 6 L = _____ qt
8. 3 in. = _____ mm
9. 100 cm = _____ yd
10. 300 mL = _____ fl oz
11. 7.5 yd = _____ m
12. 80 g = _____ oz

Solve. Round each answer to the nearest tenth, if rounding is necessary.

13. An electric motor is to be installed on a shelf that has a maximum safe load of 150 lb. The mass of the motor is 75 kg.
    (a) What is the weight of the motor in pounds?
    (b) Is it safe to install this motor on the shelf?

14. A floor box for a wood floor has the following dimensions: $3\frac{3}{4}''$ (length), 3" (width), and $3\frac{1}{2}''$ (height). Convert each of these dimensions to centimeters.

15. A 2.5-oz bottle of liquid flux cost $2.95.
    (a) How many milliliters are in the bottle?
    (b) What is the cost per milliliter?

16. **Challenge:** Art is placing a light every 48 inches along a walkway that is 50 meters long. He places the first light at the beginning of the walkway. How many lights can he place? Where does he place the last light? (Hint: Use the exact conversion factor noted at the top of the table.)

CONVERTING MEASURES OF TEMPERATURE

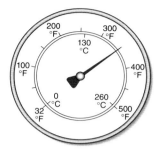

There are two temperature scales in the metric system—the **Celsius (C) scale** (measured in degrees) and the **Kelvin (K) scale** (measured in kelvins). Water freezes at 0°C and boils at 100°C, so there are 100 degrees between the freezing point and the boiling point of water on the Celsius scale. Likewise, there are 100 kelvins between the freezing point and the boiling point of water on the Kelvin scale, but the freezing point is approximately 273 K and the boiling point is approximately 373 K. The lowest temperature on the Kelvin scale is zero kelvins, which is called *absolute zero*. Unlike the degree Fahrenheit and degree Celsius, the kelvin is not referred to as a "degree," nor is it typeset with a degree symbol; that is, it is written K and not °K.

The kelvin is the base unit of temperature in the metric system, because the Kelvin scale is useful for scientific purposes. The Celsius scale, however, is more commonly used than the Kelvin scale.

In the customary system, we measure temperature on the Fahrenheit scale. On the Fahrenheit scale, there are 212 degrees between the freezing point and the boiling point of water. The table below summarizes some important temperatures on all three scales.

Temperature Scales			
Temperature	**Fahrenheit**	**Celsius**	**Kelvin**
boiling point of water	212°	100°	373.15 K
freezing point of water (melting point of ice)	32°	0°	273.15 K
absolute zero	−459.67°	−273.15°	0 K

To convert between the Celsius and the Fahrenheit scale, use one of the formulas below.

To convert *from* Celsius (°C) *to* Fahrenheit (°F):

$$F = \frac{9}{5}C + 32$$

To convert *from* Fahrenheit (°F) *to* Celsius (°C):

$$C = \frac{5}{9}(F - 32)$$

The temperature of a grill is 350°F. What is that temperature on the Celsius scale?

Convert 350°F to degrees Celsius. Round to the nearest whole degree.

Use the formula $C = \frac{5}{9}(F - 32)$. Substitute 350 for F.

$$C = \frac{5}{9}(F - 32) = \frac{5}{9}(350 - 32) = \frac{5}{9}(318) = 176.\overline{6} \approx 177°C$$

350°F is approximately 177°C.

EXAMPLE A

Convert 65°F to degrees Celsius. Round your answer to the nearest whole degree Celsius.

Use the formula $C = \frac{5}{9}(F - 32)$. Substitute 65 for F. Remember to subtract within the parentheses before multiplying by $\frac{5}{9}$.

$C = \frac{5}{9}(F - 32)$ ◄— Use the formula for converting from degrees Fahrenheit to degrees Celsius.

$= \frac{5}{9}(65 - 32)$ ◄— Substitute 65 for F.

$= \frac{5}{9}(33)$ ◄— Subtract within the parentheses.

$= 18.\overline{3}$ ◄— Multiply.

≈ 18 ◄— Round to the nearest whole number.

65°F is approximately 18°C.

EXAMPLE B

Convert 90°C to degrees Fahrenheit.

Use the formula $F = \frac{9}{5}C + 32$. Substitute 90 for C. There are no parentheses in this formula, so multiply first.

$F = \frac{9}{5}C + 32$ ◄— Use the formula for converting from degrees Celsius to degrees Fahrenheit.

$= \frac{9}{5}(90) + 32$ ◄— Substitute 90 for C.

$= 162 + 32$ ◄— Multiply.

$= 194$ ◄— Add.

90°C is equal to 194°F.

EXAMPLE C

Convert −5°C to degrees Fahrenheit.

$F = \frac{9}{5}C + 32$ ◄— Use the formula for converting from degrees Celsius to degrees Fahrenheit.

$= \frac{9}{5}(-5) + 32$ ◄— Substitute −5 for C.

$= -9 + 32$ ◄— Multiply.

$= 23$ ◄— Add.

−5°C is equal to 23°F.

EXERCISES 8-4

Convert each of the following temperatures as indicated. Round each answer to the nearest tenth, if rounding is necessary.

1. 125°C = _____°F

2. 0°F = _____°C

3. 44.5°C = _____°F

4. 10°C = _____°F

5. 100°F = _____°C

6. −2°F = _____°C

7. $50°C =$ _____ $°F$ **8.** $5°C =$ _____ $°F$

9. $68°F =$ _____ $°C$ (comfortable room temperature)

10. $98.6°F =$ _____ $°C$ (normal body temperature)

11. $451°F =$ _____ $°C$ (temperature at which paper ignites)

Solve. Round your answer to the nearest tenth of a degree as needed.

12. The melting point of copper is 2,723°F. Convert that temperature to degrees Celsius.

13. The body temperature for a person with the flu is 103.5°F. What is this temperature in degrees Celsius?

14. The minimum, ideal, and maximum operating temperatures for a certain electric motor are −46°C, 22°C, and 60°C. Express each of these temperatures in degrees Fahrenheit, to the nearest degree.

15. Convert −40°C to degrees Fahrenheit.

16. What is the only temperature at which the Fahrenheit and Celsius temperatures are equal?

17. **What's the Error?** To convert 54°F to degrees Celsius, a student showed the following work. Explain what was done incorrectly, and give the correct answer.
$$C = \frac{5}{9}(F - 32) = \frac{5}{9} \cdot \frac{54}{1} - 32 = 30 - 32 = -2$$

CONVERTING MEASURES OF AREA AND VOLUME

Area

2.5 cm

←—2.2 cm—→

Area is the number of square units on a surface. To find a conversion factor for converting an area measure, write a conversion factor using the appropriate units of length and then square that conversion factor.

The area of a certain postage stamp is 5.5 cm². Convert the area of the stamp to mm².

In this case, $\frac{10 \text{ mm}}{1 \text{ cm}}$ is the appropriate conversion factor to convert from centimeters to millimeters. Square this conversion factor to get the conversion factor for area.

$$\left(\frac{10 \text{ mm}}{1 \text{ cm}}\right)^2 = \left(\frac{10 \text{ mm}}{1 \text{ cm}}\right)\left(\frac{10 \text{ mm}}{1 \text{ cm}}\right) = \frac{100 \text{ mm}^2}{1 \text{ cm}^2}$$

$5.5 \text{ cm}^2 = \frac{5.5 \text{ cm}^2}{1}$ ← Write the quantity as a fraction.

$5.5 \text{ cm}^2 = \frac{5.5 \text{ cm}^2}{1} \times \frac{100 \text{ mm}^2}{1 \text{ cm}^2}$ ← Multiply by the conversion factor for area.

$5.5 \text{ cm}^2 = \frac{5.5 \text{ cm}^2}{1} \times \frac{100 \text{ mm}^2}{1 \text{ cm}^2} = 550 \text{ mm}^2$ ← Cancel units and simplify.

The area of the postage stamp is 550 mm².

EXAMPLE A

A park has an area of 3.5 square kilometers. Convert the area of the park to square meters.

Square the conversion factor for length, $\frac{1000 \text{ m}}{1 \text{ km}}$, to get the conversion factor for area.

$$\left(\frac{1000 \text{ m}}{1 \text{ km}}\right)^2 = \left(\frac{1000 \text{ m}}{1 \text{ km}}\right)\left(\frac{1000 \text{ m}}{1 \text{ km}}\right) = \frac{1\,000\,000 \text{ m}^2}{1 \text{ km}^2}$$

$$3.5 \text{ km}^2 = \frac{3.5 \text{ km}^2}{1}$$ ← Write the quantity as a fraction.

$$3.5 \text{ km}^2 = \frac{3.5 \text{ km}^2}{1} \times \frac{1\,000\,000 \text{ m}^2}{1 \text{ km}^2}$$ ← Multiply by the conversion factor for area.

$$3.5 \text{ km}^2 = \frac{3.5 \text{ km}^2}{1} \times \frac{1\,000\,000 \text{ m}^2}{1 \text{ km}^2} = 3\,500\,000 \text{ m}^2$$ ← Cancel units and simplify.

The park has an area of 3 500 000 square meters.

EXAMPLE B

Convert 8000 square centimeters to square meters.

Square the conversion factor for length, $\frac{1 \text{ m}}{100 \text{ cm}}$, to get the conversion factor for area.

$$\left(\frac{1 \text{ m}}{100 \text{ cm}}\right)^2 = \left(\frac{1 \text{ m}}{100 \text{ cm}}\right)\left(\frac{1 \text{ m}}{100 \text{ cm}}\right) = \frac{1 \text{ m}^2}{10\,000 \text{ cm}^2}$$

$$8000 \text{ cm}^2 = \frac{8000 \text{ cm}^2}{1}$$ ← Write the quantity as a fraction.

$$8000 \text{ cm}^2 = \frac{8000 \text{ cm}^2}{1} \times \frac{1 \text{ m}^2}{10\,000 \text{ cm}^2}$$ ← Multiply by the conversion factor for area.

$$8000 \text{ cm}^2 = \frac{8000 \text{ cm}^2}{1} \times \frac{1 \text{ m}^2}{10\,000 \text{ cm}^2} = 0.8 \text{ m}^2$$ ← Cancel units and simplify.

8000 square centimeters is equal to 0.8 square meter.

EXAMPLE C

An office requires 1,250 square feet of carpet. About how many square yards of carpet is that? Round your answer to the nearest whole number.

Square the conversion factor for length, $\frac{1 \text{ yd}}{3 \text{ ft}}$, to get the conversion factor for area.

$$\left(\frac{1 \text{ yd}}{3 \text{ ft}}\right)^2 = \left(\frac{1 \text{ yd}}{3 \text{ ft}}\right)\left(\frac{1 \text{ yd}}{3 \text{ ft}}\right) = \frac{1 \text{ yd}^2}{9 \text{ ft}^2}$$

$1,250 \text{ ft}^2 = \dfrac{1,250 \text{ ft}^2}{1}$ ← Write the quantity as a fraction.

$1,250 \text{ ft}^2 = \dfrac{1,250 \text{ ft}^2}{1} \times \dfrac{1 \text{ yd}^2}{9 \text{ ft}^2}$ ← Multiply by the conversion factor for area.

$1,250 \text{ ft}^2 = \dfrac{1,250 \cancel{\text{ ft}^2}}{1} \times \dfrac{1 \text{ yd}^2}{9 \cancel{\text{ ft}^2}} \approx 139 \text{ yd}^2$ ← Cancel units and simplify.

There are approximately 139 square yards in 1,250 square feet.

The Table of Customary Measures includes some facts about area that can be used directly for conversion within the customary system. The fact 144 sq in. = 1 sq ft is used in Example D.

EXAMPLE D

How many square inches are there in a tabletop whose area is 6 square feet?

$6 \text{ sq ft} = \dfrac{6 \text{ sq ft}}{1}$ ← Write the quantity as a fraction.

$6 \text{ sq ft} = \dfrac{6 \text{ sq ft}}{1} \times \dfrac{144 \text{ sq in.}}{1 \text{ sq ft}}$ ← Multiply by the conversion factor.

$6 \text{ sq ft} = \dfrac{6 \cancel{\text{ sq ft}}}{1} \times \dfrac{144 \text{ sq in.}}{1 \cancel{\text{ sq ft}}} = 864 \text{ sq in.}$ ← Cancel units and simplify.

There are 864 square inches in the tabletop.

Language Box

Square feet were written as ft² in Example C and as sq ft in Example D. Both are correct. In the metric system, units of area are written using exponents, so square meter is written m², as in Example B.

Volume

Volume is the number of cubic units enclosed by or occupied by a three-dimensional figure. To find a conversion factor for converting a volume measure, write a conversion factor using appropriate units of length and then cube that conversion factor.

How many cubic feet are there in $2\frac{1}{2}$ cubic yards of mulch?

Cube the appropriate conversion factor for length to get the conversion factor for volume.

$$\left(\frac{3 \text{ ft}}{1 \text{ yd}}\right)^3 = \left(\frac{3 \text{ ft}}{1 \text{ yd}}\right)\left(\frac{3 \text{ ft}}{1 \text{ yd}}\right)\left(\frac{3 \text{ ft}}{1 \text{ yd}}\right) = \frac{27 \text{ ft}^3}{1 \text{ yd}^3}$$

$$2\frac{1}{2} \text{ yd}^3 = \frac{5 \text{ yd}^3}{2}$$ ← Write the quantity as a fraction.

$$2\frac{1}{2} \text{ yd}^3 = \frac{5 \text{ yd}^3}{2} \times \frac{27 \text{ ft}^3}{1 \text{ yd}^3}$$ ← Multiply by the conversion factor for volume.

$$2\frac{1}{2} \text{ yd}^3 = \frac{5 \text{ yd}^3}{2} \times \frac{27 \text{ ft}^3}{1 \text{ yd}^3} = \frac{135 \text{ ft}^3}{2} = 67\frac{1}{2} \text{ ft}^3$$ ← Cancel units and simplify.

There are $67\frac{1}{2}$ cubic feet in $2\frac{1}{2}$ cubic yards of mulch.

EXAMPLE E

Convert 4500 cubic millimeters to cubic centimeters.

Cube the conversion factor for length, $\frac{1 \text{ cm}}{10 \text{ mm}}$, to get the correct conversion factor for volume.

$$\left(\frac{1 \text{ cm}}{10 \text{ mm}}\right)^3 = \left(\frac{1 \text{ cm}}{10 \text{ mm}}\right)\left(\frac{1 \text{ cm}}{10 \text{ mm}}\right)\left(\frac{1 \text{ cm}}{10 \text{ mm}}\right) = \frac{1 \text{ cm}^3}{1000 \text{ mm}^3}$$

$$4500 \text{ mm}^3 = \frac{4500 \text{ mm}^3}{1}$$ ← Write the quantity as a fraction.

$$4500 \text{ mm}^3 = \frac{4500 \text{ mm}^3}{1} \times \frac{1 \text{ cm}^3}{1000 \text{ mm}^3}$$ ← Multiply by the conversion factor for volume.

$$4500 \text{ mm}^3 = \frac{4500 \text{ mm}^3}{1} \times \frac{1 \text{ cm}^3}{1000 \text{ mm}^3} = 4.5 \text{ cm}^3$$ ← Cancel units and simplify.

4500 cubic millimeters equals 4.5 cubic centimeters.

EXAMPLE F

Convert 9 cubic meters to cubic centimeters.

Follow the same steps as in previous examples.

$$\left(\frac{100 \text{ cm}}{1 \text{ m}}\right)^3 = \left(\frac{100 \text{ cm}}{1 \text{ m}}\right)\left(\frac{100 \text{ cm}}{1 \text{ m}}\right)\left(\frac{100 \text{ cm}}{1 \text{ m}}\right) = \frac{1\,000\,000 \text{ cm}^3}{1 \text{ m}^3}$$

$$9 \text{ m}^3 = \frac{9 \text{ m}^3}{1}$$

$$9 \text{ m}^3 = \frac{9 \text{ m}^3}{1} \times \frac{1\,000\,000 \text{ cm}^3}{1 \text{ m}^3}$$

$$9 \text{ m}^3 = \frac{9 \text{ m}^3}{1} \times \frac{1\,000\,000 \text{ cm}^3}{1 \text{ m}^3} = 9\,000\,000 \text{ cm}^3$$

There are 9 000 000 cubic centimeters in 9 cubic meters.

EXAMPLE G

Convert 2 liters to cubic centimeters.

In the metric system, 1 milliliter is equal to 1 cubic centimeter and also to 0.001 liter. Therefore, it is possible to convert a measure of capacity, given in liters, to a measure of volume.

$$\begin{array}{ccccc}
0.001\ \text{L} & = & 1\ \text{mL} & = & 1\ \text{cm}^3 \\
\times\quad 2000 & & \times\ 2000 & & \times\ 2000 \\
\hline
2\ \text{L} & = & 2000\ \text{mL} & = & 2000\ \text{cm}^3
\end{array}$$

← Write the basic fact.
← Multiply each measure by 2000 because there are 2000 mL in 2 L.

So, 2 liters can also be expressed as 2000 milliliters or 2000 cubic centimeters.

Conversions of Area and Volume between Systems

Use the table below to convert measures of area and volume between the customary system and the metric system.

Table of Customary/Metric Equivalents
(All entries are approximations.)

Area
1 square foot (ft²) = 0.093 square meter (m²)
1 square yard (yd²) = 0.836 square meter (m²)
1 square meter (m²) = 10.764 square feet (ft²)
1 square meter (m²) = 1.20 square yards (yd²)

Volume
1 cubic inch (in.³) = 16.387 cubic centimeters (cm³)
1 cubic foot (ft³) = 0.028 cubic meter (m³)
1 cubic yard (yd³) = 0.765 cubic meter (m³)
1 cubic centimeter (cm³) = 0.061 cubic inch (in.³)
1 cubic meter (m³) = 35.315 cubic feet (ft³)
1 cubic meter (m³) = 1.308 cubic yards (yd³)

EXAMPLE H

Convert 2.25 square meters to square feet. Round your answer to the nearest hundredth.

To write a conversion factor to convert square meters to square feet, choose the fact that relates 1 square meter to square feet: 1 square meter = 10.764 square feet.

$$2.25\ \text{m}^2 = \frac{2.25\ \text{m}^2}{1}$$

← Write the quantity as a fraction.

$$2.25\ \text{m}^2 = \frac{2.25\ \text{m}^2}{1} \times \frac{10.764\ \text{ft}^2}{1\ \text{m}^2}$$

← Multiply by the conversion factor.

$$2.25\ \text{m}^2 = \frac{2.25\ \cancel{\text{m}^2}}{1} \times \frac{10.764\ \text{ft}^2}{1\ \cancel{\text{m}^2}} \approx 24.22\ \text{ft}^2$$

← Cancel units and round your answer.

2.25 square meters is approximately 24.22 square feet.

EXAMPLE 1

Convert 2 cubic yards to cubic meters.

To write a conversion factor to convert cubic yards to cubic meters, choose the fact that relates 1 cubic yard to cubic meters: 1 cubic yard = 0.765 cubic meter.

$$2 \text{ yd}^3 = \frac{2 \text{ yd}^3}{1}$$ ← Write the quantity as a fraction.

$$2 \text{ yd}^3 = \frac{2 \text{ yd}^3}{1} \times \frac{0.765 \text{ m}^3}{1 \text{ yd}^3}$$ ← Multiply by the conversion factor.

$$2 \text{ yd}^3 = \frac{2 \cancel{\text{ yd}^3}}{1} \times \frac{0.765 \text{ m}^3}{1 \cancel{\text{ yd}^3}} = 1.53 \text{ m}^3$$ ← Cancel units.

2 cubic yards is approximately 1.53 cubic meters.

EXERCISES 8-5

Convert each of the following measures of area to the unit indicated. Round each answer to the nearest tenth, if rounding is necessary.

1. 3 sq mi = _____ acres

2. 1.5 sq yd = _____ sq in.

3. 1.2 m² = _____ cm²

4. 54 700 m² = _____ km²

5. 100 sq yd = _____ sq ft

6. 0.5 m² = _____ cm²

Helpful Hint

Use the Table of Measures shown earlier in this chapter as a reference for customary units.

Convert each of the following measures of volume to the unit indicated. Round each answer to the nearest tenth, if rounding is necessary.

7. 2,000 cu in. = _____ cu ft

8. 1.5 cu yd = _____ cu in.

9. 450 cm³ = _____ mm³

10. 0.8 m³ = _____ cm³

11. 5 cu ft = _____ cu in.

12. 0.000 01 m³ = _____ cm³

Convert each of the following measures to the unit indicated. Round each answer to the nearest tenth, if rounding is necessary.

13. 20 cm³ = _____ in.³

14. 15 m³ = _____ yd³

15. 100 ft² = _____ m²

16. 1 yd² = _____ m²

17. 100 yd³ = _____ m³

18. 0.6 m³ = _____ ft³

Helpful Hint

All conversion factors, including those for area and volume, have the same form:

$$\frac{\text{new units}}{\text{old units}}$$

Solve.

19. A bottle of liquid flux contains 90 mL. What is this quantity in cubic centimeters?

20. Chris's driveway has an area of 20 square yards. The driveway sealer he is purchasing describes coverage in square feet. What is the area of Chris's driveway in square feet?

21. A battery has a volume of 32.5 cu in. What is this volume in cubic centimeters? Round your answer to the nearest tenth.

22. Challenge: What is the capacity in quarts of a tank that holds 24 000 cm³ of water?

USING THE MIL

Mil

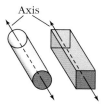

Axis

A cross-section of a solid is the surface exposed when the solid is cut at a right angle to its axis.

A **mil** is a unit of measure equal to one thousandth of an inch. The cross-sectional dimensions of a conductor are often given in mils. The thickness of a material such as plastic can also be given in mils.

1 mil = 0.001 inch = 10^{-3} inch

1,000 mils = 1 inch

To convert mils to inches, use the conversion factor $\dfrac{1 \text{ in.}}{1{,}000 \text{ mils}}$.

Convert 1,750 mils to inches.

$$1{,}750 \text{ mils} = \frac{1{,}750 \text{ mils}}{1} \times \frac{1 \text{ in.}}{1{,}000 \text{ mils}} = \frac{1{,}750 \text{ in.}}{1{,}000} = 1.75 \text{ in.}$$

EXAMPLE A

Convert 0.5 inch to mils.

To convert inches to mils, use the conversion factor $\dfrac{1{,}000 \text{ mils}}{1 \text{ in.}}$.

$$0.5 \text{ in.} = \frac{0.5 \text{ in.}}{1} \times \frac{1{,}000 \text{ mils}}{1 \text{ in.}} = 500 \text{ mils}$$

0.5 inch equals 500 mils.

EXAMPLE B

Convert $\frac{1}{16}$ inch to mils.

$$\frac{1}{16} \text{ in.} = \frac{1 \text{ in.}}{16} \times \frac{1{,}000 \text{ mils}}{1 \text{ in.}} = \frac{1{,}000 \text{ mils}}{16} = 62.5 \text{ mils}$$

$\frac{1}{16}$ inch equals 62.5 mils.

EXAMPLE C

Convert 42.5 mils to inches.

$$42.5 \text{ mils} = \frac{42.5 \text{ mils}}{1} \times \frac{1 \text{ in.}}{1{,}000 \text{ mils}} = \frac{42.5 \text{ in.}}{1{,}000} = 0.0425 \text{ in.}$$

42.5 mils equals 0.0425 inch.

Square Mil

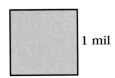

1 mil

1 mil

(Larger than actual size)

A **square mil** is the area of a square measuring 1 mil on each side. The abbreviation for square mil is sq mil.

Square mils can be used to measure the area of a cross-section of a square or rectangular conductor. The area of a square or rectangle is found by multiplying the length and the width.

Find the area in square mils of a cross-section of a conductor if the cross-section is a rectangle with width 0.375 inch and length 4 inches.

0.375 in. = 375 mils   ← Express the dimensions of the
4 in. = 4,000 mils   cross-section in mils.

$A = l \times w = 375$ mils $\times 4,000$ mils $= 1,500,000$ sq mils

The area of a cross-section of the conductor is 1,500,000 square mils.

EXAMPLE D

Find the area in square mils of a cross-section of a conductor if the cross-section is a square with each side measuring 0.75 inch.

0.75 in. = 750 mils ← Express the length of the side of the square in mils.

$A = l \times w = 750$ mils $\times 750$ mils $= 562,500$ sq mils

The area of a cross-section of the conductor is 562,500 square mils.

Circular Mil

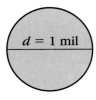

$d = 1$ mil

(Larger than actual size)

A **circular mil** is the area of a circle that has a diameter of 1 mil. If a wire has a diameter of 1 mil, then its cross-sectional area is 1 circular mil.

The relationship between the area of a circle in circular mils and the diameter of the circle in mils is given by the equation:

$A_{\text{cmil}} = d^2{}_{\text{mil}}$

The following table shows the diameter and the area in circular mils for various cross-sections.

Diameter in mils	1 mil	2 mils	3 mils	4 mils	5 mils
Area in cmil	1 cmil	4 cmil	9 cmil	16 cmil	25 cmil

Find the circular mil area of a conductor whose cross-section is a circle with a diameter of 45 mils.

$A_{\text{cmil}} = d^2{}_{\text{mil}}$

$= 45^2$ ← Substitute 45 for the diameter d.

$= 2,025$ cmil

The circular mil area of a conductor with a diameter of 45 mils is 2,025 cmil.

EXAMPLE E

Find the circular mil area of a conductor whose cross-section is a circle with a diameter of 350 mils.

$$A_{cmil} = d^2_{mil}$$
$$= 350^2 \quad \leftarrow \text{Substitute 350 for the diameter } d.$$
$$= 122,500 \text{ cmil}$$

The circular mil area of a conductor with a diameter of 350 mils is 122,500 cmil.

EXAMPLE F

Find the circular mil area of a conductor whose cross-section is a circle with a diameter of $\frac{3}{8}$ inch.

$\frac{3}{8}$ in. = 0.375 in. $\leftarrow$ Express the diameter as a decimal.

$\frac{3}{8}$ in. = 0.375 in. = 375 mils $\leftarrow$ Express the diameter in mils.

$$A_{cmil} = d^2_{mil}$$
$$= 375^2 \qquad \leftarrow \text{Substitute 375 for } d.$$
$$= 140,625 \text{ cmil}$$

The circular mil area of a conductor with a diameter of $\frac{3}{8}$ inch is 140,625 cmil.

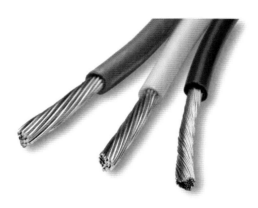

EXAMPLE G

A conductor has a circular cross-section with an area of 81 cmil. Find the diameter of the cross-section in mils and in inches.

$$A_{cmil} = d^2_{mil}$$
$$81 = d^2_{mil} \quad \leftarrow \text{Substitute 81 for } A_{cmil}.$$
$$9 = d \qquad \leftarrow \text{Find the square root of 81.}$$

To convert 9 mils to inches, use the conversion factor $\frac{0.001 \text{ in.}}{1 \text{ mil}}$.

9 mils = $\frac{9 \text{ mils}}{1} \times \frac{0.001 \text{ in.}}{1 \text{ mil}} = 0.009$ in.

The diameter of the cross-section is 9 mils, or 0.009 inch.

Relationship between Square Mil and Circular Mil

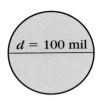

d = 100 mil

(Larger than actual size)

If the area of a circular cross-section is found using the formula $A_{cmil} = d^2_{mil}$, the result is in circular mils. But if the area of a circular cross-section is found using the formula $A = \frac{\pi}{4}d^2$, the result is in square mils. Consider a circle whose diameter d is 100 mils. The area of the circle is found below using both formulas. Use 3.1416 for π.

$$A_{cmil} = d^2_{mil} \qquad\qquad A = \frac{\pi}{4}d^2$$

$$= 100^2 \qquad\qquad\qquad \approx \frac{3.1416}{4} \times 10,000$$

$$= 10,000 \text{ circular mils} \qquad = 7,854 \text{ square mils}$$

The calculations above illustrate the following (approximate) relationship between circular mils and square mils.

$$10,000 \text{ cmil} = 7,854 \text{ sq mil}$$

This relationship leads to two conversion formulas.

- To convert an area from circular mils to square mils, use $A = 0.7854A_{cmil}$, where A is in sq mil.

- To convert an area from square mils to circular mils, use $A_{cmil} = \frac{A}{0.7854}$, where A is in sq mil.

Helpful Hint

The traditional formula for the area of a circle is $A = \pi r^2$, where r is the radius. An equivalent formula is $A = \frac{\pi}{4}d^2$, where d is the diameter.

EXAMPLE H

A 10-gauge wire has a diameter of 102 mils. What is the cross-sectional area in circular mils and in square mils?

Calculate the area in *circular mils.*

$$A_{cmil} = d^2$$

$$A_{cmil} = (102)^2 \quad \longleftarrow \text{ Substitute 102 for } d.$$

$$A_{cmil} = 10,404 \quad \longleftarrow \text{ This is the circular mil area.}$$

Convert the circular mil area to square mils. Use A to represent the area in square mils.

$$A = 0.7854A_{cmil} = 0.7854 \times 10,404 = 8,171 \text{ sq mil}$$

The cross-sectional area of the wire is 10,404 circular mils, or 8,171 square mils.

EXERCISES 8-6

Convert each of the following to the unit indicated.

1. 3,000 mils = _____ in.
2. $2\frac{1}{4}$ in. = _____ mils
3. 12.375 in. = _____ mils
4. 60.3 mils = _____ in.
5. 850 mils = _____ in.
6. 0.875 in. = _____ mils
7. 20 mils = _____ in.
8. $\frac{1}{4}$ ft = _____ mils

Solve.

9. Find the area in square mils of a cross-section of a conductor if the cross-section is a square with a side of 0.25 inch.

10. Find the circular mil area of a conductor whose cross-section is a circle with a diameter of 1,200 mils.

11. Find the area in square mils of a cross-section of a conductor for which the cross-section is a rectangle a with width of 0.5 inch and a length of 4 inches.

12. A conductor has a circular cross-section with an area of 3,600 cmil. Find the diameter of the cross-section in mils and in inches.

Complete the table below. Round your answer for area to the nearest whole number, if rounding is necessary.
(s = side, w = width, l = length, and d = diameter)

	Cross-Section	Dimension(s)	Square Mil Area	Circular Mil Area
13.	square	s = 80 mils		
14.	circle			640,000 cmil
15.	rectangle	w = 0.375 in. l = 2 in.		
16.	circle	d = 0.125 in.		
17.	square	$s = \dfrac{3}{16}$ in.		

Solve.

18. A stereo wire has a cross-sectional area of 200 square mils. What is the circular mil cross-sectional area of the wire?

19. A wire has a circular mil area of 9.5 cmil. What is the cross-sectional area of the wire in square mils?

20. **Challenge:** A strand of cable is made up of 55 round wires, each with a diameter of 0.0004 in. What is the total cross-sectional area of the cable in circular mils?

CHAPTER 8 REVIEW EXERCISES

Write the conversion factor or factors needed for each conversion.

1. Centimeters to meters
2. Feet to inches
3. Inches to millimeters
4. Kilograms to pounds

Convert each measurement to the indicated unit. If rounding is necessary, round each answer to the nearest tenth.

5. 1.5 ft = ___ in.
6. 2.5 lb = ___ oz
7. 2.25 ft^2 = ___ in.^{2}
8. 35 cm = ___ mm
9. 775 mL = ___ L
10. 5 lb = ___ kg
11. 0.65 kA = ___ A
12. 13.87 kV = ___ V
13. 40.25 MV = ___ kV
14. 5 fl oz = ___ mL
15. 32.5 ft^2 = ___ m^2
16. 0.45 m^3 = ___ cm^3
17. 4.75 in. = ___ mils
18. 2,575 mils = ___ in.
19. 6.25 mA = ___ μA
20. 0.095 MΩ = ___ Ω

21. Kayla cut a 10.814 m long wire into eight pieces. If each cut was 2 mm wide and there was no waste, what was the length of each piece of wire in

    (a) meters

    (b) centimeters?

22. Number 16 aluminum wire weighs 2.37 lb/1,000 ft.

    (a) What is the weight, in pounds, of 45.75 yards of #16 wire?

    (b) Express the answer to (a) in ounces to the nearest tenth.

23. America operates about 157,000 miles of high voltage (>230 kV) electric transmission lines.

    (a) Convert this distance to kilometers.

    (b) Convert 230 kV to volts.

24. A 100-meter reel of Kevlar-reinforced Ethernet cable weights 25 pounds.

    (a) What is this length in feet?

    (b) What is the weight of this reel in kilograms?

25. The minimum, ideal, and maximum operating temperatures for a certain electric motor are $-50°F$, $72°F$, and $150°F$. Express each of these temperatures to the nearest degree Celsius.

26. Multi-mode fiber optic cable has diameters in the 50- to 100-micron range. A micron is 10^{-6} m.

(a) What are these diameters in mils? Round to the nearest tenth.

(b) What is the area in circular mils of a cable with a diameter of 50 microns?

(c) What is the area in circular mils of a cable with a diameter of 100 microns?

27. A cable has an area of 42 cmil. What is this area in square mils?

Building a Foundation in Mathematics

Introducing Algebra

Using Properties of Addition and
Multiplication

Simplifying Algebraic Expressions

Overview

Until now, our work has been with numbers and we
have limited ourselves to the basic operations with
numbers, powers, roots, and absolute values. We
have also learned how to write and work with numbers in scientific and engineering notations. In this
chapter we begin our work with algebra.

We start with the basic definitions, a review of
the order of operations, and a discussion of how to
evaluate expressions. Next, we look at the Associative
and Commutative Properties of Addition and Multiplication, the Distributive Property of Multiplication

over Addition, and the Properties of Inverses, Zeros,
and One. We end the chapter by demonstrating how
to simplify algebraic expressions by combining like
terms.

Throughout this text, we have shown how you can
make your work easier by using a scientific calculator.
Calculators use electricity to operate, and people who
design calculators rely on their knowledge of algebra
and electronics to ensure that the calculators operate
correctly.

Algebra Essentials

Objectives

After completing this chapter, you will be able to:

- Substitute values in an expression and evaluate the expression

- Rewrite expressions using the Associative, Commutative, and Distributive Properties

- Use the Identity and Inverse Properties for addition and multiplication

- Use the order of operations to evaluate or simplify expressions

- Simplify combined operations of literal term expressions

INTRODUCING ALGEBRA

Variables and Expressions

The name *Algebra* comes from the title of a book written in Arabic by Al-Khwarizimi in the 9th century A.D. Taken literally, it means *reunion of broken parts*.

Problems in arithmetic involve operations with numbers only. The study of algebra follows naturally from the study of arithmetic. **Algebra** is the study of operations and relationships among numbers, often using symbols to represent numbers. Symbols that represent numbers are called **variables,** and they are usually letters such as x and y. An **expression** is a number, a variable, or a combination of numbers, variables, and operations. To evaluate an expression means to find the value of it.

Evaluate $x + 7$, if $x = 5$.

$x + 7 =$ ← Write the expression.

$5 + 7 =$ ← Substitute 5 for x.

    $= 12$ ← Add.

If $x = 5$, the value of the expression $x + 7$ is 12.

EXAMPLE A

Evaluate $9x$, if $x = -3$.

$9x =$ ← Write the expression.

$9(-3) =$ ← Substitute -3 for x.

    $= -27$ ← Multiply.

If $x = -3$, the value of the expression $9x$ is -27.

EXAMPLE B

Find the value of the expression $3 + x + 4$, if $x = 2$.

$3 + x + 4 =$ ← Write the expression.

$3 + 2 + 4 =$ ← Substitute 2 for x.

    $= 9$ ← Add.

If $x = 2$, the value of the expression $3 + x + 4$ is 9.

EXAMPLE C

Find the value of the expression $8x + y$, if $x = 5$ and $y = 1$.

$8x + y =$ ← Write the expression.

$8 \cdot 5 + 1 =$ ← Substitute 5 for x and 1 for y.

$40 + 1 =$ ← Multiply.

    $= 41$ ← Add.

If $x = 5$ and $y = 1$, the value of the expression $8x + y$ is 41.

Using the Order of Operations

We discussed the order of operations earlier. To review, the **order of operations** is a list of steps to be followed for finding the value of an expression. If a step does not apply to an expression, go to the next step.

Order of Operations

- Complete all calculations within grouping symbols. Grouping symbols include parentheses, brackets, and absolute value signs.
- Complete all calculations indicated by exponents.
- Perform all multiplications and divisions in order from left to right.
- Perform all additions and subtractions in order from left to right.

Evaluate $3 + (4 + 1)^2 - 10 \div 5$.

$3 + (4 + 1)^2 - 10 \div 5 = \longleftarrow$ Identify the operation within parentheses.

$3 + (5)^2 - 10 \div 5 = \longleftarrow$ Complete the calculation inside the parentheses.

$3 + 25 - 10 \div 5 = \longleftarrow$ Complete the calculation indicated by the exponent.

$3 + 25 - 2 = \longleftarrow$ Perform division before addition or subtraction.

$28 - 2 = \longleftarrow$ Perform addition and subtraction in order from left to right.

$= 26$

The value of the expression $3 + (4 + 1)^2 - 10 \div 5$ is 26.

When a fraction has an expression in the numerator or denominator, parentheses around the expression are implied since the fraction bar is a grouping symbol, even if they are not shown. This is demonstrated in Example D.

EXAMPLE D

Evaluate
$20 - \dfrac{16 - 4}{4 + 2} + 5.$

$20 - \dfrac{16 - 4}{4 + 2} + 5 = \longleftarrow$ Think of the fraction as $\dfrac{(16 - 4)}{(4 + 2)}$.

$20 - \dfrac{12}{6} + 5 = \longleftarrow$ Evaluate the numerator and the denominator of the fraction.

$20 - 2 + 5 = \longleftarrow$ Perform division before addition or subtraction.

$18 + 5 = \longleftarrow$ Perform addition and subtraction in order from left to right.

$= 23$

The value of $20 - \dfrac{16 - 4}{4 + 2} + 5$ is 23.

A scientific calculator, such as the TI-30X IIS, will follow the order of operations to evaluate an expression. However, you must be careful to type in all sets of parentheses, both stated and implied. To use a scientific calculator for the expression in Example D, press 20 .

Beware that a nonscientific calculator will not follow the order of operations.

EXAMPLE E

Evaluate
$2^3 + 2^4 - 6 \times 3$.

$2^3 + 2^4 - 6 \times 3 =$ ← Identify the calculations with exponents.

$8 + 16 - 6 \times 3 =$ ← Complete the calculations indicated by exponents.

$8 + 16 - 18 =$ ← Perform multiplication before addition or subtraction.

$24 - 18 =$ ← Perform addition and subtraction in order from left to right.

$= 6$

The value of $2^3 + 2^4 - 6 \times 3$ is 6.

EXAMPLE F

Evaluate
$-8(10^2 \div 25) + 12$.

$-8(10^2 \div 25) + 12 =$ ← Identify the operations within parentheses.

$-8(100 \div 25) + 12 =$ ← Begin the calculations inside the parentheses. Do the calculation indicated by the exponent first.

$-8(4) + 12 =$ ← Complete the calculations inside the parentheses.

$-32 + 12 =$ ← Perform multiplication before addition.

$= -20$

The value of $-8(10^2 \div 25) + 12$ is -20.

Helpful Hint

Multiplication can be indicated by a times sign, a raised dot, an asterisk, or parentheses. For example, 3×16, $3 \cdot 16$, $3 * 16$, and $3(16)$ all indicate the product of 3 and 16.

An expression such as $(2 + (10 - 4)) \div 8$ has two sets of parentheses. To make it easier to read, the same expression can be written with brackets: $[2 + (10 - 4)] \div 8$. To evaluate an expression like this, work from the inside out. Perform the calculation inside the parentheses first, and then perform the calculation inside the brackets. This is shown in Example G. Your calculator does not have bracket keys, so you should use parentheses whenever you see brackets.

EXAMPLE G

Evaluate
$[2 + (10 - 4)] \div 8$.

$[2 + (10 - 4)] \div 8 =$ ← Identify the operation within parentheses.

$[2 + 6] \div 8 =$ ← Perform the calculation inside the parentheses.

$8 \div 8 =$ ← Perform the calculation inside the brackets.

$= 1$ ← Divide.

The value of $[2 + (10 - 4)] \div 8$ is 1.

Evaluating Expressions

$25n - 5p$

↑ variable   ↑ variable

An **algebraic expression** contains one or more variables. When values are given to use for the variables, it is possible to evaluate an algebraic expression. Substitute the values given for the variables, and follow the order of operations.

Evaluate the expression $25n - 5p$ for $n = 3$ and $p = 2$.

$25n - 5p =$

$25(3) - 5(2) =$ ← Substitute 3 for n and 2 for p.

$75 - 10 =$ ← Perform multiplication from left to right.

$= 65$ ← Subtract.

For $n = 3$ and $p = 2$, the value of the expression $25n - 5p$ is 65.

EXAMPLE H

Evaluate the expression $5n^2$ for $n = -4$.

$5n^2 =$

$5(-4)^2 =$ ← Substitute -4 for n.

$5(16) =$ ← Complete the calculation indicated by the exponent.

$= 80$ ← Multiply.

For $n = -4$, the value of the expression $5n^2$ is 80.

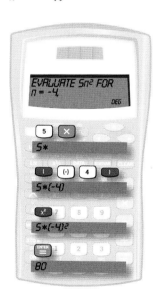

EXAMPLE I

Evaluate the expression $27 - 2(n + 6)$ for $n = 8$.

$27 - 2(n + 6) =$

$27 - 2(8 + 6) =$ ← Substitute 8 for n.

$27 - 2(14) =$ ← Complete the calculation inside the parentheses.

$27 - 28 =$ ← Multiply.

$= -1$ ← Subtract.

For $n = 8$, the value of the expression $27 - 2(n + 6)$ is -1.

EXAMPLE J

Evaluate the
expression
$(5 + n)^2 + 12$
for $n = -2$.

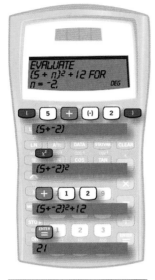

$(5 + n)^2 + 12 =$

$(5 + (-2))^2 + 12 =$ ⟵ Substitute -2 for n.

$(3)^2 + 12 =$ ⟵ Complete the calculation inside the parentheses.

$9 + 12 =$ ⟵ Complete the calculation indicated by the exponent.

$= 21$ ⟵ Add.

For $n = -2$, the value of the expression $(5 + n)^2 + 12$ is 21.

EXAMPLE K

Evaluate the expression $10m^2 + 5n + 25$ for $m = 2$ and $n = 3$.

$10m^2 + 5n + 25 =$

$10(2)^2 + 5(3) + 25 =$ ⟵ Substitute 2 for m and 3 for n.

$10(4) + 5(3) + 25 =$ ⟵ Complete the calculation indicated by the exponent.

$40 + 15 + 25 =$ ⟵ Perform multiplication from left to right.

$= 80$ ⟵ Add.

For $m = 2$ and $n = 3$, the value of the expression $10m^2 + 5n + 25$ is 80.

Helpful Hint

If more than one operation must be performed inside parentheses, follow the order of operations to complete. For example, to evaluate $(4 + 6^2)$, square 6 first, and then add the result to 4.

EXERCISES 9-1

Evaluate each of the following expressions. Show each step of your work.

1. $2 \times 3^2 - 9$

2. $92 + 3(7 - 3)^2$

3. $\dfrac{2 \times 9}{20 - 2} - 5$

4. $32 \div \dfrac{8}{9 - 5}$

5. $8^2 \div 4(9 - 7)^2$

6. $[(14 \div 7)^2 + 1] + 6$

7. $54 + 14 - 5 \times 2^3$

8. $(4 + 6^2) \div 10 - 9$

9. $5^2 + 9^2 - 6$

10. $-2(40 - 2 \times 3^2)$

11. $45 \div [20 - (10 + 1)]$

12. $[28 + (5 + 2)^2] \div 11$

Evaluate each expression for the given values of the variables.

13. $12k - 8$ for $k = 3$

14. $2a^2 + 3a$ for $a = 5$

15. $8x + 4y$ for $x = -2$ and $y = 3$

16. $5y^3 + 2$ for $y = -1$

17. $4rs$ for $r = 2$ and $s = -5$

18. $-4(a - 9) + a$ for $a = 5$

19. $\frac{6 + b}{4} + 5$ for $b = 2$

20. $\frac{x + y}{x - y}$ for $x = 5$ and $y = 4$

21. $2n^2 \div 5$ for $n = 5$

22. $7(2m + 3n)$ for $m = 6$ and $n = 0$

Solve.

23. Explain how to apply the order of operations when using the formula $C = \frac{5}{9}(F - 32)$ to convert degrees Fahrenheit to degrees Celsius.

24. Evaluate $I = \frac{V}{R}$ when $V = 24$ and $R = 20$.

25. Evaluate $P = I^2R$ when $I = 2.5$ and $R = 1.6$.

26. Evaluate $V = \sqrt{P \cdot R}$ when $P = 15$ and $R = 10$.

27. Insert operation symbols between the digits below so that the value of the expression is zero.

$$9 \quad 6 \quad 3 \quad 1$$

28. Challenge: Evaluate $18 - 5^2 - [2(3 + 12) + (-3)^2]$.

USING PROPERTIES OF ADDITION AND MULTIPLICATION

Commutative Properties

The properties of addition and multiplication are applied in algebra to simplify expressions or to rewrite expressions in different forms.

This lesson presents the properties of addition and multiplication and gives examples of each, beginning with the Commutative Properties of Addition and Multiplication.

The Commutative Property of Addition states that changing the order of addends does not change the sum. The Commutative Property of Multiplication states that changing the order of factors does not change the product.

Commutative Properties

For all numbers a and b: $a + b = b + a$

$$ab = ba$$

Examples:

$4 + 5 = 5 + 4$ $2 \cdot 3 = 3 \cdot 2$

$x + 7 = 7 + x$ $9 \cdot x = x \cdot 9$

$(3 + a) + r = r + (3 + a)$ $3(x + y) = (x + y)3$

Note that the order of the addends $(3 + a)$ and r is changed.

Note that the order of the factors 3 and $(x + y)$ is changed.

Alert! There are no commutative properties for subtraction and division. Changing the order in a subtraction or a division expression usually does change the result. For example, $5 - 2 \ne 2 - 5$ and $8 \div 4 \ne 4 \div 8$.

Complete the following statement. Name the property applied.

$-3 + c = c + $ _____

$-3 + c = c + \underline{(-3)}$ by the Commutative Property of Addition.

Language Box

To remember the meaning of the Commutative Property, think of the root word *commute,* meaning to change locations.

EXAMPLE A

Complete the following statement. Name the property applied.
$4 \cdot x = $ ___ $\cdot 4$

$4 \cdot x = \underline{x} \cdot 4$ by the Commutative Property of Multiplication.

EXAMPLE B

Complete the following statement. Name the property applied.
$(x + y) + 1 = 1 + $ ___

$(x + y) + 1 = 1 + \underline{(x + y)}$ by the Commutative Property of Addition.

Sometimes a property is used on only part of an expression. This is shown in Example C.

EXAMPLE C

Complete the following statement. Name the property applied.
$6 + 3 + 7 = 6 + 7 + $ ___

$6 + 3 + 7 = 6 + 7 + $ ___ ← The position of the 7 is changed.

$6 + 3 + 7 = 6 + 7 + \underline{3}$ ← Apply the Commutative Property of Addition.
$3 + 7 = 7 + 3$

$6 + 3 + 7 = 6 + 7 + 3$ by the Commutative Property of Addition.

EXAMPLE D

Complete the statement
$-4 \cdot 2 \cdot b = 2 \cdot $ ___ $\cdot b$. Name the property applied.

$-4 \cdot 2 \cdot b = 2 \cdot $ ___ $\cdot b$ ← The position of the 2 is changed.

$-4 \cdot 2 \cdot b = 2 \cdot \underline{(-4)} \cdot b$ ← Apply the Commutative Property of Multiplication. $-4 \cdot 2 = 2 \cdot (-4)$

$-4 \cdot 2 \cdot b = 2 \cdot (-4) \cdot b$ by the Commutative Property of Multiplication.

Associative Properties

John Q. Adams and Associates *Painting Experts*				
205 Rose Avenue 814-438-5000			20 years experience	
Quantity	Description		Rate	Total

The Associative Property of Addition states that changing the grouping of addends does not change the sum. The Associative Property of Multiplication states that changing the grouping of factors does not change the product.

Associative Properties

For all numbers a, b, and c: $(a + b) + c = a + (b + c)$
$$(ab)c = a(bc)$$

Examples:

$(3 + 5) + 2 = 3 + (5 + 2)$ $(8 \cdot 3) \cdot 2 = 8 \cdot (3 \cdot 2)$

$(-1 + 6) + x = -1 + (6 + x)$ $(-5 \cdot x) \cdot 9 = -5 \cdot (x \cdot 9)$

$(w + 9) + z = w + (9 + z)$ $(10a)b = 10(ab)$

Note that only the placement of the parentheses is different on the left and the right side of each equation.

Complete the following statement. Name the property applied.
$(\underline{} \cdot 3) \cdot (-7) = 6 \cdot (3 \cdot (-7))$

$(6 \cdot 3) \cdot (-7) = 6 \cdot (3 \cdot (-7))$ by the Associative Property of Multiplication.

EXAMPLE E

Complete the statement $(8 + a) + b = 8 + (\underline{} + b)$. Name the property applied.

$(8 + a) + b = 8 + (\underline{a} + b)$ by the Associative Property of Addition.

Distributive Property

The Distributive Property involves the operations of addition and multiplication.

The Distributive Property of Multiplication over Addition states that multiplying a sum by a number gives the same result as multiplying each addend by the number and then adding the products. The word *distribute* means to give out or deliver. In the case of the Distributive Property, multiplication is "delivered" to each addend in parentheses.

Distributive Property

For all numbers a, b, and c: $a(b + c) = ab + ac$

The Distributive Property is illustrated in the equation below. The expression on each side of the equation is evaluated according to the order of operations.

$$6(4 + 3) = 6 \cdot 4 + 6 \cdot 3$$

$6(7)$	$24 + 18$
42	42

Use the Distributive Property to rewrite the expression $5(x + 4)$.

$$5(x + 4) = 5x + 5 \cdot 4$$

EXAMPLE F

Use the Distributive Property to rewrite the expression $a(y + 3)$.

$$a(y + 3) = ay + a \cdot 3$$

The Distributive Property can also be stated $ab + ac = a(b + c)$. This form is used in Example G.

EXAMPLE G

Use the Distributive Property to rewrite the expression $2 \cdot 8 + 2 \cdot 3$.

$$2 \cdot 8 + 2 \cdot 3 = 2(8 + 3)$$

Properties of Zero and One

Several properties involving operations with zero and one are listed and illustrated below.

The Identity Property for Addition states that the sum of zero and any number is that number.

Identity Property for Addition

For any number a: $a + 0 = a$ and $0 + a = a$

Examples:

$$7 + 0 = 0 + 7 = 7$$
$$\frac{2}{3} + 0 = 0 + \frac{2}{3} = \frac{2}{3}$$
$$-2.5 + 0 = 0 + (-2.5) = -2.5$$

The Identity Property for Multiplication states that the product of any number and one is that number.

Identity Property for Multiplication

For any number a: $1 \cdot a = a$ and $a \cdot 1 = a$

Examples:

$6 \cdot 1 = 1 \cdot 6 = 6$

$-3.75 \cdot 1 = 1 \cdot (-3.75) = -3.75$

$\frac{4}{5} \cdot 1 = 1 \cdot \frac{4}{5} = \frac{4}{5}$

The Multiplication Property of Zero states that the product of any number and zero is zero.

Multiplication Property of Zero

For any number a: $a \cdot 0 = 0$ and $0 \cdot a = 0$

Examples:

$-2 \cdot 0 = 0 \cdot (-2) = 0$

$-4\frac{3}{8} \cdot 0 = 0 \cdot \left(-4\frac{3}{8}\right) = 0$

$2.3 \cdot 0 = 0 \cdot (2.3) = 0$

Complete the statements.

$6 + \underline{} = 6 \quad \longrightarrow \quad 6 + \underline{0} = 6$

$1 \cdot \underline{} = 4 \quad \longrightarrow \quad 1 \cdot \underline{4} = 4$

$(-10) \cdot \underline{} = 0 \longrightarrow (-10) \cdot \underline{0} = 0$

> ### Language Box
>
> Zero is called the identity element for addition. One is called the *identity element* for multiplication.

EXAMPLE H

Use the properties of zero and one to complete the statements.

$\underline{} \cdot (3) = 0$

$9 = 9 + \underline{}$

$(4.35) \cdot 1 = \underline{}$

$14 \cdot \underline{} = 0$

$\underline{} \left(-\frac{1}{5}\right) = -\frac{1}{5}$

$0 + \underline{} = 16$

$\underline{0} \cdot (3) = 0$

$9 = 9 + \underline{0}$

$(4.35) \cdot 1 = \underline{4.35}$

$14 \cdot \underline{0} = 0$

$\underline{1} \left(-\frac{1}{5}\right) = -\frac{1}{5}$

$0 + \underline{16} = 16$

Properties of Inverses

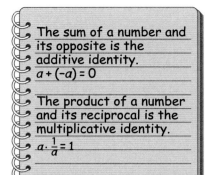

The sum of a number and its opposite is the additive identity.
$a + (-a) = 0$

The product of a number and its reciprocal is the multiplicative identity.
$a \cdot \frac{1}{a} = 1$

The **opposite** of a number is called the **additive inverse** of the number. The opposite of a number is formed by multiplying that number by -1. For example, the opposite of 10 is -10.

The **reciprocal** of any number except zero is called the **multiplicative inverse** of the number. The reciprocal of a number is formed by writing the number as a fraction and then inverting the fraction. For example, to form the reciprocal of 3, first write $3 = \frac{3}{1}$, and then invert the fraction. The reciprocal of 3 is $\frac{1}{3}$.

The Property of Additive Inverses states that the sum of a number and its opposite is zero.

Property of Additive Inverses

For any number a: $a + (-a) = 0$ and $-a + a = 0$

Examples:

$7 + (-7) = -7 + 7 = 0$

$\frac{1}{4} + \left(-\frac{1}{4}\right) = -\frac{1}{4} + \frac{1}{4} = 0$

$-8.4 + 8.4 = 8.4 + (-8.4) = 0$

The Property of Multiplicative Inverses states that the product of any number, except zero, and its reciprocal is 1.

Property of Multiplicative Inverses

For any nonzero number a: $a \cdot \frac{1}{a} = 1$ and $\frac{1}{a} \cdot a = 1$

Examples:

$9 \cdot \frac{1}{9} = \frac{1}{9} \cdot 9 = 1$

$\left(-\frac{3}{2}\right) \cdot \left(-\frac{2}{3}\right) = \left(-\frac{2}{3}\right) \cdot \left(-\frac{3}{2}\right) = 1$

$-5 \cdot \left(-\frac{1}{5}\right) = \left(-\frac{1}{5}\right)(-5) = 1$

$4 \cdot 4^{-1} = 4\left(\frac{1}{4}\right) = 1$

Complete the following statements.

$4 + \underline{} = 0 \longrightarrow 4 + \underline{-4} = 0$

$\underline{\phantom{\frac{1}{2}}} \cdot 2 = 1 \longrightarrow \underline{\frac{1}{2}} \cdot 2 = 1$

EXAMPLE I

Use the properties of inverses to complete the statements.

$\frac{1}{4} \cdot \underline{\hspace{1em}} = 1$

$-6 + \underline{\hspace{1em}} = 0$

$0 = 4.8 + \underline{\hspace{1em}}$

$-12 \cdot \underline{\hspace{1em}} = 1$

$1 + (-1) = \underline{\hspace{1em}}$

$16 \cdot \frac{1}{16} = \underline{\hspace{1em}}$

$\left(-\frac{2}{3}\right)^{-1} \left(-\frac{2}{3}\right) = \underline{\hspace{1em}}$

$\frac{1}{4} \cdot \underline{4} = 1$

$-6 + \underline{6} = 0$

$0 = 4.8 + \underline{(-4.8)}$

$-12 \cdot \left(-\frac{1}{12}\right) = 1$

$1 + (-1) = \underline{0}$

$16 \cdot \frac{1}{16} = \underline{1}$

$\left(-\frac{2}{3}\right)^{-1} \left(-\frac{2}{3}\right) = \underline{1}$

EXERCISES 9-2

Complete each of the following statements. Name the property used.

1. $9 + \underline{\hspace{1em}} + 2 = 9 + 2 + 6$

2. $5(8 + 3) = 5 \cdot 8 + 5 \cdot \underline{\hspace{1em}}$

3. $-4 \cdot (6 \cdot 7) = (-4 \cdot \underline{\hspace{1em}}) \cdot 7$

4. $\underline{\hspace{1em}} + 7 = 0$

5. $18 \cdot \underline{\hspace{1em}} = 18$

6. $1 \cdot 9 = \underline{\hspace{1em}}$

7. $\left(2\frac{3}{4}\right) \cdot \underline{\hspace{1em}} = 1$

8. $\underline{\hspace{1em}} + 10 + 4 = 10 + 5 + 4$

9. $\underline{\hspace{1em}} \cdot 12 = 1$

10. $6 + (-6) = \underline{\hspace{1em}}$

11. $(3 + 7) + 1 = 3 + (7 + \underline{\hspace{1em}})$

12. $9(a + b) = \underline{\hspace{1em}} + 9b$

13. $(xy)z = \underline{\hspace{1em}}(yz)$

14. $w \cdot \underline{\hspace{1em}} = 1 \; (w \neq 0)$

15. $(2 + \underline{\hspace{1em}}) + b = 2 + (a + b)$

16. $-y + \underline{\hspace{1em}} = 0$

17. $8w + 8y = 8(w + \underline{\hspace{1em}})$

18. $14x \cdot \underline{\hspace{1em}} = 0$

19. $7 + b = b + \underline{\hspace{1em}}$

20. $\frac{n}{21} \cdot \underline{\hspace{1em}} = 1$

Solve.

21. Which two properties must be applied to explain the statement $(3 + 5) + 7 = (7 + 3) + 5$?

22. If you know that x has a value of 6 in the statement $4xy = 0$, what is the value of y? On which property did you base your answer?

23. The total power in two circuits is given by $12 (3I_a^2 + 4I_b^2)$. Use the Distributive Property to rewrite this expression.

24. At a certain time after the switch is closed, the current in a circuit containing a resistor R in series with a capacitor is given by $\frac{V}{0.3679R} - \frac{V_0}{0.3679R}$. Use the Distributive Property to rewrite this expression.

25. The power output of a generator is given by $VI - RI^2$. Use the Distributive Property to rewrite this expression.

26. **Challenge:** Explain why the Property of Multiplicative Inverses does not apply to zero.

SIMPLIFYING ALGEBRAIC EXPRESSIONS

Like terms have the same variables raised to the same powers.

Many expressions consist of terms. A **term** can be a number, a variable, or a product of numbers and variables. Some expressions have only one term; other expressions have more than one term. In an expression, terms are separated by addition or subtraction signs. For example, in the expression $4x^2 + 3x + 8$, there are three terms separated by addition signs.

$$4x^2 \qquad + \qquad 3x \qquad + \qquad 8$$

$4x^2$ is a term. $3x$ is a term.   8 is a term.

The numerical factor in a term is called a **coefficient** In the term $4x^2$, 4 is the coefficient. In the term $3x$, 3 is the coefficient.

Terms with variables are **like terms** if they have the same variables raised to the same powers. Also, all numbers are like terms. Note the differences between the like terms and the unlike terms in the following lists.

Like terms:	Unlike terms:
$8a$ and $7a$	$8a$ and 7
$3n^2$ and $5n^2$	$3n$ and $5n^2$
$2xy$ and $-2xy$	$2xy$ and $-2x$
$-n$ and n	-1 and n
-5 and $12\frac{1}{2}$	-5 and $5x$
$3xy^2$ and $6xy^2$	$3xy^2$ and $3x^2y$

Language Box

When an algebraic expression has been simplified, it is said to be in **simplest form**.

To simplify an algebraic expression means to combine the like terms in the expression. Group like terms by writing them next to each other. To combine like terms, add or subtract the coefficients.

Simplify the expression $8x + 3y - 2x + 5$.

$8x + 3y - 2x + 5 =$ ← Identify the like terms $8x$ and $-2x$.

$8x - 2x + 3y + 5 =$ ← Use the Associative Property to group the like terms together.

$(8 - 2)x + 3y + 5 =$ ← Apply the Distributive Property.

$6x + 3y + 5$ ← Subtract 2 from 8 to combine the like terms.

In simplest form, $8x + 3y - 2x + 5$ is $6x + 3y + 5$.

EXAMPLE A

Simplify
$9a + 2c + 3a - 6c$.

$9a + 2c + 3a - 6c =$ $\longleftarrow$ Identify the like terms $9a$ and $3a$, $2c$ and $-6c$.

$9a + 3a + 2c - 6c =$ $\longleftarrow$ Group the like terms together.

$(9 + 3)a + (2 - 6)c =$ $\longleftarrow$ Apply the Distributive Property.

$= 12a - 4c$ $\longleftarrow$ Combine the like terms.

In simplest form, $9a + 2c + 3a - 6c$ is $12a - 4c$.

EXAMPLE B

Simplify
$x^2 - 9 + 7x + 2x + 14$.

$x^2 - 9 + 7x + 2x + 14 =$ $\longleftarrow$ Identify the like terms $7x$ and $2x$, -9 and 14.

$x^2 + 7x + 2x - 9 + 14 =$ $\longleftarrow$ Group the like terms together.

$x^2 + (7 + 2)x - 9 + 14 =$ $\longleftarrow$ Apply the Distributive Property.

$= x^2 + 9x + 5$ $\longleftarrow$ Combine the like terms.

In simplest form, $x^2 - 9 + 7x + 2x + 14$ is $x^2 + 9x + 5$.

When an expression contains parentheses, use the Distributive Property to remove the parentheses before combining like terms. This is shown in Example C.

EXAMPLE C

Simplify
$4a + 5(a - 1) - 2a$.

$4a + 5(a - 1) - 2a =$

$4a + 5a - 5 - 2a =$ $\longleftarrow$ Apply the Distributive Property, then identify the like terms $4a$, $5a$, and $-2a$.

$4a + 5a - 2a - 5 =$ $\longleftarrow$ Group the like terms together.

$(4 + 5 - 2)a - 5 =$ $\longleftarrow$ Apply the Distributive Property.

$= 7a - 5$ $\longleftarrow$ Combine the like terms.

In simplest form, $4a + 5(a - 1) - 2a$ is $7a - 5$.

EXAMPLE D

Simplify
$3a + 6ab + 2a - ab$.

$3a + 6ab + 2a - ab =$ ← Identify the like terms $3a$ and $2a$, $6ab$ and $-ab$.

$3a + 2a + 6ab - ab =$ ← Group the like terms together.

$(3 + 2)a + (6 - 1)ab =$ ← Apply the Distributive Property.

$= 5a + 5ab$ ← Combine the like terms.

In simplest form, $3a + 6ab + 2a - ab$ is $5a + 5ab$.

Helpful Hint

In the term $-ab$, the coefficient -1 is understood.

$6ab - ab =$
$6ab + (-1ab) = 5ab$

EXAMPLE E

Simplify
$5n^2 + 5n - 3n + 2 + n^2$.

$5n^2 + 5n - 3n + 2 + n^2 =$ ← Identify the like terms $5n^2$ and n^2, $5n$ and $-3n$.

$5n^2 + n^2 + 5n - 3n + 2 =$ ← Group the like terms together.

$(5 + 1)n^2 + (5 - 3)n + 2 =$ ← Apply the Distributive Property.

$= 6n^2 + 2n + 2$ ← Combine the like terms.

In simplest form, $5n^2 + 5n - 3n + 2 + n^2$ is $6n^2 + 2n + 2$.

EXERCISES 9-3

Simplify.

1. $8n - 5 - 6n$
2. $3(x + 9) + 1$
3. $a^2 + 5a + 16 + 3a$
4. $4m - 3n + 6m + n$
5. $-a + 9(a + 4)$
6. $4w + w^2 + 7w - w^2 + w$
7. $8ab - 6a - 9ab$
8. $10a + 2(a + 9) - a$
9. $3b^2 + 2b + 18b + 12$
10. $-6w - 2wz + 2w + 4$
11. The current in a circuit is given by $2t^2 - 3t + 10t - 15$, where t is time. Simplify this expression.
12. The current in a circuit is given by $2t^2 + 5t - 8t - 40$, where t is time. Simplify this expression.
13. **Challenge:** Simplify the expression
$5xy + 10y + 5x^2 - 15y - 5x(x + y) + 5y$.

CHAPTER 9 REVIEW EXERCISES

Evaluate each expression for the given values of the variables.

1. $12.5p - (2 + p)$ for $p = 2.5$

2. $a^2 - \frac{3 - a}{a + 2}$ for $a = -4$

3. $\frac{x^2 - 9}{x + 3} + \sqrt{x - 9}$ when $x = 25$

4. $2t^3 + |3t + 5|$ when $t = -1.5$

Complete each of the following statements. Name the property used.

5. $(12 + 6)(-5) = 12 \cdot \underline{\quad} + 6(-5)$

6. $-3p + \underline{\quad} = 0$

Simplify.

7. $11x - 6 + 7x$

8. $5p^2 + 19p - 5p + 10 + p$

9. $12V + 2(5 - 6V)$

10. $6R - 2R^2 + 3R + 5R^2 - 7$

Solve.

11. Evaluate $I = \frac{E}{R + r}$, when $E = 3.85$, $R = 1.5$, and $r = 0.04$.

12. Evaluate I $= \sqrt{\frac{P}{R}}$ when $P = 2.5$ and $R = 1.6$.

13. The total power in two circuits is given by $15I_a^2 + 12I_b^2$. Use the Distributive Property to rewrite this expression.

14. The current in a circuit is given by $5t^2 + 3t + 4t^2 - 4t + 9$, where t is time. Simplify this expression.

Building a Foundation in Mathematics

Writing Equations

Solving One-Step Equations

Solving Multi-Step Equations

Using Formulas

Using Equations to Solve Word
Problems

Writing Inequalities

Solving Inequalities

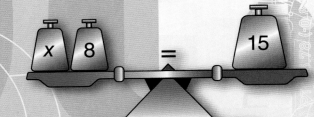

Overview

In the last chapter you learned the essential aspects and properties of algebra. In this chapter you will continue that work and learn how to write equations and inequalities for situations that are described in words. Next, you will learn how to solve both simple and complex equations, including those requiring several steps to reach solutions. You will also learn how to solve simple and multi-step inequalities.

Meter readings often provide the basic information for solving an algebraic problem.

Equations, Formulas, and Inequalities

Objectives

After completing this chapter, you will be able to:

- Express word problems as equations or inequalities
- Solve equations using the fundamental principles of equality
- Solve problems by writing equations or inequalities and determining the values of unknowns
- Substitute values in formulas and solve for the unknowns
- Solve inequalities using the fundamental principles of inequalities

WRITING EQUATIONS

Operation Signs

+ increased by, added to

− decreased by, less than

× times, product of

÷ divided into equal groups

An **equation** is a mathematical sentence that states that two expressions are equal. Many problems can be solved by writing and solving equations. To write an equation, you may need to translate words or phrases into symbols. The symbols will usually be operation signs and variables to represent unknown quantities. Remember that variables are often represented by letters and that any letter can be used. Although the letters x and y are most often used, some of the following examples use other letters.

Write this statement as an equation.
Some number increased by 5 is equal to 9.

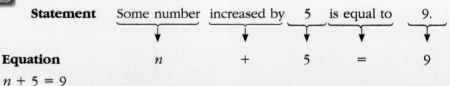

$n + 5 = 9$

EXAMPLE A

Write this statement as an equation: *7 is 5 more than some number.*

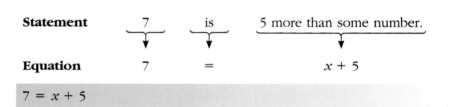

$7 = x + 5$

EXAMPLE B

Write this statement as an equation: *10 is 3 less than some number.*

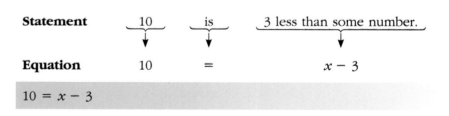

$10 = x - 3$

EXAMPLE C

Write this statement as an equation: *A number decreased by 8 is equal to 4.*

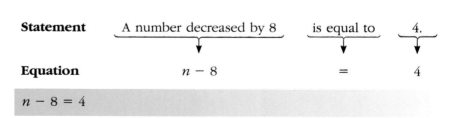

$n - 8 = 4$

EXAMPLE D

Write this statement as an equation: *7 times a number is 98.*

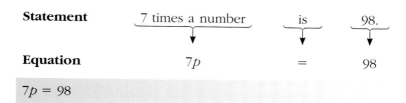

$$7p = 98$$

EXAMPLE E

Write this statement as an equation: *15 divided by some number is 3.*

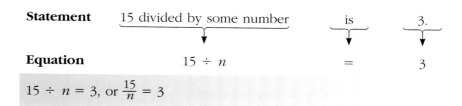

$$15 \div n = 3, \text{ or } \frac{15}{n} = 3$$

EXAMPLE F

Write this statement as an equation: *Some number squared is equal to 144.*

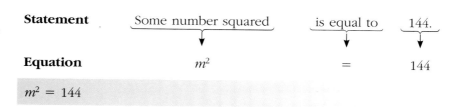

$$m^2 = 144$$

EXAMPLE G

Write this statement as an equation: *The square root of some number is 6.*

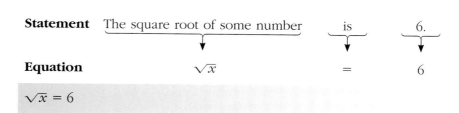

$$\sqrt{x} = 6$$

To translate a statement with more than one operation, you may need to use parentheses to make the order of operations clear. This is shown in Examples H and I.

EXAMPLE H

Write this statement as an equation: *3 times the sum of a number and 7 is 51.*

Statement

| 3 | times | the sum of a number and 7 | is | 51. |

Equation

| 3 | $\times$ | $(n + 7)$ | = | 51 |

$$3 \times (n + 7) = 51, \text{ or } 3(n + 7) = 51$$

EXAMPLE I

Write this statement
as an equation:
*5 divided by the
difference of 6 and
some number is 1.*

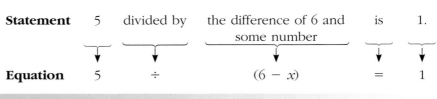

$$5 \div (6 - x) = 1, \text{ or } \frac{5}{6-x} = 1$$

In Example J, it is helpful to rewrite the sentence before translating it into an equation. The sentence contains the phrase *6 less than,* which means that 6 is subtracted from another term. Rewriting the sentence will help you write the symbols in the correct order.

EXAMPLE J

Write this statement
as an equation:
*6 less than the
product of 8 and a
number is 2.*

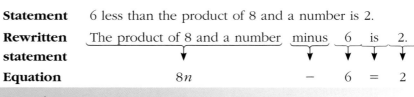

$$8n - 6 = 2$$

EXERCISES 10-1

Write each statement as an equation.

1. Some number divided by 6 is 7.

2. Two times the sum of a number and 9 is 20.

3. A number divided by 7 is −35.

4. The square root of some number is equal to 20 plus 5.

5. Twelve decreased by some number is 8.

6. Three times the sum of 8 and a number is 30.

7. To find the area, *A,* of a circle, multiply the square of the radius, *r,* by π.

8. One third the sum of 45 and a number is 16.

9. If you multiply a number by 6 and then add 1, the result is 55.

10. Three more than the square of a number is 52.

11. Eight less than some number is −2.

12. One less than the quotient of 25 and a number is 4.

13. The square root of the difference between a certain voltage, *V,* and 9 is 4.

14. Seven more than one half of the square of resistance, *R,* is 15.

15. A current of a 15.3 A is the square root of the fraction formed by the power, *P,* divided by a resistance of 12.25 Ω.

Answer the following.

16. Does the statement "Some number times the sum of 7 plus 5 is equal to 52" have the same meaning as $n \times 7 + 5 = 52$? Explain.

17. Which statement corresponds to $x^2 = y^2 + z^2$? Explain your choice.

    (a) One number squared is equal to the square of the sum of a second and third number.

    (b) One number squared is equal to a second number squared plus the square of a third number.

18. Challenge: Nine increased by 6 times some number is 10 less than one half the number. Write this statement as an equation.

SOLVING ONE-STEP EQUATIONS

Addition and Subtraction Equations

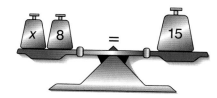

An equation is similar to a balanced scale. The expression on one side of the equal sign equals the expression on the other side of the equal sign. You must keep the equation in balance as you solve it.

To solve an equation means to find the value(s) of the variable that makes the equation true. This is done by isolating the variable. A value that makes the equation true is called a **solution.**

If a variable is part of an addition expression, you can add an opposite to isolate the variable. For example, to isolate the variable x in $x + 8 = 15$, add the opposite of 8 (-8) to both sides. The same method works for subtraction because every subtraction can be expressed as an addition. If a number is added to (or subtracted from) one side of an equation, then the same number must be added to (or subtracted from) the other side of the equation.

To check your solution, substitute it for the variable in the original equation. Verify that the solution makes the equation true.

Solve $x + 8 = 15$. Then check your solution.

Solve:

$$x + 8 = 15 \qquad \longleftarrow \text{8 is added to } x.$$
$$x + 8 + (-8) = 15 + (-8) \qquad \longleftarrow \text{Add } -8 \text{ to both sides of the equation.}$$
$$x + 0 = 7 \qquad \longleftarrow \text{Do addition.}$$
$$x = 7$$

Check:

$$x + 8 = 15 \longleftarrow \text{Write the original equation.}$$
$$7 + 8 \overset{?}{=} 15 \longleftarrow \text{Substitute your solution for } x.$$
$$15 = 15 \checkmark$$

The solution is 7.

Helpful Hint

Recall that you can subtract a number by adding its opposite. For example,
$3 - 5 = 3 + (-5)$
$= -2.$

EXAMPLE A

Solve $5 = x + 10$. Then check your solution.

Solve:

$$5 = x + 10 \quad \longleftarrow \text{10 is added to } x.$$
$$5 - 10 = x + 10 - 10 \quad \longleftarrow \text{Add } -10 \text{ to both sides of the equation.}$$
$$-5 = x$$

Check:

$$5 = x + 10 \quad \longleftarrow \text{Write the original equation.}$$
$$5 \stackrel{?}{=} -5 + 10 \quad \longleftarrow \text{Substitute your solution for } x.$$
$$5 = 5 \checkmark$$

The solution is -5.

EXAMPLE B

Solve $a - 9 = 14$. Then check your solution.

Solve:

$$a - 9 = 14 \quad \longleftarrow \text{Think of this as } a + (-9) = 14.$$
$$a - 9 + 9 = 14 + 9 \quad \longleftarrow \text{Add 9 to both sides of the equation.}$$
$$a = 23$$

Check:

$$a - 9 = 14 \quad \longleftarrow \text{Write the original equation.}$$
$$23 - 9 \stackrel{?}{=} 14 \quad \longleftarrow \text{Substitute your solution for } a.$$
$$14 = 14 \checkmark$$

The solution is 23.

EXAMPLE C

Solve $-7 + n = -12$. Then check your solution.

Solve:

$$-7 + n = -12 \quad \longleftarrow n \text{ is added to } -7.$$
$$-7 + n + 7 = -12 + 7 \quad \longleftarrow \text{Add 7 to both sides of the equation.}$$
$$n = -5$$

Check:

$$-7 + n = -12 \quad \longleftarrow \text{Write the original equation.}$$
$$-7 + (-5) \stackrel{?}{=} -12 \quad \longleftarrow \text{Substitute your solution for } n.$$
$$-12 = -12 \checkmark$$

The solution is -5.

Multiplication and Division Equations

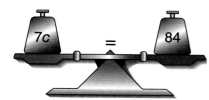

If a variable is multiplied by a number, you can divide by that number (or multiply by the reciprocal) to isolate the variable. If a variable is divided by a number, multiply by that number. For example, to isolate the variable c in $7c = 84$, divide both sides by 7.

Solve $7c = 84$. Then check your solution.

Solve:

$7c = 84$ ← c is multiplied by 7.

$\dfrac{7c}{7} = \dfrac{84}{7}$ ← Divide both sides of the equation by 7.

$\dfrac{\not{7}c}{\not{7}} = \dfrac{\not{7} \times 12}{\not{7}}$

$c = 12$

Check:

$7c = 84$ ← Write the original equation.

$7 \times 12 \stackrel{?}{=} 84$ ← Substitute your solution for c.

$84 = 84$ ✓

The solution is 12.

EXAMPLE D

Solve $24 = -6n$. Then check your solution.

Solve:

$24 = -6n$ ← n is multiplied by -6.

$\dfrac{24}{-6} = \dfrac{-6n}{-6}$ ← Divide both sides of the equation by -6.

$-4 = n$

Check:

$24 = -6n$ ← Write the original equation.

$24 \stackrel{?}{=} -6 \times (-4)$ ← Substitute your solution for n.

$24 = 24$ ✓

The solution is -4.

EXAMPLE E

Solve $\dfrac{n}{-8} = 9$. Then check your solution.

Solve:

$\dfrac{n}{-8} = 9$ ← n is divided by -8.

$-8 \cdot \dfrac{n}{-8} = -8 \cdot 9$ ← Multiply both sides of the equation by -8.

$n = -72$

Check:

$\dfrac{n}{-8} = 9$ ← Write the original equation.

$\dfrac{-72}{-8} \stackrel{?}{=} 9$ ← Substitute your solution for n.

$9 = 9$ ✓

The solution is -72.

EXAMPLE F

Solve $5c = -\frac{15}{16}$. Then check your solution.

Solve:

$$5c = -\frac{15}{16} \quad \longleftarrow \; c \text{ is multiplied by 5.}$$

$$\left(\tfrac{1}{5}\right)(5c) = \left(\tfrac{1}{5}\right)\left(-\tfrac{15}{16}\right) \quad \longleftarrow \; \text{Multiply both sides of the equation by the reciprocal of 5.}$$

$$c = -\frac{3}{16}$$

Check:

$$5c = -\frac{15}{16} \quad \longleftarrow \; \text{Write the original equation.}$$

$$5\left(-\tfrac{3}{16}\right) \stackrel{?}{=} -\tfrac{15}{16} \quad \longleftarrow \; \text{Substitute your solution for } c.$$

$$-\frac{15}{16} = -\frac{15}{16} \; \checkmark$$

The solution is $-\dfrac{3}{16}$.

EXAMPLE G

Solve $-\frac{2}{3}x = 10$. Then check your solution.

Solve:

$$-\frac{2}{3}x = 10 \quad \longleftarrow \; x \text{ is multiplied by } -\tfrac{2}{3}.$$

$$\left(-\tfrac{3}{2}\right)\left(-\tfrac{2}{3}\right)x = \left(-\tfrac{3}{2}\right)10 \quad \longleftarrow \; \text{Multiply both sides of the equation by the reciprocal of } -\tfrac{2}{3}.$$

$$x = -15$$

Check:

$$-\frac{2}{3}x = 10 \quad \longleftarrow \; \text{Write the original equation.}$$

$$-\frac{2}{3}(-15) \stackrel{?}{=} 10 \quad \longleftarrow \; \text{Substitute your solution for } x.$$

$$10 = 10 \; \checkmark$$

The solution is -15.

Equations with Powers

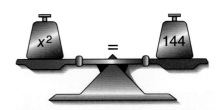

If a power of a variable appears in a one-step equation, you can use what you know about roots to isolate the variable. Every equation has the same number of roots as the exponent. But, not all of these roots can be written as either a rational or irrational number, so we will use the following rule: If the variable is raised to an even power, the equation will have two distinct solutions (or no solution). If the equation is raised to an odd power, the equation will have one solution.

Solve $x^2 = 144$. Then check your solution.

Solve:

$$x^2 = 144 \quad \longleftarrow \; \text{The power of } x \text{ is } x \text{ squared. Use the square root of 144 to find the solutions: } \sqrt{144} = 12.$$

$$x = 12 \text{ or } x = -12 \quad \longleftarrow \; \text{There are two solutions because } 12^2 = 144 \text{ and } (-12)^2 = 144.$$

Checks:

$$x^2 = 144 \qquad\qquad x^2 = 144$$
$$12^2 \overset{?}{=} 144 \qquad\qquad (-12)^2 \overset{?}{=} 144$$
$$12 \times 12 \overset{?}{=} 144 \qquad (-12) \times (-12) \overset{?}{=} 144$$
$$144 = 144 \ \checkmark \qquad\qquad 144 = 144 \ \checkmark$$

The solutions are 12 and -12.

EXAMPLE H

Solve $n^2 = 100$. Then check your solution.

Solve:

$n^2 = 100$ ← The power of n is n *squared.* Use the *square root* of 100 to find the solutions: $\sqrt{100} = 10$.

$n = 10$ or $n = -10$ ← There are two solutions because $10^2 = 100$ and $(-10)^2 = 100$.

Checks:

$$n^2 = 100 \qquad\qquad n^2 = 100$$
$$10^2 \overset{?}{=} 100 \qquad\qquad (-10)^2 \overset{?}{=} 100$$
$$10 \times 10 \overset{?}{=} 100 \qquad (-10) \times (-10) \overset{?}{=} 100$$
$$100 = 100 \ \checkmark \qquad\qquad 100 = 100 \ \checkmark$$

The solutions are 10 and -10.

EXAMPLE I

Solve $x^3 = 64$. Then check your solution.

Solve:

$x^3 = 64$ ← The power of x is x *cubed.* Use the *cube root* of 64 to find the solution: $\sqrt[3]{64} = 4$.

$x = 4$ ← There is only one solution because only $4^3 = 64$.

Check:

$$x^3 = 64 \quad \text{← Write the original equation.}$$
$$4^3 \overset{?}{=} 64 \quad \text{← Substitute your solution for } x.$$
$$4 \times 4 \times 4 \overset{?}{=} 64$$
$$64 = 64 \ \checkmark$$

The solution is 4.

EXAMPLE J

Solve $a^3 = -1,000$. Then check your solution.

Solve:

$a^3 = -1,000$ ⬅ The power of a is a *cubed*. Use the *cube root* of $-1,000$ to find the solution: $\sqrt[3]{-1,000} = -10$.

$a = -10$ ⬅ There is only one solution because only $(-10)^3 = -1,000$.

Check:

$a^3 = -1,000$ ⬅ Write the original equation.

$(-10)^3 \stackrel{?}{=} -1,000$ ⬅ Substitute your solution for a.

$(-10)(-10)(-10) \stackrel{?}{=} -1,000$

$-1,000 = -1,000$ ✓

The solution is -10.

EXAMPLE K

Solve $x^4 = \frac{1}{16}$. Then check your solution.

Helpful Hint

Remember that taking any root of the number 1 gives a result of 1.

Solve:

$x^4 = \frac{1}{16}$ ⬅ The power of x is the *fourth power*. Use the *fourth root* of $\frac{1}{16}$ to find the solutions: $\sqrt[4]{\frac{1}{16}} = \frac{1}{2}$.

$x = \frac{1}{2}$ or $x = -\frac{1}{2}$ ⬅ There are two solutions because $\left(\frac{1}{2}\right)^4 = \frac{1}{16}$ and $\left(-\frac{1}{2}\right)^4 = \frac{1}{16}$.

Checks:

$$x^4 = \frac{1}{16} \qquad\qquad x^4 = \frac{1}{16}$$

$$\left(\frac{1}{2}\right)^4 \stackrel{?}{=} \frac{1}{16} \qquad\qquad \left(-\frac{1}{2}\right)^4 \stackrel{?}{=} \frac{1}{16}$$

$$\left(\frac{1}{2}\right)\left(\frac{1}{2}\right)\left(\frac{1}{2}\right)\left(\frac{1}{2}\right) \stackrel{?}{=} \frac{1}{16} \qquad \left(-\frac{1}{2}\right)\left(-\frac{1}{2}\right)\left(-\frac{1}{2}\right)\left(-\frac{1}{2}\right) \stackrel{?}{=} \frac{1}{16}$$

$$\frac{1}{16} = \frac{1}{16} \checkmark \qquad\qquad \frac{1}{16} = \frac{1}{16} \checkmark$$

The solutions are $\frac{1}{2}$ and $-\frac{1}{2}$.

EXAMPLE L

Solve $x^2 = -36$.

Solve:

$x^2 = -36$ ⬅ There is no number that can be squared to get a negative number.

There is no solution to the equation.

Equations with Roots

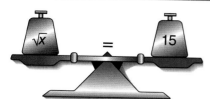

If a root of a variable appears in an equation, you can use what you know about powers to isolate the variable. You want to write the variable without a radical. Therefore, to isolate the variable x in $\sqrt[3]{x} = 10$, raise both sides of the equation to the third power. Then you can write $x = 1,000$.

A radical symbol with an even index, such as $\sqrt{}$ or $\sqrt[4]{}$, can represent only a nonnegative root. A radical symbol with an odd index, such as $\sqrt[3]{}$ or $\sqrt[5]{}$, can represent a positive, a negative, or zero root.

Solve $\sqrt[3]{x} = 10$. Then check your solution.

$\sqrt[3]{x} = 10$ ← The root of x is a *cube root*.

$x = 1,000$ ← The solution is 10 *cubed*: $10^3 = 1,000$.

Check:

$\sqrt[3]{x} = 10$ ← Write the original equation.

$\sqrt[3]{1,000} \overset{?}{=} 10$ ← Substitute your solution for x.

$10 = 10$ ✓

The solution is 1,000.

EXAMPLE M

Solve $\sqrt[3]{x} = -2$. Then check your solution.

Solve:

$\sqrt[3]{x} = -2$ ← The root of x is a *cube root*.

$x = -8$ ← The solution is (-2) *cubed*: $(-2)^3 = -8$.

Check:

$\sqrt[3]{x} = -2$ ← Write the original equation.

$\sqrt[3]{-8} \overset{?}{=} -2$ ← Substitute your solution for x.

$-2 = -2$ ✓

The solution is -8.

EXAMPLE N

Solve $\sqrt[3]{n} = \frac{3}{4}$. Then check your solution.

Solve:

$\sqrt[3]{n} = \frac{3}{4}$ ← The root of n is a *cube root*.

$n = \frac{27}{64}$ ← The solution is $\frac{3}{4}$ *cubed*: $\left(\frac{3}{4}\right)^3 = \frac{27}{64}$.

Check:

$\sqrt[3]{n} = \frac{3}{4}$ ← Write the original equation.

$\sqrt[3]{\frac{27}{64}} \overset{?}{=} \frac{3}{4}$ ← Substitute your solution for n.

$\frac{3}{4} = \frac{3}{4}$ ✓

The solution is $\frac{27}{64}$.

EXAMPLE O

Solve $\sqrt[4]{x} = \frac{1}{2}$. Then check your solution.

Solve:

$\sqrt[4]{x} = \frac{1}{2}$ ← The root of x is a *fourth root*.

$x = \frac{1}{16}$ ← The solution is $\frac{1}{2}$ to the *fourth power*: $\left(\frac{1}{2}\right)^4 = \frac{1}{16}$.

Check:

$\sqrt[4]{x} = \frac{1}{2}$ ← Write the original equation.

$\sqrt[4]{\frac{1}{16}} \overset{?}{=} \frac{1}{2}$ ← Substitute your solution for x.

$\frac{1}{2} = \frac{1}{2}$ ✓

The solution is $\frac{1}{16}$.

EXAMPLE P

Solve $\sqrt[4]{x} = 5$. Then check your solution.

Solve:

$\sqrt[4]{x} = 5$ ← The root of x is a *fourth root*.

$x = 625$ ← The solution is 5 to the *fourth power*: $5^4 = 625$.

Check:

$\sqrt[4]{x} = 5$ ← Write the original equation.

$\sqrt[4]{625} \overset{?}{=} 5$ ← Substitute your solution for x.

$5 = 5$ ✓

The solution is 625.

EXAMPLE Q

Solve $\sqrt{p} = 0$.

Solve:

$\sqrt{p} = 0$ ← The only number that can be squared to get zero as a result is zero.

$p = 0$

The solution is 0.

EXAMPLE R

Solve $\sqrt{x} = -4$.

Solve:

$\sqrt{x} = -4$ ← The symbol $\sqrt{}$ cannot represent a negative number. There is no number whose square is -4.

There is no solution to the equation.

EXERCISES 10-2

Solve each equation.

1. $n - 15 = -3$

2. $-5y = 20$

3. $\sqrt{d} = 8$

4. $27 = a - 18$

5. $3x = \frac{6}{7}$

6. $\frac{k}{6} = 7$

7. $b^2 = 81$

8. $\sqrt{m} = 30$

9. $\frac{c}{-15} = 4$

10. $3c = \frac{1}{3}$

11. $x^3 = 1,000$

12. $100 = n^2$

13. $-108 = -12n$

14. $11 = \sqrt{c}$

15. $\sqrt{w} = 25$

16. $\sqrt{n} = \frac{5}{6}$

17. $-6 + a = -36$

18. $\sqrt[3]{n} = -3$

19. $80 = 4n - 16$

20. $\frac{n}{24} = -2$

21. $\sqrt{n} = \frac{4}{9}$

22. $n^3 = -125$

23. $w^4 = \frac{1}{10,000}$

24. $\sqrt[5]{x} = 1$

25. $x^4 = \frac{1}{256}$

26. $5 + y = 4$

27. $\frac{x}{-7} = -7$

28. $\frac{-y}{5} = 10$

Solve.

29. The current, I, in a certain circuit is $12.7I = 50.8$, where I is in amps. Solve this equation for I.

30. What is the voltage, V, in a certain circuit if $V^2 = 12,100$.

31. Manuel says that he can solve the equation $3n = 21$ by multiplying both sides by $\frac{1}{3}$. Explain why he is correct.

32. Explain why there is no solution to the equation $n^2 = -9$.

33. **What's the Error?** A student wrote the following equation and solution:
$$\sqrt{x} = \frac{9}{16}; \; x = \frac{3}{4}.$$
Explain the error and give the solution.

SOLVING MULTI-STEP EQUATIONS

Two-Step Equations

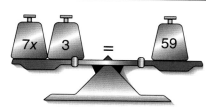

Many equations have more than one operation. To solve these equations you will need to apply more than one operation to isolate the variable.

In the equation $7x + 3 = 59$, there are two operations in the variable expression. The expression $7x + 3$ indicates that the variable x is multiplied by 7, and then 3 is added to the result. To isolate the variable, subtract 3, and then divide by 7.

Solve $7x + 3 = 59$. Then check your solution.

Solve:

$$7x + 3 = 59 \qquad \longleftarrow \text{3 is added to the variable term.}$$

$$7x + 3 - 3 = 59 - 3 \qquad \longleftarrow \text{Add } -3 \text{ to both sides of the equation.}$$

$$7x = 56 \qquad \longleftarrow x \text{ is multiplied by 7.}$$

$$\frac{7x}{7} = \frac{56}{7} \qquad \longleftarrow \text{Divide both sides of the equation by 7.}$$

$$x = 8$$

Check:

$$7x + 3 = 59 \qquad \longleftarrow \text{Write the original equation.}$$

$$7(8) + 3 \overset{?}{=} 59 \qquad \longleftarrow \text{Substitute your solution for } x.$$

$$56 + 3 \overset{?}{=} 59$$

$$59 = 59 \checkmark$$

The solution is 8.

EXAMPLE A

Solve $2a - 7 = 17$. Then check your solution.

Solve:

$$2a - 7 = 17 \qquad \longleftarrow -7 \text{ is added to the variable term.}$$

$$2a - 7 + 7 = 17 + 7 \qquad \longleftarrow \text{Add 7 to both sides of the equation.}$$

$$2a = 24 \qquad \longleftarrow a \text{ is multiplied by 2.}$$

$$\frac{2a}{2} = \frac{24}{2} \qquad \longleftarrow \text{Divide both sides of the equation by 2.}$$

$$a = 12$$

Check:

$$2a - 7 = 17 \qquad \longleftarrow \text{Write the original equation.}$$

$$2(12) - 7 \overset{?}{=} 17 \qquad \longleftarrow \text{Substitute your solution for } a.$$

$$24 - 7 \overset{?}{=} 17$$

$$17 = 17 \checkmark$$

The solution is 12.

EXAMPLE B

Solve $\frac{c}{4} + 5 = 29$. Then check your solution.

Solve:

$$\frac{c}{4} + 5 = 29 \qquad \longleftarrow \text{5 is added to the variable term.}$$

$$\frac{c}{4} + 5 - 5 = 29 - 5 \qquad \longleftarrow \text{Add } -5 \text{ to both sides of the equation.}$$

$$\frac{c}{4} = 24 \qquad \longleftarrow c \text{ is divided by 4.}$$

$$4 \cdot \frac{c}{4} = 4 \cdot 24 \qquad \longleftarrow \text{Multiply both sides of the equation by 4.}$$

$$c = 96$$

Check:

$\dfrac{c}{4} + 5 = 29$ ◄— Write the original equation.

$\dfrac{96}{4} + 5 \stackrel{?}{=} 29$ ◄— Substitute your solution for c.

$24 + 5 \stackrel{?}{=} 29$

$29 = 29$ ✓

The solution is 96.

EXAMPLE C

Solve $10 - 2n = -16$. Then check your solution.

Solve:

$10 - 2n = -16$ ◄— 10 is added to the variable term.

$10 - 2n - 10 = -16 - 10$ ◄— Add -10 to both sides of the equation.

$-2n = -26$ ◄— n is multiplied by -2.

$\dfrac{-2n}{-2} = \dfrac{-26}{-2}$ ◄— Divide both sides of the equation by -2.

$n = 13$

Check:

$10 - 2n = -16$ ◄— Write the original equation.

$10 - 2(13) \stackrel{?}{=} -16$ ◄— Substitute your solution for n.

$10 - 26 \stackrel{?}{=} -16$

$-16 = -16$ ✓

Helpful Hint

When only $-2n$ remains on the left side of the equation, think of the coefficient as negative 2.

The solution is 13.

EXAMPLE D

Solve $6n^2 = 150$. Then check your solution.

Solve:

$6n^2 = 150$ ◄— n^2 is multiplied by 6.

$\dfrac{6n^2}{6} = \dfrac{150}{6}$ ◄— Divide both sides of the equation by 6.

$n^2 = 25$ ◄— Use $\sqrt{25}$ to find the solutions: $\sqrt{25} = 5$.

$n = 5$ or $n = -5$ ◄— There are two solutions because $5^2 = 25$ and $(-5)^2 = 25$.

Checks:

$6n^2 = 150$ $\qquad\qquad$ $6n^2 = 150$

$6(5)^2 \stackrel{?}{=} 150$ $\qquad\qquad$ $6(-5)^2 \stackrel{?}{=} 150$

$6 \times 5 \times 5 \stackrel{?}{=} 150$ $\qquad$ $6 \times (-5) \times (-5) \stackrel{?}{=} 150$

$150 = 150$ ✓ $\qquad\qquad$ $150 = 150$ ✓

The solutions are 5 and -5.

EXAMPLE E

The current, I (in amps), in a circuit with a resistance of 120 Ω and power 30 W is given by $30 = I^2 \cdot 120$. Solve this equation.

$30 = I^2 \cdot 120$ ← I^2 is multiplied by 120.

$\dfrac{30}{120} = \dfrac{I^2 \cdot 120}{120}$ ← Divide both sides by 120.

$\dfrac{1}{4} = I^2$ ← Use $\sqrt{\dfrac{1}{4}}$ to find the solutions: $\sqrt{\dfrac{1}{4}} = \dfrac{1}{2}$.

$I = \dfrac{1}{2}$ or $I = -\dfrac{1}{2}$ ← There are two solutions because $\left(\dfrac{1}{2}\right)^2 = \dfrac{1}{4}$ and $\left(-\dfrac{1}{2}\right)^2 = \dfrac{1}{4}$.

The current in this circuit is either $-\dfrac{1}{2}$ A $= -0.5$ A or $\dfrac{1}{2}$ A $= 0.5$ A.

Equations Requiring More Than Two Steps

Proceed one step at a time to isolate the variable.

When an equation contains parentheses or has more than one variable term, it may take more than two steps to isolate the variable. Use the following steps in the order given to solve equations of this nature.

- When there are parentheses in an equation, use the Distributive Property first to remove the parentheses.
- If like terms appear on the same side of the equation, combine the like terms.
- If there are variable terms on both sides of the equation, add the opposite of one of those terms to both sides to collect all the variable terms on the same side.
- Isolate the variable in the equation.

Solve $3(x + 5) = -42$.

$3(x + 5) = -42$

$3x + 15 = -42$ ← Use the Distributive Property to remove the parentheses.

$3x + 15 - 15 = -42 - 15$ ← Add -15 to both sides of the equation.

$3x = -57$

$\dfrac{3x}{3} = -\dfrac{57}{3}$ ← Divide both sides of the equation by 3.

The solution is -19.

Although the check steps are not shown for the examples that follow, you should continue to check your solutions by substitution.

EXAMPLE F

Solve
$5(b - 7) = 2b - 29$.

$$5(b - 7) = 2b - 29$$
$$5b - 35 = 2b - 29$$ ← Use the Distributive Property to remove the parentheses.

$$5b - 35 - 2b = 2b - 29 - 2b$$ ← Add $-2b$ to both sides of the equation.
$$3b - 35 = -29$$
$$3b - 35 + 35 = -29 + 35$$ ← Add 35 to both sides of the equation.
$$3b = 6$$
$$\frac{3b}{3} = \frac{6}{3}$$ ← Divide both sides of the equation by 3.
$$b = 2$$

The solution is 2.

EXAMPLE G

Solve $4n + 12 + n$
$= -3n + 17$.

$$4n + 12 + n = -3n + 17$$
$$5n + 12 = -3n + 17$$ ← Combine like terms on the left side of the equation.

$$5n + 12 + 3n = -3n + 17 + 3n$$ ← Add $3n$ to both sides of the equation.
$$8n + 12 = 17$$
$$8n + 12 - 12 = 17 - 12$$ ← Add -12 to both sides of the equation.

$$8n = 5$$
$$\frac{8n}{8} = \frac{5}{8}$$ ← Divide both sides of the equation by 8.

$$n = \frac{5}{8}$$

The solution is $\frac{5}{8}$.

EXAMPLE H

Solve $5w^2 + 6 = 51$.

$$5w^2 + 6 = 51$$
$$5w^2 + 6 - 6 = 51 - 6$$ ← Add -6 to both sides of the equation.
$$5w^2 = 45$$
$$\frac{5w^2}{5} = \frac{45}{5}$$ ← Divide both sides of the equation by 5.
$$w^2 = 9$$ ← Find the square root of both sides of the equation.
$$w = \sqrt{9}$$
$$w = 3 \text{ or } w = -3$$

The solutions are 3 and -3.

EXERCISES 10-3

Solve each of the following equations.

1. $12b + 1 = -71$

2. $\dfrac{a + 4}{8} = 2$

3. $45 = 6x - 9$

4. $35 - 3c = -10$

5. $2x^2 = 32$

6. $-10 - 5c = 20$

7. $8x + 4 = -14 + 2x$

8. $5 + 9a = 7a + 11$

9. $\dfrac{b}{-7} + 1 = -1$

10. $9(2n + 3) = 18$

11. $-4b + 2 = -98$

12. $10 = 3(6 + 8n)$

13. $2q + 10 = -5q + 3$

14. $\dfrac{r + 5}{3} = 17$

15. $x^2 - 5 = 44$

16. $12 = -6(c + 1)$

17. $3a^2 - 4 = 104$

18. $5 + \dfrac{c}{3} = -10$

Solve.

19. According to Kirchhoff's voltage law, the current, I, in a certain circuit is $6.4I + 9.2I = 50.7$, where I is in amps. Solve this equation for I.

20. The voltage of 240 V in a circuit is the square root of the product of the power of 3600 W and the resistance, R.

    (a) Write this statement as an equation.

    (b) Solve this equation for resistance, in ohms. Round your answer to the nearest ohm.

21. The power in a certain circuit of 4.25 W is the quotient of the square of the voltage, V, and a resistance of 2850 Ω.

    (a) Write this statement as an equation.

    (b) Solve this equation for the voltage in volts. Round your answer to the nearest volt.

22. The impedance of 10 Ω in a certain circuit is the square root of the sum of the square of the resistance of 4.5 Ω and the square of the reactance, X, in ohms.

    (a) Write this statement as an equation.

    (b) Solve this equation for the reactance, X, in ohms. Round your answer to the nearest ohm.

23. **Challenge:** Solve $\dfrac{c + 11}{3c} = 4$.

USING FORMULAS

$P = 4s$

$A = \frac{1}{2}bh$

$P = 2(l + w)$

$A = \pi r^2$

Some equations contain more than one variable. A **formula** is an equation that describes a relationship among quantities, and it usually contains more than one variable.

To solve an equation (or a formula) for a given variable means to isolate that variable in the equation. For example, the equation $x = a + b$ is already solved for x. It is also possible to solve the equation for the variable a or the variable b.

Solve the equation $x = a + b$ for a.

$x = a + b$

$x - b = a + b - b$ ← Add the opposite of b to both sides of the equation.

$x - b = a$

$a = x - b$ ← It is customary to rewrite the equation with the desired variable on the left.

When the equation is solved for a, the result is $a = x - b$.

EXAMPLE A

The area of a triangle is given by the formula $A = \frac{1}{2}bh$. Solve this formula for h.

$A = \frac{1}{2}bh$

$2 \cdot A = 2 \cdot \frac{1}{2}bh$ ← Multiply both sides of the equation by 2.

$2A = bh$

$\frac{2A}{b} = \frac{bh}{b}$ ← Divide both sides of the equation by b.

$\frac{2A}{b} = h$

$h = \frac{2A}{b}$ ← Rewrite the equation with the desired variable on the left.

$h = \frac{2A}{b}$

EXAMPLE B

The area of a circle is given by the formula $A = \pi r^2$. Solve this formula for r.

$A = \pi r^2$

$\frac{A}{\pi} = \frac{\pi r^2}{\pi}$ ← Divide both sides of the equation by π.

$\frac{A}{\pi} = r^2$

$\sqrt{\frac{A}{\pi}} = r$ ← Take the square root of both sides. This is permitted because all quantities in the formula are positive.

$r = \sqrt{\frac{A}{\pi}}$ ← Rewrite the equation with the desired variable on the left.

$r = \sqrt{\frac{A}{\pi}}$

EXAMPLE C

Solve the perimeter formula $P = 2(l + w)$ for w.

$$P = 2(l + w)$$

$$P = 2l + 2w$$ ← Use the Distributive Property first.

$$P - 2l = 2l + 2w - 2l$$ ← Add the opposite of $2l$ to both sides of the equation.

$$P - 2l = 2w$$

$$\frac{P - 2l}{2} = \frac{2w}{2}$$ ← Divide both sides of the equation by 2.

$$\frac{P - 2l}{2} = w$$

$$w = \frac{P - 2l}{2}$$ ← Rewrite the equation with the desired variable on the left.

$$w = \frac{P - 2l}{2}$$

Once a variable has been isolated, its value can be determined based on other known quantities. In the formula $d = r \cdot t$, d represents distance, r represents rate of speed, and t represents time. In Example D, the formula $d = r \cdot t$ is solved for r, and then values are substituted for d and t.

EXAMPLE D

The formula relating distance, rate, and time is $d = r \cdot t$. Solve the formula for r. Find the rate if $d = 260$ miles and $t = 4$ hours.

$$d = r \cdot t$$

$$\frac{d}{t} = \frac{r \cdot t}{t}$$ ← Divide both sides of the equation by t.

$$\frac{d}{t} = r$$

$$r = \frac{260 \text{ miles}}{4 \text{ hours}}$$ ← Substitute the given values.

$$r = 65 \text{ miles per hour}$$

When $d = 260$ miles and $t = 4$ hours, the rate of speed, r, is 65 miles per hour.

EXAMPLE E

The formula relating current (I), voltage (E), and resistance (R) is given by the formula $I = \frac{E}{R}$. Solve the formula for R. Find the resistance if $E = 110$ volts and $I = 44$ amperes.

$$I = \frac{E}{R}$$

$$R \cdot I = R \cdot \frac{E}{R}$$ ← Multiply both sides of the equation by R.

$$R \cdot I = E$$

$$\frac{R \cdot I}{I} = \frac{E}{I}$$ ← Divide both sides of the equation by I.

$$R = \frac{E}{I}$$

$$R = \frac{110}{44}$$ ← Substitute the given values.

$$R = 2.5 \text{ ohms}$$

When $E = 110$ volts and $I = 44$ amperes, $R = 2.5$ ohms.

10-4 EXERCISES

Solve each equation or formula for the variable specified. Assume that no variables are equal to zero.

1. $A = lw$ for l

2. $C = \pi d$ for d

3. $V = \pi r^2 h$ for r

4. $m = \dfrac{a + b}{2}$ for b

5. $P = I \times E$ for I

6. $P = \dfrac{E^2}{R}$ for E

7. $x = ay + t$ for y

8. $E = mc^2$ for m

9. $C = \dfrac{q}{V}$ for V

10. $S = 6s^2$ for s

11. $P = I^2 R$ for I

12. $X_C = \dfrac{1}{2\pi f C}$ for C

13. $f = \dfrac{1}{2\pi\sqrt{LC}}$ for L

14. $Z = \sqrt{X^2 + R^2}$ for X

Solve each equation or formula for the variable specified. Then substitute the given values to find the value of the specified variable.

15. $R = \dfrac{K \cdot L}{A}$ for K

Find K if:

$R = 2,500$

$L = 100$

$A = 5$

16. $y = mx + b$ for b

Find b if:

$y = 8$

$m = -2$

$x = 1$

17. $V = lwh$ for h

Find h if:

$V = 450$ cubic meters

$l = 5$ meters

$w = 6$ meters

18. $V = s^3$ for s

Find s if:

$V = 1,000$ cubic feet

19. $Z = \sqrt{X^2 + R^2}$

Find R to the nearest tenth ohm if:

$Z = 45\,875 \; \Omega$

$X = 25 \; k\Omega$

20. $\dfrac{1}{R} = \dfrac{1}{R_1} + \dfrac{1}{R_2}$

Find R_1 to the nearest tenth ohm if:

$R = 9 \; \Omega$

$R_2 = 16 \; \Omega$

21. $R = \dfrac{1}{\dfrac{1}{R_1} + \dfrac{1}{R_2}} + R_3$

Find R_3 to the nearest tenth ohm if:

$R = 34 \; \Omega$

$R_1 = 40 \; \Omega$

$R_2 = 60 \; \Omega$

22. Challenge: The formula $A = p + prt$ gives the amount A in an account when p dollars are invested at an annual rate of simple interest r for t years. Solve the formula for r. Find the value of r when $A = 6,200$, $p = 5,000$, and $t = 6$.

USING EQUATIONS TO SOLVE WORD PROBLEMS

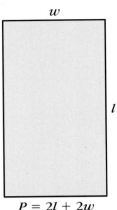

w

l

$P = 2l + 2w$

Sometimes the best way to solve a word problem is with an equation. Solving word problems in this way involves several steps. For word problems that can be solved using one variable, the steps are:

- Make a sketch if appropriate.
- Use a variable to represent the unknown quantity or quantities.
- Write an equation that represents a relationship among the quantities.
- Solve the equation for the variable.
- Use the value of the variable to answer the question in the word problem.

The length of a rectangle is three times its width. The perimeter, P, of the rectangle is 56 inches. What are the width and length of the rectangle?

- ☐ x ← Make a sketch.
  $3x$

- Let x = width of the rectangle ← Use a variable to represent the width
  $3x$ = length of the rectangle   and the length of the rectangle.

- $P = 2 \times$ length $+ 2 \times$ width ← Write an equation using your
  $56 = \quad 2(3x) \quad + \quad 2(x)$ variable that represents the
                                         relationship among quantities.

- $56 = 2(3x) + 2(x)$ ← Solve the equation.
  $56 = 6x + 2x$
  $56 = 8x$
  $7 = x$

- $x = 7$ (width of rectangle) ← Answer the question in the problem.
  $3x = 21$ (length of rectangle)

The width of the rectangle is 7 inches; the length of the rectangle is 21 inches.

EXAMPLE A

José is 4 years older than Shasta. The sum of their ages is 34. What are José's and Shasta's ages?

- Let s = Shasta's age   ← Use a variable to represent both
  $s + 4$ = José's age   ages.

- Shasta's age + José's age = 34 ← Write an equation using your
  $s \qquad + (s + 4) = 34$ variable to represent the
                                     relationship among quantities.

- $s + (s + 4) = 34$ ← Solve the equation.
  $2s + 4 = 34$
  $2s = 30$
  $s = 15$

- $s = 15$ (Shasta's age)   ← Answer the question in the
  $s + 4 = 19$ (José's age)   problem.

Shasta is 15 years old; José is 19 years old.

EXAMPLE B

The sum of three consecutive integers is 129. What are the integers?

Let n = first integer
$n + 1$ = second integer
$n + 2$ = third integer

← Use a variable to represent the integers.

$$\text{sum of the integers} = 129$$
$$n + (n + 1) + (n + 2) = 129$$

← Write an equation using your variable to represent the relationship among quantities.

$$n + n + 1 + n + 2 = 129$$
$$3n + 3 = 129$$
$$3n = 126$$
$$n = 42$$

← Solve the equation.

$n = 42$ (first integer)
$n + 1 = 43$ (second integer)
$n + 2 = 44$ (third integer)

← Answer the question in the problem.

The three consecutive integers are 42, 43, and 44.

EXAMPLE C

The sum of two consecutive odd integers is 76. What are the integers?

Helpful Hint

Two consecutive odd integers are always 2 apart. For example, 5 and 7 are consecutive odd integers. If $x = 5$, then $x + 2 = 7$.

Let n = first integer
$n + 2$ = second integer

← Use a variable to represent the integers.

$$\text{sum of the integers} = 76$$
$$n + (n + 2) = 76$$

← Write an equation using your variable to represent the relationship among quantities.

$$n + n + 2 = 76$$
$$2n + 2 = 76$$
$$2n = 74$$
$$n = 37$$

← Solve the equation.

$n = 37$ (first integer)
$n + 2 = 39$ (second integer)

← Answer the question in the problem.

The two consecutive odd integers are 37 and 39.

KEY IDEA

Did you notice that the sum of two consecutive odd integers is an even integer? This is always true! If you add three consecutive odd integers, is the sum odd or even? Can you show that your answer is correct?

Coins are a frequent topic for word problems. One way to solve a word problem about a collection of coins is to let the variable represent the number of one type of coin. It is best to use the value of each type of coin in pennies to write the equation. This is shown in Example D.

EXAMPLE D

In a collection of coins, there are 4 times as many nickels as quarters and 2 times as many pennies as quarters. The coins are worth $1.41. How many of each coin are there?

Let q = the number of quarters ← Use a variable to represent the
$4q$ = the number of nickels   number of each type of coin.
$2q$ = the number of pennies

$25(q) + 5(4q) + 1(2q) = 141$ ← Write an equation for the value
of the collection of coins.

$$25(q) + 5(4q) + 1(2q) = 141$$ ← Solve the equation.
$$25q + 20q + 2q = 141$$
$$47q = 141$$
$$q = 3$$

$q = 3$ (3 quarters) ← Answer the question in the problem.
$4q = 12$ (12 nickels)
$2q = 6$ (6 pennies)

There are 3 quarters, 12 nickels, and 6 pennies in the collection of coins.

EXERCISES 10-5

Solve each word problem.

1. Five less than twice some number is equal to 21. What is the number?

2. If the perimeter of a certain rectangle is 36 feet and the length is 2 times the width, what is the width?

3. Rita is twice as old as her cousin Lenny, who is 4 years older than their cousin Tony. The sum of all three ages is 72. How old is Rita?

4. A handful of coins is made up of dimes and nickels. There are 2 more dimes than nickels. The coins have a value of $1.85. How many of each coin are there?

5. The sum of two consecutive even integers is -22. What are the two integers?

6. Twice some number decreased by 5 is equal to 1 more than that number. What is the number?

7. The sum of three consecutive integers is 102. What are the three consecutive integers?

8. Xavier has one $10 bill, some $1 bills, and some $5 bills for a total of $31. There are twice as many $1 bills as $5 bills. How many $1 bills are there?

9. The sum of two consecutive odd integers is zero. What are the two odd integers?

10. Hilkka purchases a digital multimeter and a portable power monitor. Together they cost $417.94. The power monitor costs $44.04 more than the multimeter. Find the cost of each tool.

11. Mladen is paid $14.75 an hour. He gets paid time-and-a-half for each overtime hour he works over 40 hours a week. One week he makes $722.75. How many hours of overtime did he work?

12. Challenge: Ten years from now, Jane will be three times as old as she was four years ago. How old is Jane now?

WRITING INEQUALITIES

To write an **inequality,** write a statement, comparing two quantities, using one of the symbols given below.

Symbol	**Meaning**
>	is greater than
<	is less than
≥	is greater than or equal to
≤	is less than or equal to
≠	is not equal to

To translate a written statement to an inequality, translate each part of the statement into mathematical symbols just as you did to write an equation. Then use one of the symbols shown above instead of an = symbol.

Look at the street sign above. Write an inequality to describe a possible fine for parking.

Write a sentence describing the situation:
A fine is less than or equal to $100.

Translate the sentence into mathematical symbols.

Sentence	A fine	is less than or equal to	$100.
	↓	↓	↓
Inequality	f	≤	100

$f ≤ 100$

EXAMPLE A

Use the following statement to write an inequality: *The Star Tool Company produces at least 750 hammers each day.*

Write a sentence that can be translated into an inequality:
The number of hammers produced is greater than or equal to 750.

Translate the sentence into an inequality.

Sentence	The number of hammers produced	is greater than or equal to	750.
	↓	↓	↓
Inequality	h	≥	750

$h ≥ 750$

EXAMPLE B

Use the following statement to write an inequality: The price of 12 packages of outlets was less than $150.

Think of the statement as this: The product of 12 and a number is less than 150.

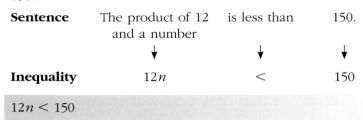

$12n < 150$

EXAMPLE C

Use the following statement to write an inequality: *The number chosen in the raffle is not 219.*

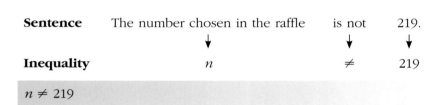

$n \neq 219$

EXAMPLE D

Write this statement as an inequality: *3 times a number divided by 5 is greater than −18.*

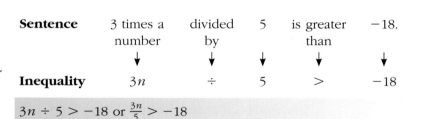

$3n \div 5 > -18$ or $\frac{3n}{5} > -18$

EXAMPLE E

Use the following statement to write an inequality: *At one dealership, the number of cars sold this month is at least 45.*

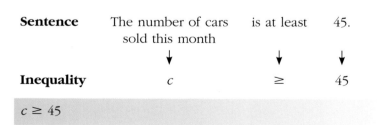

$c \geq 45$

EXERCISES 10-6

Write each statement as an inequality.

1. A number divided by 6 is less than 24.

2. Twelve is greater than or equal to the sum of a number and 7.

3. The product of −4 and a number is less than 16.

4. Four more than 2 times a number is greater than −10.

5. Three times the sum of a number and 5 is greater than or equal to 20.

Write an inequality using each of the following facts.

6. The fixture is designed for use with a lightbulb whose wattage does not exceed 60 watts.

7. The population of the town exceeds 250,000 this year.

8. The balance in Jack's bank account is not $325.

9. Pablo can spend no more than $875 for supplies.

10. The amount of time that it took to complete the wiring job is less than half of the budgeted 64 hours.

11. Challenge: Does the inequality $x \geq 0$ have the same meaning as the statement *x is not a negative number?* Explain.

SOLVING INEQUALITIES

One-Step Inequalities

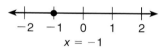

A **solution** of an inequality is any value of the variable that makes the inequality true. The **solution set** of an inequality is the set of all the values that make the statement true.

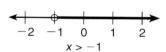

There are many solutions to the inequalities $x > -1$ and $x < -1$. The solution sets to these inequalities are indicated by the bold portions of the number lines shown at the left.

Solving an inequality is similar to solving an equation. However, there is one important difference. *When you multiply or divide both sides of an inequality by a negative number, you must reverse the inequality symbol.*

To understand why this is necessary, consider the two inequalities below.

8 < 12		**8 < 12**	
$-2(8) > -2(12)$	Multiply both sides by -2.	$\dfrac{8}{-2} > \dfrac{12}{-2}$	Divide both sides by -2.
$-16 > -24$	The inequality symbol *must be reversed* to make the statement true.	$-4 > -6$	The inequality symbol *must be reversed* to make the statement true.

Solve $-4n < 24$. Then check your solution set.

Solve:

$-4n < 24$

$\dfrac{-4n}{-4} > \dfrac{24}{-4}$ ← Divide both sides of the inequality by -4. Reverse the inequality symbol.

$n > -6$

The solution set is shown by the bolded portion of the number line below.

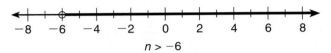

$n > -6$

Check:

$-4n < 24$ ← Write the original inequality.

$-4(-5) \overset{?}{<} 24$ ← Substitute a number greater than -6 for n.

$20 < 24$ ✓

The solution set of $-4n < 24$ is given by $n > -6$.

Language Box

In the inequality $n > -6$, the number -6 is the boundary value of the set. The open circle on the graph shows that -6 is not included in the solution set.

There really is no way to completely check a solution set of an inequality, because there are an infinite number of solutions. However, we will continue to use the direction "check your solution set" and know that we are really only checking one solution.

EXAMPLE A

Solve $x - 10 \geq 14$. Then check your solution set.

Solve:

$x - 10 \geq 14$

$x - 10 + 10 \geq 14 + 10$ ← Add 10 to both sides of the inequality.

$x \geq 24$

The solution set is shown on the number line below.

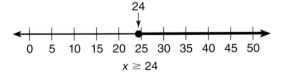

$x \geq 24$

Helpful Hint

To show that a boundary value is included in the solution set, use a filled-in circle.

Check:

$x - 10 \geq 14$ ← Write the original inequality.

$30 - 10 \overset{?}{\geq} 14$ ← Substitute a number greater than or equal to 24 for x.

$20 \geq 14$ ✓

The solution set of $x - 10 \geq 14$ is given by $x \geq 24$.

EXAMPLE B

Solve $5w < -45$. Then check your solution set.

Solve:

$5w < -45$

$\dfrac{5w}{5} < \dfrac{-45}{5}$ ⟵ Divide both sides of the inequality by 5.

$w < -9$

The solution set is shown on the number line below.

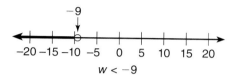

$w < -9$

Check:

$5w < -45$ ⟵ Write the original inequality.

$5(-10) \overset{?}{<} -45$ ⟵ Substitute a number less than -9 for w.

$-50 < -45$ ✓

The solution set of $5w < -45$ is given by $w < -9$.

Helpful Hint

To check the solution set to an inequality, choose a number close to, but not equal to, the boundary value.

EXAMPLE C

Solve $2 \geq \dfrac{a}{-7}$. Then check your solution set.

Solve:

$2 \geq \dfrac{a}{-7}$

$-7 \cdot 2 \leq -7 \cdot \dfrac{a}{-7}$ ⟵ Multiply both sides of the inequality by -7. Reverse the inequality symbol.

$-14 \leq a$

$a \geq -14$ ⟵ This is equivalent to $-14 \leq a$.

The solution set is shown on the number line below.

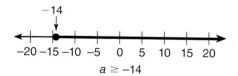

$a \geq -14$

Check:

$2 \geq \dfrac{a}{-7}$ ⟵ Write the original inequality.

$2 \overset{?}{\geq} \dfrac{-7}{-7}$ ⟵ Substitute a number greater than -14 for a.

$2 \geq 1$ ✓

The solution set of $2 \geq \dfrac{a}{-7}$ is given by $a \geq -14$.

Multi-Step Inequalities

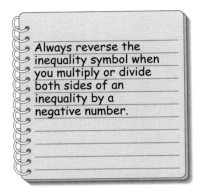

Always reverse the inequality symbol when you multiply or divide both sides of an inequality by a negative number.

Suppose that a contractor charges a flat fee of $600 plus $80 per hour for a project. He wants to do a job for $3,000 or less. What is the maximum number of hours he can work to be able to charge $3,000 or less?

Solve the inequality $600 + 80h \leq 3,000$. Then check your solution set.

Solve:

$$600 + 80h \leq 3,000$$

$$600 + 80h - 600 \leq 3,000 - 600 \quad \longleftarrow \text{Add } -600 \text{ to both sides of the inequality.}$$

$$80h \leq 2,400$$

$$\frac{80h}{80} \leq \frac{2,400}{80} \quad \longleftarrow \text{Divide both sides of the inequality by 80.}$$

$$h \leq 30$$

Check:

$$600 + 80h \leq 3,000 \quad \longleftarrow \text{Write the original inequality.}$$

$$600 + 80(28) \leq 3,000 \quad \longleftarrow \text{Substitute a number less than 30 for } h.$$

$$600 + 2,240 \stackrel{?}{\leq} 3,000$$

$$2,840 \leq 3,000 \checkmark$$

The solution set is given by $h \leq 30$. The contractor can work up to 30 hours and charge $3,000 or less.

EXAMPLE D

Solve $-3x - 7 < 23$. Then check your solution set.

Solve:

$$-3x - 7 < 23$$

$$-3x - 7 + 7 < 23 + 7 \quad \longleftarrow \text{Add 7 to both sides of the inequality.}$$

$$-3x < 30$$

$$\frac{-3x}{-3} > \frac{30}{-3} \quad \longleftarrow \text{Divide both sides of the inequality by } -3. \text{ Reverse the inequality symbol.}$$

$$x > -10$$

Check:

$$-3x - 7 < 23 \quad \longleftarrow \text{Write the original inequality.}$$

$$-3(-9) - 7 \stackrel{?}{<} 23 \quad \longleftarrow \text{Substitute a number greater than } -10 \text{ for } x.$$

$$27 - 7 \stackrel{?}{<} 23$$

$$20 < 23 \checkmark$$

The solution set of $-3x - 7 < 23$ is given by $x > -10$.

EXAMPLE E

Solve $-2 + \frac{n}{10} \leq -6$. Then check your solution set.

Solve:

$$-2 + \frac{n}{10} \leq -6$$

$$-2 + \frac{n}{10} + 2 \leq -6 + 2 \quad \longleftarrow \text{ Add 2 to both sides of the inequality.}$$

$$\frac{n}{10} \leq -4$$

$$10 \cdot \frac{n}{10} \leq 10 \cdot (-4) \quad \longleftarrow \text{ Multiply both sides of the inequality by 10.}$$

$$n \leq -40$$

Check:

$$-2 + \frac{n}{10} \leq -6 \quad \longleftarrow \text{ Write the original inequality.}$$

$$-2 + \frac{-50}{10} \overset{?}{\leq} -6 \quad \longleftarrow \text{ Substitute a number less than } -40 \text{ for } n.$$

$$-2 - 5 \overset{?}{\leq} -6$$

$$-7 \leq -6 \checkmark$$

The solution set is given by $n \leq -40$.

EXAMPLE F

Solve $-4(w + 2) < 4$. Then check your solution set.

Solve:

$$-4(w + 2) < 4$$

$$-4w - 8 < 4 \quad \longleftarrow \begin{array}{l}\text{Use the Distributive Property to} \\ \text{remove parentheses.}\end{array}$$

$$-4w - 8 + 8 < 4 + 8 \quad \longleftarrow \text{ Add 8 to both sides of the inequality.}$$

$$-4w < 12$$

$$\frac{-4w}{-4} > \frac{12}{-4} \quad \longleftarrow \begin{array}{l}\text{Divide both sides of the inequality} \\ \text{by } -4. \text{ Reverse the inequality symbol.}\end{array}$$

$$w > -3$$

Check:

$$-4(w + 2) < 4 \quad \longleftarrow \text{ Write the original inequality.}$$

$$-4(0 + 2) \overset{?}{<} 4 \quad \longleftarrow \text{ Substitute a number greater than } -3 \text{ for } w.$$

$$-4(2) \overset{?}{<} 4$$

$$-8 < 4 \checkmark$$

The solution set is given by $w > -3$.

EXAMPLE G

Solve $a + 5 \geq -2a - 1$. Then check your solution set.

Solve:

$$a + 5 \geq -2a - 1$$

$$a + 5 + 2a \geq -2a - 1 + 2a \quad \leftarrow \text{Add } 2a \text{ to both sides of the inequality.}$$

$$3a + 5 \geq -1$$

$$3a + 5 - 5 \geq -1 - 5 \quad \leftarrow \text{Add } -5 \text{ to both sides of the inequality.}$$

$$3a \geq -6$$

$$\frac{3a}{3} \geq \frac{-6}{3} \quad \leftarrow \text{Divide both sides of the inequality by 3.}$$

$$a \geq -2$$

Check:

$$a + 5 \geq -2a - 1 \quad \leftarrow \text{Write the original inequality.}$$

$$-1 + 5 \overset{?}{\geq} -2(-1) - 1 \quad \leftarrow \text{Substitute a number greater than } -2 \text{ for } a.$$

$$4 \overset{?}{\geq} 2 - 1$$

$$4 > 1 \checkmark$$

The solution set is given by $a \geq -2$.

EXERCISES 10-7

Solve each inequality.

1. $4x > -32$

2. $8n - 13 \geq 11$

3. $\frac{w}{9} < 3$

4. $\frac{n}{2} + 3 < -1$

5. $x - 4 \leq 16 + 5x$

6. $3(w - 1) > -12$

7. $9 > n + 10$

8. $4x + 8 \leq 0$

9. $-7n \geq -21$

10. $18 - 3w < 6 - w$

11. $30 > \frac{n}{-2}$

12. $9(n + 1) < -27$

13. $-x + 6 < 30$

14. $90 - n \geq -10n$

15. $-7x + 1 \geq -55$

16. $1 < \frac{2x}{8}$

Solve.

17. Julio bought a new watt/VAR transducer for $957.50. He paid $145 down and will pay $32.50 a month until it is fully paid for.

 (a) Write an inequality to describe how many months it will take before he has fully paid for the transducer.

 (b) How many months will it take before he has fully paid for the transducer?

18. Four times the sum of a number and 3 is less than 28. What numbers satisfy this condition?

19. Challenge: The perimeter of a square is greater than 64 inches but less than 152 inches. What are the possible lengths for the side of the square?

CHAPTER 10 REVIEW EXERCISES

1. A resistance, R, decreased by 25 Ω is 37 Ω.

   (a) Write an equation to describe this statement.

   (b) Determine the resistance.

2. Two times the difference of a certain voltage and 12 V is 216 V.

   (a) Write an equation to describe this statement.

   (b) Determine the voltage.

3. A power of 40 W is the square of the current in amps times a resistance of 250 Ω.

   (a) Write an equation to describe this statement.

   (b) Determine the voltage.

4. The diameter, d, in mils, of 240 feet of 10-gauge copper wire is given by $0.25 = \frac{(10.8)(240)}{d^2}$. Solve this for d to the nearest tenth.

5. Solve the equation $I = \frac{E}{R + r}$ for r.

6. Solve the equation $I = \frac{E_x - E_c}{R}$ for E_c.

7. If $R = \frac{KL}{d^2}$, find L to the nearest tenth foot if $R = 2.5$ Ω, $K = 10.8$, and $d = 85$ mil. (Note, the K is a constant and does not have any units.)

8. Suppose $R = R_1 + \frac{R_2 R_3}{R_2 + R_3}$.

   (a) Solve this equation for R_1.

   (b) Determine R_1 if $R = 70$ Ω, $R_2 = 20$ Ω, and $R_3 = 30$ Ω.

9. Suppose $R = R_1 + \frac{R_2 R_3}{R_2 + R_3}$. Solve this equation for R_2 if $R = 70$ Ω, $R_1 = 58$ Ω, and $R_3 = 20$ Ω.

10. Pavel is paid time-and-a-half for each hour he works over 40 hours a week. One week he worked 12 hours of overtime and made $962.80.

    (a) Write an equation for this statement using w as his hourly wage.

    (b) What is his hourly wage?

11. Write an inequality that describes the fact that it will take no more than 1,750 feet of cable to wire a certain house.

12. Julio needed a transducer to replenish the supplies in his electrical shop. The cost of the supplies is $8,500. He paid $1,245 down and will pay $420.50 a month until it is fully paid for.

    (a) Write an inequality to describe how many months it will take before he has fully paid for the transducer.

    (b) How many months will it take before he has fully paid for the transducer?

Building a Foundation in Mathematics

Writing and Simplifying Ratios

Writing Unit Rates

Identifying and Solving Proportions

Solving Direct Proportion Problems

Solving Inverse Proportion Problems

Overview

The ability to solve applied problems using ratio and proportion is vital to electricians. Electrical resistance and wire sizes are two examples that use proportions. In this chapter you will learn how to express relationships as ratios or rates. You will also learn how to multiply the means and extremes in order to solve a proportion. Finally, you will learn how to determine when two comparisons vary directly or inversely and how to set up and solve direct and inverse proportions.

Ratios, Rates, and Proportions

Objectives

After completing this chapter, you will be able to:

- Write comparisons as ratios or rates

- Solve applied ratio problems

- Solve for the missing terms of given proportions

- Solve proportion problems by substituting values in formulas

- Analyze problems to determine whether they are direct or inverse proportions, set up proportions, and solve for unknowns

Chapter 11

WRITING AND SIMPLIFYING RATIOS

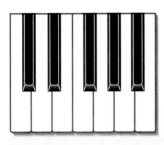

A **ratio** is a comparison of two like quantities by division. In the diagram at the left, the ratio of white keys to black keys is 7 to 5. This ratio can also be expressed with a colon, 7:5, or as a fraction, $\frac{7}{5}$.

The order in which the ratio is written is important. Note that the ratio of 5 black keys to 7 white keys is written 5 to 7, 5:7, or $\frac{5}{7}$.

Express the ratio of black keys to total keys in three ways.

There are 5 black keys and 12 keys in all. The ratio of black keys to total keys can be written:

5 to 12, 5:12, or $\frac{5}{12}$.

EXAMPLE A

Use the figures below to write the ratio of ◆s to □s.

There are 2 ◆s and 3 □s.

The ratio of ◆s to □s is 2 to 3, 2:3, or $\frac{2}{3}$.

When the numbers compared in a ratio have a common factor, you can write the ratio as an equivalent ratio in simplest form. This is shown in Example B.

EXAMPLE B

At a meeting of 54 people, there are 12 left-handed people. Write the ratio of the number of left-handed people to the total number of people.

$\frac{12}{54}$ ← Write the ratio of the number of left-handed people to the total number of people.

$\frac{12}{54} = \frac{2}{9}$ ← Simplify the fraction.

The ratio of the number of left-handed people to the total number of people is 2 to 9, 2:9, or $\frac{2}{9}$.

Both terms of a ratio must have the same unit. If the quantities have different units of measure, they can be compared as a ratio, provided it is possible to relate one unit to the other. For example, inches and feet are different units of length, but they are related by the fact 12 in. = 1 ft. We use a fact such as 12 in. = 1 ft to write a conversion factor. By using an appropriate conversion factor, we can cancel the units of measure and get a ratio that has only numbers. In Example C, the conversion factor is $\frac{12 \text{ in.}}{1 \text{ ft}}$.

EXAMPLE C

What is the ratio of 2 feet to 6 inches in simplest form?

$\dfrac{2 \text{ ft}}{6 \text{ in.}} = \dfrac{2 \text{ ft}}{6 \text{ in.}} \times \dfrac{12 \text{ in.}}{1 \text{ ft}}$ ← Write the ratio; multiply by the conversion factor.

$= \dfrac{2 \cancel{\text{ ft}}}{6 \cancel{\text{ in.}}} \times \dfrac{12 \cancel{\text{ in.}}}{1 \cancel{\text{ ft}}}$ ← Cancel the units of measure.

$= \dfrac{24}{6}$ ← Multiply.

$= \dfrac{4}{1}$ ← Simplify.

The ratio of 2 feet to 6 inches is 4 to 1, 4:1 or $\frac{4}{1}$.

It is impossible to express two quantities as ratios if the terms have unlike units of measure that cannot be represented as like units. For example, inches and pounds cannot be compared as ratios.

EXAMPLE D

What is the ratio of 3 hours to 15 minutes in simplest form?

Use the conversion factor $\frac{60 \text{ min}}{1 \text{ hr}}$ to cancel the units of measure in the ratio.

$\frac{3 \text{ hr}}{15 \text{ min}} = \frac{3 \text{ hr}}{15 \text{ min}} \times \frac{60 \text{ min}}{1 \text{ hr}}$ ← Write the ratio; multiply by the conversion factor.

$= \frac{3 \text{ hr}}{15 \text{ min}} \times \frac{60 \text{ min}}{1 \text{ hr}}$ ← Cancel the units.

$= \frac{180}{15}$ ← Multiply.

$= \frac{12}{1}$ ← Simplify.

The ratio of 3 hours to 15 minutes is 12 to 1, 12:1, or $\frac{12}{1}$.

When an object is represented in a drawing or in a model, a scale is used to show the dimensions of the object proportionally. A **scale** is a ratio that compares the dimensions in the drawing or the model to the actual dimensions. A scale is given as a ratio with the names of the units included. This is shown in Example E.

EXAMPLE E

A distance of 90 miles is represented by 3 inches on a map. Write the scale of the map in simplest form.

$\frac{3 \text{ in.}}{90 \text{ mi}}$ ← Write the ratio.

$\frac{3 \text{ in.}}{90 \text{ mi}} = \frac{3 \text{ in.} \div 3}{90 \text{ mi} \div 3} = \frac{1 \text{ in.}}{30 \text{ mi}}$ ← Divide to get a numerator of 1; keep the units in the fraction.

1 in.:30 mi   ← A scale may be written as a ratio using a colon.

The scale of the map is 1 in.:30 mi.

EXAMPLE F

A wall 20 feet long is represented by an 8-inch segment on a blueprint. What is the scale of the blueprint, written as an equation?

$\frac{8 \text{ in.}}{20 \text{ ft}}$ ← Write the ratio.

$\frac{8 \text{ in.}}{20 \text{ ft}} = \frac{8 \text{ in.} \div 8}{20 \text{ ft} \div 8} = \frac{1 \text{ in.}}{2.5 \text{ ft}}$ ← Divide to get a numerator of 1; keep the units in the fraction.

1 in. = 2.5 ft   ← A scale is sometimes written as an equation. This scale indicates that 1 inch on the blueprint represents an actual length of 2.5 feet.

The scale of the blueprint is given by the equation 1 in. = 2.5 ft.

EXERCISES 11-1

Express each of the following ratios in three ways. Write each ratio in simplest form.

1. Of 48 vehicles sold last month at a dealership, 16 were hybrids. What is the ratio of the number of hybrids sold to the total number of cars sold?

2. What is the ratio of 60 volts to 8 volts?

3. A transformer has 1,128 primary turns and 16,848 secondary turns. Express the ratio of primary turns to secondary turns in lowest terms.

4. A contractor spent 90 minutes preparing an estimate for a job and 24 hours doing the job. What is the ratio of the time spent preparing the estimate to the time spent doing the work on the job?

5. What is the ratio of 35 millimeters to 35 meters?

6. In a handful of coins, there are 4 nickels, 4 dimes, and 12 quarters. What is the ratio of the number of dimes to the total number of coins?

Write each scale in simplest form and as an equation.

7. Two cities that are actually 350 kilometers apart are 7 centimeters apart on a map. What is the scale of the map?

8. A drawing of a solar heating panel on a blueprint is 2 centimeters wide. The actual panel is 3 meters wide. What is the scale of the drawing?

9. The length of a room on a drawing is 3.25 inches. The actual room will be 15 feet $8\frac{1}{2}$ inches. What is the scale of the drawing?

10. **Challenge:** What is the ratio in simplest form of 80 centimeters to 2 kilometers?

WRITING UNIT RATES

A **rate** is a type of ratio that compares two unlike quantities. For example, the rate at which cars are passing through a tollbooth could be expressed as 50 cars per 30-minute interval. In this case, the unlike quantities are cars and minutes.

A **unit rate** is a rate in which the denominator is 1 unit. A speed limit of 65 on a U.S. interstate means that the drivers should not be driving any faster than 65 miles per hour. The unit rate 65 miles per hour can be written $\frac{65 \text{ mi}}{1 \text{ hr}}$. Note that the denominator is 1 hour.

Express the ratio $\frac{330 \text{ mi}}{6 \text{ hr}}$ as a unit rate.

$$\frac{330 \text{ mi}}{6 \text{ hr}} = \frac{330 \text{ mi} \div 6}{6 \text{ hr} \div 6}$$ ← Write the ratio. Divide both parts of the ratio by the numerical part of the denominator.

$$= \frac{55 \text{ mi}}{1 \text{ hr}}$$ ← Simplify.

The unit rate is $\frac{55 \text{ mi}}{1 \text{ hr}}$, or 55 miles per hour.

EXAMPLE A

Express the rate $\frac{\$8.75}{2.5\text{ lb}}$ as a unit rate.

$$\frac{\$8.75}{2.5\text{ lb}} = \frac{\$8.75 \div 2.5}{2.5\text{ lb} \div 2.5}$$ ← Write the rate. Divide both parts of the rate by the numerical part of the denominator.

$$= \frac{\$3.50}{1\text{ lb}}$$ ← Simplify.

The unit rate is $\frac{\$3.50}{1\text{ lb}}$, or $3.50 per pound.

EXAMPLE B

A case of 12 quarts of oil costs $10.20. What is the unit cost of the oil?

$$\frac{\$10.20}{12\text{ qt}} = \frac{\$10.20 \div 12}{12\text{ qt} \div 12}$$ ← Write the rate. Divide both parts of the rate by the numerical part of the denominator.

$$= \frac{\$0.85}{1\text{ qt}}$$ ← Simplify.

Language Box

For the cost of an item, unit rate is called **unit cost**.

The unit cost is $\frac{\$0.85}{1\text{ qt}}$, or $0.85 per quart.

EXAMPLE C

A tourist received 43.80 British pounds in exchange for $60 in U.S. money. How many pounds did he receive for each U.S. dollar?

$$\frac{43.80\text{ pounds}}{\$60} = \frac{43.80\text{ pounds} \div 60}{\$60 \div 60}$$ ← Write the rate. Divide both parts of the rate by the numerical part of the denominator.

$$= \frac{0.73\text{ pounds}}{\$1}$$ ← Simplify.

He received 0.73 British pound for each U.S. dollar.

EXERCISES 11-2

Express each of the following rates as a unit rate.

1. $\frac{576\text{ mi}}{18\text{ gal}}$

2. $\frac{\$4.56}{12\text{ oz}}$

3. $\frac{\$73.50}{6\text{ shirts}}$

4. $\frac{640\text{ km}}{8\text{ hr}}$

5. $\frac{5,000\text{ people}}{50\text{ mi}^2}$

6. $\frac{\$15.60}{24\text{ lightbulbs}}$

7. $\frac{54\text{ ft}^3}{2\text{ yd}^3}$

8. $\frac{15,4000\text{ ft}}{5\text{ min}}$

Helpful Hint

To find a unit rate, divide by the number of the quantity that follows the word *per*.

Answer each of the following.

9. If 70 feet of flat electrical cable can be purchased for $164.50, what is the price per foot?

10. The resistance of a wire depends on its length. If a wire that is 16 feet long has a resistance of 0.4 Ω, what is the resistance per foot?

11. Paul received $62 in Canadian dollars in exchange for $50 in U.S. money. What was the exchange rate for each U.S. dollar?

12. Population density is the number of residents per square mile. In 2010, there were 75,000 residents living in a certain county with an area of 213 square miles. What was the population density of the county? Round your answer to the nearest whole number.

13. **Challenge:** Which product has the lower unit cost: product A at 48 ounces for $3.84 or product B at 64 ounces for $5.44? Explain your answer.

IDENTIFYING AND SOLVING PROPORTIONS

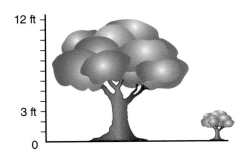

12 ft

3 ft

0

A **proportion** is an equation that states that two ratios are equal. Proportions can be written several ways. The proportion $\frac{1}{4} = \frac{3}{12}$ can also be written 1:4 = 3:12. It is read *1 is to 4 as 3 is to 12.*

In the proportion $a:b = c:d$, the terms a and d are the **extremes,** and the terms b and c are the **means.** This proportion could also be written $\frac{a}{b} = \frac{c}{d}$.

means

$$a \div b = c \div d$$

extremes

When one term in a proportion is not known, a variable may be used in its place. For example, in the proportion $\frac{3}{8} = \frac{x}{40}$, the value of the term represented by x is unknown. Cross multiplication can be used to find the value of x. Do this by setting the product of the means equal to the product of the extremes and solving the equation for the variable.

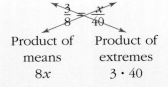

Product of   Product of
means   extremes
8x 3 · 40

The product of the means and the product of the extremes are called simply the *cross products.*

Language Box

Finding the value of a variable in a pro-portion is sometimes called *solving the proportion*.

Solve the proportion $\frac{3}{8} = \frac{x}{40}$.

Solve:

$\frac{3}{8} = \frac{x}{40}$ ← Write the proportion.

$8x = 3 \cdot 40$ ← Set the product of the means equal to the product of the extremes.

$8x = 120$

$\frac{8x}{8} = \frac{120}{8}$ ← Divide by 8 to isolate the variable.

$x = 15$

Check:

$\frac{3}{8} = \frac{x}{40}$ ← Write the original proportion.

$\frac{3}{8} \stackrel{?}{=} \frac{15}{40}$ ← Substitute your solution for x.

$8 \cdot 15 \stackrel{?}{=} 3 \cdot 40$

$120 = 120$ ✓ ← Verify that the cross products are equal.

$x = 15$

Helpful Hint

When a proportion is correct, the cross products are equal.

EXAMPLE A

Solve the proportion $\frac{30}{x} = \frac{120}{100}$.

Solve:

$\frac{30}{x} = \frac{120}{100}$ ← Write the proportion.

$120x = 30 \cdot 100$ ← Set the product of the means equal to the product of the extremes.

$120x = 3,000$

$\frac{120x}{120} = \frac{3,000}{120}$ ← Divide by 120 to isolate the variable.

$x = 25$

Check:

$\frac{30}{x} = \frac{120}{100}$ ← Write the original proportion.

$\frac{30}{25} \stackrel{?}{=} \frac{120}{100}$ ← Substitute your solution for x.

$\frac{6}{5} = \frac{6}{5}$ ✓ ← Verify that the ratios are equal.

$x = 25$

Helpful Hint

When a proportion is true, both ratios of the proportion are the same when written in simplest form.

EXAMPLE B

Solve the proportion $\frac{n}{0.6} = \frac{5}{3}$.

Solve:

$\dfrac{n}{0.6} = \dfrac{5}{3}$ ← Write the proportion.

$3n = 0.6 \cdot 5$ ← Set the product of the extremes equal to the product of the means.

$3n = 3$ ← Solve the equation for n.

$n = 1$

Check:

$\dfrac{n}{0.6} = \dfrac{5}{3}$ ← Write the original proportion.

$\dfrac{1}{0.6} \stackrel{?}{=} \dfrac{5}{3}$ ← Substitute your solution for n.

$1 \cdot 3 \stackrel{?}{=} 0.6 \cdot 5$

$3 = 3$ ✓   ← Verify that the cross products are equal.

$n = 1$

EXAMPLE C

Solve the proportion $\frac{25}{10a} = \frac{20}{36}$.

Solve:

$\dfrac{25}{10a} = \dfrac{20}{36}$ ← Write the proportion.

$20 \cdot 10a = 25 \cdot 36$ ← Set the product of the means equal to the product of the extremes.

$200a = 900$

$\dfrac{200a}{200} = \dfrac{900}{200}$ ← Divide both sides of the equation by 200.

$a = 4.5$

Check:

$\dfrac{25}{10a} = \dfrac{20}{36}$ ← Write the original proportion.

$\dfrac{25}{10(4.5)} \stackrel{?}{=} \dfrac{20}{36}$ ← Substitute your solution for a.

$\dfrac{25}{45} \stackrel{?}{=} \dfrac{20}{36}$

$\dfrac{5}{9} = \dfrac{5}{9}$ ✓   ← Verify that the ratios are equal.

$a = 4.5$

A variable may appear in more than one term in a proportion. In that case, write the proportion and cross multiply as you would with any proportion. Then use the steps you have learned for solving equations to isolate the variable and find the value of the variable. This is shown in Example D.

EXAMPLE D

Solve the proportion
$$\frac{5x - 3}{4} = \frac{5x + 3}{6}.$$

Solve:

$$\frac{5x - 3}{4} = \frac{5x + 3}{6}$$ ← Write the proportion.

$$6(5x - 3) = 4(5x + 3)$$ ← Set the product of the extremes equal to the product of the means.

$$30x - 18 = 20x + 12$$ ← Use the Distributive Property to remove parentheses.

$$10x - 18 = 12$$ ← Subtract $20x$ from both sides.

$$10x = 30$$ ← Add 18 to both sides.

$$\frac{10x}{10} = \frac{30}{10}$$ ← Divide by 10 to isolate the variable.

$$x = 3$$

Check:

$$\frac{5x - 3}{4} = \frac{5x + 3}{6}$$ ← Write the original proportion.

$$\frac{5(3) - 3}{4} \overset{?}{=} \frac{5(3) + 3}{6}$$ ← Substitute your solution for x.

$$\frac{15 - 3}{4} \overset{?}{=} \frac{15 + 3}{6}$$

$$\frac{12}{4} \overset{?}{=} \frac{18}{6}$$

$$3 = 3 \checkmark$$ ← Verify that the ratios are equal.

$$x = 3$$

EXERCISES 11-3

Solve each proportion.

1. $\frac{5}{4} = \frac{75}{n}$

2. $\frac{a}{36} = \frac{1}{9}$

3. $\frac{2}{3} = \frac{8}{2x}$

4. $\frac{90}{64} = \frac{n}{32}$

5. $\frac{4w - 9}{5} = \frac{w + 3}{3}$

6. $\frac{3a}{7} = \frac{6}{1}$

7. $\frac{x}{35} = \frac{16}{40}$

8. $\frac{6x - 2}{7} = \frac{5x + 7}{8}$

9. $\frac{352}{11} = \frac{w}{0.4}$

10. $\frac{3}{14} = \frac{3n}{126}$

11. $\frac{11}{n} = \frac{77}{14}$

12. $\frac{6}{5w} = \frac{4.8}{9}$

13. Six is to 15 as some number x is to 45. Set up a proportion and solve for x.

14. The ratio of 10 to 0.5 is the same as the ratio of 40 to some number x. What is the value of x?

15. The resistance of a wire is proportional to its length. If a 24-foot wire has a resistance of 0.6 Ω, what is the resistance of 42 feet of the same wire?

16. The ratio of the length to the width of a printed circuit board is 4.5 : 3.2. The maximum width of a circuit board is 406.4 millimeters. What is the maximum length?

17. **Challenge:** If you multiply both sides of the proportion $\frac{a}{b} = \frac{c}{d}$ by bd, what equation do you get? What does the equation represent?

SOLVING DIRECT PROPORTION PROBLEMS

Servings	Oats	Water
2	1 cup	2 cups
4	2 cups	4 cups
8	4 cups	8 cups

Two variables, x and y, are **directly proportional** if the ratio of y to x is a nonzero constant. This can be expressed by the equation $\frac{y}{x} = k$, where k is a nonzero number. For example, the ratio of y to x may be 4 to 8. In this case, $\frac{y}{x} = \frac{4}{8}$, and the constant ratio in simplest form is $\frac{1}{2}$. To set up a direct proportion, compare the quantities in the same way in each ratio of the proportion. Consider the amounts of oats and water for various servings of oatmeal shown at the left.

For two servings of oatmeal, the ratio of oats to water is $\frac{1 \text{ cup oats}}{2 \text{ cups water}}$, or 1 to 2. For four servings, the ratio of oats to water is $\frac{2 \text{ cups oats}}{4 \text{ cups water}}$, or 1 to 2. The same ratio holds true for eight servings: $\frac{4 \text{ cups oats}}{8 \text{ cups water}}$, or 1 to 2. We say that the amount of water used is directly proportional to the amount of oats used. That is, as one increases, so does the other.

Language Box

In the equation $\frac{y}{x} = k$ describing a direct proportion, k is called the **constant of variation** or *constant of proportionality*.

How many cups of water are used with 3 cups of oats to make the same type of oatmeal?

Let x represent the number of cups of water needed. Solve the proportion $\frac{3 \text{ cups oats}}{x \text{ cups water}}$ or $\frac{3}{x} = \frac{1}{2}$.

Solve:

$\frac{3}{x} = \frac{1}{2}$ ← Write the proportion.

$3 \cdot 2 = 1 \cdot x$ ← Cross multiply to solve the proportion.

$6 = x$

Check:

$\frac{3}{x} = \frac{1}{2}$ ← Write the original proportion.

$\frac{3}{6} \stackrel{?}{=} \frac{1}{2}$ ← Substitute your solution for x.

$\frac{1}{2} = \frac{1}{2}$ ✓   ← Verify that the ratios are equal.

For 3 cups of oats, 6 cups of water are needed.

EXAMPLE A

If 3 spools of wire of equal weight weigh a total of 54 pounds, how much would 7 of the same spools weigh?

Solve:

Let w represent the weight of the 7 spools.

$$\frac{3 \text{ spools}}{54 \text{ pounds}} = \frac{7 \text{ spools}}{w \text{ pounds}}$$ ← Compare the quantities in the same way in each ratio.

$$\frac{3}{54} = \frac{7}{w}$$ ← Write the proportion with only numbers and variables.

$$3w = 7 \cdot 54$$ ← Cross multiply and then solve the equation.

$$\frac{3w}{3} = \frac{378}{3}$$

$$w = 126$$

Check:

$$\frac{3}{54} = \frac{7}{w}$$ ← Write the original proportion.

$$\frac{3}{54} \stackrel{?}{=} \frac{7}{126}$$ ← Substitute your solution for w.

$$\frac{1}{18} = \frac{1}{18} \checkmark$$ ← Verify that the ratios are equal.

Seven of the same spools of wire would weigh 126 pounds.

A unit rate may be given as part of a direct proportion. This is shown in Example B. Continue to check the solutions to the remaining examples in the lesson on your own.

EXAMPLE B

A certain car gets an average of 22.5 miles to a gallon of gas. How many gallons will the car use for a 900-mile trip?

Let n represent the number of gallons.

$$\frac{22.5 \text{ miles}}{1 \text{ gallon}} = \frac{900 \text{ miles}}{n \text{ gallons}}$$ ← Compare the quantities in the same way in each ratio.

$$\frac{22.5}{1} = \frac{900}{n}$$ ← Write the proportion with only numbers and variables.

$$900 = 22.5n$$ ← Cross multiply and then solve the equation.

$$\frac{900}{22.5} = \frac{22.5n}{22.5}$$

$$40 = n$$

The car will use 40 gallons of gas for a 900-mile trip.

Frequently, subscripts are used with variables to identify the terms in a proportion. If variables x and y are directly proportional, a subscript of 1 is used in the first ratio and a subscript of 2 is used in the second ratio. The direct proportion is written $\frac{y_1}{x_1} = \frac{y_2}{x_2}$. This is shown in Example C.

EXAMPLE C

Variables x and y are directly proportional. If $x_1 = 5$, $y_1 = 9$, and $x_2 = 15$, determine the value of y_2.

$\frac{y_1}{x_1} = \frac{y_2}{x_2}$ ⟵ Set up the direct proportion.

$\frac{9}{5} = \frac{y_2}{15}$ ⟵ Substitute the given values.

$5y_2 = 9 \cdot 15$ ⟵ Cross multiply and then solve the equation.

$\frac{5y_2}{5} = \frac{135}{5}$

$y_2 = 27$

$y_2 = 27$

You may wonder if there is more than one way to set up a direct proportion for a set of numbers. The answer is yes. A proportion is true if the cross products are equal. For example, $\frac{2}{3} = \frac{6}{9}$ is true because $2 \cdot 9 = 6 \cdot 3 = 18$. For any true proportion, you can arrange terms to get another true proportion as long as the cross product remains the same. The following are some of the true proportions that are equivalent to $\frac{2}{3} = \frac{6}{9}$. Notice that they all have the same cross product, 18.

$\frac{2}{6} = \frac{3}{9} \qquad \frac{9}{3} = \frac{6}{2} \qquad \frac{6}{2} = \frac{9}{3}$

EXAMPLE D

Variables x and y are directly proportional. If $x_1 = 3$, $y_1 = 5$, and $y_2 = 20$, determine the value of x_2 using three different proportions.

METHOD 1	METHOD 2	METHOD 3
$\frac{x_1}{y_1} = \frac{x_2}{y_2}$	$\frac{x_1}{x_2} = \frac{y_1}{y_2}$	$\frac{y_2}{y_1} = \frac{x_2}{x_1}$
$\frac{3}{5} = \frac{x_2}{20}$	$\frac{3}{x_2} = \frac{5}{20}$	$\frac{20}{5} = \frac{x_2}{3}$
$5x_2 = 3(20)$	$5x_2 = 3(20)$	$5x_2 = 20(3)$
$5x_2 = 60$	$5x_2 = 60$	$5x_2 = 60$
$x_2 = 12$	$x_2 = 12$	$x_2 = 12$

Because the proportions are all equivalent, the solution, $x_2 = 12$, is the same in all three cases.

EXERCISES 11-4

Assume that the variables x and y are directly proportional. Determine the missing value in each case.

1. If $x_1 = 5$, $y_1 = 125$, and $y_2 = 25$, determine the value of x_2.

2. If $y_1 = 6$, $x_2 = 15$, and $y_2 = 18$, determine the value of x_1.

3. If $x_1 = 2$, $y_1 = 16$, and $x_2 = 3.5$, determine the value of y_2.

4. If $x_1 = 32$, $y_1 = 80$, and $y_2 = 45$, determine the value of x_2.

5. If $x_1 = 45$, $x_2 = 0.5$, and $y_2 = 2$, determine the value of y_1.

6. If $x_1 = \$11.00$, $y_1 = 2$, and $y_2 = 3$, determine the value of x_2.

7. If $y_1 = 560$, $x_2 = 5$, and $y_2 = 100$, determine the value of x_1.

8. If $x_1 = 1$, $y_1 = 120$, and $x_2 = 1.75$, determine the value of y_2.

Solve each of the following problems.

9. The resistance in a wire is directly proportional to the length of the wire. If 1,000 feet of #18 copper wire has a resistance of 6.5 ohms, what is the resistance of 1,200 feet of #18 copper wire?

10. When the resistance is constant, the current, I, varies directly as the voltage, V, varies.

    (a) Write the proportion for this problem.

    (b) If $I = 4.25$ A when $V = 110$ V, what is the current when the voltage is 60 V? Round your answer to the nearest hundredth.

11. George's car gets an average of 21 miles to a gallon of gas. At this rate, how many gallons will the car use on a 378-mile trip?

12. In an inductive-resistive circuit, the inductive reactance varies directly as the frequency changes. In a given circuit, a frequency of 2500 Hz produces an inductive reactance of 650.4 Ω. What frequency would produce an inductive reactance of 1740 Ω? Round your answer to the nearest tenth.

13. In a transformer, the voltage across the windings is directly proportional to the number of turns in the windings. If the primary windings have 100 turns and 60 volts across them, what is the voltage in the secondary windings with 150 turns?

14. **Challenge:** Using a copier, an office assistant reduces a document 8 inches wide using a 75% ratio. Then she enlarges the new document using a 150% ratio. What is the width of the final document?

SOLVING INVERSE PROPORTION PROBLEMS

rate (mph)	×	time (hr)	=	200 (miles)
20	×	10	=	200
25	×	8	=	200
50	×	4	=	200
80	×	2.5	=	200

When two variables x and y are **inversely proportional,** the product of x and y is a nonzero constant. This can be expressed by the equation $xy = k$, where k is a nonzero number. For example, the product xy may be 200. In this case, $xy = 200$ and the constant of variation k is 200.

Recall that if x and y are directly proportional, then $\frac{x_1}{y_1} = \frac{x_2}{y_2}$ and $x_1 y_2 = x_2 y_1$. Two variables x and y are inversely proportional if $\frac{x_1}{y_2} = \frac{x_2}{y_1}$ and $x_1 y_1 = x_2 y_2$.

The table at the left shows several different rates of speed and corresponding times that all result in the same distance traveled, 200 miles. Note that as the rate increases, the time decreases, but the distance is the same in every case. In this situation, we say that the time is inversely proportional to the rate. That is, as one increases, the other decreases, but the product rt (xy) equals 200 (k) in each case.

How long would it take a train to go 200 miles at an average rate of 40 miles per hour?

$rt = d$ ← Write the formula *rate × time = distance*.

$40t = 200$ ← Substitute 40 for r and 200 for d.

$\dfrac{40t}{40} = \dfrac{200}{40}$ ← Solve.

$t = 5$

It would take 5 hours to go 200 miles at 40 miles per hour.

Recall that the area of a rectangle is found using the formula area = length · width. For a given area, length is inversely proportional to width. This formula is used in Example A.

EXAMPLE A

Complete the table to show some different lengths and widths of rectangles that have an area of 80 cm².

Area	=	length	·	width		width
80 cm²	=	2 cm	·	_____		40 cm
80 cm²	=	4 cm	·	_____		20 cm
80 cm²	=	5 cm	·	_____		16 cm
80 cm²	=	10 cm	·	_____		8 cm
80 cm²	=	20 cm	·	_____		4 cm

Area = 80 cm²

To solve an inverse proportion problem involving x_1, y_1, x_2, and y_2, you can begin with the equation $x_1 y_1 = x_2 y_2$.

EXAMPLE B

Variables x and y are inversely proportional. If $x_1 = 2$, $y_1 = 24$, and $x_2 = 16$, determine the value of y_2.

Solve:

$x_1 y_1 = x_2 y_2$ ← Write the equation.

$2 \cdot 24 = 16 \cdot y_2$ ← Substitute the given values.

$48 = 16 y_2$ ← Solve the equation.

$\dfrac{48}{16} = \dfrac{16 y_2}{16}$

$3 = y_2$

Check:

$2 \cdot 24 = 16 \cdot y_2$ ← Write the original equation.

$2 \cdot 24 \overset{?}{=} 16 \cdot 3$ ← Substitute your solution.

$48 = 48$ ✓   ← Verify that the products are equal.

$y_2 = 3$

Language Box

The expressions *indirect proportion* and *inverse proportion* are both used to describe the relationship $x_1 y_1 = x_2 y_2$.

EXAMPLE C

The capacitive reactance in a circuit varies inversely as the capacitance changes. A circuit with a capacitance 0.05 μF produces a capacitive reactance of 19 980 Ω. If the capacitance changes to 0.075 μF, find the new capacitive reactance.

Solve:

We will let $C_1 = 0.05$ μF, $X_{C1} = 19\,980$ Ω, and $C_2 = 0.075$ μF. We want to solve for X_{C2} in ohms.

$C_1 X_{C1} = C_2 X_{C2}$ ← Write the equation.

$(0.05)(19\,980) = 0.075 X_{C2}$ ← Substitute the given values.

$999 = 0.075 X_{C2}$ ← Solve the equation.

$\dfrac{999}{0.075} = \dfrac{0.075 X_{C2}}{0.075}$

$13\,320 = X_{C2}$

Check.

$(0.05)(19\,980) = 0.075 X_{C2}$ ← Write the original equation.

$999 \stackrel{?}{=} (0.075)(13\,320)$ ← Substitute your solution.

$999 = 999$ ✓ ← Verify that the products are equal.

The new capacitive reactance is 13 320 Ω.

EXAMPLE D

The time to do a certain job is indirectly proportional to the number of people working on the job. If 6 people can do the job in 90 hours, how many people are required to do the job in 30 hours?

Solve:

Let $x_1 = 6$, $y_1 = 90$, and $y_2 = 30$. Solve for x_2, the number of people required to do the job in 30 hours.

$x_1 y_1 = x_2 y_2$ ← Write the equation.

$6 \cdot 90 = x_2 \cdot 30$ ← Substitute the given values.

$540 = 30 x_2$ ← Solve the equation.

$\dfrac{540}{30} = \dfrac{30 x_2}{30}$

$18 = x_2$

So, 18 people are required to do the job in 30 hours.

COMMON ERROR

ALERT

If the given values are not substituted into the proportion correctly, an error will occur. Write the equation for an indirect proportion first: $x_1 y_1 = x_2 y_2$. Then substitute the given values. Check the subscripts carefully as you substitute the values.

EXERCISES 11-5

Assume that the variables x and y are inversely proportional. Determine the missing value in each case.

1. If $x_1 = 7$, $x_2 = 4$, and $y_2 = 21$, determine the value of y_1.

2. If $x_1 = 3$, $y_1 = 12$, and $x_2 = 9$, determine the value of y_2.

3. If $y_1 = 1.5$, $x_2 = 2.5$, and $y_2 = 3$, determine the value of x_1.

4. If $x_1 = 25$, $x_2 = 5$, and $y_2 = 125$, determine the value of y_1.

5. If $x_1 = \frac{1}{2}$, $y_1 = 6$, and $y_2 = 9$, determine the value of x_2.

6. If $y_1 = 1$, $x_2 = 4$, and $y_2 = 7$, determine the value of x_1.

7. If $x_1 = 10$, $x_2 = 2$, and $y_2 = 85$, determine the value of y_1.

8. If $x_1 = 45$, $y_1 = 0.2$, and $y_2 = 9$, determine the value of x_2.

Solve each of the following problems.

9. If variables x and y are inversely proportional, with $x_1 = 5$ and $y_1 = 4$, what is the constant of variation?

10. Variables x and y are inversely proportional with a constant of variation of 2.5. If $x_1 = 6.25$, what is y_1?

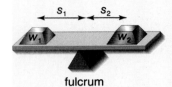

fulcrum

11. If the fulcrum is at the midpoint of a lever and the weights w_1 and w_2 are inversely proportional to distances s_1 and s_2 from the fulcrum, respectively, then the lever is balanced. If w_1 equals 120 grams, s_1 is 9 centimeters, w_2 equals 135 grams, and the lever is balanced, what is the distance s_2?

12. In a transformer, the ratio of the currents is inversely proportional to the ratio of turns. Use N_1 for the number of turns in the primary winding, N_2 for the number of turns in the secondary winding, I_1 for the primary current, and I_2 for the secondary current.

(a) Write the proportion for this problem.

(b) Suppose a transformer has 100 turns in its primary winding and 800 turns in its secondary winding. If the primary current is 5 A, determine the secondary current.

13. In a parallel circuit, the currents are inversely proportional to the resistances. Use I_1 and I_2 for the currents and R_1 and R_2 for the resistances.

(a) Write the proportion for this problem.

(b) If $I_1 = 0.5$ A, $I_2 = 0.35$ A, and $R_2 = 175$ Ω, determine R_1.

14. **What's the Error?** In a transformer, the currents in the primary and secondary windings are inversely proportional to the voltages in those windings. Given that the primary voltage E_p is 120 volts, the primary current I_p is 5 amps, and the secondary voltage E_s is 480 volts, a student solved a proportion for I_s as shown below. Describe the error and give the solution.

$$\frac{E_p}{I_p} = \frac{E_s}{I_s}; \frac{120}{5} = \frac{480}{I_s}; 120I_s = 5 \cdot 480; I_s = 20 \text{ amps}$$

CHAPTER 11 REVIEW EXERCISES

1. An amplifier has an input power of 1.5 W and an output power of 42.6 W. Write this ratio in three ways, with each ratio written in simplest form.

2. On an architectural drawing, one side of a building measures $6\frac{1}{4}''$. The actual side will be 50 feet. Write the scale in simplest form and as an equation.

3. Write $\dfrac{486.4 \text{ mi}}{15.2 \text{ gal}}$ as a unit rate.

4. If 24.2 gallons of diesel fuel can be purchased for $59.27, what is the price per gallon? Round the answer to the nearest tenth cent.

5. Solve $\dfrac{12}{56} = \dfrac{M}{21}$.

6. Solve $\dfrac{52}{d + 4} = \dfrac{18}{2.25}$.

7. Solve $\dfrac{2p + 5}{3p - 5} = \dfrac{32.4}{36.1}$.

8. Solve $\dfrac{7}{12} = \dfrac{21}{m^2}$.

9. When the resistance is constant, the square of the voltage, V, varies directly as the power, P, changes.

   (a) Write the proportion for this problem.

   (b) If $P = 120$ W when $V = 125$ V, what is the power when the voltage is 150 V?

10. When the resistance is constant, the power, P, is directly proportional to the square of the current, I.

    (a) Write the proportion for this problem.

    (b) If $P = 250$ W when $I = 4.25$ A, what is the current to the nearest tenth watt when the power is 175 W?

11. When the voltage is constant, the resistance, R, varies inversely with the current, I.

    (a) Write the proportion for this problem.

    (b) If $R = 60\ \Omega$ when $I = 2.5$ A, what is the current when the resistance is 48 Ω?

12. For a given length of wire, the electrical resistance, R, is inversely proportional to the square of its diameter, d.

    (a) Write the proportion for this problem.

    (b) If a certain length of wire with a diameter of 64.1 mils has a resistance of 2.525 kΩ, what is the resistance of the same length and type of wire with a diameter of 50.8 mils?

Building a Foundation in Mathematics

Converting Fractions, Decimals, and Percents

Solving Percent Problems with Proportions

Solving Percent Problems with Equations

Solving Word Problems

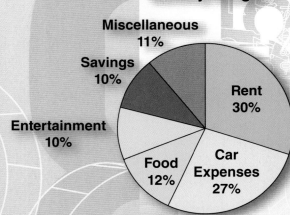

Monthly Budget

- Miscellaneous 11%
- Savings 10%
- Entertainment 10%
- Food 12%
- Car Expenses 27%
- Rent 30%

Overview

In this chapter you will learn about percentages and how to solve problems that involve percents.

Each day, people are faced with various kinds of percentage problems to solve. Sales taxes, savings interest, loan payments, insurance premiums, and income tax payments are all based on percents.

Percentages are widely used in both business and nonbusiness fields. Selling prices and discounts, wage deductions, and equipment depreciations are determined by percentages. Business profit and loss are often expressed as percents.

Percents

Objectives

After completing this chapter, you will be able to:

- Express decimal fractions and common fractions as percents
- Express percents as decimal fractions and common fractions
- Solve percent problems using proportions
- Solve percent problems using equations

CONVERTING FRACTIONS, DECIMALS, AND PERCENTS

A **percent** is a special ratio that compares a number to 100 using the symbol %. Sebastian answered 78 of the 100 multiple-choice questions correctly on an exam.

Express the fraction $\frac{78}{100}$ as a decimal and as a percent.

The fraction $\frac{78}{100}$ is equal to the decimal 0.78; both $\frac{78}{100}$ and 0.78 are read *seventy-eight hundredths*. The fraction $\frac{78}{100}$ can be written as 78% because *percent* means *hundredths*.

$$\frac{78}{100} = 0.78 = 78\%$$

Notice that you can convert between decimals and percents by just moving the decimal point.

$0.78 = 078.\% = 78\%$ ← Move the decimal point 2 places to the right, and insert the % sign.

$78\% = 78.\% = 0.78$ ← Move the decimal point 2 places to the left, and delete the % sign

Fractions, decimals, and percents can all represent parts of a whole. The table below shows the steps to convert a number from one form to another form.

Types of Conversion	Examples	Steps
Percent to Decimal	75% = 0.75 95% = 0.95 6.75% = 0.0675	Move the decimal point 2 places to the left.
Percent to Fraction	$75\% = \frac{75}{100} = \frac{3}{4}$ $42\% = \frac{42}{100} = \frac{21}{50}$	Simplify the fraction when possible.
Decimal to Percent	0.75 = 75% 0.23 = 23% 0.001 = 0.1% 1.50 = 150%	Move the decimal point 2 places to the right.
Decimal to Fraction	$0.75 = \frac{75}{100} = \frac{3}{4}$ $0.7 = \frac{7}{10}$ $0.125 = \frac{125}{1,000} = \frac{1}{8}$	Simplify the fraction when possible.
Fraction to Decimal	$\frac{3}{4} \rightarrow 4\overline{)3.00}^{\,0.75}$ $\frac{3}{4} = 0.75$	Divide the numerator by the denominator.
Fraction to Percent	$\frac{3}{4} = 0.75 = 75\%$ $\frac{3}{8} \rightarrow 8\overline{)3.000}^{\,0.375}$ $\frac{3}{8} = 0.375 = 37.5\%$	Convert to a decimal, and then to a percent.

EXAMPLE A

In a training class, 60% of the students are male. Express 60% as a fraction in simplest form.

$60\% = \dfrac{60}{100}$ ← Write a fraction in hundredths.

$= \dfrac{3}{5}$ ← Simplify the fraction.

$60\% = \dfrac{3}{5}$

Helpful Hint

In many cases, it helps to remember that *percent* means *hundredths*.

A percent greater than 100% can be converted to an improper fraction with a denominator of 100. The improper fraction can then be written as a mixed number. This is shown in Example B.

EXAMPLE B

A contractor saw an increase of 150% in the cost of materials for a job. Express 150% as a mixed number.

$150\% = \dfrac{150}{100}$ ← Write a fraction in hundredths.

$= \dfrac{3}{2}$ ← Simplify the fraction.

$= 1\dfrac{1}{2}$ ← Write the improper fraction as a mixed number.

$150\% = 1\dfrac{1}{2}$

EXAMPLE C

A wiring job is 45% complete. Express 45% as a decimal.

$45\% = 0.45$ ← Drop the % symbol; move the decimal point 2 places to the left.

$45\% = 0.45$

EXAMPLE D

The percent of acid in a certain solution is 7.5%. Express 7.5% as a decimal.

$7.5\% = 0.075$ ← Drop the % symbol; move the decimal point 2 places to the left.

$7.5\% = 0.075$

EXAMPLE E

A technician reported that a tank was filled to 0.875 of capacity. Express 0.875 as a percent.

$0.875 = 87.5\%$ ← Move the decimal point 2 places to the right; include a percent symbol.

$0.875 = 87.5\%$

EXAMPLE F

An order was placed for 16 electrical wall plates. Eleven of the plates were received, and the rest were back ordered. What percent were back ordered?

Since 11 of the 16 plates were received, $16 - 11 = 5$ plates were back ordered. We need to convert $\frac{5}{16}$ to a percent by dividing 5 by 16.

$$16\overline{)5.0000} \quad \begin{array}{l}0.3125\end{array}$$

← Convert the fraction to a decimal.

$0.3125 = 31.25\%$ ← Move the decimal point 2 places to the right; include a percent symbol.

31.25% of the electrical wall plates were back ordered.

EXAMPLE G

Convert $\frac{4}{9}$ to a percent. Round your answer to the nearest tenth of a percent.

$$9\overline{)4.0000} \quad \begin{array}{l}0.4444\end{array}$$

← Convert the fraction to a decimal.

$0.444\overline{4}... \approx 44.4\%$ ← Move the decimal point 2 places to the right. Round to the nearest tenth of a percent; include a percent symbol.

$\frac{4}{9} \approx 44.4\%$

EXAMPLE H

Calvin received only 3.2% of the vote in an election. Write 3.2% as a fraction.

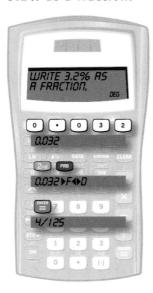

$3.2\% = 0.032$ ← Write the percent as a decimal by dropping the % symbol and moving the decimal point 2 places to the left.

$= \frac{32}{1,000}$ ← Write the decimal as a fraction.

$= \frac{4}{125}$ ← Simplify the fraction.

$3.2\% = \frac{4}{125}$

A table containing frequently used percents, decimals, and fractions is given in Appendix D. It is helpful to commit as many of these as possible to memory because they can serve as benchmarks for other values. For example, 27% is close to 25%, which is the same as the fraction $\frac{1}{4}$. So 27% is close to $\frac{1}{4}$.

EXERCISES 12-1

Convert each of the following percents to a fraction or mixed number in simplest form.

1. 23%	**2.** 75%	**3.** 30%
4. 1.2%	**5.** 8%	**6.** 120%
7. 48%	**8.** 225%	**9.** 3%

Rewrite each of the following percents as a decimal.

10. 43%	**11.** 82%	**12.** 10.5%
13. 128%	**14.** 6%	**15.** 0.2%
16. 21.9%	**17.** 31%	**18.** 95%

Express each of the following decimals as a percent.

19. 0.37 **20.** 0.091 **21.** 0.65

22. 2.48 **23.** 0.8 **24.** 0.09

25. 0.77 **26.** 1.3 **27.** 0.111

Express each of the following decimals as a fraction or mixed number in simplest form.

28. 0.44 **29.** 0.03 **30.** 1.65

31. 0.001 **32.** 3.06 **33.** 2.1

34. 2.75 **35.** 0.8 **36.** 0.007

Convert the following fractions to percents.

37. $\frac{55}{100}$ **38.** $\frac{7}{10}$ **39.** $\frac{1}{8}$

40. $\frac{12}{30}$ **41.** $\frac{5}{8}$ **42.** $\frac{8}{6}$

43. $\frac{3}{40}$ **44.** $\frac{52}{50}$ **45.** $\frac{75}{1,000}$

46. Complete the table below.

	Percent	Decimal	Fraction or Mixed Number
a)	15%	0.15	
b)	4%		$\frac{1}{25}$
c)		1.35	$1\frac{7}{20}$
d)	72.5%		
e)		0.4	$\frac{2}{5}$
f)	90%		
g)		2.5	$2\frac{1}{2}$

47. Under full load, an AC motor rotates at 85% of its rated speed. Write this percentage as a decimal and a fraction in lowest terms.

48. An electric motor is on sale for 18% off. Write this percentage as a decimal and a fraction in lowest terms.

49. Incandescent lightbulbs produce about 290% more heat than compact fluorescent bulbs. Write this percentage as a decimal and a fraction in lowest terms.

50. The taxes Luis paid this year were 108% of what he paid last year. Write this percentage as a decimal and a fraction in lowest terms.

SOLVING PERCENT PROBLEMS WITH PROPORTIONS

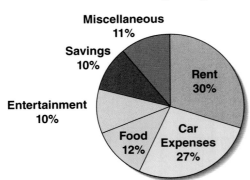

Monthly Budget

Miscellaneous 11%

Savings 10%

Entertainment 10%

Rent 30%

Food 12%

Car Expenses 27%

Juanita used a computer program to create a circle graph of her monthly budget. The graph, shown at the left, indicates the percent of her take-home pay budgeted for each type of expense. How can you use a proportion to find the amount budgeted for car expenses if her take-home pay is $2,100 per month?

Find 27% of $2,100.

STEP 1

Write the percent as a fraction.

$27\% = \frac{27}{100}$

STEP 2

Let n represent the amount budgeted for car expenses. Set up a proportion comparing *part* to *whole* in both ratios.

$$\text{Part} \longrightarrow \frac{27}{100} = \frac{n}{2,100} \longleftarrow \text{Part}$$
$$\text{Whole} \longrightarrow \qquad\qquad \longleftarrow \text{Whole}$$

STEP 3

Solve the proportion.

$\frac{27}{100} = \frac{n}{2,100}$ ← Write the proportion.

$100n = 27(2,100)$ ← Cross multiply.

$100n = 56,700$

$n = 567$ ← Solve for n.

The amount budgeted for car expenses is $567.

EXAMPLE A

Use a proportion to find 30% of 15.

STEP 1

Write the percent as a fraction.

$30\% = \frac{30}{100} = \frac{3}{10}$

STEP 2

Let x represent the unknown value. Set up a proportion comparing *part* to *whole* in both ratios.

$\frac{3}{10} = \frac{x}{15}$

STEP 3

Solve the proportion.

$\dfrac{3}{10} = \dfrac{x}{15}$ ← Write the proportion.

$10x = 3(15)$ ← Cross multiply.

$10x = 45$

$x = 4.5$ ← Solve for n.

So, 30% of 15 is 4.5.

A proportion can also be used to find the *whole* amount. This is shown in Example B.

EXAMPLE B

The sale price of an electric motor is $51. This is 85% of the regular, or presale, price. What was the regular price?

The problem can be restated as this: 51 is 85% of what number?

In this problem, 51 is the *part* and the *whole* is unknown. Let n represent the unknown number.

$\dfrac{85}{100} = \dfrac{51}{n}$ ← Set up a proportion.

$85n = 100(51)$ ← Cross multiply.

$85n = 5{,}100$

$n = 60$ ← Solve for n.

So, 51 is 85% of 60.

The regular price was $60.

EXAMPLE C

What percent of 200 is 125?

In this problem, 125 is the *part* and 200 is the *whole*. Let p represent the unknown percent.

$\dfrac{p}{100} = \dfrac{125}{200}$ ← Set up a proportion.

$200p = 100(125)$ ← Cross multiply.

$200p = 12{,}500$

$p = 62.5$ ← Solve for p.

125 is 62.5% of 200.

Sometimes a mixed number is used to express a percent. This is shown in Example D.

EXAMPLE D

What percent of 30 is 20?

In this problem, the *part* is 20 and the *whole* is 30. Let p represent the unknown percent.

$$\frac{p}{100} = \frac{20}{30}$$

$$30p = 100(20)$$

$$30p = 2,000$$

$$p = \frac{2,000}{30}$$

$$p = 66\frac{2}{3}$$

$$66\frac{20}{30} = 66\frac{2}{3}$$

$$\begin{array}{r} 30\overline{)\,2,000} \\ -\ 180 \\ \hline 200 \\ -\ 180 \\ \hline 20 \end{array}$$

20 is $66\frac{2}{3}\%$ of 30.

Sometimes it may be necessary to round an answer in a percent problem. This may be stated in the directions, or it may be dictated by the situation. These situations are shown in Examples E and F.

EXAMPLE E

Jorge bought some electrical supplies that were priced at $17.52. He had to pay a sales tax of 6.5%. How much sales tax did Jorge have to pay? Round your answer to the nearest cent.

The problem can be restated as this: What is 6.5% of 17.52?

Write 6.5% as $\frac{6.5}{100}$. Let n represent the unknown amount.

$$\frac{6.5}{100} = \frac{n}{17.52}$$

$$100n = 6.5(17.52)$$

$$100n = 113.88$$

$$n = 1.1388$$

$$n \approx 1.14$$

Rounded to the nearest cent, Jorge had to pay $1.14 in sales tax.

EXAMPLE F

On average, 75% of the cameras produced on a certain shift pass inspection. If 255 cameras are produced on the shift, about how many would you expect to pass inspection?

Let x represent the number of cameras you expect to pass inspection.

$$\frac{75}{100} = \frac{x}{255}$$

$$100x = 75(255)$$

$$100x = 19,125$$

$$x = 191.25$$

$$x \approx 191$$

You would expect about 191 cameras to pass inspection.

EXERCISES 12-2

Solve each of the following problems.

1. Find 38% of 65.
2. Find 98% of 50.
3. What percent of 48 is 42?
4. What percent of 90 is 3?
5. Find 4% of $29.00.
6. Find 125% of 360.
7. What percent of 20 is 0.1?
8. What percent of 96 is 12?
9. 20 is 4% of what number?
10. 18 is 150% of what number?
11. What is 140% of 45?
12. What is 0.2% of 30?
13. What percent of 18 is 6?
14. What percent of 26 is 26?

15. The regular price of an electric motor is $98.95. During a sale, the price was reduced by 18%.
    (a) To the nearest cent, how much was the price of the motor reduced?
    (b) What was the sale price?

16. Luis charged $425 for a wiring job. He figured that materials for the job were 55% of the total cost and the rest of the money was for his labor.
    (a) What was the cost, to the nearest cent, of the materials?
    (b) How much did Luis receive for his labor?

17. A printed circuit board has 53 resistors, 32 capacitors, 12 transistors, and 14 integrated circuits. Express each type of component as a percent of the total number of components. Round your answer to the nearest tenth percent.

18. The power delivered by a transformer is 4275 W. This is 95% of the power supplied. What was the power supplied to the transformer?

19. **What's the Error?** To find 2.5% of 18, a student performed the following calculation.
$$\frac{0.025}{100} = \frac{x}{18}$$
$$100x = 0.025(18)$$
$$100x = 0.45$$
$$x = 0.0045$$
Find the error in the calculation and give the solution.

Helpful Hint

Whenever a percent is given in a problem, write the number in front of the percent sign over 100 in a fraction. For example, write 4% as $\frac{4}{100}$; write 150% as $\frac{150}{100}$.

SOLVING PERCENT PROBLEMS WITH EQUATIONS

In the last lesson, you solved percent problems using proportions. In this lesson, you will learn to solve percent problems with equations.

Wes is planning to purchase a new car. The approximate cost of the car is $18,000. If the tax rate is 5% in his state, about how much will Wes pay in taxes for this car?

Find 5% of $18,000.

$n = 5\%$ of 18,000   ⟵   Let n represent the unknown amount; write an equation.

$n = 0.05 \times 18,000$ ⟵   Write the percent as a decimal and replace *of* with a multiplication symbol.

$n = 900$

For a car costing $18,000, Wes will pay $900 in taxes.

EXAMPLE A

What is 82% of 350?

$n = 82\%$ of 350 ← Let n represent the unknown amount; write an equation.

$n = 0.82 \times 350$ ← Write the percent as a decimal and replace *of* with a multiplication symbol.

$n = 287$

82% of 350 is 287.

When a percent is greater than 100%, the percent of a number is greater than the number. This is shown in Example B.

EXAMPLE B

Find 150% of 80.

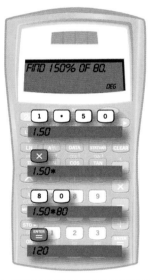

$n = 150\%$ of 80 ← Let n represent the unknown amount; write an equation.

$n = 1.50 \times 80$ ← Write the percent as a decimal and replace *of* with a multiplication symbol.

$n = 120$

150% of 80 is 120.

You can use the benchmarks of 100%, 10%, 1%, and 0.1% of a number to help judge the reasonableness of your answers to percent problems. The table below shows various percents of the number 235. Notice that the decimal point in the original number moves 1 place to the left for 10%, 2 places to the left for 1%, and 3 places to the left for 0.1%.

Percent	100%	10%	1%	0.1%
Percent as a decimal	1.00	0.1	0.01	0.001
Percent of 235	100% of 235 $1.00 \times 235 = \mathbf{235}$	10% of 235 $0.1 \times 235 = \mathbf{23.5}$	1% of 235 $0.01 \times 235 = \mathbf{2.35}$	0.1% of 235 $0.001 \times 235 = \mathbf{0.235}$

EXAMPLE C

Find 0.5% of 34.

$n = 0.5\%$ of 34   ← Let n represent the unknown amount; write an equation.

$n = 0.005 \times 34$ ← Write the percent as a decimal and replace *of* with a multiplication symbol.

$n = 0.17$

Use the benchmark of 1% and the fact that 0.5% is less than 1% to judge the reasonableness of your answer.

Compare your answer to 1% of 34:
1% of 34 $= 0.01 \times 34 = 0.34$.
Because 0.17 is less than 0.34, the answer is reasonable.

0.5% of 34 is 0.17.

When the percent is asked for in a problem, find the fractional part and then convert the fraction to a percent. This is shown in Examples D and E.

EXAMPLE D

What percent of 25 is 20?

What percent of 25 is 20?

What part of 25 is 20?   ← Restate the question, replacing *percent* with *part*.

$x \cdot 25 = 20$ ← Let x represent the unknown part; replace *of* with a multiplication symbol.

$25x = 20$

$\dfrac{25x}{25} = \dfrac{20}{25}$ ← Divide both sides by 25 to isolate the variable, x.

$x = \dfrac{20}{25}$

$x = 0.8$ ← Divide the numerator by the denominator.

$x = 80\%$ ← Move the decimal point 2 places to the right and include a percent sign.

20 is 80% of 25.

COMMON ERROR

ALERT

As we have seen before, a decimal percent such as 2.6% has the decimal number equivalent 0.026. Do not confuse 2.6% with the decimal number 2.6.

EXAMPLE E

1.5 is what percent of 40?

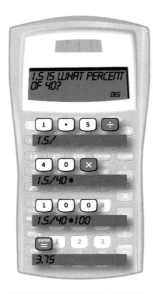

1.5 is what percent of 40?

1.5 is what part of 40?   ← Restate the question, replacing *percent* with *part*.

$1.5 = x \cdot 40$ ← Let x represent the unknown part; replace *of* with a multiplication symbol.

$1.5 = 40x$

$\dfrac{1.5}{40} = \dfrac{40x}{40}$ ← Divide both sides by 40 to isolate the variable, x.

$\dfrac{1.5}{40} = x$

$0.0375 = x$ ← Divide 1.5 by 40.

$3.75\% = x$ ← Move the decimal point 2 places to the right and include a percent sign.

1.5 is 3.75% of 40.

Sometimes you need to find a number when a percent of it is known. This is shown in Examples F and G.

EXAMPLE F

You received a 30% discount on a bill. If the discount was $6, how much was the original bill?

Helpful Hint

Multiply the given percent, 30%, by your solution for the whole amount to check your work.

$6 \stackrel{?}{=} 0.30 \times 20$

$6 = 6 \checkmark$

The problem can be restated as this: 6 is 30% of what number?

6 is 30% of what number?

$6 = 30\%$ of n ← Let n represent the unknown number; write an equation.

$6 = 0.30 \cdot n$ ← Write the percent as a decimal and replace *of* with a multiplication symbol.

$6 = 0.30n$

$\dfrac{6}{0.30} = \dfrac{0.30n}{0.30}$ ← Divide both sides by 0.30 to isolate the variable, n.

$\dfrac{6}{0.30} = n$

$20 = n$ ← Divide 6 by 0.30.

6 is 30% of 20.

The original bill was $20.

EXAMPLE G

12 is 2.25% of what number? Round your answer to the nearest hundredth.

12 is 2.25% of what number?

$12 = 2.25\%$ of x ← Let x represent the unknown number; write an equation.

$12 = 0.0225 \cdot x$ ← Write the percent as a decimal and replace *of* with a multiplication symbol.

$12 = 0.0225x$

$\dfrac{12}{0.0225} = \dfrac{0.0225x}{0.0225}$ ← Divide both sides by 0.0225 to isolate the variable, x.

$\dfrac{12}{0.0225} = x$

$533.33\overline{3} = x$ ← Use a calculator to divide 12 by 0.0225.

$533.33 \approx x$ ← Round to the nearest hundredth.

To the nearest hundredth, 12 is 2.25% of 533.33.

EXERCISES 12-3

Solve each of the following.

1. Find 8% of 80.
2. Find 2% of 3,250.
3. What percent of 210 is 31.5?
4. What percent of 5 is 0.6?
5. 10% of what number is 34?
6. 5% of what number is 2?
7. What is 75% of 200?
8. What is 350% of 25?
9. What percent of 95 is 38?
10. What percent of 1,000 is 1?
11. 750 is 15% of what number?
12. 4.72 is 59% of what number?
13. The voltage drop in a line supplied by a 220 V generator is 2.5% of the generator voltage.

    (a) Determine the voltage drop.

    (b) Determine the voltage supplied to the load.

14. A GCFI circuit breaker is rated at 20 A. If it takes 21.25 A to trip the breaker, what percentage of its rating is this?

15. One week, Luis earned $615. He figured that he was going to have $14.90 plus 15% withheld in federal income tax for everything he earned that was over $200. How much was withheld for his federal income tax?

16. **Challenge:** What is 25% of 75% of 100?

SOLVING WORD PROBLEMS

Word problems involving percents can be solved using proportions or equations. The problem given below will be solved with a proportion and an equation.

A job calls for installing 270 lighting fixtures. To date, 81 have been installed. What percent of the job is complete?

What percent of 270 is 81?

Use a proportion.	Use an equation.
In this problem, 81 is the *part* and 270 is the *whole*. Let p represent the percent.	Let x represent the unknown part.
	What percent of 270 is 81?
	What *part* of 270 is 81?

Use a proportion.

In this problem, 81 is the *part* and 270 is the *whole*. Let p represent the percent.

$$\frac{p}{100} = \frac{81}{270}$$

$$270p = 100(81)$$

$$270p = 8,100$$

$$p = \frac{8,100}{270}$$

$$p = 30$$

30% of the job is complete.

Use an equation.

Let x represent the unknown part.

What percent of 270 is 81?

What *part* of 270 is 81?

$$x \cdot 270 = 81$$

$$270x = 81$$

$$x = \frac{81}{270}$$

$$x = 0.30$$

$$x = 30\%$$

30% of the job is complete.

Examples A and B will be solved using proportions.

EXAMPLE A

In a survey, 52% of the people said they had purchased a high definition television (HDTV) in the past year. If 39 people said they had purchased an HDTV, how many people were surveyed?

39 is 52% of what number?

In this problem, 39 is the *part* of the group surveyed that had purchased HDTVs. The number of people in the *whole* group is unknown.

Let n represent the number of people in the whole group.

$$\frac{52}{100} = \frac{39}{n}$$ ← Set up a proportion.

$$52n = 100(39)$$ ← Cross multiply.

$$52n = 3,900$$

$$n = 75$$ ← Solve for n.

There were 75 people surveyed.

EXAMPLE B

The voltage drop on a feeder is 3% of the 240 V source voltage. Determine the voltage drop on this feeder.

To determine the voltage drop, we need to find 3% of 240.

Write the percent as a fraction. Let x represent the voltage drop in the feeder.

$\dfrac{3}{100} = \dfrac{x}{240}$ ← Set up a proportion.

$100x = 3(240)$ ← Cross multiply.

$100x = 720$

$x = 7.2$ ← Solve for x.

The voltage drop on this feeder is 7.2 V.

The rest of the examples will be solved using equations.

EXAMPLE C

Dave paid $4.47 in sales tax on a taxable item costing $74.50. What was the sales tax rate?

What percent of $74.50 is $4.47?

What *part* of 74.50 is 4.47?   ← Restate the question, replacing *percent* with *part*.

$x \cdot 74.50 = 4.47$ ← Let x represent the part (sales tax rate); replace *of* with a multiplication symbol.

$74.50x = 4.47$

$\dfrac{74.50x}{74.50} = \dfrac{4.47}{74.50}$ ← Divide both sides by 74.50 to isolate the variable, x.

$x = \dfrac{4.47}{74.50}$

$x = 0.06$ ← Divide the numerator by the denominator.

$x = 6\%$ ← Move the decimal point 2 places to the right and include a percent sign.

The sales tax rate is 6%.

EXAMPLE D

A salesman at a certain dealership predicts that 65% of the customers who test drive a car this month will buy a car this month. If he is correct, about how many of the 182 customers will buy a car?

Find 65% of 182.

$n = 65\%$ of 182   ← Let n represent the unknown amount; write an equation.

$n = 0.65 \times 182$ ← Write the percent as a decimal and replace *of* with a multiplication symbol.

$n = 118.3$ ← Multiply.

$n \approx 118$

About 118 customers will buy a car this month.

EXAMPLE E

Roy purchased 16 pounds of burgers for a picnic and used 14.5 pounds of them. What percent of the burgers were used? Round your answer to the nearest whole percent.

What percent of 16 is 14.5?

What *part* of 16 is 14.5?   ← Restate the question, replacing *percent* with *part*.

$x \cdot 16 = 14.5$ ← Let x represent the unknown part; replace *of* with a multiplication symbol.

$16x = 14.5$

$\dfrac{16x}{16} = \dfrac{14.5}{16}$ ← Divide both sides by 16 to isolate the variable, x.

$x = 0.90625$ ← Use a calculator to divide 14.5 by 16.

$x = 90.625\%$ ← Convert the decimal to a percent.

$x \approx 91\%$ ← Round to the nearest whole percent.

About 91% of the burgers were used.

EXAMPLE F

An estimate for a home improvement job includes $1,440 for materials. If the amount for materials is 40% of the total estimate, what is the total estimate for the job?

1,440 is 40% of what number?

$1,440 = 40\%$ of n ← Let n represent the unknown number; write an equation.

$1,440 = 0.40 \cdot n$ ← Write the percent as a decimal and replace *of* with a multiplication symbol.

$1,440 = 0.40n$

$\dfrac{1,440}{0.40} = \dfrac{0.40n}{0.40}$ ← Divide both sides by 0.40 to isolate the variable, n.

$\dfrac{1,440}{0.40} = n$

$3,600 = n$ ← Divide 1,440 by 0.40.

The total estimate for the job is $3,600.

EXERCISES 12-4

Solve each of the following problems.

1. In a small town, about 25% of the families own a generator. If there are 6,245 families in the town, about how many families own a generator?

2. A motor is rated at 125 hp. It is actually developing 8% more horse-power than its rating.

   (a) How much horsepower over its rating is the motor developing?

   (b) What horsepower is the motor developing?

Helpful Hint

Solve a problem using one method and then check your work using the other method. For example, use an equation to solve Exercise 2 and then check your work using a proportion.

$n = 0.08(125)$

$n = 10$

Check:

$\dfrac{8}{100} \stackrel{?}{=} \dfrac{10}{125}$;

$1,000 = 1,000$ ✓

3. Seventy percent of the shingles on a roofing job are already installed. If 2,100 shingles are already installed, how many shingles will be on this roof altogether?

4. Ricardo has $24.05 deducted from his gross pay each week for health insurance. If his gross pay is $740, what percent is deducted for health insurance?

5. A 2,200 ohm resistor accounts for 40% of the total resistance in a circuit. What is the total resistance in this particular circuit?

6. Alexa earned $16.90 an hour. She received a 4.5% increase. What was her new hourly pay?

7. Claire was making $15.25 an hour. She received a 55¢/hour raise. What was the percent increase?

8. A circuit breaker carried 16.2 A. If this is 108% of the circuit breaker's rating, what is its rating?

9. The table below shows the sizes and the quantities of shirts purchased for a company picnic.

Small	Medium	Large	X-Large
10	15	25	35

To the nearest whole percent, what percent of the shirts purchased were size large? Explain how you found your answer.

10. During a sale, Molly will be able to buy a television for 80% of its regular cost, which is $239.99. How much will the television cost during the sale?

11. The director of sales for a company said that she expects the dollar amount of sales for the current year to be 115% of the prior year's sales. If sales in the prior year were $480,000, what does she expect sales to be in the current year?

12. In the past two weeks, Juan, who is married, earned $1,638. His federal tax rate is $55 + 15% of his earnings over $858. He must also pay 6.20% for Social Security and 1.45% for Medicare.

(a) How much was withheld from his paycheck for Social Security?

(b) How much was withheld from his paycheck for Medicare?

(c) How much was withheld from his paycheck for federal income tax?

13. Challenge: The length of the track at the Bristol Motor Speedway is 0.533 mile. If a driver has completed 25% of the required 500 laps, how much farther does he have to go? Round your answer to the nearest tenth of a mile.

CHAPTER 12 REVIEW EXERCISES

1. Convert each of the following percents to a fraction or mixed number in simplest form.

   (a) 15%

   (b) 5.25%

   (c) 112.5%

2. Rewrite each of the following percents as a decimal.

   (a) 37.5%

   (b) $7\frac{1}{2}$%

   (c) 253%

3. Express each of the following decimals as a percent.

   (a) 0.25

   (b) 1.5

   (c) 0.0063

4. Express each of the following decimals as a fraction or a mixed number.

   (a) 0.72

   (b) 1.22

   (c) 0.072

5. Convert each of the following fractions to percents.

   (a) $\frac{53}{100}$

   (b) $\frac{21}{1,000}$

   (c) $\frac{5}{4}$

6. Solve each of the following problems.

   (a) Find 12% of 85.

   (b) What percent is 45 of 36?

   (c) 84 is 12% of what number?

7. A 30-ampere circuit breaker carries a temporary 8% overload. How many amps of current flow through the circuit breaker during the overload?

8. An electronics technician tests a resistor identified as 130 Ω. The resistance is actually 128 Ω. What percent of the identified resistance is the actual resistance? Round the answer to the nearest tenth percent.

9. Carlos earns $16.75 an hour. He is given a $4\frac{1}{2}$% raise. What is his new hourly wage to the nearest cent?

10. An office remodeling takes an electrician $5\frac{3}{4}$ days to complete. The contractor estimated that it would take only $4\frac{1}{2}$ days to finish the job. What percent of the estimated time is the actual time? Round the answer to the nearest tenth percent.

Building a Foundation in Mathematics

- **Plotting Points on a Coordinate Plane**

- **Graphing Linear Equations Using a Table of Values**

- **Graphing Linear Equations Using Intercepts**

- **Finding the Slope of a Line**

- **Graphing Linear Equations Using Slope-Intercept Form**

- **Writing Equations of Lines**

Overview

A graph is a picture that shows the relationship between sets of quantities. Newspapers, magazines, books, manuals, and television news often contain graphs. Graphs are used throughout the electrical industry to illustrate relationships and ideas. Because they are used in electricity and in everyday living, it is important to be able to interpret and construct basic types of graphs. It is also important to be able to develop a formula from a graph so that you can predict what will happen at points not displayed on the graph.

The Cartesian Plane

Objectives

After completing this chapter, you will be able to:

- Plot points on a rectangular, or Cartesian, coordinate system
- Graph an equation with two variables
- Determine the intercepts of a graph or a linear equation
- Give the meaning of the slope of a line
- Determine the slope of a line from its graph
- Determine the slope of a linear equation
- Determine a linear equation given the slope and y-intercept
- Determine a linear equation given the slope and one point on the line
- Determine a linear equation given two points on the line

PLOTTING POINTS ON A COORDINATE PLANE

Upper East Side, New York City

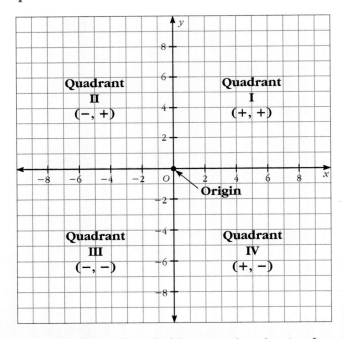

Many of the streets in New York City around Central Park are laid out on a grid. The streets intersect at right angles. Suppose you are standing at the corner of 86th Street and 3rd Avenue, and someone asks for directions to the corner of 96th Street and 1st Avenue. Using the grid shown at the left, you can direct them to go 2 blocks right and 10 blocks up. Compare the grid of the city streets to the coordinate plane below.

A **coordinate plane** is formed by two number lines intersecting at right angles. The horizontal number line is the *x*-**axis**; the vertical number line is the *y*-**axis.** The point of intersection of the *x*- and *y*-axes is called the **origin.** The *x*- and *y*-axes divide the plane into four sections called **quadrants.**

Every point in the plane can be identified by an **ordered pair** of coordinates, **(*x, y*).** All the points located in Quadrant I have positive *x*- and *y*-coordinates. This is indicated by (+, +) in Quadrant I on the coordinate plane shown above. The signs of the *x*- and *y*-coordinates of points in other quadrants are indicated by the + and − signs in the quadrants. To plot a point, begin at the origin and use the coordinates. First use the *x*-coordinate and then the *y*-coordinate.

For the *x*-coordinate: move left for negative numbers and right for positive numbers.

For the *y*-coordinate: move down for negative numbers and up for positive numbers.

Plot the point (−5, 7) in the coordinate plane.

Begin at the origin. Move 5 units to the left on the *x*-axis. From there, move up 7 units and mark the point.

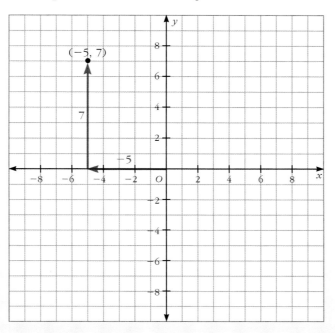

EXAMPLE A

Plot (2, 4), (−4, 3), (−3, −3), and (4, −2) on separate coordinate planes.

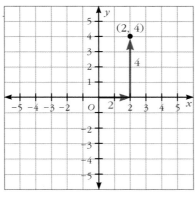

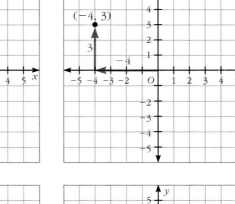

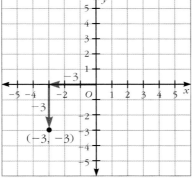

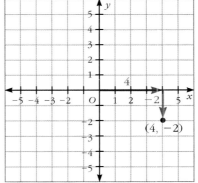

As you can see from this example, when the
  x-coordinate is positive, move right;
  x-coordinate is negative, move left;
  y-coordinate is positive, move up;
  y-coordinate is negative, move down.

The points (2, 4), (−4, 3), (−3, −3), and (4, −2) are shown on separate coordinate planes.

Some points are not located in a quadrant; rather, they are located on one of the axes. Any point with zero as a coordinate is on an axis. This is shown in Example B.

EXAMPLE B

**Plot the points
(4, 0), (0, 2),
(−4, 0), and
(0, −2) on the
coordinate plane.**

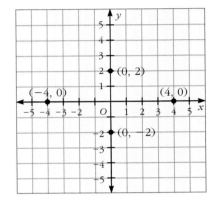

When the y-coordinate is 0, the point is on the x-axis.
When the x-coordinate is 0, the point is on the y-axis.

The points (4, 0), (0, 2), (−4, 0), and (0, −2) are shown on the graph above.

Capital letters are frequently used to name points. This will be done in the examples that follow.

EXAMPLE C

**Give the coordinates
of points C, D, E,
and F on the
coordinate plane
at the right. Name
the two points that
have the same
x-coordinate.**

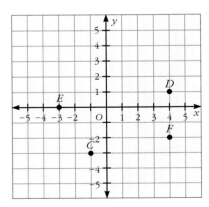

The coordinates are: $C(-1, -3)$, $D(4, 1)$, $E(-3, 0)$, and $F(4, -2)$. Points D and F have the same x-coordinate, which is 4.

EXAMPLE D

Graph the ordered pairs $G(10, 20)$, $H(30, 40)$, $I(25, 20)$, and $J(30, 35)$. Name the two points that have the same y-coordinate.

The coordinates are all positive, so only Quadrant I is needed.

The axes are numbered with multiples of 10 for convenience.

Use the tick marks for numbers such as 25 and 35.

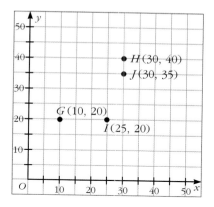

Points G and I have the same y-coordinate, which is 20.

EXAMPLE E

Give the coordinates of the points shown on the graph.

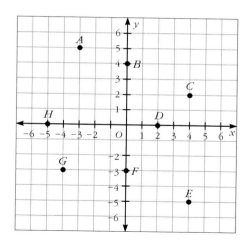

The coordinates of the points are:

$A(-3, 5)$ $B(0, 4)$ $C(4, 2)$ $D(2, 0)$

$E(4, -5)$ $F(0, -3)$ $G(-4, -3)$ $H(-5, 0)$

EXAMPLE F

The current in a 20 Ω circuit carries the following resistances.

Notice that the variables are not x and y but I and R. We will use the horizontal axis for the first row of the table, I, and the vertical axis for the R-values. This gives the following ordered pairs: $A(2, 10)$, $B(4, 5)$, $C(5, 4)$, $D(8, 2.5)$, and $E(10, 2)$. These points are plotted below.

Current, I	2	4	5	8	10
Resistance, R	10	5	4	2.5	2

Plot the points in the table.

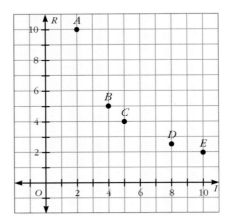

EXERCISES 13-1

For Exercises 1–6, give the coordinates of each point and name the quadrant (if any) in which it is located.

1. K
2. L
3. M
4. N
5. P
6. Q

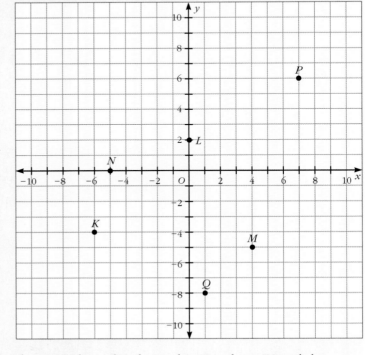

7. Explain why the point $(2, -5)$ is located in Quadrant IV and the point $(-5, 2)$ is located in Quadrant II.

8. True or False? An ordered pair (x, y) can correspond to more than one point in the coordinate plane. Explain your answer.

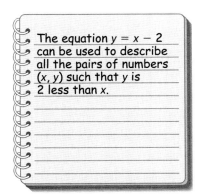

9. Use the coordinate plane at the left to name the point(s) described.
   (a) The points are on the negative y-axis.
   (b) The x- and y-coordinates are the same.
   (c) The x-coordinate is -2.
   (d) The y-coordinate is 0.

10. The current used by a 2 Ω resistor produces the following powers:

Current, I	0	2	4	6	8	10
Power, P	0	2.8	4	4.9	5.7	6.3

Plot the points in the table with I on the horizontal axis.

11. **What's the Error?** A student was asked to plot the points $A(0, 2)$ and $B(-4, 0)$. The results are shown on the grid below.

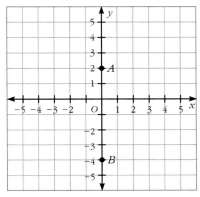

Explain the error that was made and how to correct it.

GRAPHING LINEAR EQUATIONS USING A TABLE OF VALUES

One way to graph an equation is to create a table of values and then plot the ordered pairs from the table on a coordinate plane.

Graph the equation $y = x - 2$ using a table of values.

> The equation $y = x - 2$ can be used to describe all the pairs of numbers (x, y) such that y is 2 less than x.

STEP 1

Set up a table as shown at the right.

Any values may be chosen for the x-variable.

x	$y = x - 2$
-2	
-1	
0	
1	
2	

STEP 2

For each x-value in the table, find the corresponding y-value.

Write the ordered pairs.

x	$y = x - 2$	(x, y)
-2	$-2 - 2 = -4$	$(-2, -4)$
-1	$-1 - 2 = -3$	$(-1, -3)$
0	$0 - 2 = -2$	$(0, -2)$
1	$1 - 2 = -1$	$(1, -1)$
2	$2 - 2 = 0$	$(2, 0)$

STEP 3

Plot the points.

Connect the points to form a line.

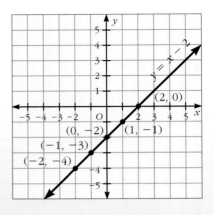

When a set of points lies in a straight line on a coordinate plane, we say that the relationship between the variables is a linear relationship. In the example above, the equation $y = x - 2$ is a **linear equation** because its graph is a line.

EXAMPLE A

Graph the equation $y = x + 3$ using a table of values.

x	$y = x + 3$	(x, y)
-5	$-5 + 3 = -2$	$(-5, -2)$
-3	$-3 + 3 = 0$	$(-3, 0)$
-1	$-1 + 3 = 2$	$(-1, 2)$
1	$1 + 3 = 4$	$(1, 4)$
3	$3 + 3 = 6$	$(3, 6)$

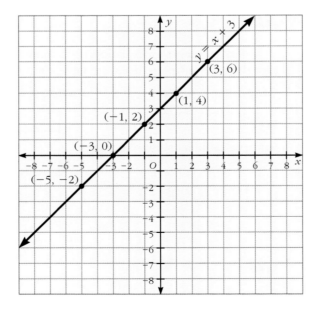

The equation $y = x + 3$ is graphed on the coordinate plane above using the ordered pairs in the table.

For the equations shown so far, five points were plotted. In reality, only two points are needed to graph a line. It is a good idea to plot a third point just to be sure that our calculations are correct. This will be done in the examples that follow. Notice that 0 is often used as one of the x-values because calculations with 0 are usually easy.

EXAMPLE B

Graph the equation $y = -\frac{1}{2}x$ using a table of values.

x	$y = -\frac{1}{2}x$	(x, y)
-2	$-\frac{1}{2}(-2) = 1$	$(-2, 1)$
0	$-\frac{1}{2}(0) = 0$	$(0, 0)$
2	$-\frac{1}{2}(2) = -1$	$(2, -1)$

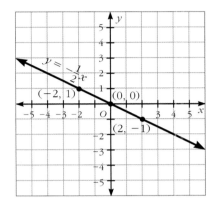

Notice that multiples of 2 were used as x-values to make it easy to multiply by $-\frac{1}{2}$.

The equation $y = -\frac{1}{2}x$ is graphed on the coordinate plane above using the ordered pairs in the table.

EXAMPLE C

Graph the equation $y = 2x - 4$ using a table of values.

x	$y = 2x - 4$	(x, y)
0	$2(0) - 4 = -4$	$(0, -4)$
2	$2(2) - 4 = 0$	$(2, 0)$
4	$2(4) - 4 = 4$	$(4, 4)$

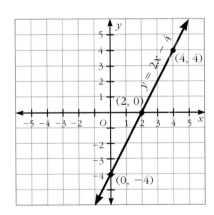

The equation $y = 2x - 4$ is graphed on the coordinate plane above using the ordered pairs in the table.

EXAMPLE D

Graph the equation 2x + 3y = 12 using a table of values.

When an equation is not solved for y, it is best to do this before setting up a table of values. For example, to graph the equation $2x + 3y = 12$, use algebra to solve the equation for y as shown.

$$2x + 3y = 12 \qquad \longleftarrow \text{Write the original equation.}$$

$$2x + 3y - 2x = 12 - 2x \qquad \longleftarrow \text{Subtract } 2x \text{ from both sides.}$$

$$3y = -2x + 12 \qquad \longleftarrow \text{Simplify and rearrange terms on the right side.}$$

$$\frac{3y}{3} = \frac{-2x}{3} + \frac{12}{3} \qquad \longleftarrow \text{Divide each term by 3.}$$

$$y = -\frac{2}{3}x + 4 \qquad \longleftarrow \text{Simplify.}$$

This form of the equation, $y = -\frac{2}{3}x + 4$, is used in the table below.

Select convenient values for x to make it easy to multiply by $-\frac{2}{3}$. That is, $-\frac{2}{3}(3) = -2$ and $-\frac{2}{3}(6) = -4$.

x	$y = -\frac{2}{3}\, x + 4$	(x, y)
0	$-\frac{2}{3}(0) + 4 = 4$	$(0, 4)$
3	$-\frac{2}{3}(3) + 4 = 2$	$(3, 2)$
6	$-\frac{2}{3}(6) + 4 = 0$	$(6, 0)$

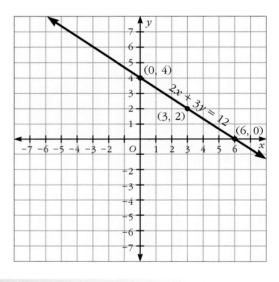

The equation $2x + 3y = 12$ is graphed on the coordinate grid above using the ordered pairs in the table.

KEY IDEA

Only two points are needed to graph a linear equation. However, we often use at least three points to reduce graphing errors.

An equation may represent a situation in which only positive values are meaningful. When it does, sometimes only Quadrant I of the coordinate plane is shown. This is the case in Example E.

EXAMPLE E

Every package of lightbulbs contains 4 lightbulbs. Write an equation that shows this relationship. Then graph the equation.

If x represents the number of packages and y represents the number of lightbulbs, the equation $y = 4x$ represents the situation.

x	$y = 4x$	(x, y)
1	$4(1) = 4$	$(1, 4)$
2	$4(2) = 8$	$(2, 8)$
3	$4(3) = 12$	$(3, 12)$

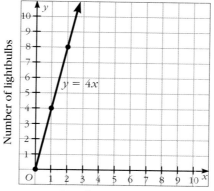

The equation showing the relationship between number of packages and number of lightbulbs is graphed on the coordinate plane above.

EXERCISES 13-2

Answer each of the following.

1. Complete the table of values below. Then graph the equation. Label the coordinates of each point.

Helpful Hint

For accuracy, as well as neatness, always use graph paper and a straightedge to graph a line.

x	$y = 2x - 4$	(x, y)
-2	$2(-2) - 4 = -8$	$(-2, -8)$
-1		
0		
1		
2		
3		

Graph each of the following linear equations by first creating a table of values. For some exercises, x-values are suggested. For other exercises, choose convenient x-values.

2. $y = x + 5$; use the x-values -1, 0, and 1 in your table.

3. $y = x - 1$

4. $y = \frac{1}{3}x$; use the x-values -6, 0, and 6 in your table.

5. $y = \frac{1}{3}x + 2$

6. $y = \frac{3}{4}x - 1$; use the x-values -4, 0, and 4 in your table.

7. $y = -\frac{3}{4}x$

8. $y = -2x + 5$

9. $x - y = 8$

10. $y + 2 = \frac{1}{2}x$

11. $3x + y = -5$

12. $3x + 2y = 6$

13. The current through a 0.5 Ω resistor is given by $I = \frac{V}{0.5} = 2\ V$, where V is the voltage in volts and I is in amps.

    (a) Create a table of I-values when V is 0, 2, 4, 6, 8, and 10.

    (b) Graph this linear equation.

14. The inductive reactance of a 0.1 H coil is given by $X_L = 0.628f$, where f is the frequency in hertz and X_L is in ohms.

    (a) Create a table of values for X_L with f-values of 0, 2, 4, 6, 8, and 10.

    (b) Graph this linear equation.

15. An electrical supply store needs to order some CFL lightbulbs. The cost, C, in dollars, of the bulbs is $C = 6.50n + 8.76$, where n is the number of bulbs purchased and 8.76 is the postage needed.

    (a) Create a table of values for C with n-values of 1, 2, 5, and 8.

    (b) Graph this linear equation.

16. To play a game of "Penny Poker," Bill exchanged some nickels for pennies. Make a table that shows the relationship between x, the possible number of nickels exchanged, and y, the number of pennies obtained. Then graph the relationship.

17. A friend asks you to do the following:
    (a) Choose any number from 0 to 10.
    (b) Divide the number by -2.
    (c) Add 8 to the result.

    Draw a graph to represent the relationship between the number selected and the result. Let x represent the number selected and let y represent the result.

18. **Challenge:** Write an equation to describe the linear relationship between the variables x and y shown in the table.

x	0	1	10	100	1,000
y	0	0.1	1	10	100

GRAPHING LINEAR EQUATIONS USING INTERCEPTS

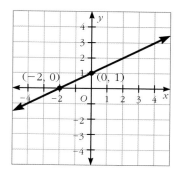

It is often useful to know the coordinates of the point where a line crosses an axis. The x-coordinate of the point where a line crosses the x-axis is called the **x-intercept.** The y-coordinate of the point where a line crosses the y-axis is called the **y-intercept.** What are the x- and y-intercepts of the line graphed at the left?

Identify the x-intercept and the y-intercept on the graph at the left.

x-intercept	y-intercept
The line crosses the x-axis at the point $(-2, 0)$.	The line crosses the y-axis at the point $(0, 1)$.
The x-intercept is -2.	The y-intercept is 1.

The x- and y-intercepts of a line may be used to graph that line. This is shown in Example A.

EXAMPLE A

Graph the line whose x-intercept is 3 and whose y-intercept is -4. Name the points where the line crosses the axes.

The x-intercept is 3, so plot the point $(3, 0)$.

The y-intercept is -4, so plot the point $(0, -4)$.

Draw a line through the two points.

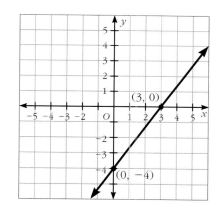

The line crosses the axes at $(3, 0)$ and $(0, -4)$.

EXAMPLE B

Name the x-intercept and the y-intercept of the graph shown.

The line crosses the x-axis at $(4, 0)$, so the x-intercept is 4.

The line crosses the y-axis at $(0, 2)$, so the y-intercept is 2.

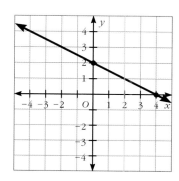

The x-intercept is 4, and the y-intercept is 2.

When the equation of a line is given, you can use the equation to find the intercepts. Use 0 as the y-value in the equation to find the x-intercept. Use 0 as the x-value in the equation to find the y-intercept. This method is shown in Example C.

EXAMPLE C

Graph the equation $y = -2x - 2$. Use the intercepts to graph the equation. Name the points where the line crosses the axes.

STEP 1

Find the x-intercept.
Let $y = 0$ in the equation and solve for x.

$$y = -2x - 2$$
$$0 = -2x - 2$$
$$0 + 2 = -2x - 2 + 2$$
$$2 = -2x$$
$$\frac{2}{-2} = \frac{-2x}{-2}$$
$$-1 = x$$

The x-intercept is -1.

STEP 2

Find the y-intercept.
Let $x = 0$ in the equation and solve for y.

$$y = -2x - 2$$
$$y = -2(0) - 2$$
$$y = 0 - 2$$
$$y = -2$$

The y-intercept is -2.

STEP 3

Use the intercepts to graph the equation.

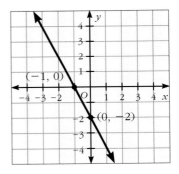

Language Box

An ordered pair contains an x- and a y-value. However, an intercept is most often considered a single number. That is the meaning of intercept in this book.

The line crosses the axes at $(-1, 0)$ and $(0, -2)$.

When the x-intercept and the y-intercept are both zero, the line passes through $(0, 0)$, the origin. In this case, you must find another point on the line in order to graph it. This is shown in Example D.

EXAMPLE D

Graph the equation $y = -5x$. Name the point where the line crosses the axes.

STEP 1

Find the x-intercept.
Let $y = 0$ in the equation and solve for x.

$$y = -5x$$
$$0 = -5x$$
$$0 = x$$

The x-intercept is 0.

STEP 2

Find the y-intercept.
Let $x = 0$ in the equation and solve for y.

$$y = -5x$$
$$y = -5(0)$$
$$y = 0$$

The y-intercept is 0.

STEP 3

Both intercepts are 0, so find another point on the line.

Let $x = 1$ in the equation.

$y = -5x$

$y = -5(1)$

$y = -5$

The point $(1, -5)$ lies on the line.

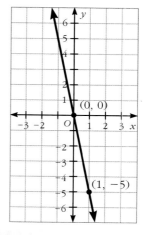

The line crosses both axes at $(0, 0)$, the origin.

EXERCISES 13-3

Answer each of the following.

1. Name the x-intercept and the y-intercept of the line shown.

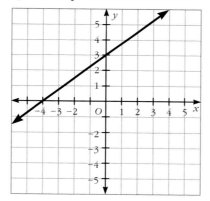

2. Name the points where the line shown crosses the axes.

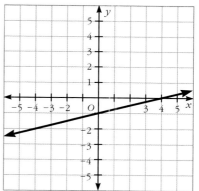

3. A line has an x-intercept of 2 and a y-intercept of -5. Graph the line.

4. A line passes through the points $(-1, 0)$ and $(0, 4)$. Graph the line.

5. Use the intercepts to graph the equation $y = -3x + 6$. Name the points where the line crosses the axes.

6. Graph the equation $y = \frac{1}{4}x$.

7. The equation of a certain line is $y = 3x + 9$. Explain why substituting $y = 0$ into the equation gives the x-intercept. Name the x-intercept of the line.

8. An electrical supply store needs to order some CFL lightbulbs. The cost, C, in dollars, of the bulbs is $C = 6.50n + 8.76$, where n is the number of bulbs purchased and 8.76 is the postage needed.

    (a) What is the y-intercept?

    (b) Does this y-intercept make sense in this situation? Explain your answer.

9. **Challenge:** The y-intercept of the line $y = mx + b$ is b. What is the x-intercept of the line $y = mx + b$?

FINDING THE SLOPE OF A LINE

From Graphs

A boat ramp must be constructed according to certain guidelines. If the ramp is too steep, boaters will damage their trailers and boats when they launch their boats. In general, slope refers to the steepness of an incline. In mathematics, the **slope** of a line is defined as the ratio of the vertical change to the horizontal change.

The graph below shows the dimensions for the construction of a safe boat ramp. What is the slope of the line representing the ramp?

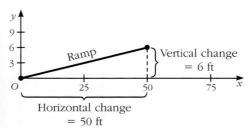

Use the formula below to calculate the slope of the boat ramp.

$$\text{Slope} = \frac{\text{vertical change}}{\text{horizontal change}}$$

Substitute the values for vertical change and horizontal change into the formula. Simplify the fraction.

$$\text{Slope} = \frac{\text{vertical change}}{\text{horizontal change}} = \frac{6}{50} = \frac{3}{25}$$

The slope of the boat ramp is $\frac{3}{25}$.

Sometimes the slope of a line is described as the ratio $\frac{rise}{run}$. This is another way of describing the ratio of vertical change (rise) to horizontal change (run). These terms are used in Examples A and B.

EXAMPLE A

Find the slope of the line at the right.

Use the points $(-3, -1)$ and $(1, 4)$ to find the rise and the run.

$$\text{Slope} = \frac{\text{vertical change}}{\text{horizontal change}} = \frac{\text{rise}}{\text{run}} = \frac{5}{4}$$

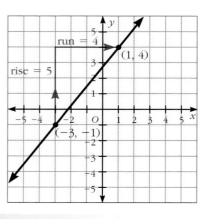

The slope of the line is $\frac{5}{4}$.

Using two given points on a line, there are two ways to calculate the slope, depending on which point is selected as the starting point. For the given points on the line in Example B, the slope is calculated in two ways.

EXAMPLE B

Use two ways to find the slope of the line at the right.

In the top graph, the *run* is a negative number because the horizontal movement is to the left.

$$\text{Slope} = \frac{\text{rise}}{\text{run}} = \frac{3}{-4} = -\frac{3}{4}$$

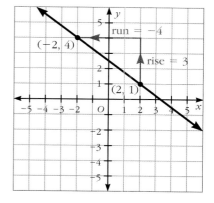

In the bottom graph, the *rise* is a negative number because the vertical movement is down.

$$\text{Slope} = \frac{\text{rise}}{\text{run}} = \frac{-3}{4} = -\frac{3}{4}$$

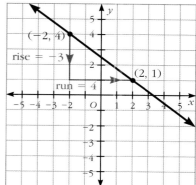

The slope of the line is $-\frac{3}{4}$

You can use any two points on a line to find the slope of the line. The slope will be the same no matter which two points you select. This is shown in Example C, using two pairs of points on the same line.

EXAMPLE C

Calculate the slope of the line shown using different pairs of points.

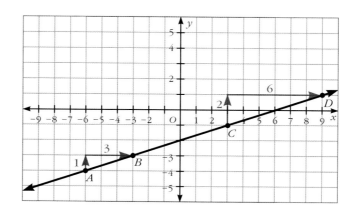

Calculate the slope using points *A* and *B*.

From *A*, count up 1 unit; that is the change in *y*. Then count 3 units to the right; that is the change in *x*.

$$\text{Slope} = \frac{\text{change in } y}{\text{change in } x} = \frac{1}{3}$$

The slope of the line is $\frac{1}{3}$.

Calculate the slope using points *C* and *D*.

From *C*, count up 2 units; that is the change in *y*. Then count 6 units to the right; that is the change in *x*.

$$\text{Slope} = \frac{\text{change in } y}{\text{change in } x} = \frac{2}{6} = \frac{1}{3}$$

From Points

As skiers improve, they are able to ski on steeper slopes.

You do not always have to rely on a graph to calculate the slope of a line. When two points on a nonvertical line are known, a formula may be used to calculate the slope of the line that contains those two points. The letter *m* is often used to represent the slope of a line. The formula for slope *m* through two points (x_1, y_1) and (x_2, y_2) is given below.

$$m = \frac{\text{change in } y}{\text{change in } x} = \frac{y_2 - y_1}{x_2 - x_1}$$

When two points are given, either point may be thought of as (x_1, y_1) in the slope formula. The result will be the same, although the work will look different.

What is the slope of the line that contains the points (1, 5) and (3, 9)?

You may think of (1, 5) as (x_1, y_1) and (3, 9) as (x_2, y_2). Substitute the numbers into the formula and simplify the fraction.

$$m = \frac{\text{change in } y}{\text{change in } x} = \frac{y_2 - y_1}{x_2 - x_1} = \frac{9 - 5}{3 - 1} = \frac{4}{2} = 2$$

The slope of the line is 2.

EXAMPLE D

Use the slope formula to find the slope of the line containing the points (2, −2) and (−5, 1).

Let (2, −2) be (x_1, y_1); let (−5, 1) be (x_2, y_2). Substitute the numbers into the formula.

$$m = \frac{\text{change in } y}{\text{change in } x} = \frac{y_2 - y_1}{x_2 - x_1} = \frac{1 - (-2)}{-5 - 2} = \frac{3}{-7} = -\frac{3}{7}$$

The slope of the line is $-\frac{3}{7}$.

There are two special cases to consider with respect to slope of a line. One is the horizontal line, and the other is the vertical line. The slope of any horizontal line is zero, and the slope of any vertical line is undefined. The reasons for this are explained in Example E.

EXAMPLE E

Find the slope of the horizontal line and the slope of the vertical line shown on the graph.

Find the slope of horizontal line a through $(0, 3)$ and $(2, 3)$.

$$m = \frac{y_2 - y_1}{x_2 - x_1} = \frac{3 - 3}{2 - 0} = \frac{0}{2} = 0$$

Any fraction with 0 in the numerator is **0.**

Try to find the slope of vertical line b through $(-3, -1)$ and $(-3, 4)$.

$$m = \frac{y_2 - y_1}{x_2 - x_1} = \frac{4 - (-1)}{-3 - (-3)} = \frac{5}{0}$$

Any fraction with 0 in the denominator is undefined.

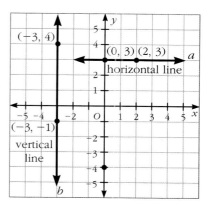

The slope of any horizontal line is zero.
The slope of any vertical line is undefined.

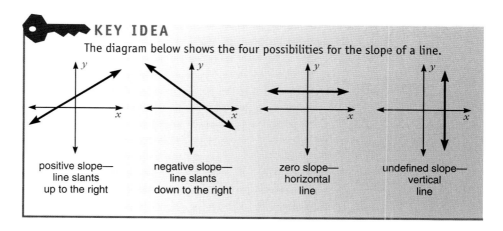

🔑 **KEY IDEA**

The diagram below shows the four possibilities for the slope of a line.

positive slope—
line slants
up to the right

negative slope—
line slants
down to the right

zero slope—
horizontal
line

undefined slope—
vertical
line

From Equations

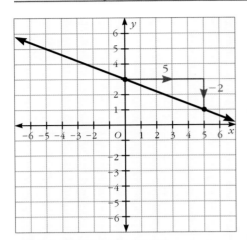

Another way of finding the slope of a line is to use the equation of that line. When a linear equation is written in the form $y = mx + b$, it is in **slope-intercept form.** The slope of the line is m; the y-intercept is b.

When you know the slope and the y-intercept of a line, you have enough information to graph the line. This will be done in the next section.

The line shown at the left is given by the equation $y = -\frac{2}{5}x + 3$. What is the slope of the line?

Name the slope and y-intercept of the line given by the equation

$$y = -\frac{2}{5}x + 3.$$

$$y = mx + b$$
$$\updownarrow$$
$$y = -\frac{2}{5}x + 3$$
$$\text{Slope} = m = -\frac{2}{5}$$

$$y\text{-intercept} = b = 3$$

When an equation is written in slope-intercept form, the coefficient of the variable x is the slope.

The slope of the line $y = -\frac{2}{5}x + 3$ is $-\frac{2}{5}$.

When an equation is written in slope-intercept form, the constant, b, is the y-intercept.

Language Box

Recall that a **coefficient** is the numerical part of a term containing a variable. In the term $-\frac{2}{5}x$, the coefficient is $-\frac{2}{5}$.

EXAMPLE F

Name the slope and y-intercept of the line given by the equation $y = 4x - 7$.

The equation is written in slope-intercept form.

The coefficient of the x-variable is 4.

The slope of the line $y = 4x - 7$ is 4.

The y-intercept of the line $y = 4x - 7$ is -7.

EXAMPLE G

Name the slope and y-intercept of the line given by the equation $y = x + 3$.

Although it is not written, the coefficient of the x-variable is 1.

In slope-intercept form, the equation is $y = 1x + 3$.

The slope of the line $y = x + 3$ is 1.

The y-intercept of the line $y = x + 3$ is 3.

Recall from the last section that a horizontal line has a slope of zero. But what does the equation of a horizontal line look like? The equation of a horizontal line has no *x*-variable. This is shown in Example H.

EXAMPLE H

Identify the slope and *y*-intercept of the line given by the equation $y = -2$.

When you write the equation $y = -2$ in slope-intercept form, you can see that the slope is zero.

$$y = -2$$

$$y = 0 \cdot x - 2$$

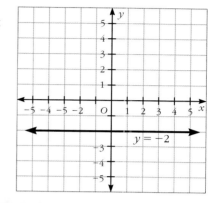

The slope of the horizontal line $y = -2$ is 0.

The *y*-intercept of the line $y = -2$ is -2.

EXAMPLE I

What is the slope and *y*-intercept of the line given by the equation $4x + 8y = 16$?

STEP 1

Write the equation in slope-intercept form by solving for *y*.

$$4x + 8y = 16$$

$4x + 8y - 4x = 16 - 4x$ ← Subtract $4x$ from both sides.

$8y = -4x + 16$ ← Simplify and rearrange terms on the right side.

$\dfrac{8y}{8} = \dfrac{-4}{8}x + \dfrac{16}{8}$ ← Divide each term by 8.

$y = -\dfrac{1}{2}x + 2$ ← Simplify.

Helpful Hint

To write an equation in slope-intercept form, isolate the variable *y* on the left side.

STEP 2

Identify the slope from the equation.

In $y = -\dfrac{1}{2}x + 2$, the slope is $-\dfrac{1}{2}$.

Identify the *y*-intercept from the simplified equation.

In the line $y = -\dfrac{1}{2}x + 2$, the y-intercept is 2.

The slope of the line given by $4x + 8y = 16$ is $-\dfrac{1}{2}$ and its *y*-intercept is 2.

The slope of a line can be found in several ways, depending on the information that is given. These ways are summarized in the table below.

Finding the Slope of a Line

Information Given	Method
Graph of a line	Choose two points on the line and use the formula: $m = \dfrac{\text{vertical change}}{\text{horizontal change}} = \dfrac{\text{rise}}{\text{run}}$
Two points, (x_1, y_1) and (x_2, y_2), on a line	Substitute the coordinates into the formula: $m = \dfrac{y_2 - y_1}{x_2 - x_1}$
Linear equation	Write the equation in slope-intercept form $y = mx + b$. The coefficient of x is the slope of the line.

EXERCISES 13-4

Answer each of the following.

1. What is the slope of the line shown?

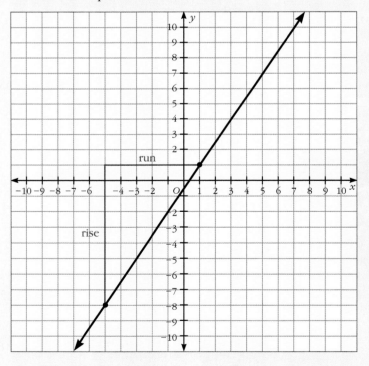

2. On a graph, a line has a vertical change of 28 units and a horizontal change of 7 units. What is the slope of the line?

3. Calculate the slope of the line shown.

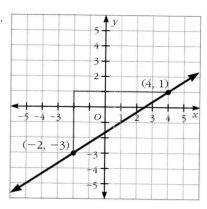

4. Calculate the slope of the line shown.

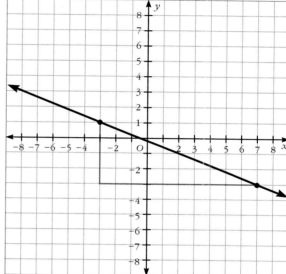

5. What is the slope of the line shown?

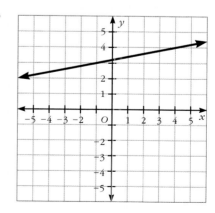

6. Find the slope of the line through the points $(0, -1)$ and $(2, 5)$.

7. What is the slope of the line through the points $(3, -10)$ and $(10, 0)$?

8. All the points on a certain line have the same y-coordinate but different x-coordinates. What is the slope of the line? Explain how you found your answer.

In Exercises 9–14, identify the slope and y-intercept of the line represented by the equation.

9. $y = 3x + 9$ **10.** $y = -x + 4$

11. $y = \frac{1}{2}x - \frac{1}{2}$ **12.** $6x + y = 0$

13. $y = -3$ **14.** $y = \frac{1}{7}x - 7$

In Exercises 15–20, write the equation in slope-intercept form, and then identify the slope and y-intercept of the line represented by the equation.

15. $-3x + y = 2$ **16.** $-4x - y = 5$

17. $y + 1 = 0$ **18.** $15x + 10y = 10$

19. $y = 8 - x$ **20.** $2x - 5y = 3$

21. The current through a 0.5 Ω resistor is given by $I = \frac{V}{0.5} = 2V$, where V is the voltage in volts and I is in amps.

(a) What is the slope of this line?

(b) What is the y-intercept?

22. The inductive reactance of a 0.1 H coil is given by $X_L = 0.628f$, where f is the frequency in hertz and X_L is in ohms.

(a) What is the slope of this line?

(b) What is the y-intercept?

23. A electrical supply store needs to order some CFL lightbulbs. The cost, C, in dollars, of the bulbs is $C = 6.50n + 8.76$, where n is the number of bulbs purchased and 8.76 is the postage needed.

(a) What is the slope of this line?

(b) What is the y-intercept?

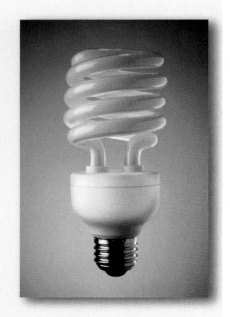

24. Challenge: A line through the points (2, 8) and (−2, k) has a slope of $\frac{3}{4}$. What is the value of k?

GRAPHING LINEAR EQUATIONS USING SLOPE-INTERCEPT FORM

$y = mx + b$

$m = \textbf{slope}$

$b = y\text{-intercept}$

When an equation is written in slope-intercept form, the slope and y-intercept are easily identified. For example, the line given by the equation $y = -3x + 4$ has a slope of -3 and a y-intercept of 4. The line given by the equation $y = \frac{2}{5}x - 1$ has a slope of $\frac{2}{5}$ and a y-intercept of -1.

To graph an equation that is in slope-intercept form, use the y-intercept to plot the point where the line crosses the y-axis and then use the slope to plot a second point.

Graph the line given by the equation $y = -3x + 4$.

- The y-intercept is 4. Plot the point $(0, 4)$.

- The slope is -3. Write the slope as the ratio of vertical change to horizontal change: $-3 = \frac{-3}{1}$.

- From the point $(0, 4)$, move down 3 units and 1 unit to the right. Plot a point there.

- Draw a line through the points.

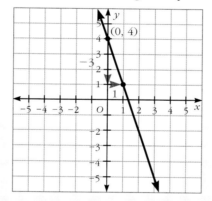

EXAMPLE A

Graph the line whose equation is $y = 3x - 2$. Identify the slope and the y-intercept.

The y-intercept is -2. Plot the point $(0, -2)$.

The slope is 3, or $\frac{3}{1}$. From $(0, -2)$, move up 3 units and 1 unit to the right. Plot a second point.

Draw a line through the points.

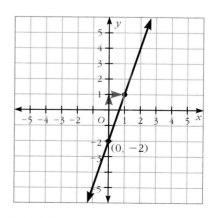

The slope of the line is 3, and the y-intercept is -2.

EXAMPLE B

Graph the line whose equation is $y = -\frac{2}{3}x + 4$. Identify the slope and the y-intercept.

The y-intercept is 4. Plot the point $(0, 4)$.

The slope is $-\frac{2}{3}$, or $\frac{-2}{3}$. From $(0, 4)$, move down 2 units and 3 units to the right.

Draw a line through the points.

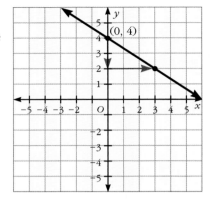

Helpful Hint

A negative fraction has three forms. For example,
$-\frac{2}{3} = \frac{-2}{3} = \frac{2}{-3}$.
To count spaces on a grid for $-\frac{2}{3}$, you can use either $\frac{-2}{3}$ or $\frac{2}{-3}$.

The slope of the line is $-\frac{2}{3}$, and the y-intercept is 4.

EXAMPLE C

Graph the line whose equation is $y = -2x$. Identify the slope and the y-intercept.

The equation in slope-intercept form is $y = -2x + 0$, so the y-intercept is 0. Plot the point $(0, 0)$.

The slope is -2, or $\frac{-2}{1}$. From $(0, 0)$, move down 2 units and 1 unit to the right.

Draw a line through the points.

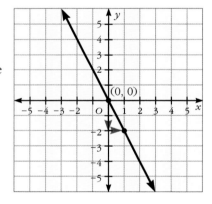

The slope of the line is -2, and the y-intercept is 0.

EXAMPLE D

Graph the line whose equation is $x + 4y = -8$. Identify the slope and the y-intercept.

When a linear equation is not written in slope-intercept form, solve the equation for y to get the equation in slope-intercept form. Then follow the steps to graph the line.

$$x + 4y = -8 \quad \longleftarrow \text{ Write the original equation.}$$

$$x + 4y - x = -8 - x \quad \longleftarrow \text{ Subtract } x \text{ from both sides.}$$

$$4y = -x - 8 \quad \longleftarrow \text{ Simplify and rearrange terms on the right side.}$$

$$\frac{4y}{4} = \frac{-x}{4} - \frac{8}{4} \quad \longleftarrow \text{ Divide each term by 4.}$$

$$y = -\frac{1}{4}x - 2 \quad \longleftarrow \text{ Simplify.}$$

The y-intercept is -2. Plot the point $(0, -2)$.

The slope is $-\frac{1}{4}$. Move down 1 unit and 4 units to the right.

Draw a line through the two points.

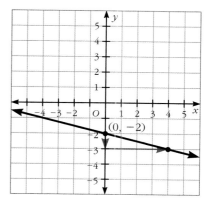

The slope of the line is $-\frac{1}{4}$, and the y-intercept is -2.

EXAMPLE E

Graph the line whose equation is $x - y = 0$. Identify the slope and the y-intercept.

$x - y = 0$ ← Write the original equation.

$x - y - x = 0 - x$ ← Subtract x from both sides.

$-y = -x$ ← Simplify.

$y = x$ ← Multiply both sides by -1.

The equation in slope-intercept form is $y = 1x + 0$, so the y-intercept is 0. Plot the point $(0, 0)$.

The slope is 1. Move up 1 unit and 1 unit to the right. Plot the point.

Draw a line through the points.

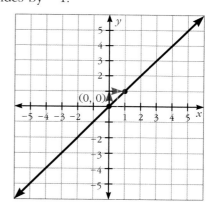

The slope of the line is 1, and the y-intercept is 0.

EXERCISES 13-5

In Exercises 1-6, identify the slope and the y-intercept. Express each slope as a fraction when appropriate.

1. $y = \frac{4}{5}x + 1$ **2.** $y = -x - 7$

3. $y = -8x + 2$ **4.** $4x + y = 9$

5. $x - y = 5$ **6.** $3x + 6y = 18$

Graph each of the following linear equations using slope-intercept form.

7. $y = \frac{3}{4}x + 1$ **8.** $y = 4x$

9. $x - 2y = 6$ **10.** $y = -x$

11. The resistance of a material depends on the temperature and can be described by a linear equation. For a certain material, the resistance is given by the equation $R = 0.0045T + 8.54$, where T is the temperature in degrees Celsius.

  (a) What is the slope of this line?

  (b) What is the y-intercept?

12. Challenge: Graph these equations on the same coordinate plane: $y = \frac{2}{3}x$, $y = \frac{2}{3}x + 2$, and $y = \frac{2}{3}x - 5$.
Identify the slope and the y-intercept of each line. What is true about a set of lines that all have the same slope?

WRITING EQUATIONS OF LINES

Given Slope and y-Intercept

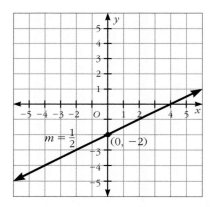

You have seen that it is possible to graph a line using certain information about the line. For example, if the slope and y-intercept are known, the line can be graphed. In this lesson, you will learn to write an equation for a line using information given about the line. The information may be given in a statement, or it may be given on a graph.

The graph at the left passes through $(0, -2)$ and has a slope of $\frac{1}{2}$. You can use this information to write an equation of the line in slope-intercept form.

Write an equation of the line shown at the left in slope-intercept form.

$y = mx + b$ ← Write the slope-intercept equation.

$y = \frac{1}{2}x + (-2)$ ← Substitute $\frac{1}{2}$ for the slope m. Substitute -2 for the y-intercept b.

$y = \frac{1}{2}x - 2$ ← Simplify.

An equation of the line in slope-intercept form is $y = \frac{1}{2}x - 2$.

EXAMPLE A

Write an equation in slope-intercept form of the line with a slope of -2 and a y-intercept of $\frac{3}{5}$.

$y = mx + b$ ← Write the slope-intercept equation.

$y = -2x + \frac{3}{5}$ ← Substitute -2 for the slope m. Substitute $\frac{3}{5}$ for the y-intercept b.

An equation of the line in slope-intercept form is $y = -2x + \frac{3}{5}$.

EXAMPLE B

Write an equation in slope-intercept form of the line with a *y*-intercept of −6 and a slope of 1.

$y = mx + b$ ⟵ Write the slope-intercept equation.

$y = 1x + (-6)$ ⟵ Substitute 1 for the slope *m*. Substitute −6 for the *y*-intercept *b*.

$y = x - 6$ ⟵ Simplify.

When the slope of a line is 1, it is customary to omit the 1 in the equation of the line.

An equation of the line in slope-intercept form is $y = x - 6$.

Given Slope and One Point

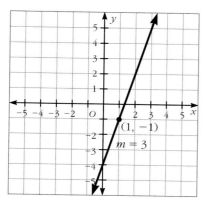

$(1, -1)$

$m = 3$

An equation of a line in variables *x* and *y* describes every point (x, y) that is on the line. If the slope of the line is *m* and you know one point on the line, then the slope between the known point and any other point is *m*.

So, if a *known* point on a line is (x_1, y_1), and (x, y) represents *any* point on the line, then $m = \frac{y - y_1}{x - x_1}$. Eliminating the fraction in this equation results in a useful form called the point-slope form.

$m = \frac{y - y_1}{x - x_1}$ ⟵ This equation represents the slope between a known point (x_1, y_1) and any other point (x, y) on the line.

$m(x - x_1) = \frac{y - y_1}{x - x_1}(x - x_1)$ ⟵ Multiply both sides by $(x - x_1)$.

$m(x - x_1) = y - y_1$ ⟵ Simplify the right side.

$y - y_1 = m(x - x_1)$ ⟵ Exchange sides of the equation.

The equation $y - y_1 = m(x - x_1)$ is called the **point-slope form** of the equation of a line. This form may be used to write an equation of a nonvertical line when the coordinates of a point (x_1, y_1) and the slope *m* are known.

Use the point-slope form to write an equation of the line shown on the graph.

STEP 1

Identify the slope and one point on the line.

The slope is 3, so $m = 3$. The known point on the line is $(1, -1)$, so $x_1 = 1$ and $y_1 = -1$.

STEP 2

Substitute the information into the point-slope form and then solve the equation for y.

$$y - y_1 = m(x - x_1)$$
$$y - (-1) = 3(x - 1)$$
$$y + 1 = 3x - 3$$
$$y + 1 - 1 = 3x - 3 - 1$$
$$y = 3x - 4$$

An equation of the line is $y = 3x - 4$.

EXAMPLE C

Write an equation of the line through (−3, 5) with a slope of −2.

Identify the slope and one point on the line. The slope is -2, so $m = -2$. The known point on the line is $(-3, 5)$, so $x_1 = -3$ and $y_1 = 5$.

Substitute the information into the point-slope form and then solve the equation for y.

$$y - y_1 = m(x - x_1)$$
$$y - 5 = -2[x - (-3)]$$
$$y - 5 = -2(x + 3)$$
$$y - 5 = -2x - 6 \qquad \longleftarrow \text{ Use the Distributive Property to remove the parentheses.}$$
$$y - 5 + 5 = -2x - 6 + 5 \longleftarrow \text{ Add 5 to both sides.}$$
$$y = -2x - 1 \qquad \longleftarrow \text{ Simplify.}$$

An equation of the line is $y = -2x - 1$.

EXAMPLE D

A line has a slope of $\frac{3}{4}$ and passes through the point (8, −4). Write an equation of the line.

Identify the slope and one point on the line. The slope is $\frac{3}{4}$, so $m = \frac{3}{4}$. The known point is $(8, -4)$, so $x_1 = 8$ and $y_1 = -4$.

Substitute the information into the point-slope form and then solve the equation for y.

$$y - y_1 = m(x - x_1)$$
$$y - (-4) = \frac{3}{4}(x - 8)$$
$$y + 4 = \frac{3}{4}(x - 8)$$
$$y + 4 = \frac{3}{4}x - 6 \qquad \longleftarrow \text{ Use the Distributive Property to remove the parentheses.}$$
$$y + 4 - 4 = \frac{3}{4}x - 6 - 4 \longleftarrow \text{ Subtract 4 from both sides.}$$
$$y = \frac{3}{4}x - 10 \qquad \longleftarrow \text{ Simplify.}$$

An equation of the line is $y = \frac{3}{4}x - 10$.

EXAMPLE E

Write the equation of a horizontal line through (3, 8).

Recall that a horizontal line has a slope of 0. Use this fact in Example E.

Identify the slope and one point on the line. The slope is 0, so $m = 0$. The known point on the line is (3, 8), so $x_1 = 3$ and $y_1 = 8$.

Substitute the information into the point-slope form. Then solve the equation for y.

$$y - y_1 = m(x - x_1)$$
$$y - 8 = 0(x - 3)$$
$$y - 8 = 0 \quad \longleftarrow \text{ The factor 0 makes the product 0.}$$
$$y - 8 + 8 = 0 + 8 \quad \longleftarrow \text{ Add 8 to both sides.}$$
$$y = 8$$

The equation of the line is $y = 8$.

Given Two Points

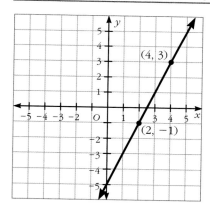

When two points on a line are known, it is a two-step process to write the equation of the line. First, calculate the slope of the line. Then use the point-slope form to write an equation of the line.

The line at the left passes through (2, −1) and (4, 3).

Write an equation of the line shown at the left.

STEP 1

Calculate the slope of the line.
$$m = \frac{y_2 - y_1}{x_2 - x_1} = \frac{3 - (-1)}{4 - 2} = \frac{4}{2} = 2$$

STEP 2

Substitute the slope and the coordinates of either point into the point-slope form. You can use (4, 3) for (x_1, y_1).

$$y - y_1 = m(x - x_1)$$
$$y - 3 = 2(x - 4)$$
$$y - 3 = 2x - 8$$
$$y - 3 + 3 = 2x - 8 + 3$$
$$y = 2x - 5$$

An equation of the line is $y = 2x - 5$.

EXAMPLE F

A line contains the points (3, 5) and (−1, 9). Write an equation of the line.

STEP 1

Calculate the slope of the line.

$$m = \frac{y_2 - y_1}{x_2 - x_1} = \frac{9 - 5}{-1 - 3} = \frac{4}{-4} = -1$$

STEP 2

Substitute the slope and the coordinates of either point into the point-slope form. You can use (3, 5) for (x_1, y_1).

$$y - y_1 = m(x - x_1)$$
$$y - 5 = -1(x - 3)$$
$$y - 5 = -x + 3$$
$$y - 5 + 5 = -x + 3 + 5$$
$$y = -x + 8$$

An equation of the line is $y = -x + 8$.

EXAMPLE G

What is an equation of the line that goes through the origin and contains the point (−6, 5)?

STEP 1

The origin is (0, 0). Calculate the slope of the line.

$$m = \frac{y_2 - y_1}{x_2 - x_1} = \frac{5 - 0}{-6 - 0} = \frac{5}{-6} = -\frac{5}{6}$$

STEP 2

Substitute the slope and the coordinates of the point into the point-slope form.

$$y - y_1 = m(x - x_1)$$
$$y - 0 = -\frac{5}{6}(x - 0)$$
$$y = -\frac{5}{6}x$$

An equation of the line is $y = -\frac{5}{6}x$.

13-6 EXERCISES

Answer each of the following.

1. Write an equation in slope-intercept form for the line shown at the right.

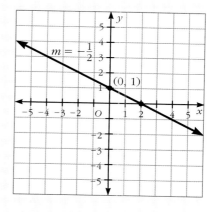

For Exercises 2–6, complete the table below.

	Slope	**y-Intercept**	**Equation of Line**
2.	-7	0	
3.	$\frac{2}{3}$	$\frac{7}{9}$	
4.	-1	10	
5.	$-\frac{5}{6}$	-2	
6.	0	9	

7. Identify the slope and one point on the line shown at the right. Then write an equation of the line.

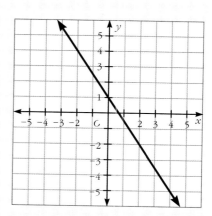

For Exercises 8–12, complete the table below.

	Slope	Point	Equation of Line
8.	−5	(6, 0)	
9.	2	(3, −2)	
10.	$\frac{1}{3}$	(−6, 3)	
11.	0	(6, −2)	
12.	−3	(4, 4)	

For Exercises 13–17, complete the table below.

	Point	Point	Equation of Line
13.	(−5, 0)	(−1, 4)	
14.	(3, 6)	(5, 10)	
15.	(0, 0)	(7, 2)	
16.	(9, 2)	(−2, −9)	
17.	(1, −6)	(0, −2)	

18. The resistance, R, of a material depends on the temperature and can be described by a linear equation. If T is the temperature in degrees Celsius, write the linear equation for a certain material if $R = 9.51$ when $T = 0$ and $R = 12.45$ when $T = 15$.

19. Carlos measures the resistance of a certain material at 6.981 Ω when the temperature is 70°F. Later that day, the temperature has risen to 87°F and the resistance is 7.644 Ω. Determine the linear equation for the resistance, R, of this material given the temperature, T, in degrees Fahrenheit.

20. Maria knows that every day her electrical shop is open, the overhead costs are $167.52. (Overhead costs include heating, air-conditioning, lighting, insurance, etc.) In addition, she has to pay $74.32 per hour in wages to her employees. You want to write an equation for her costs, C, each hour, h, of an eight-hour day.

(a) What is the C-intercept?

(b) What is the slope?

(c) What is the equation?

(d) How much must she make in a typical day in order to not lose money?

21. **Challenge:** A line has an x-intercept of −4 and a y-intercept of −3. Write an equation of the line.

CHAPTER 13 REVIEW EXERCISES

Evaluate the following.

1. When the base current of an NPN transistor is 40 μA, the reading of a collector-to-emitter voltage, V_{CE}, produces the following reading for the current collector, I_C.

V_{CE} (V)	0	0.5	1	2	4	6	8	10
I_C (mA)	0	2	2.4	2.5	2.7	2.9	3.1	3.3

Plot the points in the table with V_{CE} on the horizontal axis.

2. The following table gives the values of the current flowing through a certain resistor as the voltage is changed.

V (V)	10	20	40	60	80	100
I (A)	0.5	1.2	2.6	4.0	5.4	6.8

(a) What is the slope of the line that goes through these points?

(b) What is the equation of the line that goes through these points?

(c) What is the V-intercept of this line?

3. The current through a 1.6 Ω resistor is given by $I = \dfrac{V}{1.6}$, where V is the voltage in volts and I is in amps.

(a) Create a table of I-values when V is 0, 2, 4, 6, 8, and 10.

(b) Graph this linear equation.

(c) What is the slope of this line?

(d) What is the y-intercept?

4. The inductive reactance of a 0.75 H coil is given by $X_L = 4.712f$, where f is the frequency in hertz and X_L is in ohms.

(a) What is the slope of this line?

(b) What is the y-intercept?

5. An electronics store purchased a panel truck for $29,625. They figure that the truck will require an average of $19.65 per day in maintenance and operating costs. The total cost of owning this truck on any given day, t, can be represented by a straight line.

(a) What is the slope of this line?

(b) What is the y-intercept?

(c) Find the linear function for the total costs, C, of owning the truck after t days.

Building a Foundation in Mathematics

Solving Systems of Equations by
Graphing

Solving Systems of Equations by
Substitution

Solving Systems of Equations by
Elimination

Solving Systems of Equations for
Three Unknowns

Using Systems to Solve Word
Problems

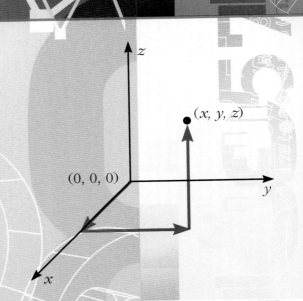

Overview

In the last chapter we saw that one equation with
two variables has an infinite number of solutions.
When you made a table of values for such an equation, you could have selected any value for one of the
variables and found the corresponding value for the
other variable.

Sometimes there is more than one relationship
between two variables. In this case, it can take two
equations to describe each of the relationships. In
this chapter, we will look at sets of two equations

with two variables and look at methods for determining the solution, if there is one. In many cases, the
two lines will intersect at a single point, and this
point is the only one that will satisfy both relationships. Other times, there will be no ordered pair that
satisfies both equations, or every ordered pair that is
a solution for one equation will be a solution of the
other. We will also see how to solve three equations
with three variables.

Systems of Equations

Objectives

After completing this chapter, you will be able to:

- Solve systems of equations graphically
- Solve systems of equations by the substitution method
- Solve systems of equations by the addition or elimination method
- Determine when systems of equations have no solution or an infinite number of solutions
- Write and solve systems of equations from word problems

SOLVING SYSTEMS OF EQUATIONS BY GRAPHING

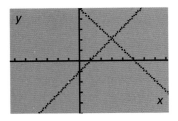

A graphing calculator can be used to graph two equations and find the point where their lines intersect. This graph shows the equations $y = x - 1$ and $y = -x + 5$.

A **system of linear equations** is two or more equations, with each equation representing a line. A **solution** of a system of two linear equations with variables x and y is an ordered pair (x, y) that satisfies both equations. If two lines intersect, the point of intersection is the solution of the system that those lines represent.

To solve a system of equations by graphing, graph the equations on the same coordinate plane. If the lines intersect in a single point, the ordered pair is the solution of the system.

The graph of the system of linear equations made up of $y = x - 1$ and $y = -x + 5$ is shown at the left and also below.

Solve the system of equations by graphing.

System of Equations

$y = x - 1$
$y = -x + 5$

Point of Intersection
$(3, 2)$

Check:
Check the ordered pair $(3, 2)$ in each equation.

$\begin{array}{ll} y = x - 1 & y = -x + 5 \\ 2 \overset{?}{=} 3 - 1 & 2 \overset{?}{=} -3 + 5 \\ 2 = 2 \checkmark & 2 = 2 \checkmark \end{array}$

Graph of the System
of Equations

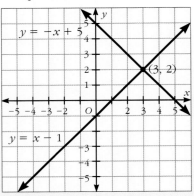

The solution of the system is the ordered pair $(3, 2)$.

EXAMPLE A

Solve the system below by graphing.
$y = 2x$
$y = -x - 3$

System of Equations

$y = 2x$
$y = -x - 3$

Point of Intersection
$(-1, -2)$

Check:
Check the ordered pair $(-1, -2)$ in each equation.

$\begin{array}{ll} y = 2x & y = -x - 3 \\ -2 \overset{?}{=} 2(-1) & -2 \overset{?}{=} -(-1) -3 \\ -2 = -2 \checkmark & -2 = -2 \checkmark \end{array}$

Graph of the System
of Equations

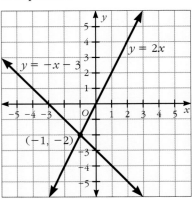

The solution of the system is the ordered pair $(-1, -2)$.

Slope-intercept form is often the easiest form of an equation for graphing. In the examples that follow, we will write equations in slope-intercept form to graph them.

EXAMPLE B

Solve the system below by graphing.
$y = 4$
$2x + y = -2$

Write $2x + y = -2$ in slope-intercept form.
$$2x + y = -2$$
$$-2x + 2x + y = -2x - 2$$
$$y = -2x - 2$$

System of Equations
$y = 4$
$2x + y = -2$

Point of Intersection
$(-3, 4)$

Check:
Check the ordered pair $(-3, 4)$ in each equation.

$y = 4$ $2x + y = -2$
$4 = 4$ ✓ $2(-3) + 4 \overset{?}{=} -2$
                $-6 + 4 \overset{?}{=} -2$
                $-2 = -2$ ✓

Graph of the System of Equations

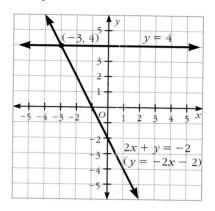

The solution of the system is the ordered pair $(-3, 4)$.

EXAMPLE C

Solve the system below by graphing.
$V = 3$
$I - 4V = -8$

Here the variables are V and I rather than x and y. We will let the horizontal axis be the V-axis and the vertical axis will be the I-axis. The answer will be an ordered pair in the form (V, I). Write $I - 4V = -8$ in slope-intercept form. The I-axis is the vertical axis, so we solve the equation for I.
$$I - 4V = -8$$
$$I - 4V + 4V = -8 + 4V$$
$$I = 4V - 8$$

System of Equations
$V = 3$
$I = 4V - 8$

Point of Intersection
$(3, 4)$

Check:
Check the ordered pair $(3, 4)$ in each equation.
$V = 3$ $I - 4V = -8$
$3 = 3$ ✓ $4 - 4(3) \overset{?}{=} -8$
                $4 - 12 \overset{?}{=} -8$
                $-8 = -8$ ✓

Graph of the System of Equations

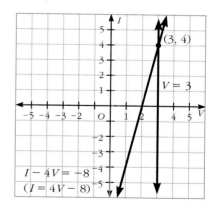

The solution of the system is the ordered pair $(3, 4)$.

EXAMPLE D

Solve the system below by graphing.
$$y = -\tfrac{1}{2}x + 1$$
$$y = -\tfrac{1}{2}x + 5$$

Language Box

Parallel lines are always the same distance apart. They will never intersect.

System of Equations
$$y = -\tfrac{1}{2}x + 1$$
$$y = -\tfrac{1}{2}x + 5$$

Point of Intersection (none)

Notice that the lines are parallel. This is because they have the same slope but different y-intercepts. There is no point of intersection.

Graph of the System of Equations

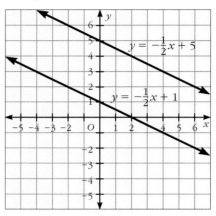

The system does not have a solution.

EXAMPLE E

Solve the system below by graphing.
$$y - 3x = 1$$
$$2y - 6x = -4$$

Write each equation in slope-intercept form.
$$y - 3x = 1$$
$$y - 3x + 3x = 1 + 3x$$
$$y = 3x + 1$$

$$2y - 6x = -4$$
$$2y - 6x + 6x = -4 + 6x$$
$$2y = 6x - 4$$
$$y = 3x - 2$$

These lines are also parallel. Writing the equations in slope-intercept form makes it easy to see that they have the same slope, 3.

Graph of the System of Equations

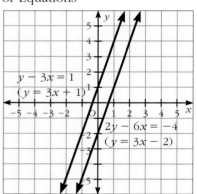

The system does not have a solution.

EXAMPLE F

Solve the system below by graphing.
$y = -3x + 1$
$6x + 2y = 2$

Write $6x + 2y = 2$ in slope-intercept form.

$$6x + 2y = 2$$
$$-6x + 6x + 2y = -6x + 2$$
$$2y = -6x + 2$$
$$y = -3x + 1$$

This is the same as the first equation in the system. The equations in the system are equivalent, so the graph of both equations is the same line. The same set of ordered pairs (x, y) will satisfy both equations.

When the equations in a system have the same line as their graph, every point on the line is a solution of the system. So there is an infinite number of solutions to the system.

Graph of the System of Equations

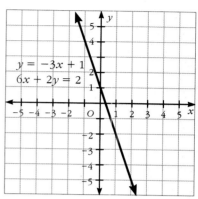

Equivalent Equations
Same Line

The solution set of the system is the set of all ordered pairs that satisfy the equation $y = -3x + 1$.

EXERCISES 14-1

Graph each system of equations. Then give the solution (if any) of the system.

1. $y = -3x$
$y = x$

2. $y = \frac{1}{4}x$
$y = \frac{1}{2}x + 1$

3. $y = -x - 2$
$y = \frac{2}{3}x + 3$

4. $y - 4x = -5$
$y - 4x = 4$

5. $y = -2$
$y = -2x + 2$

6. $x + y = 1$
$x - y = -1$

7. $V = -4$
$I = \frac{1}{4}V + 5$

8. $2V - 2I = 8$
$I = V - 4$

9. In a certain series-opposing circuit, $V_1 - V_2 = 3$ V. If the same series is series-aiding, then $V_1 + V_2 = 11$ V.

(a) Graph each line with V_1 on the horizontal axis.

(b) Determine the solution (if any) of this system.

10. Challenge: Graph the system of equations shown below.

$y = 4$

$x = -2$

$y = -x$

Explain why the system of equations does not have a solution.

SOLVING SYSTEMS OF EQUATIONS BY SUBSTITUTION

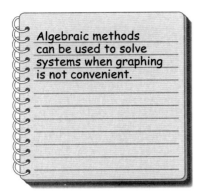

Algebraic methods can be used to solve systems when graphing is not convenient.

You can find the solution of a system of linear equations by algebraic methods. This is helpful when the equations are not easily graphed or when the solution contains a fraction. One algebraic method of solving systems of equations is called the **substitution method.** The steps are given below.

Steps for Solving a System of Two Linear Equations by Substitution

- Solve one equation for either variable.

- Substitute the expression for that variable into the other equation and solve it.

- Substitute the value you found into either original equation and solve for the other variable.

- Check the ordered pair in each original equation.

Solve the system of equations below by substitution.

$-x + y = 8$ **Equation 1 (Eq. 1)**
$x + y = 2$ **Equation 2 (Eq. 2)**

Solve Equation 1 for y.

$$-x + y = 8$$
$$x - x + y = x + 8$$
$$y = x + 8$$

Substitute $x + 8$ for y in Equation 2 and solve.

$$x + y = 2$$
$$x + x + 8 = 2$$
$$2x + 8 = 2$$
$$2x = -6$$
$$x = -3$$

Substitute -3 for x in Equation 1 and solve for y.

$$-x + y = 8$$
$$-(-3) + y = 8$$
$$3 + y = 8$$
$$y = 5$$

Check the ordered pair $(-3, 5)$ in each original equation.

$$-x + y = 8 \qquad\qquad x + y = 2$$
$$-(-3) + 5 \overset{?}{=} 8 \qquad\qquad -3 + 5 \overset{?}{=} 2$$
$$3 + 5 \overset{?}{=} 8 \qquad\qquad 2 = 2 \checkmark$$
$$8 = 8 \checkmark$$

The solution of the system is $(-3, 5)$.

When using the substitution method, you may begin by solving for either variable. Choose the variable that will be easier to isolate. In Example A, y in Equation 1 is chosen because its coefficient is 1.

EXAMPLE A

Solve the system below by substitution.

$4x + y = 20$ **Eq. 1**

$2x - y = -2$ **Eq. 2**

Solve Equation 1 for y.

$$4x + y = 20$$
$$-4x + 4x + y = -4x + 20$$
$$y = -4x + 20$$

Substitute $-4x + 20$ for y in Equation 2 and solve for x.

$$2x - y = -2$$
$$2x - (-4x + 20) = -2$$
$$2x + 4x - 20 = -2$$
$$6x - 20 = -2$$
$$6x = 18$$
$$x = 3$$

Substitute 3 for x in Equation 1 and solve for y.

$$4x + y = 20$$
$$4(3) + y = 20$$
$$12 + y = 20$$
$$y = 8$$

Check the ordered pair (3, 8) in each original equation.

$$4x + y = 20 \qquad\qquad 2x - y = -2$$
$$4(3) + (8) \stackrel{?}{=} 20 \qquad\qquad 2(3) - (8) \stackrel{?}{=} -2$$
$$12 + 8 \stackrel{?}{=} 20 \qquad\qquad 6 - 8 \stackrel{?}{=} -2$$
$$20 = 20 \checkmark \qquad\qquad -2 = -2 \checkmark$$

The graph of the system is shown at the right.

You can use the graph to verify the solution found by substitution.

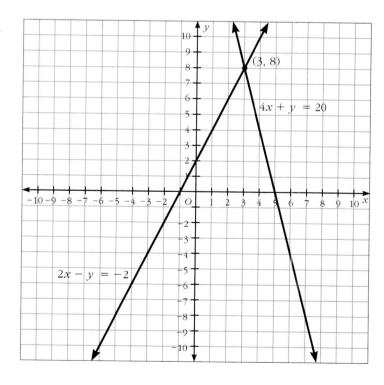

The solution of the system is (3, 8).

The solution in the next example contains fractions.

EXAMPLE B

Solve the system below by substitution.

$2x + 2y = 5$ Eq. 1
$x - y = 4$ Eq. 2

Solve Equation 2 for x.

$$x - y = 4$$
$$x - y + y = 4 + y$$
$$x = y + 4$$

Substitute $y + 4$ for x in Equation 1 and solve for y.

$$2x + 2y = 5$$
$$2(y + 4) + 2y = 5$$
$$2y + 8 + 2y = 5$$
$$4y + 8 = 5$$
$$4y = -3$$
$$y = -\frac{3}{4}$$

Substitute $-\frac{3}{4}$ for y in Equation 2 and solve for x.

$$x - y = 4$$
$$x - \left(-\frac{3}{4}\right) = 4$$
$$x + \frac{3}{4} = 4$$
$$x + \frac{3}{4} - \frac{3}{4} = 4 - \frac{3}{4}$$
$$x = \frac{16}{4} - \frac{3}{4}$$
$$x = \frac{13}{4}$$

Check the ordered pair $\left(\frac{13}{4}, -\frac{3}{4}\right)$ in each original equation.

$$2x + 2y = 5 \qquad\qquad x - y = 4$$
$$2\left(\frac{13}{4}\right) + 2\left(-\frac{3}{4}\right) \overset{?}{=} 5 \qquad \left(\frac{13}{4}\right) - \left(-\frac{3}{4}\right) = 4$$
$$\frac{13}{2} - \frac{3}{2} \overset{?}{=} 5 \qquad\qquad \frac{13}{4} + \frac{3}{4} = 4$$
$$\frac{10}{2} \overset{?}{=} 5 \qquad\qquad\qquad \frac{16}{4} \overset{?}{=} 4$$
$$5 = 5 \checkmark \qquad\qquad\qquad 4 = 4 \checkmark$$

The solution of the system is $\left(\frac{13}{4}, -\frac{3}{4}\right)$.

When one of the equations in a system has only one variable, for example, $y = 3$, you may substitute the numerical value directly into the other equation to solve the system. This is shown in Example C.

EXAMPLE C

Solve the system below by substitution.

$y = 3$ Eq. 1
$-4x + y = -1$ Eq. 2

Substitute 3 for y in Equation 2 and solve for x.

$$-4x + y = -1$$
$$-4x + 3 = -1$$
$$-4x + 3 - 3 = -1 - 3$$
$$-4x = -4$$
$$x = 1$$

Check the ordered pair (1, 3) in each original equation.

$y = 3$ $-4x + y = -1$

$3 = 3$ ✓ $-4(1) + 3 \overset{?}{=} -1$

$-4 + 3 \overset{?}{=} -1$

$-1 = -1$ ✓

The solution of the system is (1, 3).

EXAMPLE D

Solve the system below by substitution.

$2x + y = 3$ **Eq. 1**

$4x + 2y = -1$ **Eq. 2**

Solve Equation 1 for y.

$$2x + y = 3$$
$$-2x + 2x + y = -2x + 3$$
$$y = -2x + 3$$

Substitute $-2x + 3$ for y in Equation 2 and solve for x.

$$4x + 2y = -1$$
$$4x + 2(-2x + 3) = -1$$
$$4x - 4x + 6 = -1$$
$$6 = -1 \quad \longleftarrow \text{ The result is a false statement.}$$

When the algebraic steps lead to a false statement, the system has no solution.

The graph of the system is shown at the right.

You can see that the system has no solution because the lines are parallel.

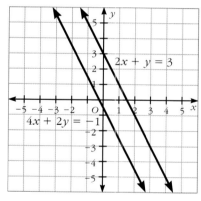

The system does not have a solution.

EXAMPLE E

Solve the system below by substitution.

$y = \frac{3}{4}x + 2$ **Eq. 1**

$4y - 3x = 8$ **Eq. 2**

Equation 1 is already solved for y: $y = \frac{3}{4}x + 2$.

Substitute $\frac{3}{4}x + 2$ for y in Equation 2 and solve.

$$4y - 3x = 8$$
$$4\left(\frac{3}{4}x + 2\right) - 3x = 8$$
$$3x + 8 - 3x = 8$$
$$8 = 8 \quad \longleftarrow \text{ The result is a true statement.}$$

When the algebraic steps lead to a true statement containing no variables, the system has an infinite number of solutions.

The graph of the system is shown at the right.

You can see that the equations are equivalent because their graphs are the same line.

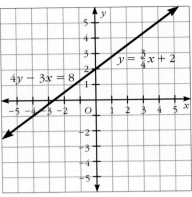

Equivalent Equations
Same Line

The solution set of the system is the set of all ordered pairs that satisfy the equation $y = \frac{3}{4}x + 2$.

EXERCISES 14-2

Solve each system of equations by substitution.

1. $-2x + y = 1$
$3x - 4y = 6$

2. $3x + 5y = 7$
$-4x + y = 6$

3. $x - 4y = 0$
$2x + 5y = 26$

4. $3x + 3y = 6$
$y = -x + 2$

5. $3x + y = -8$
$x + 2y = 4$

6. $3x - 2y = -3$
$3x + y = 3$

7. $x = 3$
$4x + 5y = 22$

8. $8x + 3y = -5$
$8x - y = 7$

9. $2x + y = 9$
$y = 1$

10. $x + 2y = 6$
$2x + 4y = 10$

11. $x + 2y = 3$
$2x - y = 1$

12. $4x + 3y = 6$
$-2x + 6y = 7$

13. $y = \frac{1}{3}x + 1$
$x = -6$

14. $-6x + 3y = -1$
$2x - y = -1$

15. In a certain series-aiding circuit, $V_1 + V_2 = 23$ V. If the same series is series-opposing, then $V_1 - V_2 = 17$ V.

This gives the system of equations.

$V_1 + V_2 = 23$

$V_1 - V_2 = 17$

Solve this system by substitution.

> **Helpful Hint**
>
> Remember to check your solution to a system in both of the original equations.

16. In a certain series-aiding circuit, $V_1 + V_2 = 72$ V. If the same series is series-opposing, then $V_1 - V_2 = 36$ V.

This gives the system of equations.

$V_1 + V_2 = 72$
$V_1 - V_2 = 36$

Solve this system by substitution.

17. Challenge: A student attempted to solve the system of equations below but decided that the system had no solution.

$x = -5$
$y = 6$

Do you agree with the student? Explain your answer.

SOLVING SYSTEMS OF EQUATIONS BY ELIMINATION

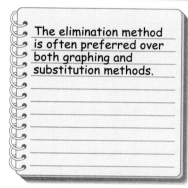

The elimination method is often preferred over both graphing and substitution methods.

Another algebraic method used to solve systems of linear equations is the **elimination method.** This method is sometimes called the *addition method* because similar terms of the equations are added to make it possible to solve for one variable at a time. The steps for using this method are given below.

Steps for Solving a System of Two Linear Equations by Elimination

- Add like terms of the equations to eliminate one variable. (You may need to multiply before adding.)
- Solve the resulting equation for the remaining variable.
- Substitute the value you found into either of the original equations and solve for the other variable.
- Check the ordered pair in each original equation.

Language Box

Eliminating a variable does not mean that the variable is gone forever. It means that you are able to work with an equation without that variable in it.

Solve the system of equations below by elimination.

$2x - y = 5$ Equation 1 (Eq. 1)
$x + y = 4$ Equation 2 (Eq. 2)

Notice that the y-terms are $1y$ and $-1y$. The coefficients of the y-terms are opposites of each other, so their sum will be 0.

Add like terms of the equations to eliminate y.

$$\begin{array}{r} 2x - y = 5 \\ x + y = 4 \\ \hline 3x = 9 \end{array}$$

Solve the resulting equation for x.

$3x = 9$
$x = 3$

Substitute 3 for x in Equation 2 and solve for y.

$x + y = 4$
$3 + y = 4$
$y = 1$

Check the ordered pair (3, 1) in each original equation.

$$\begin{array}{ll} 2x - y = 5 & x + y = 4 \\ 2(3) - 1 \overset{?}{=} 5 & 3 + 1 \overset{?}{=} 4 \\ 6 - 1 \overset{?}{=} 5 & 4 = 4 \checkmark \\ 5 = 5 \checkmark & \end{array}$$

The solution of the system is (3, 1).

EXAMPLE A

Solve the system below by elimination.

$3x - 9y = -6$ Eq. 1
$3x + 4y = 7$ Eq. 2

In this case, neither variable has opposite coefficients. However, if both sides of Equation 1 are multiplied by -1, the coefficients of x will be opposites. This multiplication is shown in the first step below.

Add similar terms of the equations to eliminate x.

Write the original equations.	Multiply both sides of Equation 1 by -1.	Add.
$3x - 9y = -6$	$-1(3x - 9y) = -1(-6)$	$-3x + 9y = 6$
$3x + 4y = 7$	$3x + 4y = 7$	$\underline{3x + 4y = 7}$
		$13y = 13$

Solve the resulting equation for y.

$13y = 13$
$y = 1$

Substitute 1 for y in Equation 2 and solve for x.

$3x + 4y = 7$
$3x + 4(1) = 7$
$3x + 4 = 7$
$3x = 3$
$x = 1$

Check the ordered pair $(1, 1)$ in each original equation.

$$3x - 9y = -6 \qquad\qquad 3x + 4y = 7$$
$$3(1) - 9(1) \stackrel{?}{=} -6 \qquad 3(1) + 4(1) \stackrel{?}{=} 7$$
$$3 - 9 \stackrel{?}{=} -6 \qquad\qquad 3 + 4 \stackrel{?}{=} 7$$
$$-6 = -6 \checkmark \qquad\qquad 7 = 7 \checkmark$$

The solution of the system is $(1, 1)$.

EXAMPLE B

Solve the system below by elimination.

$2x + 5y = 22$ Eq. 1
$x + y = 2$ Eq. 2

In this case, it will not be sufficient to multiply by -1 to get opposite coefficients. However, if both sides of Equation 2 are multiplied by -2, the coefficients of x will be opposites. This multiplication is shown in the first step below.

Add similar terms of the equations to eliminate x.

Write the original equations.	Multiply both sides of Equation 2 by -2.	Add.
$2x + 5y = 22$	$2x + 5y = 22$	$2x + 5y = 22$
$x + y = 2$	$-2(x + y) = -2(2)$	$\underline{-2x - 2y = -4}$
		$3y = 18$

Helpful Hint

Always be certain that the like terms are lined up in columns before adding the equations.

Solve the equation for y.

$3y = 18$
$y = 6$

Substitute 6 for y in Equation 2 and solve for x.

$x + y = 2$
$x + 6 = 2$
$x = -4$

Check the ordered pair $(-4, 6)$ in each original equation.

$$2x + 5y = 22 \qquad\qquad x + y = 2$$
$$2(-4) + 5(6) \overset{?}{=} 22 \qquad -4 + 6 \overset{?}{=} 2$$
$$-8 + 30 \overset{?}{=} 22 \qquad\qquad 2 = 2 \checkmark$$
$$22 = 22 \checkmark$$

The solution of the system is $(-4, 6)$.

KEY IDEA

To eliminate a variable, the coefficients of that variable must be opposites. You may need to multiply both sides of an equation by a constant, and sometimes both sides of the other equation by a different constant, to get opposite coefficients.

EXAMPLE C

Solve the system below by elimination.

$4x - 7y = 10$ Eq. 1
$5x + 4y = -13$ Eq. 2

Neither variable has coefficients that are opposites. One way to get coefficients that are opposites is to multiply Equation 1 by 4 and Equation 2 by 7. The result will have $4(-7y) = -28y$ in the first equation and $7(4y) = 28y$ in the second equation.

Add similar terms of the equations to eliminate y.

Write the original equations.

$$4x - 7y = 10$$
$$5x + 4y = -13$$

Multiply both sides of Equation 1 by 4 and both sides of Equation 2 by 7.

$$4(4x - 7y) = 4(10)$$
$$7(5x + 4y) = 7(-13)$$

Add.

$$16x - 28y = 40$$
$$\underline{35x + 28y = -91}$$
$$51x \qquad = -51$$

Solve the equation for x.

$$51x = -51$$
$$x = -1$$

Substitute -1 for x in Equation 1 and solve for y.

$$4x - 7y = 10$$
$$4(-1) - 7y = 10$$
$$-4 - 7y = 10$$
$$-7y = 14$$
$$y = -2$$

Check the ordered pair $(-1, -2)$ in each original equation.

$$4x - 7y = 10 \qquad\qquad 5x + 4y = -13$$
$$4(-1) - 7(-2) \overset{?}{=} 10 \qquad 5(-1) + 4(-2) \overset{?}{=} -13$$
$$-4 + 14 \overset{?}{=} 10 \qquad\qquad -5 - 8 \overset{?}{=} -13$$
$$10 = 10 \checkmark \qquad\qquad -13 = -13 \checkmark$$

The solution of the system is $(-1, -2)$.

EXAMPLE D

Solve the system below by elimination.

$4x - y = -3$ Eq. 1
$8x - 2y = 8$ Eq. 2

Add similar terms of the equations to eliminate x.

Write the original equations.	Multiply both sides of Equation 1 by -2.	Add.
$4x - y = -3$ $\rightarrow$	$-2(4x - y) = -2(-3)$ $\rightarrow$	$-8x + 2y = 6$
$8x - 2y = 8$	$8x - 2y = 8$	$\underline{8x - 2y = 8}$
		$0 = 14$

When the addition step results in a false statement, such as $0 = 14$, the system has no solution.

The system has no solution.

EXAMPLE E

Solve the system below by elimination.

$x - \frac{1}{4}y = 5$ Eq. 1
$-4x + y = -20$ Eq. 2

Add similar terms of the equations to eliminate x.

Write the original equations.	Multiply both sides of Equation 1 by 4.	Add.
$x - \frac{1}{4}y = 5$ $\rightarrow$	$4(x - \frac{1}{4}y) = 4(5)$ $\rightarrow$	$4x - y = 20$
$-4x + y = -20$	$-4x + y = -20$	$\underline{-4x + y = -20}$
		$0 = 0$

When the addition step results in a true statement, such as $0 = 0$, the system has an infinite number of solutions because Equation 1 and Equation 2 represent the same line.

The solution set of the system is the set of all ordered pairs that satisfy the equation $-4x + y = -20$.

EXERCISES 14-3

Solve the following systems of equations by elimination.

1. $x - y = 17$
   $x + y = 5$

2. $5x + 4y = 12$
   $3x + 4y = 4$

3. $x - y = 3$
   $5x - 5y = 15$

4. $x + y = 5$
   $5x + 6y = 8$

5. $2x - 3y = -4$
   $-4x + 3y = 6$

6. $2x + 3y = 2$
   $3x - 4y = -14$

7. $x + y = -4$
   $3x + 3y = 1$

8. $2x + 3y = 8$
   $x - y = 2$

9. $x - 2y = 5$
   $3x - y = -20$

10. $3x - 6y = 12$
    $2x + 3y = 1$

11. $y = -2$
    $x + 2y = 6$

12. $x - y = 0$
    $4x + 10y = -7$

13. $2x + 3y = 1$
    $5x + 7y = 3$

14. $-7x - 3y = 10$
    $2x + 2y = -8$

15. Applying Kirchhoff's laws to a certain circuit with three resistors satisfies the following equations:

$$I_1 = I_2 + I_3$$
$$E_1 = R_1 I_1 + R_2 I_2$$
$$E_2 = R_2 I_2 - R_3 I_3$$

If $E_1 = 8$ V, $E_2 = 5$ V, $R_1 = 3$ Ω, $R_2 = 5$ Ω, and $R_3 = 6$ Ω, use elimination to solve the system for I_1, I_2, and I_3.

Hint: Solve the first equation for I_3 and substitute this in the last equation.

16. Applying Kirchhoff's laws to a certain circuit with three resistors satisfies the following equations:

$$I_1 = I_2 + I_3$$
$$E_1 = R_1 I_1 + R_2 I_2$$
$$E_2 = R_2 I_2 - R_3 I_3$$

If $E_1 = 48$ V, $E_2 = 4$ V, $R_1 = 2$ Ω, $R_2 = 4$ Ω, and $R_3 = 8$ Ω, use elimination to solve the system for I_1, I_2, and I_3.

17. Challenge: How would you use the elimination method to solve the system of equations below?

$$\frac{1}{2}x - \frac{2}{3}y = \frac{7}{3} \qquad \text{Equation 1}$$
$$3x + 4y = -50 \qquad \text{Equation 2}$$

What is the solution of the system?

SOLVING SYSTEMS OF EQUATIONS FOR THREE UNKNOWNS

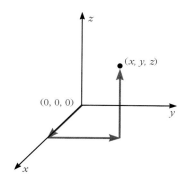

Just as an ordered pair represents a point in a plane, an ordered triple represents a point in space.

So far, you have solved systems of equations with two variables, x and y. A solution of a system with two variables, x and y, is an ordered pair (x, y) that satisfies both equations. Systems can have more than two variables. A solution of a system with three variables, x, y, and z, is an **ordered triple, (x, y, z)**, that satisfies all three equations.

One method of solving a system with three unknowns is to use substitution to reduce the original system to a system of two equations with two unknowns. After solving that system, use the two values you found to find the value of the third variable. The steps are given below.

Steps for Solving a System of Three Linear Equations

- Solve one equation for one variable. Substitute that expression for the variable into the other equations.
- Solve the resulting system of two equations.
- Substitute one or both of the values you found into any of the original equations to solve for the third variable.
- Check the ordered triple in all three original equations.

Solve the system of equations given below.

$$x + y = 6 \qquad \text{Equation 1 (Eq. 1)}$$
$$x + z = 8 \qquad \text{Equation 2 (Eq. 2)}$$
$$y + z = 10 \qquad \text{Equation 3 (Eq. 3)}$$

Solve Equation 1 for x. Substitute that expression into the other equations, where possible.

Equation 1
$$x + y = 6$$
$$x + y - y = 6 - y$$
$$x = -y + 6$$

Equation 2 $\quad (-y + 6) + z = 8 \qquad \rightarrow \qquad -y + z = 2$

Equation 3 $\qquad\qquad y + z = 10 \qquad\qquad\qquad y + z = 10$

Solve the resulting system of two equations.

$\begin{aligned} -y + z &= 2 \\ y + z &= 10 \\ \hline 2z &= 12 \\ z &= 6 \end{aligned}$ $\qquad$ This system can be solved by the elimination method.

Helpful Hint

Use a highlighter to indicate the value of each variable as you find it.

Substitute 6 for z in Equation 3 and solve for y.

$$y + z = 10$$
Equation 3 $\quad y + 6 = 10$
$$y = 4$$

Substitute 4 for y in Equation 1 and solve for x.

Equation 1 $\quad x + y = 6$
$$x + 4 = 6$$
$$x = 2$$

Check the ordered triple $(2, 4, 6)$ in all three original equations.

Equation 1	Equation 2	Equation 3
$x + y = 6$	$x + z = 8$	$y + z = 10$
$2 + 4 \overset{?}{=} 6$	$2 + 6 \overset{?}{=} 8$	$4 + 6 \overset{?}{=} 10$
$6 = 6$ ✓	$8 = 8$ ✓	$10 = 10$ ✓

The solution of the system is the ordered triple $(2, 4, 6)$.

If a system has a variable whose coefficient is 1, isolate that variable first. In Example A, the variable z in Equation 3 is isolated first because its coefficient is 1.

EXAMPLE A

Solve the system below.

$3x + 2y = 19 \quad$ **Eq. 1**
$-3x + 2z = 5 \quad$ **Eq. 2**
$-y + z = 2 \quad$ **Eq. 3**

Solve Equation 3 for z. Substitute that expression into the other equations, where possible.

Equation 3 $\qquad -y + z = 2$
$$-y + z + y = y + 2$$
$$z = y + 2$$

Equation 1 $\qquad\qquad 3x + 2y = 19 \qquad \rightarrow \qquad 3x + 2y = 19$

Equation 2 $\quad -3x + 2(y + 2) = 5 \qquad\qquad -3x + 2y = 1$

Solve the resulting system of two equations.

$\begin{aligned} 3x + 2y &= 19 \\ -3x + 2y &= 1 \\ \hline 4y &= 20 \\ y &= 5 \end{aligned}$ $\quad$ This system can be solved by the elimination method.

Substitute 5 for y in Equation 1 and solve for x.

Equation 1 $\quad 3x + 2y = 19$
$$3x + 2(5) = 19$$
$$3x + 10 = 19$$
$$3x = 9$$
$$x = 3$$

COMMON ERROR ALERT

An error in writing the ordered triple (x, y, z) will occur if you are not careful to match the values found with the correct variable.

Substitute 5 for y in Equation 3 and solve for z.

Equation 3 $\qquad -y + z = 2$

$$-(5) + z = 2$$
$$-5 + z + 5 = 2 + 5$$
$$z = 7$$

Check the ordered triple $(3, 5, 7)$ in all three original equations.

Equation 1	Equation 2	Equation 3
$3x + 2y = 19$	$-3x + 2z = 5$	$-y + z = 2$
$3(3) + 2(5) \stackrel{?}{=} 19$	$-3(3) + 2(7) \stackrel{?}{=} 5$	$-(5) + 7 \stackrel{?}{=} 2$
$9 + 10 \stackrel{?}{=} 19$	$-9 + 14 \stackrel{?}{=} 5$	$2 = 2$ ✓
$19 = 19$ ✓	$5 = 5$ ✓	

The solution of the system is the ordered triple $(3, 5, 7)$.

EXAMPLE B

Solve the system below.

$\quad x + y + z = 10$ **Eq. 1**
$2x - y + 3z = 20$ **Eq. 2**
$x + 2y - z = 5$ **Eq. 3**

Solve Equation 1 for x. Substitute that expression into the other equations.

Equation 1 $\quad x + y + z = 10$

$$x = -y - z + 10$$

Equation 2 $\quad 2(-y - z + 10) - y + 3z = 20$
Equation 3 $\quad (-y - z + 10) + 2y - z = 5 \quad \longrightarrow$

$-2y - 2z + 20 - y + 3z = 20 \quad \longrightarrow \quad -3y + z = 0$
$-y - z + 10 + 2y - z = 5 \qquad\qquad\qquad y - 2z = -5$

Solve the resulting system of two equations. This system can be solved by the elimination method. To do this, multiply both sides of $-3y + z = 0$ by 2 and then add the similar terms.

$-3y + z = 0 \quad \longrightarrow \quad 2(-3y + z) = 2(0) \quad \longrightarrow \quad -6y + 2z = 0$
$\quad y - 2z = -5 \qquad\qquad\qquad y - 2z = -5 \qquad\qquad\qquad \underline{\quad y - 2z = -5}$
$\qquad\qquad\qquad\qquad\qquad\qquad\qquad\qquad\qquad\qquad\qquad -5y \qquad = -5$
$\qquad\qquad\qquad\qquad\qquad\qquad\qquad\qquad\qquad\qquad\qquad\qquad y = 1$

Substitute 1 for y in $-3y + z = 0$ and solve for z.

$\qquad -3y + z = 0$
$-3(1) + z = 0$
$\quad -3 + z = 0$
$\qquad\qquad z = 3$

Substitute 1 for y and 3 for z in Equation 1 and solve for x.

Equation 1 $\quad x + y + z = 10$

$$x + 1 + 3 = 10$$
$$x + 4 = 10$$
$$x = 6$$

Check the ordered triple $(6, 1, 3)$ in all three original equations.

Equation 1	Equation 2	Equation 3
$x + y + z = 10$	$2x - y + 3z = 20$	$x + 2y - z = 5$
$6 + 1 + 3 \stackrel{?}{=} 10$	$2(6) - 1 + 3(3) \stackrel{?}{=} 20$	$6 + 2(1) - 3 \stackrel{?}{=} 5$
$10 = 10$ ✓	$12 - 1 + 9 \stackrel{?}{=} 20$	$6 + 2 - 3 \stackrel{?}{=} 5$
	$20 = 20$ ✓	$5 = 5$ ✓

The solution of the system is the ordered triple $(6, 1, 3)$.

If one of the equations in a system contains only one variable, solve for that variable first. This is shown in Example C.

EXAMPLE C

Solve the system below.

$$-2z = 2 \quad \text{Eq. 1}$$
$$-3y - 4z = -2 \quad \text{Eq. 2}$$
$$x + z = -1 \quad \text{Eq. 3}$$

Solve Equation 1 for z.
$$-2z = 2$$
$$\boxed{z = -1}$$

Substitute -1 for z in Equation 3 and solve for x.
$$x + z = -1$$
$$x + (-1) = -1$$
$$\boxed{x = 0}$$

Substitute -1 for z in Equation 2 and solve for y.
$$-3y - 4z = -2$$
$$-3y - 4(-1) = -2$$
$$-3y + 4 = -2$$
$$-3y = -6$$
$$\boxed{y = 2}$$

Check the ordered triple $(0, 2, -1)$ in all three original equations.

Equation 1	Equation 2	Equation 3
$-2z = 2$	$-3y - 4z = -2$	$x + z = -1$
$-2(-1) \overset{?}{=} 2$	$-3(2) - 4(-1) \overset{?}{=} -2$	$0 + (-1) \overset{?}{=} -1$
$2 = 2$ ✓	$-6 + 4 \overset{?}{=} -2$	$-1 = -1$ ✓
	$-2 = -2$ ✓	

The solution of the system is the ordered triple $(0, 2, -1)$.

EXERCISES 14-4

Solve each of the following systems of equations.

1. $x + y = 1$
    $x - z = 1$
    $-y + z = 4$

2. $3x - 2y = -2$
    $y - z = -1$
    $x + z = 2$

3. $4x + z = 11$
    $2x + 5y = -1$
    $3y + 4z = 9$

4. $x + y = 3$
    $x - y - z = -2$
    $x - y + z = 0$

5. $2x = 20$
    $x + z = 15$
    $4y - 2z = -6$

6. $y + 2z = 4$
    $6x - 2y = 2$
    $3x - 2z = -1$

7. $x + y = 9$
    $x + z = 8$
    $y + z = 7$

8. $3x + 3y = 3$
    $-3x - 2z = -11$
    $y + z = -1$

9. $2x + y - 5z = 4$
    $x + 3y + z = 5$
    $3x - y - 4z = -11$

10. $x + 8y + 2z = -24$
    $3x + y + 7z = -3$
    $4x - 3y + 6z = 9$

11. Applying Kirchhoff's laws to a certain circuit with three resistors satisfies the following equations:

$$I_1 + I_2 - I_3 = 0$$
$$3I_1 - 5I_2 = 2$$
$$5I_2 + 6I_3 = 109$$

Solve this system for I_1, I_2, and I_3.

12. Applying Kirchhoff's laws to a certain circuit with three resistors satisfies the following equations:

$$2I_1 + 4I_2 - 6I_3 = 32$$
$$5I_1 - 5I_2 + I_3 = -6$$
$$6I_1 - 6I_2 + 4I_3 = 4$$

Solve this system for I_1, I_2, and I_3.

13. Challenge: Solve the system below. Show your work.

$$y = 2x \quad \text{Equation 1}$$
$$z = 2y \quad \text{Equation 2}$$
$$x = 2z \quad \text{Equation 3}$$

USING SYSTEMS TO SOLVE WORD PROBLEMS

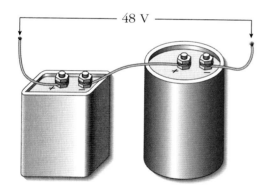
— 48 V —

Systems of equations can be used to solve some word problems. In a system of equations, each equation represents a condition placed on the variables. If there are two unknowns, you must have two variables and two equations to solve the problem.

To solve a word problem, follow these steps:

- Assign a variable to each unknown value.
- Write an equation for each condition imposed on the variables.
- Solve the system of equations.
- Check your solution.

Consider the problem below.

Two batteries are connected in series, producing a total voltage of 48 volts. The voltage of the first battery is 6 volts less than twice the voltage of the second battery.

Find the actual voltage of each battery.

Assign a variable to each unknown value.
Let x = the voltage of the first battery.
Let y = the voltage of the second battery.

Write an equation for each condition imposed on the variables.
The sum of the voltages is 48 volts. $x + y = 48$
The voltage of the first battery is
6 volts less than twice the voltage $x = 2y - 6$
of the second battery.

Helpful Hint

The substitution or the elimination method may be used to solve this system of equations. The variable x is already isolated in an equation, making the substitution method a good choice.

Solve the system by substitution. Substitute $2y - 6$ for x in the first equation.

$$x + y = 48$$
$$2y - 6 + y = 48$$
$$3y - 6 = 48$$
$$3y = 54$$
$$y = 18$$

Substitute 18 for y in the first equation.

$$x + y = 48$$
$$x + 18 = 48$$
$$x = 30$$

Check your solution.

Is the sum of the voltages 48? $30 + 18 \overset{?}{=} 48$
$$48 = 48 \checkmark$$

Is the voltage of the first battery 6 volts less than twice the voltage of the second battery?

$$30 \overset{?}{=} 2(18) - 6$$
$$30 \overset{?}{=} 36 - 6$$
$$30 = 30 \checkmark$$

The voltage of the first battery is 30 volts. The voltage of the second battery is 18 volts.

EXAMPLE A

The sum of two voltages in a series-aided circuit is 31 V. The difference between them in a series-opposed circuit is 3 V. Find the voltages.

Assign a variable to each unknown value.

Let V_1 = the first voltage.

Let V_2 = the second voltage.

Write an equation for each condition imposed on the variables.

The sum of the two voltages is 31. $V_1 + V_2 = 31$

The difference between the two voltages is 3. $V_1 - V_2 = 3$

Solve the system by elimination.

$$V_1 + V_2 = 31$$
$$V_1 - V_2 = 3$$
$$2V_1 = 32$$
$$V_1 = 17$$

Substitute 17 for V_1 in the first equation and solve for V_2.

$$V_1 + V_2 = 31$$
$$17 + V_2 = 31$$
$$V_2 = 14$$

Check the solution.

Is the sum of the numbers 31? $17 + 14 \overset{?}{=} 31$
$$31 = 31 \checkmark$$

Is the difference between the numbers 3? $17 - 14 \overset{?}{=} 3$
$$3 = 3 \checkmark$$

The voltages are 17 and 14.

EXAMPLE B

Wally mixes peanuts that cost \$2.50 per pound with cashews that cost \$5.50 per pound. He gets 3 pounds of mix that costs \$9.00. How many pounds of each type of nut are in the mix?

Assign a variable to each unknown value.
Let x = the number of pounds of peanuts.
Let y = the number of pounds of cashews.

Write an equation for each condition imposed on the variables.
He gets 3 pounds of mix. $x + y = 3$
The cost of the mix is \$9.00. $2.50x + 5.50y = 9.00$

Solve the system by substitution.
$$x + y = 3$$
$$x + y - y = 3 - y$$
$$x = 3 - y$$

Substitute $3 - y$ for x in the second equation and solve for y.
$$2.50x + 5.50y = 9.00$$
$$2.50(3 - y) + 5.50y = 9.00$$
$$7.50 - 2.50y + 5.50y = 9.00$$
$$7.50 + 3.00y = 9.00$$
$$3.00y = 1.50$$
$$y = 0.5$$

Substitute 0.5 for y in the first equation and solve for x.
$$x + y = 3$$
$$x + 0.5 = 3$$
$$x = 2.5$$

Check your solution.
Is the number of pounds 3? $2.5 + 0.5 \stackrel{?}{=} 3$
$$3 = 3 \checkmark$$

Is the cost of the mix \$9.00? $2.50(2.5) + 5.50(0.5) \stackrel{?}{=} 9.00$
$$6.25 + 2.75 \stackrel{?}{=} 9.00$$
$$9.00 = 9.00 \checkmark$$

There are 2.5 pounds of peanuts and 0.5 pound of cashews in the mix.

Helpful Hint

You could also solve this system by elimination. To eliminate x, multiply the first equation by (-2.50).

EXERCISES 14-5

Answer each of the following.

1. The sum of two numbers is 9. The greater number is 3 more than twice the lesser number. What are the numbers?

2. The length of a rectangle is 3 times the width. The difference between the length and the width is 9 feet. Determine the rectangle's length and width.

3. Ray charges \$35 per hour for his repair service. Wanda charges a flat fee of \$25 plus \$30 per hour for her repair service. For which number of hours will their charge be the same? What will that charge be?

4. Aaron deposited a total of \$2,000 in two savings accounts. One account pays 3% interest per year, and the other pays 4% per year. If Aaron earned \$72 interest in one year, how much did he deposit in each account?

Helpful Hint

For Exercise 4, use the formula
Interest = Rate × Time. Remember to write the interest rate as a decimal in the problem.

5. The sum of two voltages in a series-aided circuit is 52 V. The difference between them in a series-opposed circuit is 18 V. Find the voltages.

6. Applying Kirchhoff's laws, we find that the sum of three currents is 30. The third current is the sum of the first and second. The third current is three times the second. What are the three currents?

7. Pavel asks for some bids for transistors and resistors. One bid says that he can get 20 resistors and 20 transistors for $28.40. Another states that 30 resistors and 10 transistors will cost $22.20. What is the price of each resistor and transistor?

8. **Challenge:** Marla bought twice as many pounds of cherries as grapes. She paid $15.00 for a total of 4.5 pounds of both items. Cherries cost twice as much as grapes. How much do grapes cost per pound?

CHAPTER 14 REVIEW EXERCISES

1. Solve the following system by graphing.

   $x + y = 9$

   $y - 2x = 0$

2. Solve the following system of equations by substitution.

   $2V - 2I = 16$

   $2I = V + 9$

3. In a certain series-opposing circuit, $V_1 - V_2 = 23$ V. If the same series is series-aiding, then $V_2 + V_1 = 57$ V.

   (a) Graph each line with V_1 on the horizontal axis.

   (b) Determine the solution (if any) of this system.

4. Solve the following system of equations by elimination.

   $x + y = 32$

   $7x - 3y = 29$

5. Solve the following system of equations.

   $x + y + z = 15$

   $5x - z = 18$

   $4y + z = 19$

6. Pavel is going to buy some compact fluorescent lightbulbs (CFLs). He can purchase 35 of the 20 W CFL Spiral and 20 of the 25 W CFL Spiral bulbs for $352.15, or he can purchase 45 of the 20 W CFL Spiral and 10 of the 25 W CFL Spiral bulbs for $344.65. Determine the price of each bulb.

7. Applying Kirchhoff's laws to a certain circuit with three resistors satisfies the following equations:

$$I_1 + I_3 = I_2$$
$$5.2I_1 - 3.25I_2 = -12.35$$
$$3.25I_2 + 2.6I_3 = 137.15$$

Solve this system for I_1, I_2, and I_3.

Building a Foundation in Mathematics

Basic Concepts

Measuring and Drawing Angles

Angles, Angle Pairs, and Parallel Lines

Polygons and Their Properties

Finding Perimeter

Finding Area

Finding Circumference and Area of Circles

Congruent and Similar Triangles

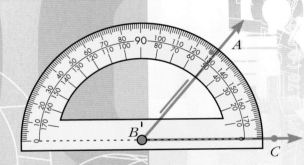

Overview

Plane geometry is the part of mathematics that deals with points, lines, angles, and various figures that are made up of points, lines, and angles. The figures all lie on a flat surface, called a plane. An example of plane geometry is a wiring diagram for a building.

Geometry is fundamental to wiring a building, so it is essential to understand the definitions and terms used in geometry. It is just as important to be able to apply the geometric principles in problem solving. In this chapter, you will learn not only the definitions and terms of geometry but also how to apply them in real-life electrical problems.

Plane Geometry

Chapter 15

Objectives

After completing this chapter, you will be able to:

- Identify types of angles and find their measure
- Use a protractor to draw an angle of a given size
- Identify types of triangles by sides or angles
- Identify various types of polygons
- Identify corresponding sides and corresponding angles of congruent or similar triangles
- Determine whether given pairs of triangles are similar
- Identify similar parts of polygons
- Compute lengths of sides and perimeters of similar polygons
- Compute areas of common polygons, given heights and bases
- Compute heights of common polygons, given areas and bases
- Compute bases of common polygons, given areas and heights
- Compute radii, diameters, circumferences, and areas of circles
- Solve applied problems involving polygons
- Solve applied problems involving circles

BASIC CONCEPTS

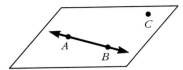

The study of geometry is based on three fundamental figures: points, lines, and planes.

A **point** is a specific location. A point is represented by a dot and is named by a single letter.

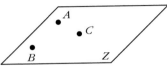

Point *P*

A **line** has no end points and extends in a straight path endlessly in both directions. A line is named either by two points that lie on the line or by a single letter.

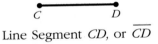

Line *AB*, $\overleftrightarrow{AB}$, or *ℓ*

A **plane** is a flat surface that extends endlessly in all directions. A plane can be named by either a single letter or by three points in the plane.

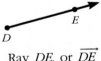

Plane *Z*, or plane *ABC*

Using points, lines, and planes, we can define line segments, rays, and angles.

A **line segment** is a portion of a line. It consists of two points of the line and all points between them. Line segments are named by their end points.

Line Segment *CD*, or $\overline{CD}$

A **ray** is the portion of a line that starts at one point and extends endlessly in one direction. Rays are named by their end point and another point on the ray, in that order.

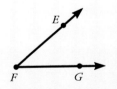

Ray *DE*, or $\overrightarrow{DE}$

An **angle** is formed by two rays that have the same end point. The rays are called the *sides* of the angle, and the common end point is called the **vertex** of the angle.

Angle *EFG*, or ∠*EFG*

There are different ways to name lines, line segments, and rays. Some examples are shown below.

$\overrightarrow{PQ}$ and $\overleftrightarrow{QP}$ name the same line.

$\overline{JK}$ and $\overline{KJ}$ name the same line segment.

$\overrightarrow{AB}$ and $\overrightarrow{AC}$ name the same ray.

There are also different ways to name angles. Some of the customary ways to name angles are shown below.

∠B or ∠ABC or ∠CBA or ∠1

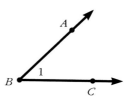

∠LMN or ∠NML or ∠2

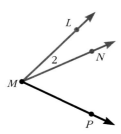

∠NMP or ∠PMN

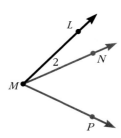

∠LMP or ∠PML

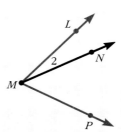

Notice that ∠B above can be named by a single letter because it is the only angle with vertex B. The angles that share vertex M each require three letters in their names, when letters are used. When three letters are used, the letter for the vertex must be in the middle.

Name each figure below in as many ways as possible.

1.

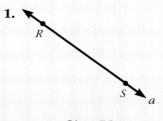

Line RS

2.

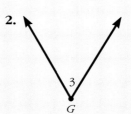

Angle 3

1. $\overleftrightarrow{RS}$, $\overleftrightarrow{SR}$, a

2. ∠3, ∠G

EXAMPLE A

Name each figure in as many ways as possible.

1.

2.

Line Segment *AF*   Ray *PQ*

1. $\overline{AF}$, $\overline{FA}$ 2. $\overrightarrow{PQ}$, $\overrightarrow{PR}$

EXAMPLE B

Name each angle in as many ways as possible.

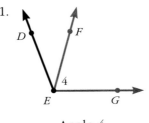

1.   2.

Angle 4   Angle *GED*

1. ∠4, ∠*FEG*, ∠*GEF*   2. ∠*GED*, ∠*DEG*

EXAMPLE C

Name the sides and vertex of angle *STU*.

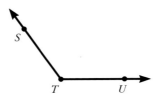

The sides are $\overrightarrow{TS}$ and $\overrightarrow{TU}$. The vertex is *T*.

EXAMPLE D

Name the point at which lines *ℓ* and *m* intersect. Name the rays that form ∠5.

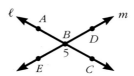

If two lines meet, we say they intersect.

Lines *ℓ* and *m* intersect at point *B*. The rays that form ∠5 are $\overrightarrow{BE}$ and $\overrightarrow{BC}$.

EXERCISES 15-1

For Exercises 1–4, use the diagram below.

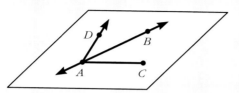

1. Name a line.

2. Name a ray.

3. Name a line segment.

4. Name the angle that has sides $\overrightarrow{AD}$ and $\overrightarrow{AB}$.

5. What is the vertex of $\angle LMN$?

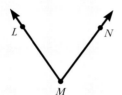

6. In $\angle POR$, what is the name of the ray on which point P lies?

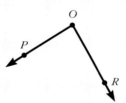

7. Give four different names for the angle shown below.

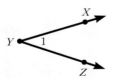

8. Name three angles in the diagram below.

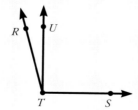

9. Challenge: Look at the diagram below.

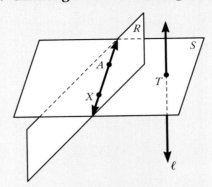

What is the intersection of planes *R* and *S*? What is the intersection of line ℓ and plane *S*?

MEASURING AND DRAWING ANGLES

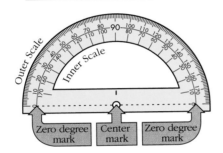

A protractor is used to measure and draw angles.

Angles are measured in **degrees.** To measure the number of degrees in an angle, or to draw an angle with a specific measure, a tool called a **protractor** is used. The diagram at the left shows a protractor. The numbers around the curved edge of the protractor are used to find the measure of an angle, in degrees, when the vertex of the angle is at the center mark.

The protractor has an inner scale with degree marks from 0° to 180° counterclockwise and an outer scale with degree marks from 0° to 180° clockwise. To measure or draw an angle, always use the scale that has 0° touching one side of the angle. If you use the 0° mark on the right side of the protractor, you will measure using the inner scale of the protractor. On the other hand, if you use the 0° mark on the left side of the protractor, you will measure using the outer scale.

Find the measure of ∠ABC.

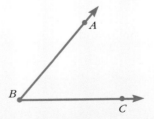

Language Box

The symbol for degrees is °. The measure of an angle is denoted by *m.* The statement *m*∠*ABC* = 50° is read "the measure of angle ABC equals 50 degrees."

Line up the vertex of the angle, point *B,* with the center mark on the protractor. Line up a side of the angle with a 0° mark.

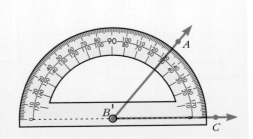

The measure of ∠*ABC* is given by the degree mark on the protractor where the other side of the angle intersects the inner scale. So, *m*∠*ABC* = 50°.

If you use the 0° mark on the left side of the protractor, you will measure using the outer scale of the protractor. This is shown in Example A.

EXAMPLE A

Find the measure of ∠*DEF* shown at the right.

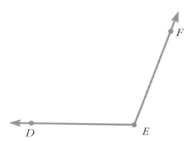

Line up the vertex of the angle, point *E*, with the center mark on the protractor. Line up a side of the angle with a 0° mark. This time, use the outer scale of the protractor.

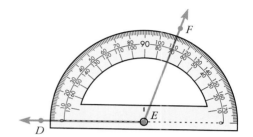

The degree mark where the other side of the angle intersects the outer scale of the protractor is 110°. So, *m*∠*DEF* = 110°.

Another way to make sure that you use the correct scale on the protractor is to first determine whether the measure of the angle is more or less than 90°. This can reassure you that you are reading the correct scale on the protractor.

A 90° angle is a **right angle.** In the diagram below, ∠*BED* is a right angle because *m*∠*BED* = 90°. Angle *CED* has a measure less than 90°; its measure is correctly read as 50°, not 130°. Angle *AED* has a measure more than 90°; its measure is correctly read as 130°, not 50°.

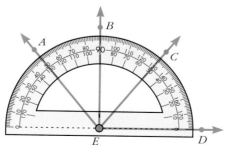

EXAMPLE B

Draw ∠LMN with a measure of 40°.

A protractor can also be used to draw an angle with a specific measure.

When drawing an angle with a specific measure, you should use the following four steps.

STEP 1

Draw one side of the angle. Label points *M* and *N*.

STEP 2

Point *M* is the vertex because it is the middle letter in the name of the angle. Line up the vertex with the center mark on the protractor. Line up the side of the angle with the 0° mark.

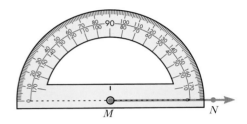

STEP 3

Find the 40° mark on the inner scale of the protractor. Make a dot at 40° and label it point *L*.

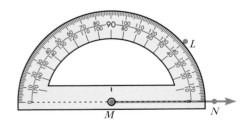

STEP 4

Draw a ray from point *M* through point *L*.

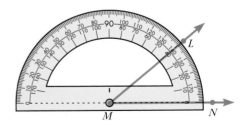

Angle *LMN* is shown above and has a measure of 40°.

By definition, an angle is formed by two rays. However, an angle can be represented in a diagram by line segments; the arrowheads used to indicate the rays are not necessary. Both diagrams below represent angle *ABC*.

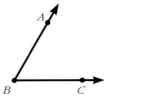

EXAMPLE C

Find the measure of ∠Q in the triangle.

In triangles and other figures, angles are often represented by line segments.

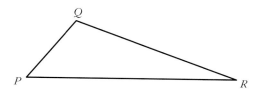

STEP 1

Rotate the triangle to make it easy to measure ∠Q.

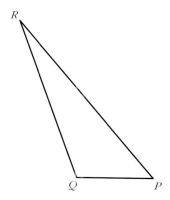

Helpful Hint

You may find it easier to rotate the protractor than to rotate the triangle.

STEP 2

Measure ∠Q.

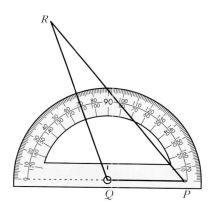

$m\angle Q = 110°$

EXERCISES 15-2

Use a protractor to measure each angle in Exercises 1–4.

1.

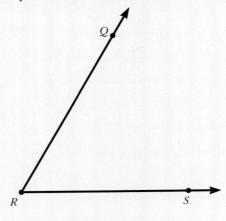

2.

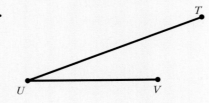

3.

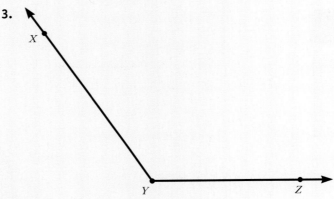

4.

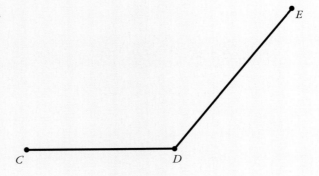

Draw and label an angle with the specified measure in Exercises 5–8.

5. $m\angle ABC = 45°$

6. $m\angle DEF = 70°$

7. $m\angle GHI = 150°$

8. $m\angle JKL = 115°$

Without using a protractor, estimate the correct size of each of the angles in Exercises 9–10 as 30°, 45°, 60°, or 120°.

9.

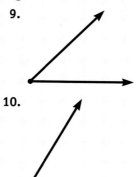

10.

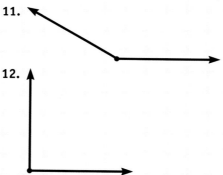

Without using a protractor, estimate the correct size of each of the angles in Exercises 11–12 as 90°, 135°, 150°, or 180°.

11.

12.

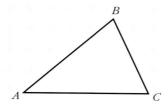

13. Find the measure of $\angle B$ in the triangle.

> ### Helpful Hint
>
> With a pencil, make the sides of $\angle B$ longer. This will make it easier to use your protractor to measure the angle.

14. Find the measure of ∠*C* in the figure below.

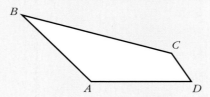

15. What's the Error? Rodney used a protractor to measure angle *RST* below and stated that *m*∠*RST* = 80°.

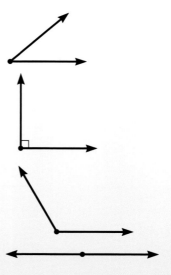

What error did Rodney make? What is the correct measure of angle *RST*?

ANGLES, ANGLE PAIRS, AND PARALLEL LINES

Classifying Angles

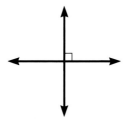

Perpendicular lines form right angles.

Angles are classified according to their measures.

An **acute angle** is an angle that measures less than 90°.

A **right angle** is an angle that measures 90°. A right angle is indicated by a small square at its vertex.

An **obtuse angle** is an angle that measures more than 90° and less than 180°.

A **straight angle** is an angle that measures 180°.

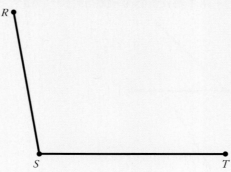

Classify each angle based on its measure. Explain.

∠B is an acute angle because its measure is less than 90°.

∠ACB is a straight angle because its measure is 180°.

EXAMPLE A

Classify each angle according to its measure. Explain.

∠X is an obtuse angle because its measure is greater than 90° and less than 180°.

∠Y is a right angle because its measure is exactly 90°.

Perpendicular lines are lines that form right angles.

EXAMPLE B

Name the right angles formed by the perpendicular lines at the right.

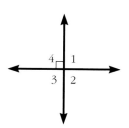

Each of the angles formed by the perpendicular lines is a right angle. So, ∠1, ∠2, ∠3, and ∠4 are all right angles.

Angle Pairs

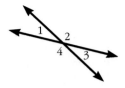

So far, you have learned how to name and classify angles and how to measure and draw angles. In this section, you will learn about certain *pairs* of angles that have special characteristics.

Vertical angles are formed by two intersecting lines; they share a common vertex but have no common side. In the figure at the left, the intersecting lines form two pairs of vertical angles: $\angle 1$ and $\angle 3$ are vertical angles; $\angle 2$ and $\angle 4$ are vertical angles. Vertical angles have equal measure, so $m\angle 1 = m\angle 3$ and $m\angle 2 = m\angle 4$.

What is the measure of $\angle 4$ in the figure below? Explain.

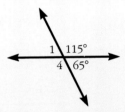

There are two intersecting lines, so two pairs of vertical angles are formed. One of these pairs is $\angle 4$ and the angle measuring 115°.

Because the measures of vertical angles are equal, $m\angle 4 = 115°$.

Two angles in the same plane that have a common vertex and a common side are called **adjacent angles.** In the diagram below, $\angle 5$ and $\angle 6$ are adjacent angles.

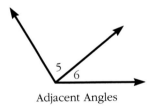

Adjacent Angles

Complementary angles are two angles whose measures have a sum of 90°. In the figure below, $\angle A$ and $\angle B$ are complementary angles because $35° + 55° = 90°$.

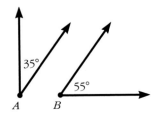

As you will see in Example C, a pair of adjacent complementary angles forms a right angle.

EXAMPLE C

In the figure at the right, name a pair of complementary angles. What is the measure of $\angle NMP$?

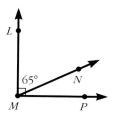

The sum of the measures of $\angle LMN$ and $\angle NMP$ is 90°, so $\angle LMN$ and $\angle NMP$ are complementary.

$m\angle NMP = 90° - 65° = 25°$

$\angle LMN$ and $\angle NMP$ are complementary, and $m\angle NMP = 25°$.

Supplementary angles are two angles whose measures have a sum of 180°. In the figure below, $\angle C$ and $\angle D$ are supplementary angles because $60° + 120° = 180°$.

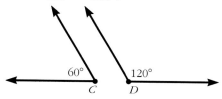

In Example D, you will see that a pair of adjacent supplementary angles forms a straight angle.

EXAMPLE D

In the figure at the right, name a pair of supplementary angles. What is the measure of $\angle ZXY$?

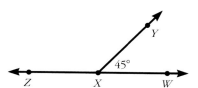

The sum of the measures of $\angle ZXY$ and $\angle WXY$ is 180°, so $\angle ZXY$ and $\angle WXY$ are supplementary.

$m\angle ZXY = 180° - 45° = 135°$

$\angle ZXY$ and $\angle WXY$ are supplementary, and $m\angle ZXY = 135°$.

We can use what we know about vertical angles and supplementary angles to help us find missing angle measures.

EXAMPLE E

Find the values of *x* and *y* in the figure at the right.

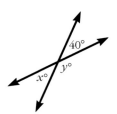

First, notice that the angle labeled *x*° and the 40° angle are vertical angles. This means that their measures are equal, and *x* = 40.

Now find the value of *y*.

$y + 40 = 180$ ← The angles labeled *y*° and 40° are supplementary.

$y = 180 - 40$ ← Subtract 40 from both sides.

$y = 140$ ← Simplify.

$x = 40$ and $y = 140$.

Parallel Lines and Angle Pairs

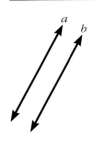

Parallel lines are lines in the same plane that do not intersect, no matter how far they are extended. In the figure at the left, lines *a* and *b* are parallel lines.

When two or more lines in the same plane are intersected by another line, the intersecting line is called a **transversal.** In the figure at the right, line *t* is a transversal.

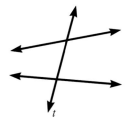

From here on, we will limit our discussion to parallel lines. Important angle relationships result when *parallel* lines are intersected by a transversal.

When a transversal intersects two parallel lines, several types of congruent angle pairs are formed. The types are described below.

∠2 and ∠6 are **corresponding angles,** so ∠2 ≅ ∠6.

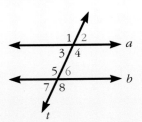

Language Box

Angles with the same measure are called congruent (≅) angles.

The following are also congruent angle pairs because they are corresponding angles: ∠1 ≅ ∠5, ∠4 ≅ ∠8, and ∠3 ≅ ∠7.

∠1 and ∠8 are **alternate exterior angles**, and ∠1 ≅ ∠8.

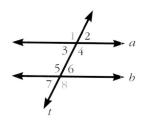

Another pair of angles are congruent because they are alternate exterior angles: ∠2 ≅ ∠7.

∠3 and ∠6 are **alternate interior angles,** and ∠3 ≅ ∠6.

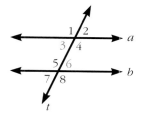

There is another pair of angles that are congruent because they are alternate interior angles: ∠4 ≅ ∠5.

Using vertical angles and the angle pairs discussed above, we know that the following sets of angles are congruent:

∠1 ≅ ∠4 ≅ ∠5 ≅ ∠8

∠2 ≅ ∠3 ≅ ∠6 ≅ ∠7

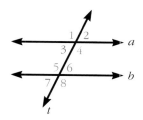

In the figure below, find $m\angle 7$ and $m\angle 8$.

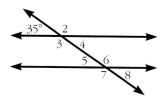

One way to find the measures of the angles is as follows:

$m\angle 8 = 35°$ ← Alternate exterior angles

$m\angle 7 + m\angle 8 = 180°$ ← Supplementary angles

$m\angle 7 + 35° = 180°$

$m\angle 7 = 145°$

So, $m\angle 7 = 145°$ and $m\angle 8 = 35°$.

EXAMPLE F

In the figure at the right, $m\angle 4 = 56°$. Find the measures of all of the angles.

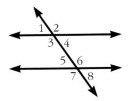

$$m\angle 2 + m\angle 4 = 180°$$ ← Supplementary angles

$$m\angle 2 + 56° = 180°$$

$$m\angle 2 = 124°$$

Now list the two sets of congruent angles:

$$\angle 1 \cong \angle 4 \cong \angle 5 \cong \angle 8$$

$$\angle 2 \cong \angle 3 \cong \angle 6 \cong \angle 7$$

So, $m\angle 1 = m\angle 4 = m\angle 5 = m\angle 8 = 56°$, and $m\angle 2 = m\angle 3 = m\angle 6 = m\angle 7 = 124°$.

EXERCISES 15-3

For Exercises 1–4, classify each angle according to its measure. Explain.

1.

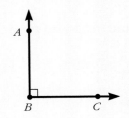

2.

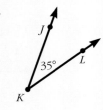

3.

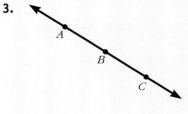

4.

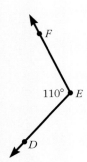

Solve the following.

5. Name the right angles that are formed by the perpendicular lines below.

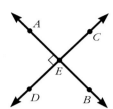

6. In the figure below, what is the measure of $\angle CEB$?

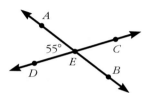

7. In the figure below, name a pair of complementary angles. What is the measure of $\angle ABD$?

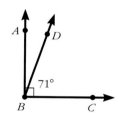

8. In the figure below, name a pair of supplementary angles. What is the measure of $\angle MNP$?

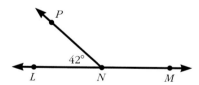

9. What are the values of x and y in the figure below?

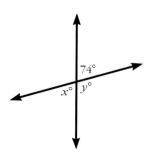

10. In the figure below, a transversal intersects two parallel lines, and $m\angle 3 = 24°$.

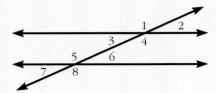

Find the measures of $\angle 5$ and $\angle 6$.

11. In the figure below, a transversal intersects two parallel lines, and $m\angle 4 = 110°$.

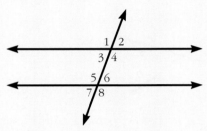

Find the measures of all of the angles.

12. Can a pair of vertical angles ever be complementary? Explain.

13. Describe a situation in which a transversal intersects two parallel lines, forming eight congruent angles.

14. An offset bend, like the one in the figure, is used to run conduit around an obstruction. In this figure, the offset angle between side AB and side BC is $43°$. What is the size of $\angle DCB$?

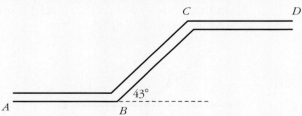

15. What's the Error? Brian stated that in *every* pair of supplementary angles, one of the angles is an obtuse angle and one of the angles is an acute angle. Give an example of two supplementary angles that shows that this statement is false.

POLYGONS AND THEIR PROPERTIES

Polygons

These are polygons.

A **polygon** is formed by three or more line segments (sides) all in the same plane, satisfying the following conditions:

* No adjacent sides lie on the same line.
* Each side intersects exactly two other sides.
* The sides intersect only at their end points (vertices).

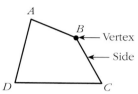

Name a polygon by giving its vertices in consecutive order. Two ways to name the polygon above are *ABCD* and *ADCB*.

A polygon is **equilateral** if all of its sides are congruent (the same length).

A polygon is **equiangular** if all of its angles are congruent (have the same measure).

A polygon is **regular** if it is both equilateral and equiangular.

Equilateral Polygon

Equiangular Polygon

Regular Polygon

We use identical marks to show that sides or angles have the same measure in a diagram. For example, in polygon *PQRS* above, $PQ = QR = RS = SP$. In polygon *ABCDEF*, $m\angle A = m\angle B = m\angle C = m\angle D = m\angle E = m\angle F$.

Is the figure below a polygon? If so, is it equilateral, equiangular, regular, or none of these?

The figure is a polygon, all of the sides are congruent, and not all of the angles are congruent.

The figure is an equilateral polygon, but it is not equiangular or regular.

EXAMPLE A

Is the figure at the right a polygon? If so, is it equilateral, equiangular, regular, or none of these?

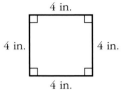

4 in.

4 in.   4 in.

4 in.

The figure is a polygon, all of the sides are congruent, and all of the angles are congruent.

The figure is a regular polygon.

EXAMPLE B

Is the figure at the right a polygon? If so, is it equilateral, equiangular, regular, or none of these?

The figure contains a curved portion, so it is not a polygon.

The figure is not a polygon.

Polygons are classified by the number of sides they have. The table below shows the names and number of sides for some basic polygons.

Name	Number of Sides
Triangle	3
Quadrilateral	4
Pentagon	5
Hexagon	6
Heptagon	7
Octagon	8
Nonagon	9
Decagon	10

Several polygons are shown below.

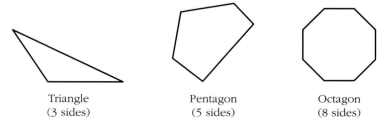

Triangle
(3 sides)

Pentagon
(5 sides)

Octagon
(8 sides)

Triangles

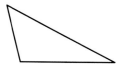

The prefix *tri-* means three.

Triangles are so important that they deserve special attention. Triangles can be classified by either sides or angles.

Classifying Triangles

By Sides

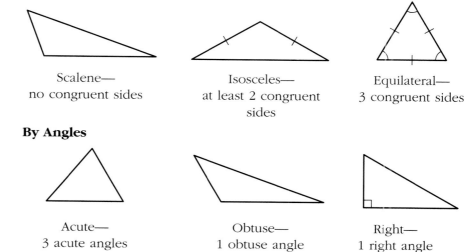

Scalene—
no congruent sides

Isosceles—
at least 2 congruent
sides

Equilateral—
3 congruent sides

By Angles

Acute—
3 acute angles

Obtuse—
1 obtuse angle

Right—
1 right angle

Which kind of triangle is shown below?

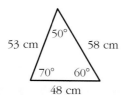

53 cm   50°   58 cm
70°   60°
48 cm

None of the sides are congruent. None of the angles are congruent, and each of the angles measures less than 90°.

The triangle is scalene and acute.

EXAMPLE C

Which kind of triangle is shown at the right?

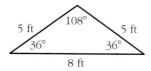

5 ft   108°   5 ft
36°   36°
8 ft

Two sides of the triangle are congruent. One angle measures more than 90°.

The triangle is isosceles and obtuse.

The sum of the measures of the angles of *any* triangle is 180°. You can use this fact to solve the problem in Example D.

EXAMPLE D

In the triangle at the right, what is the measure of ∠A?

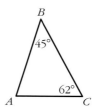

$$m\angle A + m\angle B + m\angle C = 180°$$ ← The sum of the angle measures is 180°.

$$m\angle A + 45° + 62° = 180°$$ ← Substitute the known values.

$$m\angle A + 107° = 180°$$ ← Simplify.

$$m\angle A = 73°$$

The measure of ∠A is 73°.

Quadrilaterals

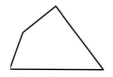

The prefix *quad-* means four.

A polygon that has four sides is called a **quadrilateral.** Special quadrilaterals and some of their properties are shown below.

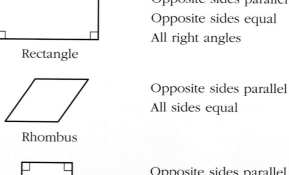

Quadrilateral	Properties
Parallelogram	Opposite sides parallel Opposite sides equal
Rectangle	Opposite sides parallel Opposite sides equal All right angles
Rhombus	Opposite sides parallel All sides equal
Square	Opposite sides parallel All sides equal All right angles

One pair of opposite sides parallel

Trapezoid

It is worth noting that rectangles, rhombuses, and squares are all parallelograms. And a square is both a rectangle and a rhombus.

Another useful property of rectangles and squares is that the diagonals are equal.

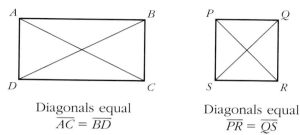

Diagonals equal
$\overline{AC} = \overline{BD}$

Diagonals equal
$\overline{PR} = \overline{QS}$

What type of quadrilateral is shown below?

All of the angles are right angles, so it is a rectangle. And because it is a rectangle, it is also a parallelogram.

The quadrilateral is both a rectangle and a parallelogram.

In Example E, it will be shown that properties of quadrilaterals can be used when building an object that must be a particular shape.

EXAMPLE E

Eric is using stakes and twine to lay out the edges of a foundation for a shed. He has the correct lengths for the edges. How can he make a rectangle with all right angles?

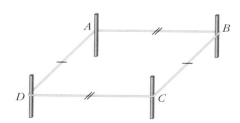

Eric already has a parallelogram because both pairs of opposite sides are equal. If he makes the diagonals $\overline{AC}$ and $\overline{BD}$ equal, the parallelogram will be a rectangle.

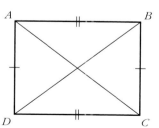

Eric can make the foundation a rectangle by making the diagonals equal.

A polygon is **convex** if the line segment that connects any two points in the interior lies entirely in the interior.

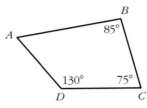

Convex   Not Convex

Most of the useful properties of polygons apply to convex polygons, including the following property.

The sum of the measures of the angles of any quadrilateral is 360°. You can use this fact to solve the problem in Example F.

EXAMPLE F

In the quadrilateral at the right, what is the measure of ∠A?

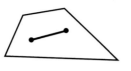

$m\angle A + m\angle B + m\angle C + m\angle D = 360°$ ← The sum of the angle measures is 360°.

$m\angle A + 85° + 75° + 130° = 360°$ ← Substitute the known values.

$m\angle A + 290° = 360°$ ← Simplify.

$m\angle A = 70°$

The measure of ∠A is 70°.

EXERCISES 15-4

For Exercises 1–5, determine whether each figure is a polygon. If so, name the polygon and tell whether it is equilateral, equiangular, regular, or none of these.

1.

2.

3.

4.

5.

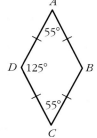

For Exercises 6–11, find the missing angle measure.

6.

7.

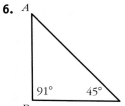

8.

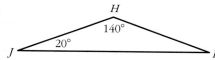

9.

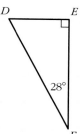

10.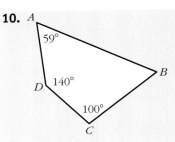

11.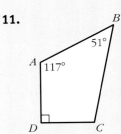

Solve the following.

12. Ryan assembled a screen door, as shown below. He made both pairs of opposite edges equal.

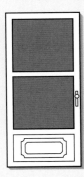

How can Ryan make sure all the corners of the door are right angles?

13. Trent says that every rhombus is also a square. Is he correct? Explain.

14. A vertical transmission tower is supported by several guy wires. One of the wires makes an angle of 57° with the level ground. What is the angle between the wire and the tower where the wire is connected to the tower?

15. Challenge: Find the value of x in the figure below.

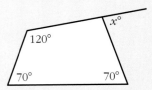

FINDING PERIMETER

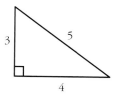

The **perimeter** of a polygon is the sum of the lengths of all of its sides.

What is the perimeter of the triangle shown at the left?

To find the perimeter, add the lengths of the sides:

$P = 3 + 4 + 5$

$\quad = 12$

The perimeter of the triangle is 12 units.

In Example A, you will see how to find the perimeter of a rectangle.

EXAMPLE A

Find the perimeter of the rectangle shown at the right.

Opposite sides of a rectangle are equal, so the unlabeled sides are 6 ft and 2 ft.

$P = 6 + 2 + 6 + 2$

$\quad = 16$

The perimeter of the rectangle is 16 feet.

Helpful Hint

You can also use the formula $P = 2l + 2w$ to find the perimeter of a rectangle.

EXAMPLE B

Betty is installing weather stripping around the outside edge of the window shown at the right. What is the total length of weather stripping she needs?

Finding the perimeter is useful when trying to find what length of a material is needed to go around a polygon.

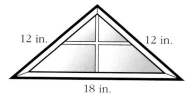

To find the total length of weather stripping, find the perimeter of the window:

$P = 12 + 12 + 18$

$\quad = 42$

Betty will need 42 inches of weather stripping.

In Example C, you will see that you can use this method to find the perimeter of *any* polygon.

EXAMPLE C

Find the perimeter of the polygon at the right.

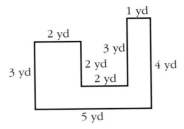

The perimeter is found by adding the lengths of the sides:

$P = 5 + 3 + 2 + 2 + 2 + 3 + 1 + 4$

$\quad = 22$

The perimeter of the polygon is 22 yards.

EXERCISES 15-5

For Exercises 1–7, find the perimeter of each polygon.

1.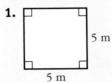

5 m

5 m

2. 2 in.

10 in.

3.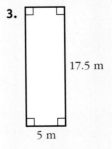

17.5 m

5 m

4.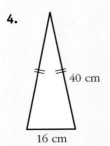

40 cm

16 cm

5.

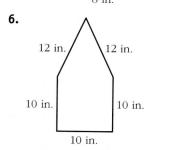

14 in.

5 in.

8 in.

6.

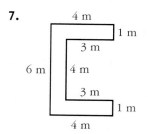

12 in. 12 in.

10 in. 10 in.

10 in.

7.

4 m

1 m

3 m

6 m 4 m

3 m

1 m

4 m

Solve the following.

8. Tony is installing cable between the three utility poles shown below.

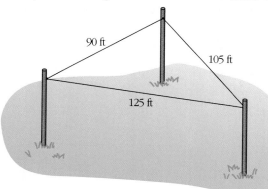

90 ft

105 ft

125 ft

How many feet of cable will Tony need?

9. Josh wants to put a fence around his garden, which is shown below.

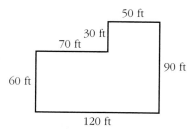

50 ft

30 ft

70 ft

60 ft

90 ft

120 ft

How many feet of fencing will he need?

10. Kevin claims that if two rectangles each have a perimeter of 26 meters, both rectangles must have the same length and the same width. Is he correct? If not, give an example.

11. A rectangular computer chip is 3.1 mm longer than it is wide. If the perimeter of the chip is 23.4 mm, what are the length and width of the chip?

12. Part of an electrical circuit is in the shape of an isosceles trapezoid. The lengths of the two parallel sides are 4.3 m and 3.8 m. The slanted sides are each 2.7 m long. What is the length of this circuit?

13. Challenge: The perimeter of the polygon below is 132 inches.

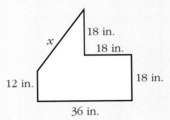

Find the value of *x*.

FINDING AREA

Area of Rectangles and Parallelograms

1 square centimeter
(1 cm²)

The **area** of a plane figure is the number of square units inside the figure. A **square unit** is a square that measures 1 unit on each side. Some square units are square centimeters (cm²), square inches (in.²), and square feet (ft²).

In the figure below, we can see that there are 8 square centimeters inside the rectangle.

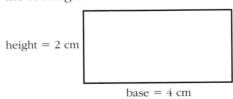

So, we say that the area of the rectangle is 8 square centimeters, abbreviated 8 cm².

In general, we can find the area *A* of a rectangle by multiplying the length of its base, *b*, by its height, *h*.

Area of a Rectangle

The area of a rectangle with base *b* and height *h* is:

$$A = bh$$

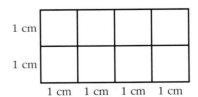

The formula for the area of a rectangle, along with every other formula in this chapter, is summarized in the appendices.

Carla's kitchen floor is a rectangle that measures 15 feet by 11 feet. What is the area of her kitchen floor?

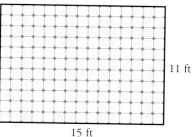

11 ft

15 ft

$A = bh$ ← Write the formula for area.

$= 15 \cdot 11$ ← Substitute known values.

$= 165$ ← Simplify.

The area of Carla's kitchen floor is 165 ft².

EXAMPLE A

Find the area of a square with side lengths of 3 m.

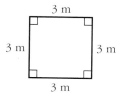

3 m

3 m   3 m

3 m

Helpful Hint

You can also find the area of a square by *squaring* its side length, *s*. For example, $A = s^2 = 3^2 = 9$.

A square is a rectangle. Both the base and the height of the square are 3 m.

$A = bh$ ← Write the formula for area.

$= 3 \cdot 3$ ← Substitute known values.

$= 9$ ← Simplify.

The area of the square is 9 m².

The area of a parallelogram is found by using the same formula as that used to find the area of a rectangle, $A = bh$. This is because a parallelogram can be formed from a rectangle, as shown below:

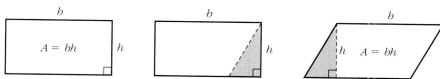

Think of forming the parallelogram from a rectangle by moving a right triangle.

Area of a Parallelogram

> The area of a parallelogram with base b and height h is:
>
> $A = bh$
>
>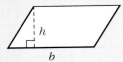

EXAMPLE B

What is the area of the parallelogram shown at the right?

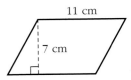

The base is 11 cm and the height is 7 cm.

$A = bh$ ← Write the formula for area.

$\quad = 11 \cdot 7$ ← Substitute known values.

$\quad = 77$ ← Simplify.

The area of the parallelogram is 77 cm².

Area of Triangles

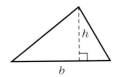

By drawing a diagonal of a parallelogram, you will create two identical triangles.

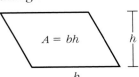   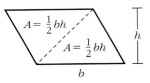

You know that the formula for the area of a parallelogram is $A = bh$. So, because the parallelogram is split into two identical triangles, the area of each triangle is half the area of the parallelogram.

Area of a Triangle

> The area of a triangle with base b and height h is:
>
> $A = \frac{1}{2}bh$
>
>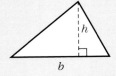

Mike has a triangular garden, as shown below. What is the area of his garden?

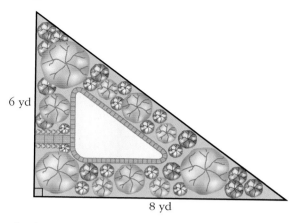

6 yd

8 yd

The base of the triangle is 8 yd, and the height of the triangle is 6 yd.

$A = \frac{1}{2}bh$ ← Write the formula for area.

$= \frac{1}{2}(8 \cdot 6)$ ← Substitute known values.

$= \frac{1}{2}(48)$ ← Simplify.

$= 24$

The area of the garden is 24 yd².

In the following example, the height is not a side of the triangle. The height is the length of the segment perpendicular to the base.

EXAMPLE C

Find the area of the triangle shown at the right.

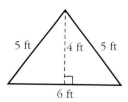

5 ft   4 ft   5 ft

6 ft

The base of the triangle is 6 feet and the height is 4 feet. The 5-foot dimension is not used to find the area.

$A = \frac{1}{2}bh$ ← Write the formula for area.

$= \frac{1}{2}(6 \cdot 4)$ ← Substitute known values.

$= \frac{1}{2}(24)$ ← Simplify.

$= 12$

The area of the triangle is 12 ft².

Area of Trapezoids

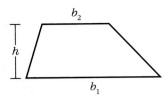

A trapezoid has exactly one pair of parallel sides. The parallel sides are called bases, and we label them b_1 and b_2. The diagram below shows that the diagonal of a trapezoid splits the trapezoid into two triangles. The bases of the triangles are b_1 and b_2, and the height of both triangles is h.

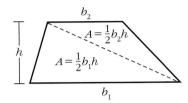

The area of the trapezoid is the sum of the areas of the two triangles:

$$A = \tfrac{1}{2}b_1h + \tfrac{1}{2}b_2h$$

Using the Distributive Property, we can rewrite the area formula as

$$A = \tfrac{1}{2}b_1h + \tfrac{1}{2}b_2h = \tfrac{1}{2}h(b_1 + b_2)$$

Area of a Trapezoid

The area of a trapezoid with bases b_1 and b_2, and height h is:

$$A = \tfrac{1}{2}h(b_1 + b_2)$$

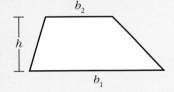

Find the area of the trapezoid below.

6 ft

3 ft

9 ft

The bases of the trapezoid are 6 ft and 9 ft, and the height is 3 ft.

$A = \tfrac{1}{2}h(b_1 + b_2)$ ⟵ Write the formula for area.

$= \tfrac{1}{2}(3)(6 + 9)$ ⟵ Substitute known values.

$= \tfrac{1}{2}(3)(15)$ ⟵ Simplify; add inside of the parentheses first.

$= \tfrac{1}{2}(45)$

$= 22\tfrac{1}{2}$

The area of the trapezoid is $22\tfrac{1}{2}$ ft^2.

Helpful Hint

In a trapezoid, it does not matter which base is b_1 and which base is b_2.

EXAMPLE D

What is the area
of the trapezoid
shown at the right?

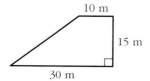

In this trapezoid, the height is one of the sides; the height is 15 m.

$A = \frac{1}{2}h(b_1 + b_2)$ ← Write the formula for area.

$= \frac{1}{2}(15)(10 + 30)$ ← Substitute known values.

$= \frac{1}{2}(15)(40)$ ← Simplify; add inside of the parentheses first.

$= \frac{1}{2}(600)$

$= 300$

The area of the trapezoid is 300 m².

Area of Other Polygons

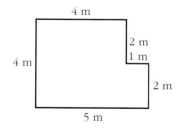

This polygon has all right angles.

To find the area of some polygons, like the one at the left, you need to separate them into polygons for which we have formulas.

What is the area of the figure at the left?

One way to find the area is to separate the figure into a square and a rectangle, as shown below.

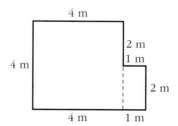

To find the area of the entire figure, add the areas of the square and the rectangle.

Area of square = 4 · 4 = 16

Area of rectangle = 1 · 2 = 2

Total area = 16 + 2 = 18

The area of the entire figure is 18 m².

In the next example, you will see that there is sometimes an extra step involved in finding a dimension of one of the polygons.

In the figure at the right, the 10-foot side is parallel to the 20-foot side. Find the total area.

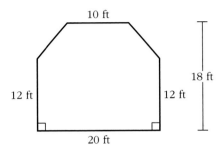

This figure can be divided into a trapezoid and a rectangle:

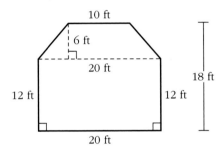

Label the figure as shown above. To find the area of the entire figure, add the areas of the trapezoid and the rectangle.

Area of trapezoid $= \frac{1}{2}h(b_1 + b_2) = \frac{1}{2}(6)(10 + 20) = \frac{1}{2}(6)(30) = 90$

Area of rectangle $= (20)(12) = 240$

Total area $= 90 + 240 = 330$

The total area of the figure is 330 ft².

EXERCISES 15-6

For Exercises 1–11, find the area of each figure.

Assume that:
* **All angles that appear to be right angles are right angles.**
* **All sides that appear to be parallel are parallel.**

1.

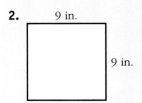

3.

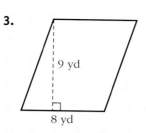

9 yd

8 yd

4.

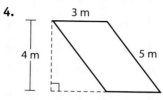

3 m

4 m

5 m

5.

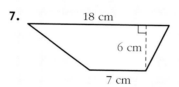

5 in.

12 in.

6.

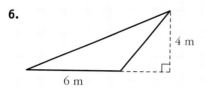

4 m

6 m

7.

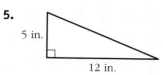

18 cm

6 cm

7 cm

8.

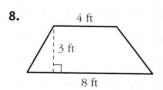

4 ft

3 ft

8 ft

9.

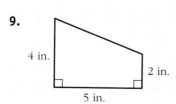

4 in.

2 in.

5 in.

10.

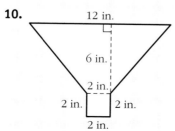

12 in.

6 in.

2 in.

2 in.   2 in.

2 in.

11.

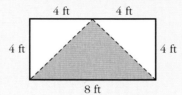

6 in.

2 in.

3 in.

3 in.

8 in.

5 in.

3 in.

8 in.

Solve the following.

12. Marta has enough paint to cover 250 ft². She needs to paint four walls, each 8 ft high and 9 ft long. Does Marta have enough paint? Explain.

13. Gerald had a 4-foot × 8-foot sheet of plywood. He cut a triangle from the sheet, as shown by the shaded region in the diagram below.

4 ft   4 ft

4 ft   4 ft

8 ft

What is the area of the triangle he cut from the sheet?

14. Olaf and Hilkka are going to install solar roof tiles on their house. Each tile measures 59" × 17" and weighs 5 lb/ft². Their roof is rectangular and measures 39 ft-4 in. long and 18 ft-5 in. wide.

(a) What is the area of their roof to the nearest tenth of a square foot?

(b) How many solar roof tiles will it take to completely cover the roof?

(c) What is the total weight of these solar roof tiles?

15. Challenge: Find the area of the figure below.

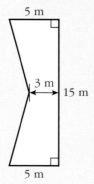

5 m

3 m

15 m

5 m

FINDING CIRCUMFERENCE AND AREA OF CIRCLES

Parts of a Circle

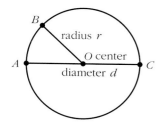

A **circle** is the set of all points in a plane that are the same distance from a single point, called the **center.** A circle is named by its center point. The circle shown at the left is called circle O.

The **radius** of a circle, r, is the distance from the center of the circle to any point on the circle. A line segment that goes from the center of a circle to any point on the circle is also called a radius. A radius is named by its end points. Every radius of the same circle has the same length.

The **diameter** of a circle, d, is the length of any line segment that connects two points on the circle and passes through the center. Any such segment itself is also called a diameter. Every diameter of the same circle has the same length.

The diameter of any circle is twice the radius:

$$d = 2r$$

Stated another way, the radius of any circle is half the diameter:

$$r = \tfrac{1}{2}d$$

In the diagram below, find the length of the diameter. Also identify the center, the diameter, and all radii shown.

Language Box

The word *radii* is the plural of *radius*.

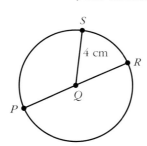

The diameter is twice the radius, so $d = 2r = 2(4) = 8$.

The length of the diameter is 8 cm.

The center is point Q. The only diameter shown is $\overline{PR}$. There are three radii shown: $\overline{QS}$, $\overline{QP}$, and $\overline{QR}$.

EXAMPLE A

In the diagram at the right, find the length of the radius. Also identify the center, the diameter, and all radii shown.

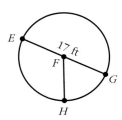

The radius is half the diameter, so $r = \frac{1}{2}d = \frac{1}{2}(17) = 8.5$.

The length of the radius is 8.5 ft.

The center is point F. The only diameter shown is $\overline{EG}$. There are three radii shown: $\overline{FE}$, $\overline{FG}$, and $\overline{FH}$.

Circumference

Circumference

The distance around the outside of a polygon is called its perimeter. We can also find the distance around a circle, which is called its **circumference, C**.

In every circle, the ratio of the circumference to the diameter is the same. The Greek letter π (pi) is used to represent this ratio:

$$\pi = \frac{C}{d}$$

Although π is a ratio of distances, it is an irrational number; it cannot be written exactly as a ratio of whole numbers or as a decimal. To perform calculations involving π, you can use 3.14, $\frac{22}{7}$, or the π key on a calculator. When you use an approximation for π, use the symbol $\approx$ (is approximately equal to) in your work. Your final answer will also be an approximation.

By multiplying both sides by d, we have a formula for the circumference of a circle:

$$C = \pi d$$

Because $d = 2r$, we can also write the formula as:

$$C = 2\pi r$$

Circumference of a Circle

For a circle with diameter d or radius r, the circumference C is given by either of the following:

$$C = \pi d \qquad \text{or} \qquad C = 2\pi r$$

Cal is applying trim to the edge of a circular table that has a diameter of 6 feet. How many feet of trim does Cal need?

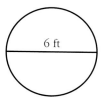

The amount of trim Cal needs is equal to the circumference of the table. The diameter is known, so use the formula $C = \pi d$.

$C = \pi d$ ⟵ Write the formula for circumference.

$\approx (3.14)(6)$ ⟵ Substitute the known value, and use 3.14 for π.

$= 18.84$ ⟵ Simplify. Use an equal sign here because $(3.14)(6)$ is exactly 18.84.

Cal needs about 18.84 ft of trim.

You can also use the circumference formula that involves radius.

Helpful Hint

Note that
$(0.84)(12)$ ft $= 10.08$ in.,
so 18.84 ft ≈ 18 ft 10 in.

EXAMPLE B

What is the circumference of a circle with a radius of 12.4 cm? Round your answer to the nearest tenth.

$C = 2\pi r$ ⟵ Write the formula for circumference.

$\approx 2(3.14)(12.4)$ ⟵ Substitute the known value.

$= 77.872$ ⟵ Simplify.

≈ 77.9 ⟵ Round to the nearest tenth.

The circle has a circumference of approximately 77.9 cm.

Given the circumference of a circle, you can work backward to find either the diameter or the radius.

EXAMPLE C

What is the diameter of a circle with a circumference of 26 ft? Round your answer to the nearest tenth.

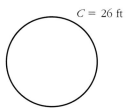

$C = \pi d$ ⟵ Write the formula for circumference.

$26 \approx 3.14d$ ⟵ Substitute the known value.

$\dfrac{26}{3.14} \approx d$ ⟵ Divide both sides by 3.14.

$8.3 \approx d$ ⟵ Simplify and round.

The circle has a diameter of about 8.3 feet.

Area

To find the area of a circle, you can use the formula shown below.

Area of a Circle

> For a circle with radius r, the area A is given by the following:
>
> $$A = \pi r^2$$

The sprinkler shown at the left waters a circular area. If the sprinkler can spray water a distance of 50 feet, what is the area of the circle that the sprinkler can water?

50 ft

Sprinkler

Because the sprinkler is in the center of the circle and can spray water 50 feet in any direction, the radius of the circle is 50 feet.

$A = \pi r^2$ ⟵ Write the formula for area.

$\approx 3.14(50)^2$ ⟵ Substitute the known value.

$= 3.14(2,500)$ ⟵ Simplify.

$= 7,850$

The sprinkler can water a circle that has an area of about 7,850 ft².

Notice that you need to know the radius of the circle to use the formula $A = \pi r^2$. Example D will show how to use the formula when given the diameter of the circle.

EXAMPLE D

What is the area of the circle shown at the right? Round your answer to the nearest tenth.

4.8 m

Because we are given the diameter of the circle, we must first find the radius.

$r = \frac{1}{2}d = \frac{1}{2}(4.8) = 2.4$

Now, find the area of the circle.

$A = \pi r^2$ ⟵ Write the formula for area.

$\approx 3.14(2.4)^2$ ⟵ Substitute the known value.

≈ 18.1 ⟵ Simplify and round.

The circle has an area of about 18.1 m².

Given the area of a circle, you can work backward to find either the diameter or the radius.

EXAMPLE E

What is the radius of a circle with an area of 100 yd²? Round your answer to the nearest tenth.

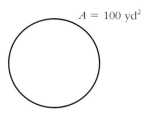

$A = 100 \text{ yd}^2$

Find the radius of the circle by solving the area formula for r.

$A = \pi r^2$ ◄─ Write the formula for area.

$100 \approx 3.14 r^2$ ◄─ Substitute the known value.

$\dfrac{100}{3.14} = r^2$ ◄─ Divide both sides by 3.14.

$\sqrt{\dfrac{100}{3.14}} = r$ ◄─ Find the positive square root.

$5.6 \approx r$ ◄─ Simplify and round.

The radius of the circle is about 5.6 yards.

EXERCISES 15-7

For Exercises 1–10, find the indicated measure for each circle. Round your answers to the nearest tenth, if rounding is necessary.

1. $r = 9$ in.

   Find d.

2. $d = 7.2$ cm

   Find r.

3. $d = 6$ yd

   Find C.

4. $r = 3.9$ cm

   Find C.

5. $C = 60$ m

   Find d.

6. $C = 27$ in.

   Find r.

7. $r = 8$ ft

   Find A.

8. $A = 300$ cm²

   Find r.

9. $d = 15$ cm

   Find A.

10. $A = 24$ yd²

    Find d.

Solve the following.

11. A trundle wheel is a tool that is used to measure distances. As the wheel is pushed, it clicks once for every full rotation.

What radius is needed so that one rotation will measure a distance of 1 yard? Round your answer to the nearest hundredth of an inch. (Remember, 1 yard = 36 inches.)

12. An electromagnet has a core that has a diameter of 13.54 mm. What is the area of one end of this core?

13. A conduit has an inside diameter of 1.54 in. What is its inside area?

14. A coil of bell wire has 42 turns. The diameter of the coil is 0.5 m.

(a) How long is the wire on this coil in meters?

(b) How long is the wire on this coil in feet?

15. Challenge: Tom wants to buy some pizza. Which will give him more pizza, three 8-inch diameter pizzas or one 16-inch diameter pizza? Explain.

CONGRUENT AND SIMILAR TRIANGLES

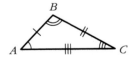 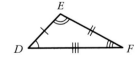

If two polygons have the same shape and the same size, we say that they are **congruent.** In congruent polygons, the corresponding angles have the same measures and the corresponding sides have the same lengths. The symbol ≅ means "is congruent to." In this lesson, we will study congruent triangles.

In the figures above, triangles *ABC* and *DEF* have the same shape and the same size, so $\triangle ABC \cong \triangle DEF$. In a congruence statement, the order of vertices is important. So, when we write $\triangle ABC \cong \triangle DEF$, we are indicating that *A* corresponds to *D, B* corresponds to *E,* and *C* corresponds to *F.* This means that:

$$\angle A \cong \angle D \qquad \overline{AB} \cong \overline{DE}$$
$$\angle B \cong \angle E \quad \text{and} \quad \overline{BC} \cong \overline{EF}$$
$$\angle C \cong \angle F \qquad \overline{AC} \cong \overline{DF}$$

Notice in the figure that congruent sides have identical marks and congruent angles have identical marks. In a diagram, angles or sides that have identical marks are congruent.

In the figure below, $\triangle ABC \cong \triangle DEF$. Find $m\angle D$ and EF.

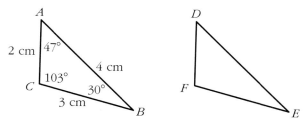

Because the triangles are congruent, corresponding angles have the same measures and corresponding sides have the same lengths.

$\angle D$ corresponds to $\angle A$, so $m\angle D = 47°$

$\overline{EF}$ corresponds to $\overline{BC}$, so $EF = 3$ cm

Therefore, $m\angle D = 47°$ and $EF = 3$ cm.

In Example A, the congruent triangles have different orientations. Be sure to use the order of vertices in the congruence statement to "match up" corresponding sides and angles correctly.

EXAMPLE A

In the figure at the right, $\triangle PQR \cong \triangle DEF$. Find $m\angle F$ and DE.

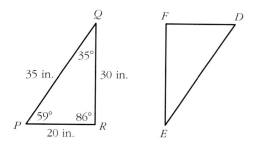

$\angle F$ corresponds to $\angle R$, so $m\angle F = 86°$

$\overline{DE}$ corresponds to $\overline{PQ}$, so $DE = 35$ in.

$m\angle F = 86°$ and $DE = 35$ in.

If two polygons have the same shape, they are **similar.** Similar polygons
may also be congruent, but often they are not congruent. In similar polygons, the corresponding angles have the same measures, and the corresponding sides are proportional. The symbol $\sim$ means "is similar to."

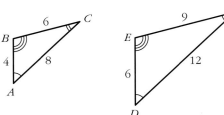

In the figure above, triangles ABC and DEF have different sizes, but they
have the same shape, so they are similar. A similarity statement that shows
the correct correspondence of vertices is $\triangle ABC \sim \triangle DEF$. This means that:

$\angle A \cong \angle D$

$\angle B \cong \angle E$ and $\dfrac{AB}{DE} = \dfrac{BC}{EF} = \dfrac{AC}{DF}$

$\angle C \cong \angle F$

Notice that by substituting the lengths of the sides, you can see that the
corresponding sides are proportional:

$$\frac{AB}{DE} = \frac{BC}{EF} = \frac{AC}{DF}$$

$$\downarrow \qquad \downarrow \qquad \downarrow$$

$$\frac{4}{6} = \frac{6}{9} = \frac{8}{12}$$

EXAMPLE B

**In the figure at the
right, $\triangle ABC \sim \triangle DEF$.
Find x.**

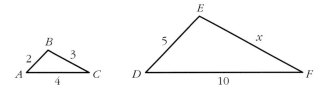

Because $\triangle ABC \sim \triangle DEF$, we know that the corresponding sides are
proportional. The variable x represents length EF, so use a proportion that
involves EF.

$\dfrac{AB}{DE} = \dfrac{BC}{EF}$ $\longleftarrow$ Write a proportion.

$\dfrac{2}{5} = \dfrac{3}{x}$ $\longleftarrow$ Substitute known values.

$2x = 3 \cdot 5$ $\longleftarrow$ Multiply both sides by $5x$.

$2x = 15$

$\dfrac{2x}{2} = \dfrac{15}{2}$ $\longleftarrow$ Divide by 2 to isolate the variable.

$x = 7.5$

$x = 7.5.$

In Example C, you will see that similar triangles can be used to solve problems in real-life situations, such as finding the height of a tall object.

EXAMPLE C

At a certain time of day, a 6-foot tall person casts a shadow that is 8 feet long. At the same time, a tree casts a shadow that is 20 feet long. How tall is the tree?

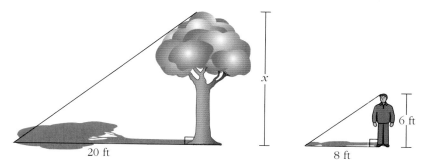

Each triangle is formed by the upright object, its shadow, and a segment that represents the sun's ray. The angle of the sun's ray is the same for both the tree and the person, so the triangles are similar. Use a proportion to find the height of the tree:

$$\frac{\text{Height of tree}}{\text{Height of person}} = \frac{\text{Length of tree's shadow}}{\text{Length of person's shadow}}$$

$$\frac{x}{6} = \frac{20}{8}$$

$$8x = 20 \cdot 6$$

$$8x = 120$$

$$\frac{8x}{8} = \frac{120}{8}$$

$$x = 15$$

The tree is 15 feet tall.

EXERCISES 15-8

Answer each of the following.

1. $\triangle ABC \cong \triangle DEF$. Find $m\angle D$ and DE.

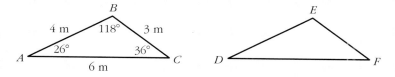

2. $\triangle STX \cong \triangle JAL$. Find $m\angle A$ and JL.

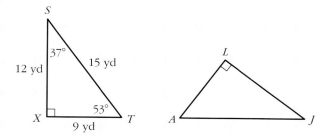

3. $\triangle NMP \cong \triangle WXY$. Find $m\angle Y$ and XY.

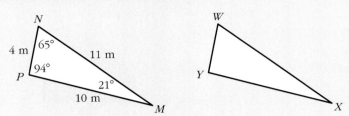

4. $\triangle RPT \cong \triangle JKL$. Find $m\angle J$ and JL.

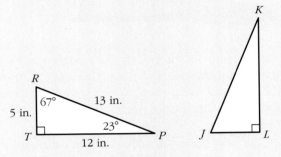

5. $\triangle ABC \sim \triangle DEF$. Find x.

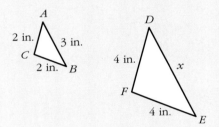

6. $\triangle RLN \sim \triangle SBT$. Find x.

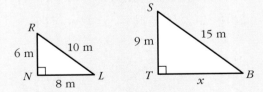

7. $\triangle PQR \sim \triangle STU$. Find x.

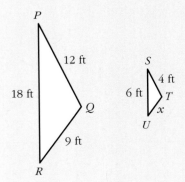

8. Elliot is 6 ft tall. At a certain time of day, he has a shadow that is 7 feet long, and a utility pole has a shadow that is 21 feet long.

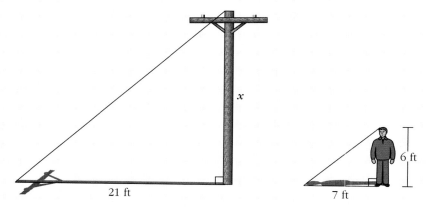

How tall is the utility pole?

9. The blueprint for a solar panel has a scale of 20 mm = 1.25 m. The length of the panel is 2.75 m, and its width is 0.48 m.

(a) What distance on the blueprint represents the length? Round your answer to the nearest millimeter.

(b) What distance on the blueprint represents the width? Round your answer to the nearest millimeter.

10. Challenge: Are the two triangles shown below similar? Explain why or why not.

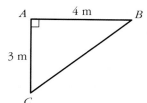
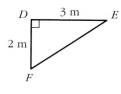

CHAPTER 15 REVIEW EXERCISES

1. Without using a protractor, estimate whether the angles in (a) and (b) are 30°, 60°, 120°, or 150°.

(a)

(b)

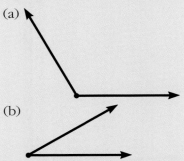

(c) Use a protractor to measure each of the above angles.

2. The outside diameter of an electrical conduit is $2\frac{1}{2}$ in. The conduit is 1/8 in. thick. What is the inside diameter of the conduit?

3. Juan is going to install some lights around a circular pond. If the lights are placed no more than 5 ft-6 in. apart and the pond had a diameter of 12 ft, how many lights will he need?

4. A switch plate cover is $4\frac{3}{16}$ in. high and $2\frac{5}{16}$ in. wide. A rectangle has been removed from the middle of the plate to allow for the switch. If the rectangle is $\frac{1}{2}$ in. wide and 1 in. high, what is the area of the plate?

5. A square surface cover is 12.5 cm on a side. A circular knockout in the center has a diameter of 1.23 cm. What is the area of the cover?

6. Felipa needed to measure the height of an inside wall of a warehouse. She does not have a ladder long enough to reach the top of the wall and, because she is inside, there is no shadow. She places a mirror on the floor, marked with an M in the figure. She then moves away from the mirror until she can see a reflection of the top of the wall in the mirror. At that time, her eyes are 5 ft-2 in. above the ground and she is 4 ft-6 in. from the mirror. If the mirror is 20 ft from the bottom of the wall, how high is the wall?

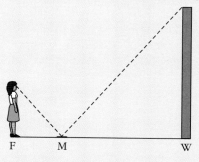

Building a Foundation in Mathematics

Identifying Solid Figures

Finding Lateral Area and Surface Area

Finding Volume

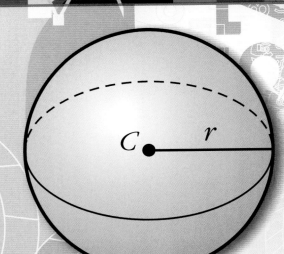

Overview

In the last chapter we studied plane geometry, the geometry of two dimensions: length and width. But, we live in a three-dimensional world, one that has length, width, and height. So, in this chapter we will study solid geometry, the study of objects in a three-dimensional world.

The study of solid geometry includes the ability to identify solid objects, create solid figures from two-dimensional figures, and determine the lateral area, surface area, and volume of the solid.

Solid Geometry

Objectives

After completing this chapter, you will be able to:

- Look at a solid geometric shape and give its name
- Look at the net of a solid geometric figure and give the name of the figure it will form
- Compute volumes of prisms, cylinders, pyramids, and cones
- Compute lateral area and surface areas of prisms, cylinders, pyramids, and cones
- Compute volumes and surface areas of spheres

Chapter 16

IDENTIFYING SOLID FIGURES

Prisms and Pyramids

The Great Pyramid at Giza is the only surviving wonder of the Seven Ancient Wonders of the World.

A **solid figure** has three dimensions and occupies space. A **polyhedron** is a solid that is formed by polygons. **Prisms** and **pyramids** are examples of polyhedrons. The polygons that make up the surface of prisms and pyramids are called **faces.** The intersection of two faces is an **edge.** The intersection of three or more edges is a **vertex.**

Prism

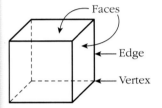

Pyramid

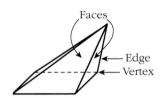

The slanting pyramid above right is called an *oblique* pyramid. The examples below are right prisms and pyramids. In a right prism, the bases are aligned, one directly above the other. In a right pyramid, the vertex at the top is directly above the center of the base. From now on, all the prisms and pyramids we study will be right prisms and pyramids.

Prisms

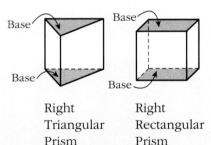

Right Triangular Prism   Right Rectangular Prism

Pyramids

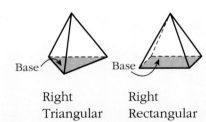

Right Triangular Pyramid   Right Rectangular Pyramid

Two of the faces of a prism are congruent polygons that lie in parallel planes. These are the **bases** of the prism. The other faces of a right prism are all rectangles.

One face of a pyramid may be any type of polygon. This is called the *base* of the pyramid. The other faces of a right pyramid are all congruent isosceles triangles.

As you can see in the figures above, a prism or a pyramid is named by the polygon that forms its base.

A **net** of a solid figure is a two-dimensional drawing that shows the shapes that will form the solid figure when the net is folded.

Which solid figure can be formed by the net shown below?

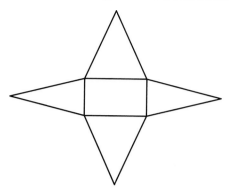

The base of the figure is a rectangle.

The other faces are triangles. This is a net for a rectangular pyramid.

EXAMPLE A

Which solid figure can be formed by the net shown?

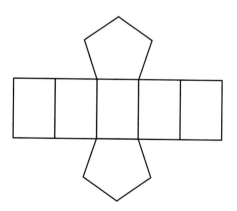

There are two congruent bases that are pentagons.

The other faces are rectangles. This is a net for a pentagonal prism.

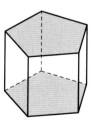

A pentagonal prism can be formed by the net shown.

EXAMPLE B

Identify the solid figure shown.

The base of the figure is a hexagon. The other faces are triangles. The figure is a hexagonal pyramid.

The figure is a hexagonal pyramid.

EXAMPLE C

Identify the solid figure shown.

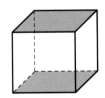

6 Congruent Square Faces

The bases of the figure are rectangles. The other faces are also rectangles. So, the figure is a rectangular prism. But because all 6 faces are also squares, it is also a cube.

The figure is a rectangular prism. It is also called a cube.

There is a formula that states a relationship involving the number of faces, edges, and vertices of any polyhedron. The mathematician Leonard Euler is credited with discovering the formula. The formula is stated below.

Euler's Formula

The number of faces (F), vertices (V), and edges (E) of any polyhedron are related by the formula $F + V = E + 2$.

Example D shows how Euler's Formula applies to several examples of polyhedrons.

EXAMPLE D

Use the figures shown to verify the table.

         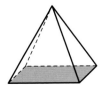

Figure 1   Figure 2   Figure 3   Figure 4

Polyhedron	Number of Faces (*F*)	Number of Vertices (*V*)	Number of Edges (*E*)
Figure 1 (cube)	6	8	12
Figure 2 (triangular prism)	5	6	9
Figure 3 (triangular pyramid)	4	4	6
Figure 4 (rectangular pyramid)	5	5	8

Notice that $F + V = E + 2$ for each figure shown earlier.

Cylinders, Cones, and Spheres

The Crystal Bridge Tropical Conservatory in Oklahoma City, OK, is a 224-foot-long cylinder that is 7 stories high.

Cylinders, cones, and **spheres** are solid figures, but they are not polyhedrons because each one has a curved surface. Examples of these solids are shown below.

Cylinder

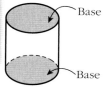
Base
Base

A cylinder has two circular, congruent bases that lie in parallel planes.

Cone

Base

A cone has one circular base.

Sphere

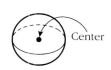
Center

A sphere is the set of all points that are equidistant from a given point, called the *center*.

In a right cylinder, the segment joining the centers of the bases is perpendicular to the bases. In a right cone, the segment joining the vertex to the center of the base is perpendicular to the base. In this chapter, we will study only right cylinders and right cones, like the ones shown above.

Name the solids that form the structure below.

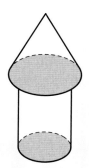

The top of the structure is a cone.
The bottom of the structure is a cylinder.

The structure is formed by a cone and a cylinder.

EXAMPLE E

Name the solids that form the structure shown.

The ends of the structure are spheres. The spheres are joined by a cylinder.

The structure is formed by two spheres and a cylinder.

EXERCISES 16-1

In Exercises 1–6, identify the solid figure shown.

1.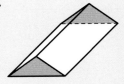

2.

3.

4.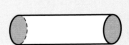

5.

6.

Solve the following problems.

7. Which solid figure can be formed by the net shown?

8. The figure below is an octagonal prism.

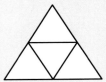

   (a) State the number of faces, vertices, and edges.

   (b) Verify that your answer in (a) satisfies Euler's Formula.

   (c) Sketch a net for this octagonal prism.

9. A prism has 8 faces and 18 edges. How many vertices does the prism have? What type of prism is it? Explain your answer.

10. The center of the sphere shown at right is C. Based on the definition of a sphere, what is true about CD, CE, and CF?

11. Challenge: A cube with vertices $ABCDEFGH$ is shown.

(a) List all edges parallel to $\overline{AE}$.

(b) List all edges parallel to $\overline{AD}$.

(c) List all edges parallel to $\overline{AB}$.

FINDING LATERAL AREA AND SURFACE AREA

Prisms and Cylinders

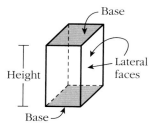

The area of a solid figure can be computed with or without the area of the base(s) included. **Lateral area (LA)** of a solid figure does not include the area of the base(s). **Surface area (SA)** is the lateral area plus the area of the base(s). The diagrams at the left show a rectangular prism and a cylinder. Although the figures have different types of lateral surfaces, both have two congruent bases.

The height of a prism or a cylinder is the distance between its bases. A right prism and a right cylinder are shown below.

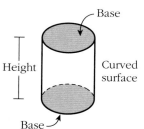

Prisms and cylinders have two congruent bases.

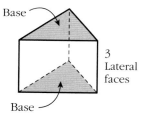

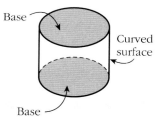

The lateral faces of a right prism are rectangles. The lateral area *(LA)* is the sum of the areas of the rectangles.

To find the surface area *(SA)* of a prism, add the areas of the bases to the lateral area.

The lateral area *(LA)* of a cylinder is the area of its curved surface. The area of this surface is the product of the circumference and the height of the cylinder.

To find the surface area *(SA)* of a cylinder, add the areas of the bases to the lateral area.

The formulas for lateral area and surface area of prisms and cylinders are given in the box that follows.

Lateral and Surface Areas of Right Prisms

$LA = ph$, where p is the perimeter of the base and h is the height of the prism

$SA = LA + 2B$, where B is the area of the base

Lateral and Surface Areas of Right Cylinders

$LA = 2\pi rh$, where r is the radius of the cylinder and h is the height of the cylinder

$SA = LA + 2B$, where B is the area of the base

Find the lateral area and the surface area of the cylinder shown.

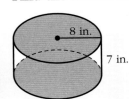

The radius of the cylinder is 8 inches; the height is 7 inches.

Helpful Hint

For a formula that contains π, you can use 3.14, $\frac{22}{7}$, or the π key on a calculator.

$LA = 2\pi rh$ ← Write the formula for lateral area.

$\approx 2(3.14)(8)(7)$ ← Substitute the known values and then simplify.

$= 351.68$

Now, to find the surface area, use the lateral area and B, the area of a base. Remember that in a cylinder, the base is a circle with area πr^2.

$SA = LA + 2B$ ← Write the formula for surface area.

$= LA + 2(\pi r^2)$ ← Substitute πr^2 for B.

$\approx 351.68 + 2(3.14)(8)^2$ ← Substitute the known values.

$= 351.68 + 401.92$

$= 753.60$

$LA \approx 351.68$ in.^{2}, $SA \approx 753.60$ in.^{2}

EXAMPLE A

Find the lateral area and the surface area of the triangular prism shown.

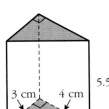

The bases of the prism have side lengths of 3, 4, and 5 centimeters. Add these lengths to find the perimeter of a base.

$p = 3 + 4 + 5$

$= 12$

$LA = ph$ ← Write the formula for lateral area.

$= 12(5.5)$ ← Substitute the known values.

$= 66$

Language Box

In the formula
$LA = ph$, h represents the height of the prism or cylinder. In the formula
$B = \frac{1}{2}bh_t$, h_t represents the height of the triangle.

Now, to find the surface area, use the lateral area and B, the area of a base. Remember that in a triangular prism, the base is a triangle with area $\frac{1}{2}bh_t$.

$SA = LA + 2B$ ← Write the formula for surface area.

$\quad = LA + 2\left(\frac{1}{2}bh_t\right)$ ← Substitute $\frac{1}{2}bh_t$ for B.

$\quad = 66 + 2\left(\frac{1}{2}\right)(4)(3)$ ← Substitute the known values.

$\quad = 66 + 12$

$\quad = 78$

$LA = 66$ cm², $SA = 78$ cm²

EXAMPLE B

The dimensions of a garage are given on the diagram shown. Find the lateral area and the surface area of the interior. (Include the area of the door and window.)

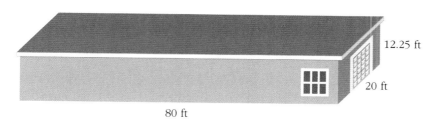

80 ft   20 ft   12.25 ft

The base of the garage is a rectangle with side lengths of 80 feet and 20 feet. Use these lengths to find the perimeter of the base.

$p = 2(80 + 20)$

$\quad = 2(100)$

$\quad = 200$

$LA = ph$ ← Write the formula for lateral area.

$\quad = 200(12.25)$ ← Substitute the known values.

$\quad = 2{,}450$

Now, to find the surface area, use the lateral area and B, the area of a base. Remember that in a rectangular prism, the base is a rectangle with area lw.

$SA = LA + 2B$ ← Write the formula for surface area.

$\quad = LA + 2lw$ ← Substitute lw for B.

$\quad = 2{,}450 + 2(80)(20)$ ← Substitute the known values.

$\quad = 2{,}450 + 3{,}200$

$\quad = 5{,}650$

$LA = 2{,}450$ ft², $SA = 5{,}650$ ft²

Regular Pyramids and Right Cones

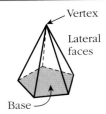

Vertex

Lateral faces

Base

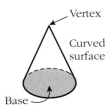

Vertex

Curved surface

Base

Pyramids and cones have one base.

A **regular pyramid** is a right pyramid whose base is a regular polygon. Recall that all sides are equal on a regular polygon and all angles are equal. The diagram below shows a regular pyramid whose base is a square. The diagram also shows a right cone.

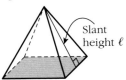

Slant height ℓ

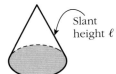

Slant height ℓ

The lateral faces of a regular pyramid are congruent isosceles triangles. The **slant height** ℓ is the height of any one of those triangles.

The **slant height** ℓ of a cone is the distance from the vertex to a point on the edge of the base.

The formulas for lateral area and surface area of regular pyramids and cones are given in the box that follows.

Lateral and Surface Areas of Regular Pyramids	**Lateral and Surface Areas of Right Cones**
$LA = \frac{1}{2}p\ell$, where p is the perimeter of the base and ℓ is the slant height $SA = LA + B$, where B is the area of the base	$LA = \pi r\ell$, where r is the radius of the base of the cone and ℓ is the slant height of the cone $SA = LA + B$, where B is the area of the base

Find the lateral area and the surface area of the regular pyramid shown.

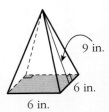

9 in.

6 in.

6 in.

The base is a square with a side length of 6 inches. The perimeter of the base is $4(6) = 24$ inches. The slant height of the pyramid is 9 inches.

$LA = \frac{1}{2}p\ell$ ← Write the formula for lateral area.

$\quad = \frac{1}{2}(24)(9)$ ← Substitute the known values.

$\quad = 108$

Now, to find the surface area, use the lateral area and B, the area of the base. Remember that in a square pyramid, the base is a square with area s^2.

$SA = LA + B$ ← Write the formula for surface area.

$\quad = LA + s^2$ ← Substitute s^2 for B.

$\quad = 108 + (6)^2$ ← Substitute the known values.

$\quad = 108 + 36$

$\quad = 144$

$LA = 108$ in.^{2}, $SA = 144$ in.^{2}

EXAMPLE C

Find the lateral area and the surface area of the cone shown.

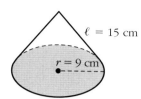

The base is a circle with radius of 9 cm. The slant height of the cone is 15 cm.

$LA = \pi r \ell$ ← Write the formula for the lateral area.

$\approx 3.14(9)(15)$ ← Substitute the known values.

$= 423.90$

Now, to find the surface area, use the lateral area and B, the area of the base. Remember that in a cone, the base is a circle with area πr^2.

$SA = LA + B$ ← Write the formula for the surface area.

$= LA + \pi r^2$ ← Substitute πr^2 for B.

$\approx 423.90 + 3.14(9)^2$ ← Substitute the known values.

$= 423.90 + 254.34$

$= 678.24$

$LA \approx 423.90 \text{ cm}^2,\ SA \approx 678.24 \text{ cm}^2$

EXAMPLE D

Find the lateral area and the surface area of the regular pyramid shown.

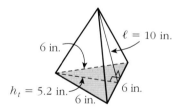

The base is an equilateral triangle with a side length of 6 inches. So, the perimeter of the base is $3(6) = 18$ inches. The slant height of the pyramid is 10 inches.

$LA = \frac{1}{2}p\ell$ ← Write the formula for lateral area.

$= \frac{1}{2}(18)(10)$ ← Substitute the known values.

$= 90$

To find the surface area, use the lateral area and B, the area of the base. Remember that in a triangular pyramid, the base is a triangle with area $\frac{1}{2}bh_t$.

$SA = LA + B$ ← Write the formula for surface area.

$= LA + \frac{1}{2}bh_t$ ← Substitute $\frac{1}{2}bh_t$ for B.

$= 90 + \frac{1}{2}(6)(5.2\ )$ ← Substitute the known values.

$= 90 + 15.6$

$= 105.6$

$LA = 90 \text{ in.}^2,\ SA = 105.6 \text{ in.}^2$

Spheres

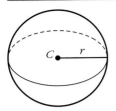

A sphere is the set of all points in space that are the same distance from one point C, called the **center** of the sphere.

The formula for the surface area of a sphere is given below.

Surface Area of a Sphere

$SA = 4\pi r^2$, where r is the radius of the sphere

Find the surface area of a sphere whose radius is 2.5 cm.

$SA = 4\pi r^2$ ← Write the formula for surface area.

$\approx 4(3.14)(2.5)^2$ ← Substitute the known value.

$= 78.5$

$SA \approx 78.5$ cm^2

EXAMPLE E

Find the surface area of a tennis ball whose radius is $1\frac{1}{4}$ inches. Round your answer to the nearest tenth.

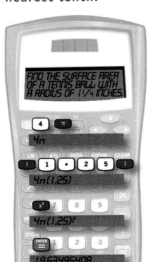

For convenience, convert the mixed number $1\frac{1}{4}$ to its decimal equivalent, 1.25.

$SA = 4\pi r^2$ ← Write the formula for surface area.

$\approx 4(3.14)(1.25)^2$ ← Substitute the known value.

$= 19.625$

≈ 19.6 ← Round your answer to the nearest tenth.

$SA \approx 19.6$ in.^{2}

The formulas used in this lesson and the next lesson are summarized in Appendix E.

EXERCISES 16-2

In the exercises that follow, round your answer to two decimal places if rounding is necessary.

In Exercises 1–4, find the lateral area and the surface area of the prism or cylinder shown.

1.

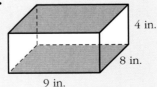

4 in.

8 in.

9 in.

2.

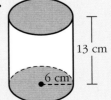

13 cm

6 cm

3.

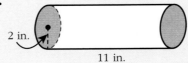

2 in.

11 in.

4.

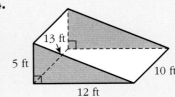

13 ft

5 ft

12 ft

10 ft

Solve the following problem.

5. The bases of the hexagonal prism shown below are regular hexagons.

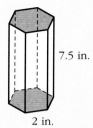

7.5 in.

2 in.

Find the lateral area of the prism.

In Exercises 6–10, find the lateral area and the surface area of the pyramid or cone shown.

6.

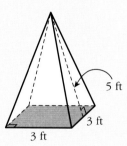

5 ft

3 ft

3 ft

7.

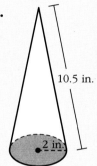

10.5 in.

2 in.

8.

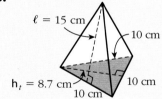

$\ell = 15$ cm

10 cm

$h_t = 8.7$ cm

10 cm

10 cm

9.

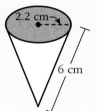

2.2 cm

6 cm

10.

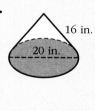

16 in.

20 in.

In Exercises 11–12, find the surface area of the sphere shown.

11.

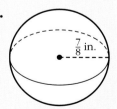

2 ft

12.

$\frac{7}{8}$ in.

Solve Exercises 13–15

13. An electrolytic capacitor is the shape of a cylinder. It is $\frac{7}{8}$ in. high and has a diameter of $\frac{23}{32}$ in.

(a) What is its lateral area?

(b) What is its surface area?

14. A tray for electronics parts is going to be made from the piece of sheet metal shown in the figure by folding the metal on the dashed lines and welding the edges that meet.

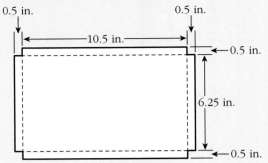

(a) What is the name of the completed figure?

(b) What is the area of the bottom of this tray?

(c) What is the surface area of the tray? (Remember, the tray does not have a top.)

15. The core of a cylindrical electromagnet is 9.25 in. long with a radius of 15.4 in. If it is entirely covered with one layer of insulation paper, what is the area of the paper?

16. Challenge: A hemisphere is half a sphere. A paperweight in the shape of a hemisphere is shown below.

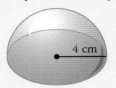

Find the surface area of the paperweight. Be sure to include the area of the flat circular base.

FINDING VOLUME

Prisms and Cylinders

Any face of a rectangular prism may serve as a base. For this box, it is natural to consider the bottom and the top to be the bases.

Volume is the amount of space that a solid figure occupies. Volume is measured in cubic units such as cubic inches (in.³) or cubic centimeters (cm³).

The volume, *V*, of a prism is the product of the area of its base and its height. Likewise, the volume, *V*, of a cylinder is the product of the area of its base and its height.

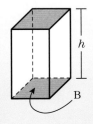

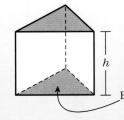

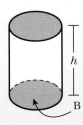

The formulas for volume of these solids are given below.

Volume of a Prism	Volume of a Cylinder
$V = Bh$, where B is the area of the base and h is the height of the prism	$V = Bh$, where B is the area of the base and h is the height of the cylinder

Find the volume of the rectangular prism (cracker box) shown below.

9.5 in.

2.5 in.

7 in.

The base of the prism is a rectangle. Use the expression lw for B, the area of the base.

$V = Bh$ ⟵ Write the formula for volume.

$\quad = lwh$ ⟵ Substitute lw for B.

$\quad = (2.5)(7)(9.5)$ ⟵ Substitute the known values and then simplify.

$\quad = 166.25$

The volume of the prism is 166.25 in.^{3}

EXAMPLE A

Find the volume of the cylinder shown.

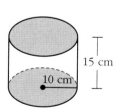

15 cm

10 cm

The base of every cylinder is a circle. Use the expression πr^2 for B, the area of the base.

$V = Bh$ ⟵ Write the formula for volume.

$\quad = \pi r^2 h$ ⟵ Substitute πr^2 for B.

$\quad \approx 3.14(10)^2(15)$ ⟵ Substitute the known values.

$\quad = 4{,}710$

$V \approx 4{,}710 \text{ cm}^3$

EXAMPLE B

Find the volume of the prism shown.

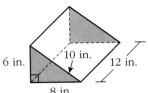

6 in.   10 in.   12 in.

8 in.

The base of the prism is a right triangle. Use the expression $\frac{1}{2}bh_t$ for B, the area of the base of the prism, where b is the base of the triangle and h_t is the height of the triangle.

$V = Bh$ ⟵ Write the formula for volume.

$\quad = \frac{1}{2}bh_t \cdot h$ ⟵ Substitute $\frac{1}{2}bh_t$ for B.

$\quad = \frac{1}{2}(8)(6)(12)$ ⟵ Substitute the known values.

$\quad = 288$

$V = 288 \text{ in.}^3$

To find the volume, V, of a cube, use the formula $V = s^3$, where s is the edge length of the cube. This formula is used in Example C.

EXAMPLE C

Find the volume of a cube whose edge is 7.5 cm. Round your answer to the nearest cubic centimeter.

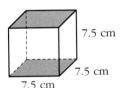

7.5 cm

7.5 cm

7.5 cm

For a cube, the formula $V = lwh$ and the formula $V = s^3$ are equivalent, because $l = w = h$ for a cube.

$V = s^3$ ⟵ Write the formula for volume of a cube.

$\quad = (7.5)^3$ ⟵ Substitute the known value.

$\quad = 421.875$

$\quad \approx 422$ ⟵ Round your answer.

$V \approx 422 \text{ cm}^3$

Pyramids and Cones

For both pyramids and cones, the volume, V, is one third the product of the area of the base and the height. In the diagram below, the base and the height of the prism and the pyramid are the same. Likewise, the base and the height of the cylinder and the cone are the same.

To find the volume of a pyramid or a cone, use the height, h, not the slant height, ℓ.

Prism   Pyramid   Cylinder   Cone

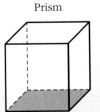

The volume of the pyramid is $\frac{1}{3}$ the volume of the prism.

The volume of the cone is $\frac{1}{3}$ the volume of the cylinder.

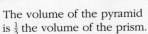

The volume formulas for pyramids and cones are given below.

Volume of a Pyramid	**Volume of a Cone**
$V = \frac{1}{3}Bh$, where B is the area of the base and h is the height	$V = \frac{1}{3}Bh$, where B is the area of the base and h is the height

Find the volume of the cone shown below.

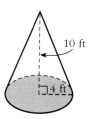

The base of every cone is a circle. Use the expression πr^2 for B, the area of the base. The radius of the cone is 4 feet; the height is 10 feet.

$V = \frac{1}{3}Bh$ ← Write the formula for volume.

$= \frac{1}{3}\pi r^2 h$ ← Substitute πr^2 for B.

$\approx \frac{1}{3}(3.14)(4)^2(10)$ ← Substitute the known values.

≈ 167.47 ← Use a calculator and round your answer.

$V \approx 167.47$ ft³

EXAMPLE D

Find the volume of the square pyramid shown.

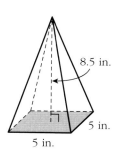

Use the expression s^2 for B, the area of the base, where s is the side of the square.

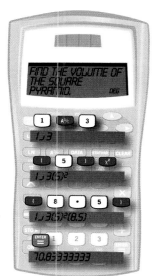

$V = \frac{1}{3}Bh$ ← Write the formula for volume.

$= \frac{1}{3}s^2 \cdot h$ ← Substitute s^2 for B.

$= \frac{1}{3}(5)^2(8.5)$ ← Substitute the known values.

≈ 70.83 ← Multiply and round your answer.

$V \approx 70.83$ in.³

Spheres

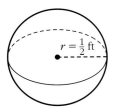

The formula for the volume of a sphere is stated below.

Volume of a Sphere

$V = \frac{4}{3}\pi r^3$, where r is the radius of the sphere

Find the volume of the sphere shown at the left.

The radius of the sphere is $\frac{1}{2}$ foot.

$V = \frac{4}{3}\pi r^3$ ← Write the formula for volume.

$\approx \frac{4}{3}(3.14)\left(\frac{1}{2}\right)^3$ ← Substitute the known value.

$= \frac{4}{3}(3.14)\left(\frac{1}{8}\right)$

≈ 0.52 ← Use a calculator and round your answer.

$V \approx 0.52 \text{ ft}^3$

EXAMPLE E

What is the volume of the hemisphere shown?

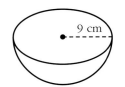

The radius of the hemisphere is 9 cm. The volume of a hemisphere is one half the volume of a sphere with the same radius.

Language Box

The prefix *hemi-* in the word *hemisphere* means "half." A hemisphere is half a sphere.

STEP 1

Use the formula for the volume of a sphere.

$V = \frac{4}{3}\pi r^3$ ← Write the formula for volume of a sphere.

$\approx \frac{4}{3}(3.14)(9)^3$ ← Substitute the known value.

$= 3{,}052.08$

STEP 2

Multiply the volume of the sphere by $\frac{1}{2}$ to find the volume of the hemisphere.

$\frac{1}{2}(3{,}052.08) = 1{,}526.04$

The volume of the hemisphere is about 1526.04 cm³.

EXERCISES 16-3

In the exercises that follow, round your answer to two decimal places if rounding is necessary.

In Exercises 1–4, find the volume of the prism or cylinder shown.

1.

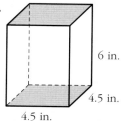

6 in.

4.5 in.

4.5 in.

2.

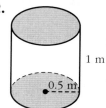

1 m

0.5 m

3.

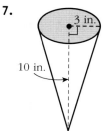

20 cm

24 cm

10 cm

26 cm

4.

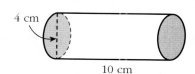

4 cm

10 cm

In Exercises 5–8, find the volume of the pyramid or cone shown.

5.

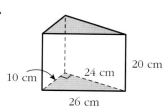

10 cm

7 cm

12 cm

6.

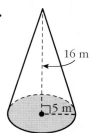

16 m

5 m

7.

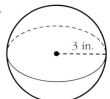

3 in.

10 in.

8.

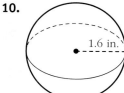

8 cm

8 cm

15 cm

In Exercises 9–11, find the volume of the figure shown.

9.

3 in.

10.

1.6 in.

11. What is the volume of the hemisphere shown?

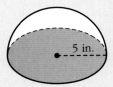

12. A spherical water tank is to have decorative lights installed. Each string will be equally spaced around the middle of the sphere and all the strings will cross at the very top and bottom of the sphere as indicated in the figure. The sphere is 50 ft in diameter. If there are 18 strings of lights, what is the total length of lighting that is needed?

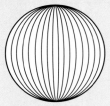

13. What is the volume of a spherical glass insulator with a diameter of 2.25 in.? Round your answer to two decimal places.

14. Challenge: The surface area of a cube is 600 in². What is the volume of the cube?

CHAPTER 16 REVIEW EXERCISES

Evaluate the following.

1. Consider the net in the figure.

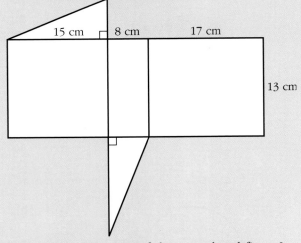

(a) What is the name of the completed figure?

(b) What is the lateral area of the figure?

(c) What is the surface area of the figure?

(d) What is the volume of this figure?

2. A coil is wrapped on a $2\frac{1}{8}$ in. diameter cylinder that is 3 in. long. What is the lateral area of the cylinder around which the wire is wrapped?

3. A solar roof tile is the shape of a rectangular prism and measures $17'' \times 59'' \times 1\frac{1}{4}''$. What is the volume of one tile?

In Exercises 4–8, find the lateral area, surface area, and volume of the figures shown.

4.

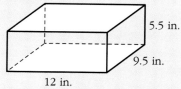

5.5 in.
9.5 in.
12 in.

5.

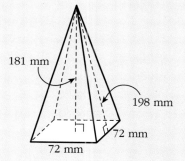

13.7 cm
5.2 cm

6.

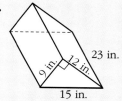

23 in.
9 in. 12 in.
15 in.

7.

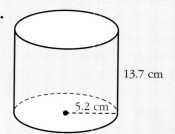

181 mm
198 mm
72 mm
72 mm

8.

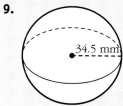

3.72 in.
3.5 in.
1.25 in.

In Exercise 9, find the surface area and volume of the figure shown.

9.

34.5 mm

Building a Foundation in Mathematics

- Using the Pythagorean Theorem

- Finding Trigonometric Function Values of Acute Angles

- Finding Trigonometric Function Values of Any Angle

- Using Inverse Trigonometric Functions

- Solving Right Triangles

- Solving Word Problems with Trigonometric Functions

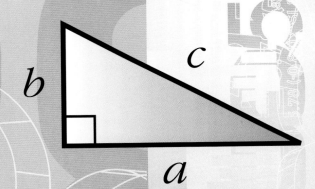

Overview

All polygons can be divided into triangles. That makes the triangle the most important plane geometric figure. Right triangles are the ones that are most often seen in construction and in building plans. Because right triangles are so important, we need to study trigonometry—the branch of mathematics that is used to compute unknown angles and sides of a triangle. This is the subject of this chapter.

We begin the chapter with the Pythagorean Theorem. This theorem allows us to determine the length of one side of any right triangle if we know the lengths of the other two sides. Then, we move on to trigonometry.

The word *trigonometry* is derived from the Greek words for triangle and measurement. With trigonometry we can determine the lengths of the sides of a right triangle if all we know is the size of one angle and the length of one side. Or, we can determine the angles based on the lengths of the sides. A knowledge of trigonometry is very valuable to an electrician in studying electrical circuits.

Trigonometry

Objectives

After completing this chapter, you will be able to:

- Compute unknown sides of a right triangle using the Pythagorean Theorem

- Identify the sides of a right triangle with reference to any angle

- State the ratios of the six trigonometric functions in relation to given triangles

- Find functions of angles given in degrees, minutes, and seconds

- Find functions of angles given in decimal degrees

- Find angles of a given trigonometric function

- Compute unknown angles of right triangles when two sides are given

- Compute unknown sides of a right triangle when an angle and a side are given

USING THE PYTHAGOREAN THEOREM

The study of trigonometry involves the relationships between the sides and angles of a triangle. Although trigonometry applies to all triangles, this chapter will focus only on **right triangles,** triangles that have a right angle. Problems in right triangle trigonometry often include calculations based on the Pythagorean Theorem. We will study the Pythagorean Theorem in this lesson and then begin the study of trigonometry in the next lesson.

In a right triangle, the side that is opposite the right angle is the **hypotenuse.** The other two sides of a right triangle are the **legs.** The hypotenuse is always the longest side of a right triangle. The **Pythagorean Theorem,** showing the relationship between the hypotenuse and the legs, is stated below.

The Pythagorean Theorem

In a right triangle, the square of the length of the hypotenuse is equal to the sum of the squares of the lengths of the legs. If c represents the length of the hypotenuse, and a and b represent the lengths of the legs, then $c^2 = a^2 + b^2$.

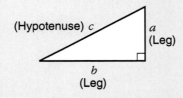

(Hypotenuse) c

a
(Leg)

b
(Leg)

Suppose the top of a building has the dimensions shown in the diagram below. What is the dimension labeled c?

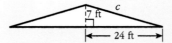

c

7 ft

|← 24 ft →|

Find the hypotenuse of a right triangle with legs of 7 feet and 24 feet.

Substitute the known values into the Pythagorean equation. Solve the equation for c.

$c^2 = a^2 + b^2$ ← Write the Pythagorean equation.

$c^2 = 7^2 + 24^2$ ← Substitute 7 and 24 for a and b.

$c^2 = 49 + 576$ ← Simplify.

$c^2 = 625$

$c = \sqrt{625}$ ← Find the positive square root.

$c = 25$

> **Helpful Hint**
>
> The equation $c^2 = 625$ has two solutions, 25 and -25. However, only the positive solution can be a length.

The hypotenuse of the triangle is 25 feet. This is the length labeled c on the diagram of the building.

EXAMPLE A

What is the length of the hypotenuse of the triangle shown?

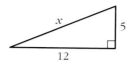

$x^2 = 5^2 + 12^2$ ← Substitute the values into the Pythagorean equation.

$x^2 = 25 + 144$ ← Simplify.

$x^2 = 169$

$x = \sqrt{169}$ ← Find the positive square root.

$x = 13$

The length of the hypotenuse is 13 units.

The Pythagorean Theorem can be used to find the length of a leg of a right triangle when the lengths of the hypotenuse and the other leg are known. This is shown in Example B.

EXAMPLE B

In a right triangle, one leg is 6 units and the hypotenuse is 10 units. Find the length of the other leg.

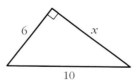

$10^2 = 6^2 + x^2$ ← Substitute the values into the Pythagorean equation.

$100 = 36 + x^2$ ← Simplify.

$64 = x^2$ ← Subtract 36 from both sides.

$\sqrt{64} = x$ ← Find the positive square root.

$8 = x$

The length of the other leg is 8 units.

EXAMPLE C

The lengths of the legs of a right triangle are 5 inches and 8 inches. What is the length of the hypotenuse? Round your answer to the nearest tenth.

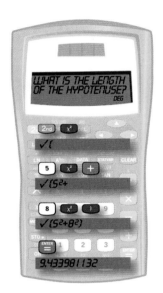

When the square root of a number is an integer, the number is a **perfect square.** For example, 169 and 64 are both perfect squares because $\sqrt{169} = 13$ and $\sqrt{64} = 8$. To find the square root of a number that is not a perfect square root, use your calculator to find an approximation.

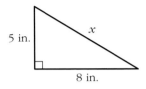

$x^2 = 5^2 + 8^2$ ← Substitute the values into the Pythagorean equation.

$x^2 = 25 + 64$ ← Simplify.

$x^2 = 89$

$x = \sqrt{89}$

$x \approx 9.4$ ← Use a calculator to find an approximation of $\sqrt{89}$.

The length of the hypotenuse is about 9.4 inches.

EXAMPLE D

Juan has a 20-ft-long cable he wants to attach to a pole. He plans to secure one end of the cable at a point on the ground 10 ft from the base of the pole. How high up the pole should the other end be placed? Round your answer to the nearest tenth.

Recall that the longest side of a right triangle is always the hypotenuse.

Draw a diagram.

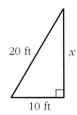

$20^2 = 10^2 + x^2$ ← Substitute the values into the Pythagorean equation.

$400 = 100 + x^2$ ← Simplify.

$300 = x^2$ ← Subtract 100 from both sides.

$\sqrt{300} = x$

$17.3 \approx x$ ← Use a calculator to find an approximation of $\sqrt{300}$.

The cable should be secured to the pole about 17.3 ft above the ground.

EXAMPLE E

What is the length of the diagonal of the square shown at the right?

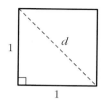

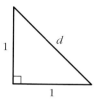

The diagonal of the square is the hypotenuse of a right triangle. Use the Pythagorean equation to find the length of the diagonal.

$d^2 = 1^2 + 1^2$ ← Substitute the values into the Pythagorean equation.

$d^2 = 1 + 1$ ← Simplify.

$d^2 = 2$

$d = \sqrt{2}$ ← Find the positive square root.

$d \approx 1.4$ ← Use a calculator to find an approximation of $\sqrt{2}$.

The length of the diagonal is about 1.4 units.

Example E contains a very useful result. A triangle formed by a diagonal of a square is a special right triangle called a 45–45–90 triangle. A 45–45–90 triangle has angle measures of 45°, 45°, and 90°. It has side lengths in the ratio 1 : 1 : $\sqrt{2}$. Two examples of 45–45–90 triangles are shown below.

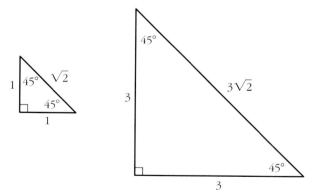

The ratio of the side lengths in any 45–45–90 triangle can be simplified to 1 : 1 : $\sqrt{2}$. For example,

$$3 : 3 : 3\sqrt{2} = \frac{3}{3} : \frac{3}{3} : \frac{3\sqrt{2}}{3} = 1 : 1 : \sqrt{2}$$

In Example F that follows, another special right triangle will be introduced—a 30–60–90 triangle.

EXAMPLE F

What is the height of the equilateral triangle shown at the right? Round your answer to the nearest tenth.

The altitude divides the equilateral triangle into two right triangles.

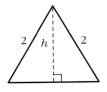

Each right triangle has a hypotenuse of length 2 and one leg of length 1. Use the Pythagorean equation to solve for h.

Language Box

The **altitude** of a triangle is a line segment. The length of the altitude is also called the *height* of the triangle.

$2^2 = h^2 + 1^2$ ← Substitute the values into the Pythagorean equation.

$4 = h^2 + 1$ ← Simplify.

$3 = h^2$ ← Subtract 1 from both sides.

$\sqrt{3} = h$ ← Find the positive square root.

$1.7 \approx h$ ← Use a calculator to find an approximation of $\sqrt{3}$.

The height of the equilateral triangle is about 1.7 units.

The special triangle formed by the altitude of an equilateral triangle is a 30–60–90 triangle. A 30–60–90 triangle has angle measures of 30°, 60°, and 90°. It has side lengths in the ratio $1 : \sqrt{3} : 2$. Two examples of 30–60–90 triangles are shown below.

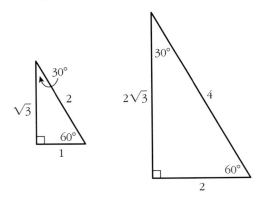

The ratio of side lengths in any 30–60–90 triangle can be simplified to $1 : \sqrt{3} : 2$

For example,

$2 : 2\sqrt{3} : 4 = \dfrac{2}{2} : \dfrac{2\sqrt{3}}{2} : \dfrac{4}{2} = 1 : \sqrt{3} : 2$

The special right triangles shown in Examples E and F are frequently used in trigonometry. Because they are so important, they are repeated in the box below. They are labeled to help you remember the ratios of their side lengths—$1 : 1 : \sqrt{2}$ and $1 : \sqrt{3} : 2$.

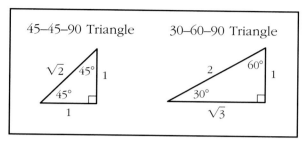

The converse of the Pythagorean Theorem can be used to determine if a triangle is a right triangle.

The Converse of the Pythagorean Theorem

If a triangle has side lengths a, b, and c such that $c^2 = a^2 + b^2$, then the triangle is a right triangle.

Language Box

For a statement in the form, "If P, then Q," its converse is "If Q, then P."

EXAMPLE G

Determine if a triangle with side lengths of 9, 12, and 15 is a right triangle.

You may assign 9 and 12 to either a or b. Assign 15 to c because it is the greatest number.

Let $a = 9$, $b = 12$, and $c = 15$.

$c^2 = a^2 + b^2$ ← Write the Pythagorean equation.

$15^2 \overset{2}{=} 9^2 + 12^2$ ← Substitute the values for a, b, and c.

$225 \overset{2}{=} 81 + 144$

$225 = 225$ ← The triangle is a right triangle because $c^2 = a^2 + b^2$.

A triangle with side lengths of 9, 12, and 15 is a right triangle.

EXAMPLE H

Determine if a triangle with side lengths of 15, 18, and 21 is a right triangle.

A triangle that has side lengths of 3, 4, and 5 is also a right triangle.

You may assign 15 and 18 to either a or b. Assign 21 to c because it is the greatest number.

Let $a = 15$, $b = 18$, and $c = 21$.

$c^2 = a^2 + b^2$ ← Write the Pythagorean Theorem.

$21^2 \overset{2}{=} 15^2 + 18^2$ ← Substitute the values for a, b, and c.

$441 \overset{2}{=} 225 + 324$

$441 \neq 549$ ← The triangle is not a right triangle because $c^2 \neq a^2 + b^2$.

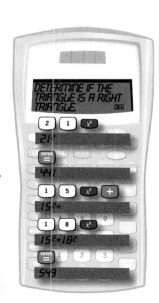

A triangle with side lengths of 15, 18, and 21 is not a right triangle.

Helpful Hint

Use the triangle shown below as a model for Exercises 1–6. Note that the triangle is not drawn to scale.

EXERCISES 17-1

In Exercises 1–6, c is the length of the hypotenuse of a right triangle, and a and b are the lengths of the legs. Find the missing side length. Round your answer to the nearest tenth, if rounding is necessary.

1. $a = 12$ $b = 16$ $c =$

2. $a =$ $b = 11$ $c = 61$

3. $a = 40$ $b =$ $c = 41$

4. $a = 5$ $b = 6$ $c =$

5. $a = 3$ $b = 3$ $c =$

6. $a = 15$ $b =$ $c = 50$

In Exercises 7–8, determine the length of the indicated side of the right triangle.

7. Find x in the triangle shown.

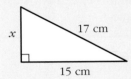

8. Find r in the triangle shown. Round your answer to the nearest tenth of a meter.

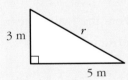

In Exercises 9–12, determine if a triangle with the given side lengths is a right triangle.

9. 30, 40, 50   **10.** 4, 6, 9

11. 0.5, 1.2, 1.3   **12.** 16, 16, $\sqrt{512}$

Helpful Hint

For Exercises 9–12, assign a letter to each length and substitute the values into the Pythagorean Theorem.

Answer each of the following.

13. A ramp at a loading dock has the dimensions shown below.

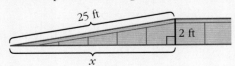

To the nearest tenth of a foot, what is the horizontal distance, x, that the ramp covers?

14. A pole 16.2 meters high is supported by a cable attached 2.3 meters from the top of the pole. What is the length of the cable required if the cable is fastened to the ground 6.3 meters from the foot of the pole? Assume the ground is level. Give the answer to the nearest hundredth meter.

15. As shown in the figure, a conduit must be bent to provide a rise of 6'-6" in a horizontal distance of 8'-9". What is the length of the conduit from *A* to *B*?

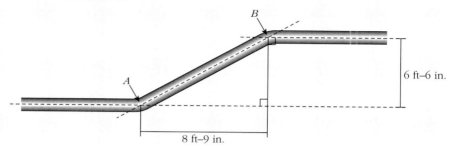

16. The hypotenuse of a certain right triangle is twice the length of the shorter leg. The shorter leg is 4 cm.

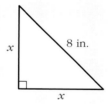

What is the length of the longer leg? Give your answer in radical form.

17. A right isosceles triangle has a hypotenuse of 8 inches.

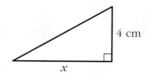

What is the length of each leg? Round your answer to the nearest tenth of an inch.

18. A triangular plot of land has side lengths of 2,000 feet, 2,100 feet, and 2,900 feet. Does the plot of land form a right triangle? Explain your reasoning.

19. Challenge: Find the lengths of $\overline{BC}$, $\overline{DE}$, and $\overline{FG}$ in the diagram below.

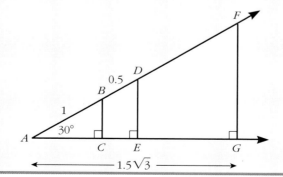

FINDING TRIGONOMETRIC FUNCTION VALUES OF ACUTE ANGLES

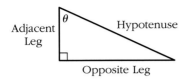

Adjacent Leg / Hypotenuse / Opposite Leg

Each acute angle of a right triangle has six ratios associated with it. Each ratio is formed by using the lengths of two of the sides of the triangle. These six ratios, called the **trigonometric ratios,** are as follows: sine (sin), cosine (cos), tangent (tan), cosecant (csc), secant (sec), and cotangent (cot).

A Greek letter is often used to represent an acute angle in a right triangle. In the diagram at the left, the letter θ (theta) is used to represent one of the acute angles. For any acute angle θ in a right triangle, the following abbreviations are used for the sides of the triangle:

- opp for the leg opposite θ
- adj for the leg adjacent to θ
- hyp for the hypotenuse

The six trigonometric ratios are defined in the box below.

Trigonometric Ratios

If θ is an acute angle of a right triangle, then:

$$\sin \theta = \frac{\text{opp}}{\text{hyp}} \qquad \csc \theta = \frac{\text{hyp}}{\text{opp}}$$

$$\cos \theta = \frac{\text{adj}}{\text{hyp}} \qquad \sec \theta = \frac{\text{hyp}}{\text{adj}}$$

$$\tan \theta = \frac{\text{opp}}{\text{adj}} \qquad \cot \theta = \frac{\text{adj}}{\text{opp}}$$

Notice that the ratios in the second column are the reciprocals of the ratios in the first column.

Write the trigonometric function values of the angle θ shown in the diagram below.

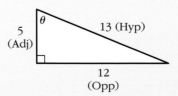

5 (Adj) / 13 (Hyp) / 12 (Opp)

$$\sin \theta = \frac{\text{opp}}{\text{hyp}} = \frac{12}{13} \qquad \csc \theta = \frac{\text{hyp}}{\text{opp}} = \frac{13}{12}$$

$$\cos \theta = \frac{\text{adj}}{\text{hyp}} = \frac{5}{13} \qquad \sec \theta = \frac{\text{hyp}}{\text{adj}} = \frac{13}{5}$$

$$\tan \theta = \frac{\text{opp}}{\text{adj}} = \frac{12}{5} \qquad \cot \theta = \frac{\text{adj}}{\text{opp}} = \frac{5}{12}$$

EXAMPLE A

Write the trigonometric function values of the angle θ shown in the diagram.

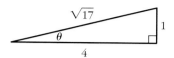

$$\sin \theta = \frac{\text{opp}}{\text{hyp}} = \frac{1}{\sqrt{17}} \qquad \csc \theta = \frac{\text{hyp}}{\text{opp}} = \frac{\sqrt{17}}{1} = \sqrt{17}$$

$$\cos \theta = \frac{\text{adj}}{\text{hyp}} = \frac{4}{\sqrt{17}} \qquad \sec \theta = \frac{\text{hyp}}{\text{adj}} = \frac{\sqrt{17}}{4}$$

$$\tan \theta = \frac{\text{opp}}{\text{adj}} = \frac{1}{4} \qquad \cot \theta = \frac{\text{adj}}{\text{opp}} = \frac{4}{1} = 4$$

The trigonometric function values of angle θ are shown above.

Helpful Hint

Trigonometric function values are often written in decimal form. For example, $\frac{4}{\sqrt{17}} \approx 0.9701$.

EXAMPLE B

Write the trigonometric function values of each acute angle in the triangle.

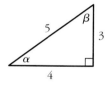

The Greek letters α (alpha) and β (beta) are also used to name angles in a triangle. These are the letters used in the triangle shown above.

Trigonometric Function Values of Angle α

$$\sin \alpha = \frac{\text{opp}}{\text{hyp}} = \frac{3}{5} \qquad \csc \alpha = \frac{\text{hyp}}{\text{opp}} = \frac{5}{3}$$

$$\cos \alpha = \frac{\text{adj}}{\text{hyp}} = \frac{4}{5} \qquad \sec \alpha = \frac{\text{hyp}}{\text{adj}} = \frac{5}{4}$$

$$\tan \alpha = \frac{\text{opp}}{\text{adj}} = \frac{3}{4} \qquad \cot \alpha = \frac{\text{adj}}{\text{opp}} = \frac{4}{3}$$

Trigonometric Function Values of Angle β

$$\sin \beta = \frac{\text{opp}}{\text{hyp}} = \frac{4}{5} \qquad \csc \beta = \frac{\text{hyp}}{\text{opp}} = \frac{5}{4}$$

$$\cos \beta = \frac{\text{adj}}{\text{hyp}} = \frac{3}{5} \qquad \sec \beta = \frac{\text{hyp}}{\text{adj}} = \frac{5}{3}$$

$$\tan \beta = \frac{\text{opp}}{\text{adj}} = \frac{4}{3} \qquad \cot \beta = \frac{\text{adj}}{\text{opp}} = \frac{3}{4}$$

The trigonometric function values of angles α and β are shown above.

In Examples C and D, you will find the trigonometric function values of the 45–45–90 triangle and the 30–60–90 triangle you studied in the last lesson.

EXAMPLE C

Use the diagram to write the sine, cosine, and tangent of a 45° angle.

$\sin 45° = \dfrac{1}{\sqrt{2}}$

$\cos 45° = \dfrac{1}{\sqrt{2}}$

$\tan 45° = \dfrac{1}{1} = 1$

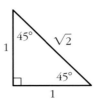

The sine, cosine, and tangent of 45° are shown above.

EXAMPLE D

Use the diagram to write the sine, cosine, and tangent of 30° and 60° angles.

$\sin 30° = \dfrac{1}{2}$ $\sin 60° = \dfrac{\sqrt{3}}{2}$

$\cos 30° = \dfrac{\sqrt{3}}{2}$ $\cos 60° = \dfrac{1}{2}$

$\tan 30° = \dfrac{1}{\sqrt{3}}$ $\tan 60° = \dfrac{\sqrt{3}}{1} = \sqrt{3}$

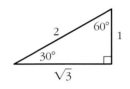

The sine, cosine, and tangent of 30° and 60° are shown above.

EXAMPLE E

Write sin 30°, cos 30°, and tan 30° in decimal form.

$\sin 30° = \dfrac{1}{2} = 0.5$

$\cos 30° = \dfrac{\sqrt{3}}{2} \approx \dfrac{1.73205}{2} \approx 0.8660$

$\tan 30° = \dfrac{1}{\sqrt{3}} \approx \dfrac{1}{1.73205} \approx 0.5774$

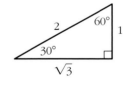

The decimal form of sin 30°, cos 30°, and tan 30° are shown above.

It is possible to find the trigonometric function values for any angle, not just a few special angles. You can find approximate trigonometric function values in a table. A sample from a table of trigonometric values is shown in Example F. A more complete table of values is given in the appendices.

EXAMPLE F

Use the table to find the sine, cosine, and tangent of 29°.

Angle	Sin	Cos	Tan
20°	0.3420	0.9397	0.3640
21°	0.3584	0.9336	0.3839
22°	0.3746	0.9272	0.4040
23°	0.3907	0.9205	0.4245
24°	0.4067	0.9135	0.4452
25°	0.4226	0.9063	0.4663
26°	0.4384	0.8988	0.4877
27°	0.4540	0.8910	0.5095
28°	0.4695	0.8829	0.5317
29°	0.4848	0.8746	0.5543
30°	0.5000	0.8660	0.5774

Locate the correct row for 29° in the table of values.

In the table, each value is rounded to four decimal places. That is why the symbol ≈ is used to state the answers.

sin 29° ≈ 0.4848, cos 29° ≈ 0.8746, tan 29° ≈ 0.5543

You may also use a scientific calculator to find trigonometric values. First, make sure that your calculator is in degree mode, by checking that the screen shows DEG in the lower right-hand corner. If it is not in degree mode, press 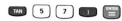 *and use the left or right arrow to move to* D E G.

EXAMPLE G

Use a calculator to find tan 57°. Round your answer to four decimal places.

Press the keys:

[TAN] [5] [7] [)] [ENTER =]

The display on the screen is 1.539864964.

$\tan 57° \approx 1.5399$

To find the cosecant, secant, or cotangent of an angle, remember that these ratios are the reciprocals of the sine, cosine, and tangent ratios, respectively. Tables of trigonometric values usually do not have entries for cosecant, secant, and cotangent. Likewise, scientific calculators do not have separate keys for cosecant, secant, and cotangent. Use the relationships in the Key Idea box that follows to find the cosecant, secant, and cotangent of an angle.

 **KEY IDEA**

For any acute angle θ in a right triangle,

$$\csc \theta = \frac{1}{\sin \theta} \qquad \sec \theta = \frac{1}{\cos \theta} \qquad \cot \theta = \frac{1}{\tan \theta}$$

EXAMPLE H

Find sec 14°.

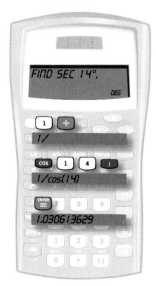

For any angle θ, sec θ is the reciprocal of cos θ. That is, $\sec \theta = \frac{1}{\cos \theta}$. So, to find sec 14°, first find cos 14°.

$\cos 14° \approx 0.9703$ ← You can use either a calculator or a table to find this value.

Now, find sec 14° by finding the reciprocal of cos 14°.

$$\sec 14° = \frac{1}{\cos 14°} \approx \frac{1}{0.9703} \approx 1.0306$$

$\sec 14° \approx 1.0306$

EXERCISES 17-2

Answer each of the following.

1. Find the missing trigonometric function values.

$$\sin \theta = \frac{\text{opp}}{\text{hyp}} = \frac{20}{29} \qquad \csc \theta = \frac{\text{hyp}}{\text{opp}} =$$

$$\cos \theta = \frac{\text{adj}}{\text{hyp}} = \qquad \sec \theta = \frac{\text{hyp}}{\text{adj}} =$$

$$\tan \theta = \frac{\text{opp}}{\text{adj}} = \qquad \cot \theta = \frac{\text{adj}}{\text{opp}} =$$

2. (a) Use the table of values in Example F to find sin 25°, cos 25°, and tan 25°.

 (b) Find csc 25°, sec 25°, and cot 25°.

3. Use a calculator to find the following values. Round your answers to four decimal places.

 (a) cos 20°   (b) sin 85°

 (c) tan 50°   (d) cos 9°

 (e) sin 14°   (f) tan 18°

4. Write the sine, cosine, and tangent of each acute angle in the triangle shown.

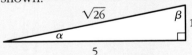

5. Using the diagram below, write sin θ, cos θ, and tan θ in both fraction form and decimal form.

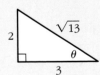

6. The cotangent of a certain angle θ is 1.7321. What is the tangent of angle θ?

7. Find the trigonometric function values of angle θ in the triangle shown below. Give your answers as decimals with four decimal places.

8. An electrician must bend a conduit to make a 6 ft-6 in. rise in an 8 ft-8 in. horizontal distance, as shown in the figure below.

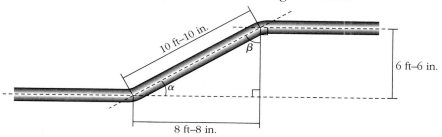

(a) Find the sin, cos, and tan of angle α in both fraction and decimal form. Simplify the fractions and round the decimal answers to four decimal places.

(b) Find the sin, cos, and tan of angle β in both fraction and decimal form. Simplify the fractions and round the decimal answers to four decimal places.

9. In an AC circuit that contains resistance, inductance, and capacitance in series, the voltage drop across the resistance V_R is $V_R = V \cos \theta$ where V is the applied voltage and θ, is the phase angle. If the phase angle is 32° and the applied voltage is 5.8V, what is the effective voltage across the resistor V_R?

10. Challenge: Explain why $\sin \theta$ and $\cos \theta$ are always less than 1 for any acute angle θ.

FINDING TRIGONOMETRIC FUNCTION VALUES OF ANY ANGLE

Angles in Standard Position

In the previous lesson, you found the trigonometric function values for various acute angles in a right triangle. To generate any angle from 0° to 360°, we will use a coordinate plane with two rays to form an angle.

The angle θ at the left is in standard position. Its vertex is at the origin. The initial ray forms one side of the angle and is placed on the positive x-axis. The terminal ray forms the other side of the angle. Think of rotating a ray from its initial position to its terminal position to form any angle.

The diagram below shows four angles in standard position, each one terminating in a different quadrant of the plane.

An angle is in standard position when its vertex is at the origin and its initial ray is on the positive x-axis.

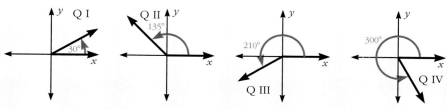

Each angle is measured from the initial ray. So, rotations less than 90° terminate in Quadrant I, rotations between 90° and 180° terminate in Quadrant II, and so on. When the rotation is counterclockwise, the angle measure is positive. When the rotation is clockwise, the angle measure is negative. Examples of negative angle measures are shown below.

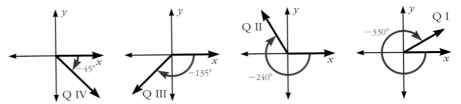

An angle that terminates on an axis is called a **quadrantal angle.** Quadrantal angles can serve as benchmarks for drawing other angles. The diagram below shows four positive rotations and four negative rotations that terminate on the axes.

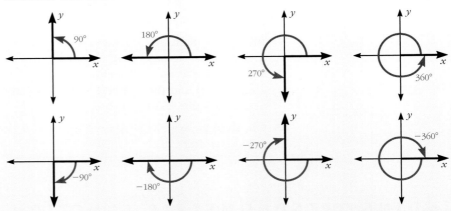

Draw an angle in standard position with a measure of 110°.

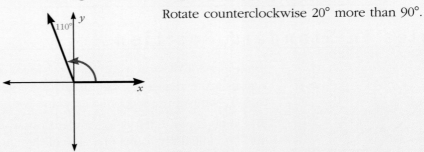

Rotate counterclockwise 20° more than 90°.

The diagram shows an angle in standard position with a measure of 110°.

EXAMPLE A

Draw an angle in standard position with a measure of 240°.

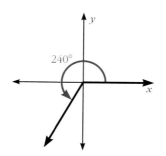

Rotate counterclockwise 60° more than 180°.

An angle of 240° in standard position is shown above.

EXAMPLE B

Draw an angle in standard position with a measure of −60°.

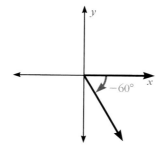

Rotate clockwise 60°.

An angle of −60° in standard position is shown above.

If a terminal ray makes one complete rotation, it forms a 360° angle. If a terminal ray makes more than one complete rotation, it forms an angle that is coterminal with another angle. Two angles in standard position are **coterminal angles** if they have the same terminal side. Coterminal angles are shown in Examples C and D.

EXAMPLE C

Draw a positive angle in standard position that is coterminal with 75°.

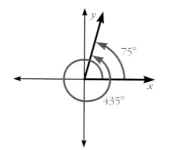

Add 360° to 75° to find one angle that is coterminal with 75°.

75° + 360° = 435°

The angle with a measure of 435° is coterminal with 75°.

EXAMPLE D

Draw a negative angle in standard position that is coterminal with 270°.

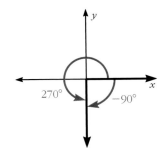

Subtract 360° from 270° to find one angle that is coterminal with 270°.

$$270° - 360° = -90°$$

The angle with a measure of −90° is coterminal with 270°.

KEY IDEA

To find an angle that is coterminal with a given angle, add or subtract any multiple of 360°.

EXAMPLE E

Give two positive angle measures and two negative angle measures that are coterminal with 10°.

Positive Angle Measures Coterminal with 10°
$$10° + 1(360°) = 10° + 360° = 370°$$
$$10° + 2(360°) = 10° + 720° = 730°$$

Negative Angle Measures Coterminal with 10°
$$10° - 1(360°) = 10° - 360° = -350°$$
$$10° - 2(360°) = 10° - 720° = -710°$$

The angle measures 370°, 730°, −350°, and −710° are coterminal with 10°.

The Unit Circle

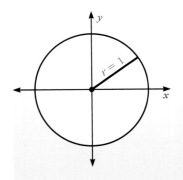

The trigonometric ratios can be defined for any angle using a unit circle. A **unit circle** is a circle whose radius is 1. The diagram at the left shows a unit circle whose center is at the origin.

The terminal ray of an angle, θ, in standard position will intersect the unit circle at a point (x, y). The trigonometric ratios of angle θ can be defined as follows.

Trigonometric Ratios in a Unit Circle

For an angle, θ, in standard position whose terminal ray intersects a unit circle at (x, y),

$$\sin \theta = \frac{y}{1} = y \qquad\qquad \csc \theta = \frac{1}{y} \text{ if } y \neq 0$$

$$\cos \theta = \frac{x}{1} = x \qquad\qquad \sec \theta = \frac{1}{x} \text{ if } x \neq 0$$

$$\tan \theta = \frac{y}{x} \text{ if } x \neq 0 \qquad\qquad \cot \theta = \frac{x}{y} \text{ if } y \neq 0$$

If $x = 0$, $\tan \theta$ and $\sec \theta$ are undefined.
If $y = 0$, $\csc \theta$ and $\cot \theta$ are undefined.

Use the unit circle and the triangle drawn below to find cos 45°.

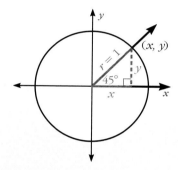

Helpful Hint

For a right triangle in a unit circle, the legs are x and y, and the hypotenuse is $r = 1$.

In a unit circle, $\cos \theta = x$, so we need to find x. The triangle is a 45–45–90 triangle, so the ratio of the side lengths is:

leg : leg : hypotenuse
 1 : 1 : $\sqrt{2}$

To write an equivalent ratio that has 1 for the hypotenuse, divide by $\sqrt{2}$:

$$1 : 1 : \sqrt{2} = \frac{1}{\sqrt{2}} : \frac{1}{\sqrt{2}} : \frac{\sqrt{2}}{\sqrt{2}} : \frac{1}{\sqrt{2}} : \frac{1}{\sqrt{2}} : 1$$

So in this triangle, $x = \frac{1}{\sqrt{2}}$, $y = \frac{1}{\sqrt{2}}$, and $r = 1$.

And $\cos 45° = x = \frac{1}{\sqrt{2}} \approx 0.7071$.

KEY IDEA

For any angle θ,

$$\sin \theta = \frac{1}{\csc \theta} \qquad \cos \theta = \frac{1}{\sec \theta} \qquad \tan \theta = \frac{1}{\cot \theta}$$

$$\csc \theta = \frac{1}{\sin \theta} \qquad \sec \theta = \frac{1}{\cos \theta} \qquad \cot \theta = \frac{1}{\tan \theta}$$

EXAMPLE F

Use the unit circle and the triangle drawn at the right to find sin 30° and cos 30°.

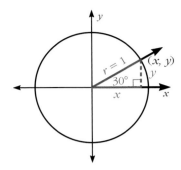

In a unit circle, $\sin \theta = y$ and $\cos \theta = x$, so we need to find y and x. The triangle is a 30–60–90 triangle, so the ratio of the side lengths is:

short leg : long leg : hypotenuse
    1   : $\sqrt{3}$:   2

To write an equivalent ratio that has 1 for the hypotenuse, divide by 2:

$$1 : \sqrt{3} : 2 = \frac{1}{2} : \frac{\sqrt{3}}{2} : \frac{2}{2} = \frac{1}{2} : \frac{\sqrt{3}}{2} : 1$$

So in this triangle, $y = \frac{1}{2}$, $x = \frac{\sqrt{3}}{2}$, and $r = 1$.

$$\sin 30° = y = \frac{1}{2} = 0.5, \quad \cos 30° = x = \frac{\sqrt{3}}{2} \approx 0.8660$$

A quadrantal angle has either an x- or a y-coordinate of 0. When a trigonometric ratio has 0 in the denominator, that particular trigonometric function value is undefined. This is shown in Example G.

EXAMPLE G

Find the trigonometric function values of 90°.

The terminal ray of 90° passes through (0, 1), so $x = 0$ and $y = 1$. Use the definitions given for the trigonometric function values to write:

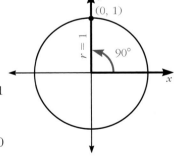

$\sin 90° = y = 1$ $\csc 90° = \frac{1}{y} = \frac{1}{1} = 1$

$\cos 90° = x = 0$ $\sec 90°$ is undefined

$\tan 90°$ is undefined $\cot 90° = \frac{x}{y} = \frac{0}{1} = 0$

The trigonometric function values of 90° are given above.

Helpful Hint

For all cases where the ratios are defined:

- $\sin \theta$ and $\csc \theta$ have the same sign.

- $\cos \theta$ and $\sec \theta$ have the same sign.

- $\tan \theta$ and $\cot \theta$ have the same sign.

The signs of trigonometric function values are determined by the signs of x and y. The diagram below shows the signs of sine, cosine, and tangent values in all the quadrants.

Signs of Function Values

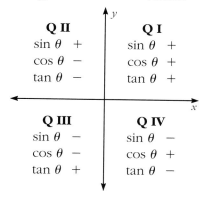

Q II	Q I
$\sin \theta$ +	$\sin \theta$ +
$\cos \theta$ −	$\cos \theta$ +
$\tan \theta$ −	$\tan \theta$ +

Q III	Q IV
$\sin \theta$ −	$\sin \theta$ −
$\cos \theta$ −	$\cos \theta$ +
$\tan \theta$ +	$\tan \theta$ −

To find trigonometric function values for any angle greater than 90°, or for any negative angle, you need to know the angle's reference angle α. The reference angle of any angle is formed by the terminal ray of the angle and the x-axis. All reference angles are positive. The diagram below shows how to find some reference angles.

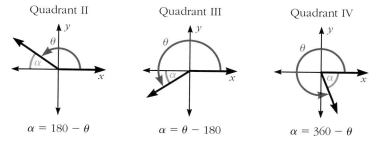

Quadrant II   Quadrant III   Quadrant IV

$\alpha = 180 - \theta$ $\alpha = \theta - 180$ $\alpha = 360 - \theta$

EXAMPLE H

Find the reference angle α for each angle θ listed at the right.

$\theta = 160°$ $\alpha = 180° - 160° = 20°$

$\theta = 210°$ $\alpha = 210° - 180° = 30°$

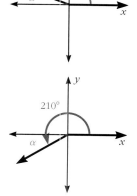

$\theta = 300°$ $\alpha = 360° - 300° = 60°$

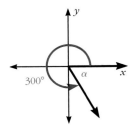

$\theta = -150°$ $360° - 150° = 210°$
            $-150°$ is coterminal with $210°$.
            $\alpha = 210° - 180° = 30°$

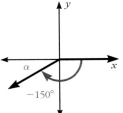

$\theta = -325$ $360° - 325° = 35°$
            $-325°$ is coterminal with $35°$.
            $\alpha = 35°$

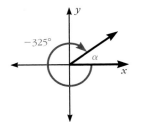

The reference angles for various values of θ are shown above.

To find a trigonometric function value for an angle greater than 90°, or for a negative angle, find the trigonometric function value of its reference angle. Then determine the sign for your answer based on the quadrant. This is shown in Examples I, J, and K that follow.

EXAMPLE I

Find cos 120° using its reference angle.

For $\theta = 120°$, the reference angle is 60°.
$\cos 60° = 0.5$

Cosine function values are negative in Quadrant II, so $\cos 120° = -0.5$.

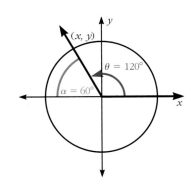

$\cos 120° = -0.5$

EXAMPLE J

Find tan (−135°) using its reference angle.

For $\theta = -135°$, the reference angle is 45°.
tan 45° = 1

Tangent function values are positive in Quadrant III, so tan (−135°) = 1.

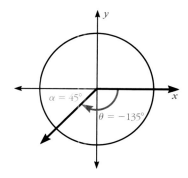

tan (−135°) = 1

EXAMPLE K

Find sin (−330°) using its reference angle.

For −330°, the reference angle is 30°.
sin 30° = 0.5

Sine function values are positive in Quadrant I, so sin (−330°) = 0.5.

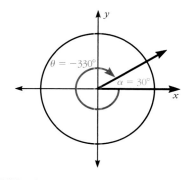

sin (−330°) = 0.5

You can also use a scientific calculator to find trigonometric function values of any angle. The calculator will assign the correct positive or negative sign to the trigonometric function value.

EXAMPLE L

Use a calculator to find tan (−134°).

tan (−134°) ≈ 1.0355

EXAMPLE M

Use a calculator to find sec 127°.

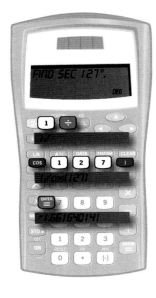

$$\sec 127° = \frac{1}{\cos 127°} \approx -1.6616$$

$\sec 127° \approx -1.6616$

EXERCISES 17-3

In Exercises 1–8, assume the angles are in standard position. Name the quadrant that contains the terminal side.

1. 95°   **2.** 350°

3. −30°   **4.** 225°

5. 390°   **6.** −300°

7. 145°   **8.** −100°

In Exercises 9–14, name one positive and one negative angle that are coterminal with the given angle.

9. 45°   **10.** 165°

11. 185°   **12.** −45°

13. 360°   **14.** −5°

In Exercises 15–16, draw an angle in standard position and give its measure.

15. $\frac{1}{3}$ of a counterclockwise rotation

16. $\frac{7}{12}$ of a clockwise rotation

Answer each of the following.

17. Use the unit circle and the triangle at the right to find cos 60°.

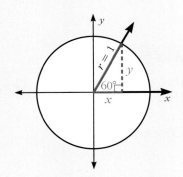

18. Use the unit circle and the triangle at the right to find tan 45° and cot 45°.

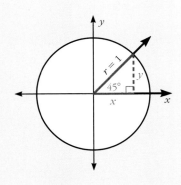

19. Use the unit circle and the triangle drawn below to find $\sin(-30°)$ and $\cos(-30°)$.

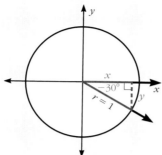

20. Find the trigonometric function values of $180°$.

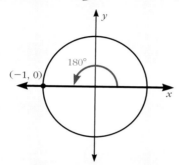

21. Find the trigonometric function values of $-90°$.

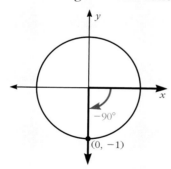

Helpful Hint

Draw each angle θ on a unit circle to help find the reference angle α.

22. Find the reference angle α for each angle θ below.

(a) $\theta = 328°$ (b) $\theta = 230°$

(c) $\theta = 188°$ (d) $\theta = -15°$

(e) $\theta = 400°$ (f) $\theta = 210°$

(g) $\theta = -270°$ (h) $\theta = 340°$

23. For each angle θ, find its reference angle α and then find $\sin \alpha$ and $\sin \theta$. (Part a has been completed.)

	θ	Reference Angle α	$\sin \alpha$	$\sin \theta$
(a)	330°	30°	$\sin 30° = \frac{1}{2}$	$\sin 330° = -\frac{1}{2}$
(b)	−135°			
(c)	120°			

24. For each angle θ, find its reference angle α and then find $\cos \alpha$ and $\cos \theta$.

	θ	Reference Angle α	$\cos \alpha$	$\cos \theta$
(a)	120°			
(b)	−30°			
(c)	−135°			

25. For each angle θ, find its reference angle α and then find $\tan \alpha$ and $\tan \theta$.

	θ	Reference Angle α	$\tan \alpha$	$\tan \theta$
(a)	−45°			
(b)	210°			
(c)	−240°			

26. Use a calculator to find each of the following values.

(a) $\cos 44°$ (b) $\tan 105°$

(c) $\cot 225°$ (d) $\sin (-150°)$

(e) $\cos (-315°)$ (f) $\sin 360°$

Answer Excercises 27–28. Round your answer to the nearest tenth.

27. The intensity, I, of a sinusoidal current in an AC circuit is given by $I = I_{max} \sin \theta$, where I_{max} is the maximum intensity of the current. Find I when $I_{max} = 32.65$ mA and $I = 132°$.

28. The potential voltage, V, of a sinusoidal current in an AC circuit is given by $V = V_{max} \sin \theta$, where V_{max} is the maximum potential. Find V when $V_{max} = 115.2$ V and $\theta = 113.4°$.

29. Challenge: Use the definitions of the trigonometric functions on a unit circle to explain why $(\sin \theta)^2 + (\cos \theta)^2 = 1$.

USING INVERSE TRIGONOMETRIC FUNCTIONS

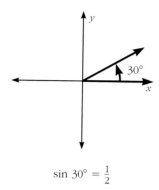

$$\sin 30° = \frac{1}{2}$$

So far in this chapter, you have found trigonometric function values for an angle θ when the value of θ was given. But is there a way to work "backward" and find θ when a value of a trigonometric function is given? Consider the problem below.

If $\sin \theta = \frac{1}{2}$, what values can θ have? Three values of θ are shown below.

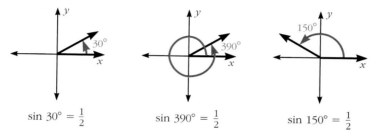

$$\sin 30° = \frac{1}{2} \qquad \sin 390° = \frac{1}{2} \qquad \sin 150° = \frac{1}{2}$$

In fact, there are many values of θ that satisfy the statement $\sin \theta = \frac{1}{2}$. When *exactly one* value of θ is needed for a statement like $\sin \theta = \frac{1}{2}$, we use an inverse trigonometric function. Three inverse trigonometric functions are defined in the table below. The notation $\sin^{-1}$ is read "inverse sine," and $\sin^{-1} t$ can be thought of as "the angle measure whose sine is t" (using the restrictions in the table to get the correct angle measure).

Inverse Trigonometric Functions

$\sin^{-1} t = \theta$ if and only if $\sin \theta = t$, and:

(Restrictions) $-1 \le t \le 1$
$-90° \le \theta \le 90°$

θ must terminate in the shaded region.

$\cos^{-1} t = \theta$ if and only if $\cos \theta = t$, and:

(Restrictions) $-1 \le t \le 1$
$0° \le \theta \le 180°$

θ must terminate in the shaded region.

$\tan^{-1} t = \theta$ if and only if $\tan \theta = t$, and:

(Restrictions) t is any real number
$-90° < \theta < 90°$

θ must terminate in the shaded region.

With a trigonometric function, you know the measure of an angle and want to find a trigonometric ratio. With an *inverse* trigonometric function, you know the trigonometric ratio and want to find the measure of the angle that corresponds to that ratio.

EXAMPLE A

What is $\cos^{-1} 0$?

$\cos^{-1} 0$ is the angle θ whose cosine is 0, such that $0° \leq \theta \leq 180°$.

$\cos \theta = x$, so if $\cos \theta = 0$, then $x = 0$. Therefore, θ must terminate on the y-axis, and $\theta = 90°$.

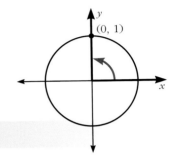

$\cos^{-1} 0 = 90°$

EXAMPLE B

What is $\tan^{-1}(-1)$?

$\tan^{-1}(-1)$ is the angle θ whose tangent is -1, such that $-90° < \theta < 90°$.

θ terminates in Q IV because $\tan \theta$ is negative.

$\tan \theta = \frac{y}{x}$, so if $\tan \theta = -1$, then $\frac{y}{x} = -1$. Therefore, $y = -x$.

Because $y = -x$, the legs of the triangle are equal, and $\theta = -45°$.

$\tan^{-1}(-1) = -45°$

You may use your calculator to solve problems involving inverse trigonometric functions. A scientific calculator is programmed with the same restrictions on angle measures for inverse trigonometric functions as those given in the table. To find $\tan^{-1}(-1)$ on your TI-30X IIS calculator, make sure you are in degree mode and then press:

$\boxed{\text{TAN}^{-1}}$ $\boxed{\text{2nd}}$ $\boxed{\text{TAN}}$ $\boxed{(\text{-})}$ $\boxed{1}$ $\boxed{)}$ $\boxed{\text{ENTER}}$.

The calculator will display $-45°$.

EXAMPLE C

What is $\sin^{-1}\frac{12}{13}$? Round your answer to the nearest whole degree.

Use the inverse sine function on your calculator. Press:

$\boxed{\text{SIN}^{-1}}$ $\boxed{\text{2nd}}$ $\boxed{\text{SIN}}$ $\boxed{1}$ $\boxed{2}$ $\boxed{÷}$ $\boxed{1}$ $\boxed{3}$ $\boxed{)}$ $\boxed{\text{ENTER}}$

The display is 67.38013505. Round your answer to $67°$.

$\sin^{-1}\frac{12}{13} \approx 67°$

EXAMPLE D

Find the measure of angle θ in the diagram at the right. Round your answer to the nearest whole degree.

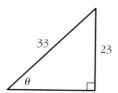

$\sin \theta = \dfrac{\text{opp}}{\text{hyp}}$ ← Write the ratio for the sine function because the leg opposite θ and the hypotenuse are given.

$\sin \theta = \dfrac{23}{33}$ ← Substitute the known values.

$\sin^{-1} \dfrac{23}{33} = \theta$ ← If $\sin \theta = \dfrac{23}{33}$, then $\sin^{-1} \dfrac{23}{33} = \theta$.

$44° \approx \theta$ ← Use a calculator to get an approximation for θ.

$\theta \approx 44°$

EXAMPLE E

Find the measure of angle θ in the diagram at the right. Round your answer to the nearest whole degree.

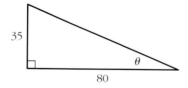

$\tan \theta = \dfrac{\text{opp}}{\text{adj}}$ ← Write the ratio for the tangent function because the leg opposite θ and the leg adjacent to θ are given.

$\tan \theta = \dfrac{35}{80}$ ← Substitute the known values.

$\tan^{-1} \dfrac{35}{80} = \theta$ ← If $\tan \theta = \dfrac{35}{80}$, then $\tan^{-1} \dfrac{35}{80} = \theta$.

$24° \approx \theta$ ← Use a calculator to get an approximation for θ.

$\theta \approx 24°$

EXAMPLE F

The phase angle, θ, in an inductive circuit between the impedance, Z, and the resistance, R, is $\theta = \cos^{-1} \dfrac{Z}{R}$. Find θ to the nearest tenth degree when $Z = 37 \ \Omega$ and $R = 40 \ \Omega$.

$\theta = \cos^{-1} \dfrac{Z}{R}$

$= \cos^{-1} \dfrac{37 \ \Omega}{40 \ \Omega}$

$= \cos^{-1} \dfrac{37}{40}$

$\approx 22.3316°$

The phase angle is about 22.3°.

EXERCISES 17-4

In Exercises 1–2, find three values of θ that satisfy the given equation.

1. $\tan \theta = 1$ **2.** $\cos \theta = -0.5$

In Exercises 3–10, find each inverse trigonometric function value. Round your answer to the nearest whole degree, if rounding is necessary.

3. $\tan^{-1}(-1)$ **4.** $\cos^{-1}(-0.5)$

5. $\sin^{-1} 0.8271$ **6.** $\tan^{-1} 1.732$

7. $\tan^{-1} 2$ **8.** $\sin^{-1}(-1)$

9. $\sin^{-1}\left(-\dfrac{3}{4}\right)$ **10.** $\tan^{-1}(-2.5)$

Answer each of the following. Round your answer to the nearest whole degree.

11. Find the measure of angle θ in the diagram.

12. Find the measure of angle θ in the diagram.

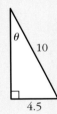

13. If the resistance is in series with the capacitive reactance in an *RC* circuit, then the phase angle, θ, between the capacitive reactance, X_C, and the resistance, R, is $\theta = \tan^{-1} \dfrac{X_C}{R}$. Find θ to the nearest tenth degree when $X_C = 33 \ \Omega$ and $R = 40 \ \Omega$.

14. In an AC circuit that contains resistance, inductance, and capacitance in series, the voltage drop across the resistance, V_R, is $V_R = V \cos \theta$, where V is the applied voltage and θ is the phase angle. What is the phase angle if the applied voltage is $V = 38.9$ V and the effective voltage across the resistor is $V_R = 12.7$ V?

15. An electrician must bend a conduit to make a 5 ft-7 in. rise in an 8 ft-8 in. horizontal distance, as shown in the figure below.

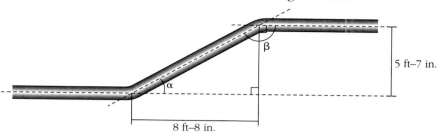

8 ft–8 in.

5 ft–7 in.

(a) Find α to four decimal places.

(b) Find β to four decimal places.

16. At two separate angles of rotation, the armature of a generator produces 15.3 V and 73.8 V. If θ is the angle of rotation, V is the peak voltage, and V_θ is the voltage when the angle of rotation is θ, then $\sin \theta = \dfrac{V_\theta}{V}$. If the peak voltage is 75 V, determine the following:

(a) These two angles of rotation

(b) The phase angle (difference) between these two voltages

17. Challenge: What is the measure of α in the diagram below?

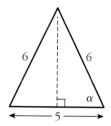

SOLVING RIGHT TRIANGLES

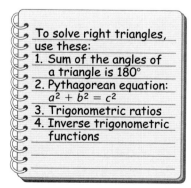

To solve right triangles, use these:
1. Sum of the angles of a triangle is 180°
2. Pythagorean equation: $a^2 + b^2 = c^2$
3. Trigonometric ratios
4. Inverse trigonometric functions

The process of finding the measures of all the unknown sides and angles of a right triangle is called *solving the triangle*. There are three angles and three sides in a triangle. Use a combination of the suggestions given at the left to solve a triangle.

For convenience, the side lengths can be represented by a, b, and c, and the angles can be named by their vertices, A, B, and C. Side a is opposite angle A, side b is opposite angle B, and side c is opposite angle C. We will let C represent the right angle in the triangles in this lesson.

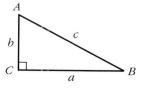

When rounding is necessary in the examples that follow, all angle measures are rounded to the nearest whole degree and all side lengths are rounded to the nearest tenth.

Solve the right triangle shown below.

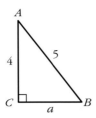

Helpful Hint

Make a chart similar to the one below.

Angles	Sides
$A =$	$a =$
$B =$	$b =$
$C =$	$c =$

Fill in the values for the angle measures and side lengths as you find them.

Because the length of one leg and the hypotenuse are already known, you can use the Pythagorean equation to find the length of the other leg, a.

$a^2 + b^2 = c^2$ ← Write the Pythagorean equation.

$a^2 + 4^2 = 5^2$ ← Substitute 4 for b and 5 for c.

$a^2 + 16 = 25$

$a^2 = 9$

$a = 3$

Now find the measures of angles A and B.

Use the inverse cosine function to find the measure of angle A.

$$\cos A = \frac{\text{adj}}{\text{hyp}}$$

$$\cos A = \frac{4}{5}$$

$$A = \cos^{-1} \frac{4}{5}$$

$$A \approx 37°$$

Use the inverse sine function to find the measure of angle B.

$$\sin B = \frac{\text{opp}}{\text{hyp}}$$

$$\sin B = \frac{4}{5}$$

$$B = \sin^{-1} \frac{4}{5}$$

$$B \approx 53°$$

It is a good idea to check that the sum of the angle measures is 180°. In this case: $37° + 53° + 90° = 180°$.

In the triangle above, $a = 3$, $A \approx 37°$, and $B \approx 53°$.

EXAMPLE A

Solve the right triangle shown.

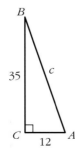

Because the lengths of two of the sides of the triangle are already known, you can use the Pythagorean equation to find the third side length, c.

$$a^2 + b^2 = c^2$$

$$35^2 + 12^2 = c^2$$

$$1{,}225 + 144 = c^2$$

$$1{,}369 = c^2$$

$$\sqrt{1{,}369} = c$$

$$37 = c$$

Now find the measures of angles A and B.

Use the inverse tangent function to find the measure of angle A.

$$\tan A = \frac{\text{opp}}{\text{adj}}$$

$$\tan A = \frac{35}{12}$$

$$A = \tan^{-1} \frac{35}{12}$$

$$A \approx 71°$$

Use the inverse tangent function to find the measure of angle B.

$$\tan B = \frac{\text{opp}}{\text{adj}}$$

$$\tan B = \frac{12}{35}$$

$$B = \tan^{-1} \frac{12}{35}$$

$$B \approx 19°$$

In the triangle on the previous page, $c = 37$, $A \approx 71°$, and $B \approx 19°$.

EXAMPLE B

Solve the right triangle shown.

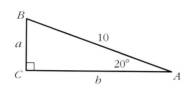

First, find the measure of angle B.

$$A + B + C = 180°$$

$$20° + B + 90° = 180°$$

$$B = 70°$$

Helpful Hint

When using a calculator, do not round off any numbers until you arrive at the final solution.

Now use trigonometric ratios to find the lengths of the sides.

Use the sine ratio to find a.

$$\sin A = \frac{\text{opp}}{\text{hyp}}$$

$$\sin 20° = \frac{a}{10}$$

$$10 \cdot \sin 20° = a$$

$$3.4 \approx a$$

Use the cosine ratio to find b.

$$\cos A = \frac{\text{adj}}{\text{hyp}}$$

$$\cos 20° = \frac{b}{10}$$

$$10 \cdot \cos 20° = b$$

$$9.4 \approx b$$

In the triangle above, $B = 70°$, $a \approx 3.4$, and $b \approx 9.4$.

EXAMPLE C

In an *RC* series circuit, the total impedance, *Z*, is related to the resistance, *R*, and the capacitive reactance, X_C, by the impedance triangle shown below. If the phase angle is 28°, $X_C = 8$ kΩ, and $Z = 17$ kΩ, solve the right triangle.

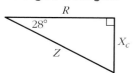

We begin by naming the unknown acute angle α and writing the known information on the triangle as shown in this figure. Notice that the legs are R and X_C and the hypotenuse is Z.

$$28° + \alpha + 90° = 180°$$

$$\alpha = 62°$$

Using the Pythagorean Theorem:

$$R^2 + X_C^2 = Z^2$$

$$R^2 + 8^2 = 17^2$$

$$R^2 + 64 = 289$$

$$R^2 = 225$$

$$R = 15$$

In the triangle above, $\alpha = 62°$ and $R = 15$ kΩ.

EXERCISES 17-5

In Exercises 1–6, solve the right triangle *ABC*. If rounding is necessary, round each angle measure to the nearest whole degree, and round each side length to the nearest tenth.

1.

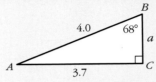

2.

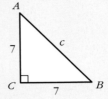

3. $A = 74°$, $a = 24$, $c = 25$

4. $a = 9$, $b = 12$, $c = 15$

5. $B = 24°$, $b = 10$

6. $A = 19°$, $b = 13$

Solve Exercises 7–10,

7. In the triangle below, $\cos A = \frac{40}{41}$. Solve the triangle. Round each angle measure to the nearest whole degree.

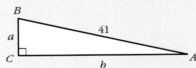

8. Solve the right triangle below. Round each angle measure to the nearest whole degree.

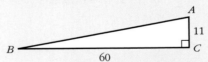

9. Given $\theta = 39.2°$ and $X_L = 139$ kΩ, solve for Z, R, and α.

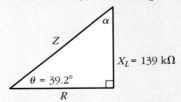

10. Given $\alpha = 18.4°$ and $Z = 59.4$ kΩ, solve for X_L, R, and β.

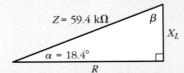

11. Challenge: In the triangle below, $\csc A = \frac{17}{15}$.

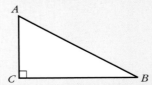

Find the measures of angles *A* and *B*. Round your answers to the nearest whole degree.

SOLVING WORD PROBLEMS WITH TRIGONOMETRIC FUNCTIONS

Trigonometry can be used to find distances when it is not practical to use a tape measure.

Some real-world problems can be modeled by right triangles and solved with trigonometry. In the problem below, the distance across a stream that cannot be measured is found using trigonometry.

Find the width x of the stream in the diagram below.

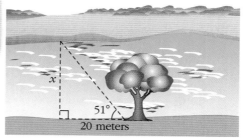

The tangent of 51° can be used to find x, the length of the opposite leg, because the length of the adjacent leg is known.

$\tan \theta = \dfrac{\text{opp}}{\text{adj}}$ ◄— Definition of tangent ratio

$\tan 51° = \dfrac{x}{20}$ ◄— Substitute the known information.

$20 \cdot \tan 51° = x$ ◄— Multiply both sides by 20.

$25 \approx x$ ◄— Evaluate using a calculator and round your answer to the nearest whole number.

The river is about 25 meters wide.

EXAMPLE A

A wire that secures a utility pole makes an angle of 20° with the utility pole. The length of the wire is 42 feet. What is the height of the utility pole? Round your answer to the nearest tenth of a foot.

Let h represent the height of the utility pole. The cosine of 20° can be used to find h, the length of the adjacent leg, because the length of the hypotenuse is known.

$\cos \theta = \dfrac{\text{adj}}{\text{hyp}}$ ◄— Definition of cosine ratio

$\cos 20° = \dfrac{h}{42}$ ◄— Substitute the known information.

$42 \cdot \cos 20° = h$ ◄— Multiply both sides by 42.

$39.5 \approx h$ ◄— Evaluate using a calculator and round your answer.

The utility pole is about 39.5 feet tall.

EXAMPLE B

A certain cell phone holder sits on a flat surface and holds the phone at an angle of 55°. How high above the flat surface is the tip of the antenna?

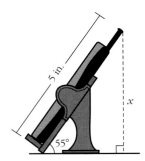

Let x represent the height of the phone above the flat surface. The sine of 55° can be used to find x, the length of the opposite leg, because the length of the hypotenuse is known.

$$\sin \theta = \frac{\text{opp}}{\text{hyp}} \quad \longleftarrow \quad \text{Definition of sine ratio}$$

$$\sin 55° = \frac{x}{5} \quad \longleftarrow \quad \text{Substitute the known information.}$$

$$5 \cdot \sin 55° = x \quad \longleftarrow \quad \text{Multiply both sides by 5.}$$

$$4.1 \approx x \quad \longleftarrow \quad \text{Evaluate using a calculator and round your answer.}$$

The tip of the antenna is about 4.1 inches above the flat surface.

EXAMPLE C

The escalator at the right connects two floors of a shopping mall. What is θ? Round your answer to the nearest degree.

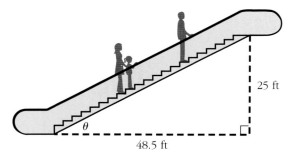

The inverse tangent function can be used to find θ.

$$\tan \theta = \frac{25}{48.5} \quad \longleftarrow \quad \text{Substitute the known information.}$$

$$\tan^{-1} \frac{25}{48.5} = \theta \quad \longleftarrow \quad \text{Definition of inverse tangent function}$$

$$27° \approx \theta \quad \longleftarrow \quad \text{Use a calculator and round your answer.}$$

$$\theta \approx 27°$$

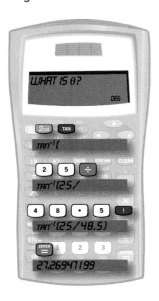

One important application of trigonometry is in the bending of conduit. An offset bend is used to run conduit around an obstruction.

In the diagram at the right, the offset angle is θ and the amount of offset is r. The center-to-center distance between bends of the conduit is x.

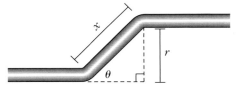

EXAMPLE D

How far apart will the center-to-center marks be on the conduit to achieve a 10-inch offset with a 30° bend?

Use the sine ratio to solve for x.

$$\sin 30° = \frac{10}{x}$$

$$x \sin 30° = 10$$

$$x = \frac{10}{\sin 30°}$$

$$x = 20$$

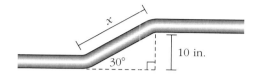

The center-to-center marks will be 20 inches apart.

EXERCISES 17-6

Solve each of the following.

1. A radio antenna is stabilized by support cables. Two of these cables are shown below.

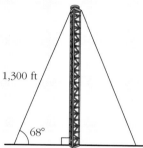

   If each of the cables is 1,300 feet long and forms an angle of 68° with the ground, how tall is the antenna? Round your answer to the nearest foot.

2. John used the ramp below to load his lawnmower into his pickup truck.

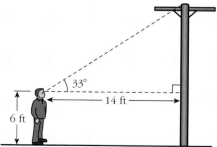

   To the nearest degree, what is the measure of angle θ, the angle between the ramp and the ground?

3. A 6-foot tall person is standing 14 feet away from a utility pole.

   If the person's angle of sight to the top of the utility pole is 33°, about how tall is the utility pole? Remember to include the person's height in your answer. Round your answer to the nearest foot.

4. Rhonda is moving some boxes using the hand truck shown below.

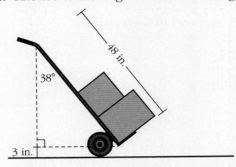

How high above the ground is the point indicated by the dashed line? Round your answer to the nearest inch.

5. How far apart will the center-to-center marks be on the conduit to achieve an 8-inch offset with a 45° bend? Round your answer to the nearest tenth of an inch.

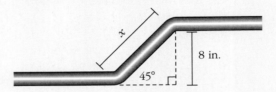

6. An 8-foot-long ladder is leaning against a wall, as shown below.

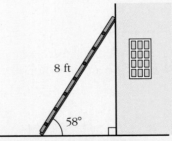

How far is the base of the wall from the base of the ladder? Round your answer to the nearest tenth of a foot.

7. Two utility poles are on opposite sides of an interstate highway. To find the distance between the poles, Ali paced off 90 ft along the highway and perpendicular to a line joining the two poles. The angle between the poles is measured at 68°. Determine the distance between the two poles to the nearest tenth foot.

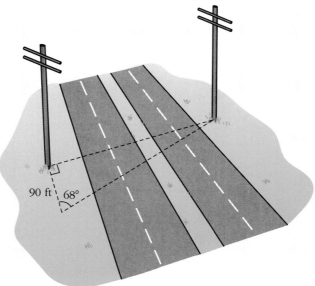

8. A security camera is positioned to record people using an ATM. The camera is located 5 ft-3 in. above and 12 ft-8 in. to the right of an average person's head.

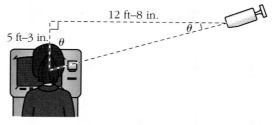

(a) How far, to the nearest inch, is the camera from the person's head?

(b) What is the angle θ to the nearest tenth degree that the camera is positioned below the horizontal?

9. Challenge: Suppose the base of the ladder in Exercise 6 has slipped, and the base of the ladder is now 5 feet from the base of the wall. To the nearest tenth of a degree, what is the angle between the ground and the ladder?

CHAPTER 17 REVIEW EXERCISES

Solve the following. Unless otherwise stated, round each answer to the nearest tenth.

1.

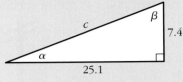

c

β

7.4

α

25.1

$c =$
$\alpha =$
$\beta =$

2.

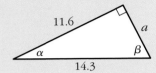

11.6

a

α   β

14.3

$a =$
$\alpha =$
$\beta =$

3.

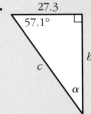

27.3

57.1°

b

c

α

$b =$
$c =$
$\alpha =$

4. Name one positive and one negative angle that are coterminal with a 137° angle.

5. If $\alpha = -112°$, find its reference angle and find sin α, cos α, and tan α.

6. An offset for a conduit is to be 4.9 m long and 1.7 m high.

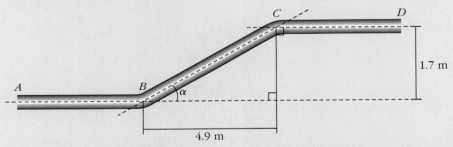

(a) What is the measure to the nearest tenth degree of the angle, $\angle\alpha$, the conduit makes with the horizontal?

(b) What is the length of the offset *BC* to the nearest tenth meter?

7. A 30m length of conduit is bent as shown in the figure below.

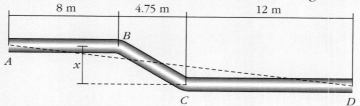

(a) What is the length of the offset *BC* to the nearest centimeter?

(b) What is the height *x* of the offset to the nearest centimeter?

(c) The straight line distance from *A* to *D* is indicated by the dotted line in the figure. What is the straight line distance from *A* to *D*?

(d) If it were possible to run a straight conduit from *A* to *D*, as indicated by the dotted line, how much conduit, to the nearest centimeter, would be saved?

8. In an RC circuit, the voltage across the resistor, V_R, is 157 V and the voltage across the capacitor, V_C, is 86 V.

(a) What is the total applied voltage V_T?

(b) What is the phase angle θ to the nearest tenth degree?

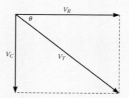

9. Challenge: A security camera is located at an ATM such that it can photograph the license plate of cars and trucks at the machine. As shown in the figure, the camera is located 8 ft-6 in. above the height of the license plate, 10 ft-9 in. behind the plate, and 4 ft-2 in. in from the plate.

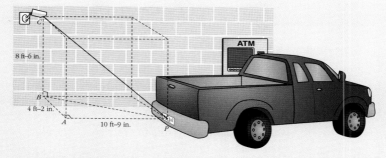

(a) Determine, to the nearest inch, the distance *CP* from the camera to the license plate.

(b) Determine, to the nearest degree, the angle ∠*BCP* makes with the wall.

Hint: First find the length *BP* in right triangle △*ABP* and then use right triangle △*BCP*.

Building a Foundation in Mathematics

Introducing Vectors

Vector Operations

Finding the Magnitude and Direction of a Vector

Finding and Using the Components of a Vector

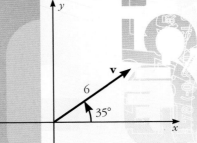

Overview

A vector is a quantity that can be described using two numbers. One number represents the magnitude and the other, the direction. This allows a vector to be drawn as a directed line segment using an arrow. In this chapter, we will look at using vectors to describe electrical situations and ways that vectors can be used to solve problems.

Vectors are important in the study of AC circuits and use the study of trigonometry from Chapter 17. Electricity, electronics, and computer graphics are some of the fields that use graphic and trigonometric applications of vectors.

Vectors

Objectives

After completing this chapter, you will be able to:

- Identify vector and scalar quantities
- Identify vectors in standard position
- Determine vector sums and differences graphically
- Determine the product of a scalar and a vector graphically
- Determine vector sums and differences using trigonometry
- Determine component vectors using trigonometry
- Solve applied problems using resultant vectors

INTRODUCING VECTORS

Directed Line Segments

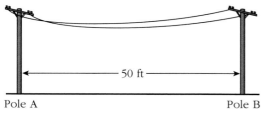

Pole A   Pole B

A physical quantity that can be described by a single number is called a **scalar quantity,** and the number that describes it is called a **scalar.** A scalar describes a **magnitude;** it is a measure that tells "how much." Some examples of scalar quantities are 4 hours, 67 feet, and 12 volts.

A physical quantity that has both a magnitude and a direction is called a **vector quantity,** and the numbers that describe it make up a **vector.** A vector is a measure that tells "how much" and "in what direction." Some examples of vector quantities are 65 miles to the south, 200 yards west, and 3.5 feet up.

65 miles to the south

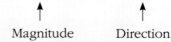

Magnitude   Direction

Two telephone poles are 50 feet apart, as shown above. Marcus walked directly from Pole A to Pole B. What is the vector that describes the path that he walked?

A vector has both magnitude and direction. The distance Marcus walked is 50 feet, and the direction is east. So, the vector that describes the path Marcus walked is *50 feet to the east.*

A vector can be represented geometrically on a coordinate plane by a **directed line segment.** The length of the directed line segment represents the magnitude of the vector and the arrowhead shows the direction.

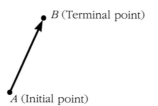

In the directed line segment $\overrightarrow{AB}$, A is called the *initial point* and B is called the *terminal point.*

EXAMPLE A

Name the initial point, the terminal point, the magnitude, and the direction of the directed line segment shown at the right.

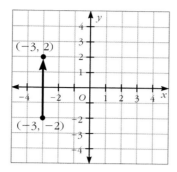

The initial point is $(-3, -2)$. The terminal point is $(-3, 2)$. The directed line segment is 4 units long and points up, so the magnitude is 4 units and the direction is up.

Directed line segments that have the same length and same direction represent **equal vectors** (or the same vector). In the diagram below, we say that $\mathbf{v}_1$, $\mathbf{v}_2$, and $\mathbf{v}_3$ are equal vectors (or represent the same vector).

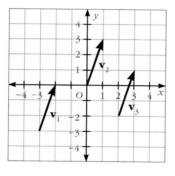

$$\mathbf{v}_1 = \mathbf{v}_2 = \mathbf{v}_3$$

Helpful Hint

In this book, vectors will be represented by a lowercase, boldface letter, such as **v.** Another notation that can be used is a lowercase letter with an arrow above it, such as $\vec{v}$. Because it is not easy to write in boldface in your own handwriting, you may wish to use the arrow notation.

Notice that the three directed line segments above have different initial points and different terminal points. A vector can be represented by *any* directed line segment that has the correct length and the correct direction.

From this point forward, we will use the word *vector* to refer to either a vector quantity or the directed line segment that represents that quantity.

EXAMPLE B

Which two vectors shown at the right are equal?

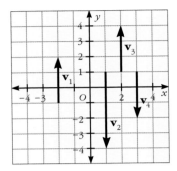

Vectors $\mathbf{v}_1$ and $\mathbf{v}_3$ are directed upward, and vectors $\mathbf{v}_2$ and $\mathbf{v}_4$ are directed downward. Vectors $\mathbf{v}_1$ and $\mathbf{v}_3$ are each 3 units long, vector $\mathbf{v}_2$ is 5 units long, and vector $\mathbf{v}_4$ is 3 units long.

Vectors $\mathbf{v}_1$ and $\mathbf{v}_3$ are equal because they have the same magnitude and direction.

Component Form

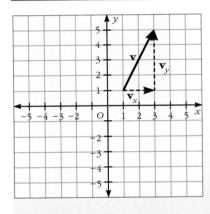

A convenient way to write vectors is in **component form.** A vector, **v,** in component form is written as $\langle v_x, v_y \rangle$, where v_x and v_y are the horizontal and vertical distances from the initial point to the terminal point of the vector. These are called the *x-component* and the *y-component,* respectively.

Write the vector shown at the left in component form.

The horizontal distance, v_x, from the initial point to the terminal point is 2 units, and the vertical distance, v_y, from the initial point to the terminal point is 4 units.

So, $v_x = 2$ and $v_y = 4$. The vector written in component form is $\langle 2, 4 \rangle$.

When writing a vector in component form, a distance that is either to the right or upward is represented by a positive component, and a distance that is either to the left or downward is represented by a negative component.

EXAMPLE C

Write vector **v** in component form.

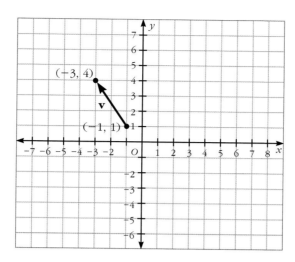

Helpful Hint

Sometimes you can find components of a vector by counting spaces on a grid.

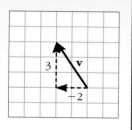

To get the components, subtract the initial coordinates from the terminal coordinates.

x-component: $v_x = -3 - (-1) = -2$
y-component: $v_y = 4 - 1 = 3$

In component form, $\mathbf{v} = \langle -2, 3 \rangle$.

EXAMPLE D

Write vector **v** in component form.

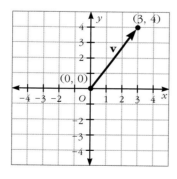

Again, subtract the initial coordinates from the terminal coordinates.

x-component: $v_x = 3 - 0 = 3$
y-component: $v_y = 4 - 0 = 4$

In component form, $\mathbf{v} = \langle 3, 4 \rangle$.

Notice that in Example D, the initial point is the origin, (0, 0). When a vector has its initial point at (0, 0), it is in standard position. This is convenient because when a vector is in standard position, the x- and y-components are the coordinates of the vector's terminal point.

In Example E, we will see that if we are given a vector in component form, it can easily be drawn in standard position.

EXAMPLE E

Draw the vector $\langle 2, 4 \rangle$ in standard position.

The vector is in standard position if the initial point is (0, 0) and the coordinates of the terminal point are the *x*- and *y*-components. That is, the terminal point is (2, 4).

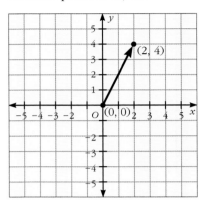

The vector $\langle 2, 4 \rangle$ is in standard position on the coordinate plane above.

In the next lesson, we will need to use the opposites of vectors. The opposite of vector **v** is written as −**v**. To find the opposite of a vector, simply find the opposite of each of its components. The vector −**v** points in the opposite direction of **v** and has the same magnitude as **v.**

EXAMPLE F

Find the opposite of the vector v = $\langle 4, 2 \rangle$, and draw both vectors in standard position.

The components of −**v** are the opposites of the components of **v.**

$$\mathbf{v} = \langle 4, 2 \rangle$$
$$-\mathbf{v} = -\langle 4, 2 \rangle = \langle -4, -2 \rangle$$

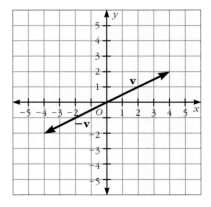

The opposite of **v** is −**v** = $\langle -4, -2 \rangle$, and both **v** and −**v** in standard position are shown above.

EXERCISES 18-1

For Exercises 1–4, tell whether each is a scalar or a vector quantity.

1. 45 minutes

2. 175 miles to the north

3. 80 feet to the west

4. 120 kilograms

Use the coordinate plane below for Exercises 5–8.

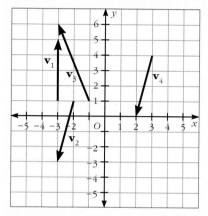

5. What are the initial and terminal points of $\mathbf{v}_4$?

6. Write each vector in component form.

7. Which vectors, if any, are equal?

8. Draw each vector in standard position on a coordinate plane.

9. If vector $\langle 5, 7 \rangle$ is drawn in standard position, what are the vector's initial and terminal points?

Use the coordinate plane below for Exercises 10–11.

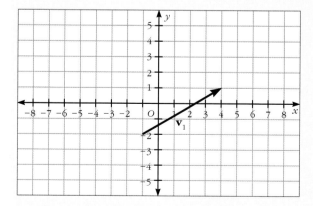

10. Write $\mathbf{v}_1$ in component form.

11. Write $-\mathbf{v}_1$ in component form, and then draw $\mathbf{v}_1$ and $-\mathbf{v}_1$ in standard position.

Answer Exercises 12–14.

12. Use the graph below to determine the vectors in component form.

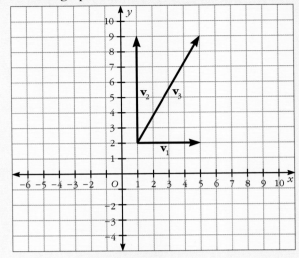

(a) $\mathbf{v}_1$

(b) $\mathbf{v}_2$

(c) $\mathbf{v}_3$

13. Use the graph below to determine the vectors in component form.

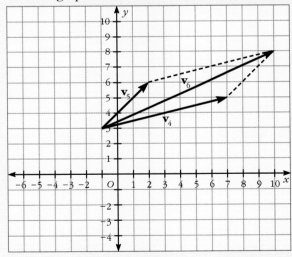

(a) $\mathbf{v}_4$

(b) $\mathbf{v}_5$

(c) $\mathbf{v}_6$

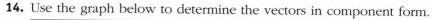

14. Use the graph below to determine the vectors in component form.

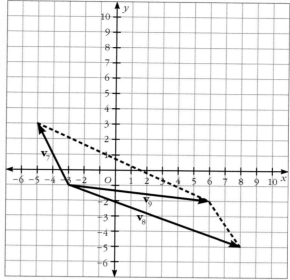

(a) $\mathbf{v}_7$

(b) $\mathbf{v}_8$

(c) $\mathbf{v}_9$

15. Challenge: The vector $-\mathbf{v}$ has an initial point of $(1, 3)$ and a terminal point of $(-2, 1)$. What is the component form of $\mathbf{v}$?

VECTOR OPERATIONS

Adding Vectors

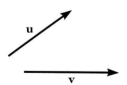

When vectors are added, the sum is a vector called the **resultant vector.** One way of adding vectors is geometrically. Recall that a given vector can be positioned anywhere on a coordinate plane. To add two vectors geometrically, we can move them as needed.

Add vectors u and v geometrically.

- Position the initial point of **v** on the terminal point of **u.**

- Draw the resultant vector **u + v** from the initial point of **u** to the terminal point of **v.**

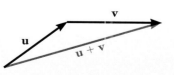

Notice that the vectors **u, v,** and the resultant vector **u + v** form a triangle. For this reason, geometric vector addition is sometimes called the *triangle method.*

We can also form a parallelogram as shown below, in which **u** + **v** is a diagonal of the parallelogram.

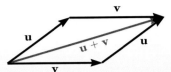

Because the resultant vector is the diagonal of a parallelogram, geometric vector addition is also sometimes called the **parallelogram method.**

EXAMPLE A

Use the triangle method to add **u** and **v** shown in the diagram.

Position the initial point of **v** on the terminal point of **u.**

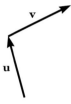

Draw the resultant vector **u** + **v** from the initial point of **u** to the terminal point of **v.**

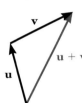

The resultant vector **u** + **v** is drawn above in red.

In Example B, we will see that when two perpendicular vectors are added, they form a right triangle.

EXAMPLE B

Use the triangle method to add the vectors shown in the diagram.

Notice that the initial point of **v** is already positioned on the terminal point of **u.** To add **u** and **v,** simply draw the vector from the initial point of **u** to the terminal point of **v.**

The resultant vector **u** + **v** is drawn above in red.

EXAMPLE C

Use the triangle method to add the vectors shown in the diagram, and then form a parallelogram whose diagonal is the resultant vector.

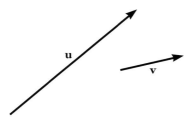

First, add the vectors by the triangle method. Then draw the parallelogram.

Position the initial point of **v** on the terminal point of **u,** and draw **u** + **v.**

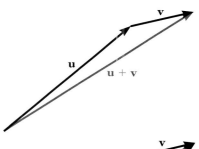

Now draw copies of **u** and **v** to form the parallelogram.

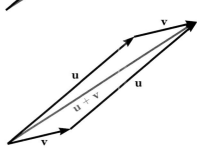

The resultant vector **u** + **v** is drawn above in red.

In Example D, we will add two vectors that point in the same direction.

EXAMPLE D

Add the vectors shown in the diagram.

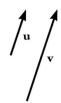

Position the initial point of **v** on the terminal point of **u.**

The resultant vector has the same direction as **u** and **v.**

The resultant vector **u** + **v** is drawn above in red.

So far, we have only added vectors geometrically. We can also add vectors by simply adding their components. The rule for adding vectors by adding their components is given in the box that follows.

Vector Addition

For any vectors $\mathbf{u} = \langle u_x, u_y \rangle$ and $\mathbf{v} = \langle v_x, v_y \rangle$, the sum $\mathbf{u} + \mathbf{v}$ can be found as follows:

$$\mathbf{u} + \mathbf{v} = \langle u_x, u_y \rangle + \langle v_x, v_y \rangle$$
$$= \langle u_x + v_x, u_y + v_y \rangle$$

EXAMPLE E

Add the vectors
u = ⟨4, 1⟩ and
v = ⟨3, 7⟩. Then
draw **u**, **v**, and
u + **v** in standard
position and show
the parallelogram.

Add.

Draw **u**, **v**, and **u** + **v** in
standard position and show
the parallelogram.

$$\mathbf{u} + \mathbf{v} = \langle 4, 1 \rangle + \langle 3, 7 \rangle$$
$$= \langle 4 + 3, 1 + 7 \rangle$$
$$= \langle 7, 8 \rangle$$

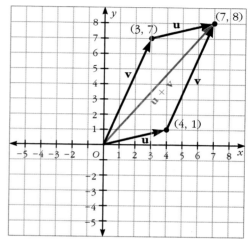

Recall that when a vector is in standard position, its components are the
coordinates of its terminal point.

The resultant vector **u** + **v** is ⟨7, 8⟩.

EXAMPLE F

Add the vectors
u = ⟨−2, 4⟩ and
v = ⟨−2, −7⟩. Then
draw **u**, **v**, and
u + **v** in standard
position and show
the parallelogram.

Add.

Draw **u**, **v**, and **u** + **v** in
standard position and
show the parallelogram.

$$\mathbf{u} + \mathbf{v} = \langle -2, 4 \rangle + \langle -2, -7 \rangle$$
$$= \langle -2 + (-2), 4 + (-7) \rangle$$
$$= \langle -4, -3 \rangle$$

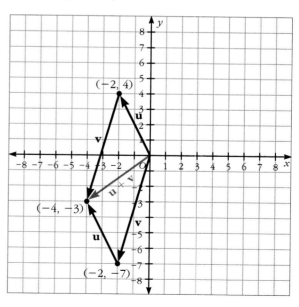

The resultant vector **u** + **v** is ⟨−4, −3⟩.

Subtracting Vectors

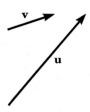

To subtract vectors, simply subtract the components.

Vector Subtraction

For any vectors $\mathbf{u} = \langle u_x, u_y \rangle$ and $\mathbf{v} = \langle v_x, v_y \rangle$, the difference $\mathbf{u} - \mathbf{v}$ can be found as follows:

$$\mathbf{u} - \mathbf{v} = \langle u_x, u_y \rangle - \langle v_x, v_y \rangle$$
$$= \langle u_x - v_x, u_y - v_y \rangle$$

Subtract $\mathbf{v} = \langle 3, 1 \rangle$ from $\mathbf{u} = \langle 5, 6 \rangle$.

$$\mathbf{u} - \mathbf{v} = \langle 5, 6 \rangle - \langle 3, 1 \rangle$$
$$= \langle 5 - 3, 6 - 1 \rangle$$
$$= \langle 2, 5 \rangle$$

So, $\mathbf{u} - \mathbf{v} = \langle 2, 5 \rangle$.

Another method of subtracting a vector is to add its opposite. Using vectors $\mathbf{u}$ and $\mathbf{v}$ from the above example, we could write:

$$\mathbf{u} - \mathbf{v} = \mathbf{u} + (-\mathbf{v}) = \langle 5, 6 \rangle + \langle -3, -1 \rangle = \langle 5 + (-3), 6 + (-1) \rangle = \langle 2, 5 \rangle$$

To show vector subtraction geometrically, use the method of adding the opposite. The diagram below shows the vector difference $\mathbf{u} - \mathbf{v}$ from the preceding example.

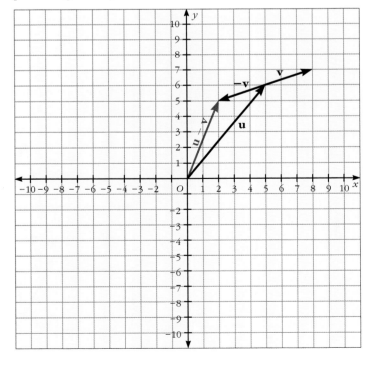

EXAMPLE G

Subtract $\mathbf{v} = \langle 2, 3 \rangle$ from $\mathbf{u} = \langle -3, 4 \rangle$, then show $\mathbf{u} - \mathbf{v}$ geometrically.

Subtract.

$$\mathbf{u} - \mathbf{v} = \langle -3, 4 \rangle - \langle 2, 3 \rangle$$
$$= \langle -3 - 2, 4 - 3 \rangle$$
$$= \langle -5, 1 \rangle$$

Show $\mathbf{u} - \mathbf{v}$ geometrically.

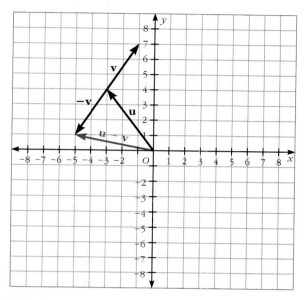

The resultant vector $\mathbf{u} - \mathbf{v}$ is $\langle -5, 1 \rangle$.

Multiplying a Vector by a Scalar

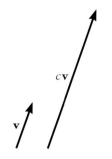

$c\mathbf{v}$ is vector $\mathbf{v}$ multiplied by scalar c.

So far, we have seen that vectors can be added or subtracted. Vectors can also be multiplied by scalars; this is called *scalar multiplication.* Any vector $\mathbf{v}$ can be multiplied by any scalar c; this is written as $c\mathbf{v}$.

The value of c can affect both the magnitude and the direction of the original vector, $\mathbf{v}$. The diagram below shows the magnitude and direction of $\frac{1}{2}\mathbf{v}$, $2\mathbf{v}$, and $-2\mathbf{v}$, all compared to the original vector $\mathbf{v}$.

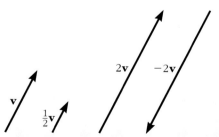

> **Helpful Hint**
>
> A scalar is just a number.

Notice that $\frac{1}{2}\mathbf{v}$ is $\frac{1}{2}$ the length of $\mathbf{v}$; $2\mathbf{v}$ and $-2\mathbf{v}$ are both 2 times the length of $\mathbf{v}$. Notice also that the direction of $-2\mathbf{v}$ is the opposite of the direction of the other vectors. When the scalar c is negative, the direction of $c\mathbf{v}$ is opposite the direction of $\mathbf{v}$.

Scalar Multiplication

> For any vectors $\mathbf{v} = \langle v_x, v_y \rangle$ and scalar c,
> $$c\mathbf{v} = c\langle v_x, v_y \rangle$$
> $$= \langle cv_x, cv_y \rangle$$

Let v = $\langle -2, 1 \rangle$. Find 3v and $-$3v. Draw v, 3v, and $-$3v.

Multiply by the scalars.

$$3\mathbf{v} = 3\langle -2, 1 \rangle = \langle 3 \cdot -2, 3 \cdot 1 \rangle = \langle -6, 3 \rangle$$
$$-3\mathbf{v} = -3\langle -2, 1 \rangle = \langle -3 \cdot -2, -3 \cdot 1 \rangle = \langle 6, -3 \rangle$$

Draw **v**, **3v**, and **–3v**.

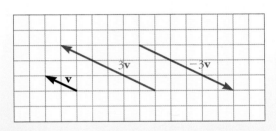

So, $3\mathbf{v} = \langle -6, 3 \rangle$ and $-3\mathbf{v} = \langle 6, -3 \rangle$.

EXAMPLE H

Let $\mathbf{u} = \langle 12, -4 \rangle$. Find $-\frac{1}{4}\mathbf{u}$, and then draw $\mathbf{u}$ and $-\frac{1}{4}\mathbf{u}$.

Multiply by the scalar.

$$-\frac{1}{4}\mathbf{u} = -\frac{1}{4}\langle 12, -4 \rangle = \langle -3, 1 \rangle$$

Draw $\mathbf{u}$ and $-\frac{1}{4}\mathbf{u}$.

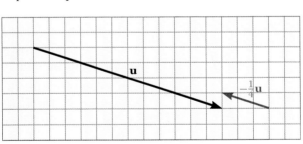

$$-\frac{1}{4}\mathbf{u} = \langle -3, 1 \rangle$$

EXAMPLE I

Let $\mathbf{v} = \langle -2, 7 \rangle$.
Find $-1\mathbf{v}$, and then
draw $\mathbf{v}$ and $-1\mathbf{v}$.

Notice in Example I that the opposite of a vector is -1 times that vector.

Multiply by the scalar. $-1\mathbf{v} = -1\langle -2, 7 \rangle = \langle 2, -7 \rangle$

Draw $\mathbf{v}$ and $-1\mathbf{v}$.

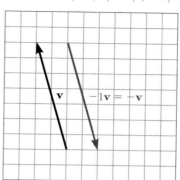

$-1\mathbf{v} = \langle 2, -7 \rangle$

EXERCISES 18-2

For Exercises 1–2, add geometrically by the triangle method and draw the resultant vector.

1.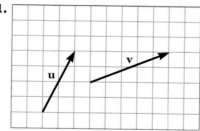

2.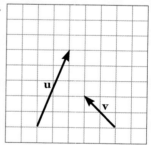

For Exercises 3–4, add $\mathbf{u}$ and $\mathbf{v}$. Then draw $\mathbf{u}$, $\mathbf{v}$, and $\mathbf{u} + \mathbf{v}$ in standard position and show the parallelogram.

3. $\mathbf{u} = \langle 1, 5 \rangle$, $\mathbf{v} = \langle 7, 1 \rangle$

4. $\mathbf{u} = \langle -4, 1 \rangle$, $\mathbf{v} = \langle 5, -6 \rangle$

5. Add the vectors in the diagram below. Draw the resultant vector $\mathbf{u} + \mathbf{v}$.

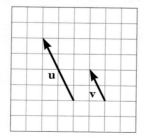

For Exercises 6–7, give the component form of the resultant vector **u** + **v**.

 6. $\mathbf{u} = \langle 4, 7 \rangle$, $\mathbf{v} = \langle -8, 1 \rangle$

 7. $\mathbf{u} = \langle -2, -5 \rangle$, $\mathbf{v} = \langle 3, -9 \rangle$

For Exercises 8–9, subtract **v** from **u**. Then show **u** − **v** geometrically.

 8. $\mathbf{u} = \langle 2, 7 \rangle$, $\mathbf{v} = \langle -3, 2 \rangle$

 9. $\mathbf{u} = \langle -4, 1 \rangle$, $\mathbf{v} = \langle 4, 4 \rangle$

For Exercises 10–11, give the component form of **u** − **v**.

10. $\mathbf{u} = \langle 0, 3 \rangle$, $\mathbf{v} = \langle -8, 6 \rangle$

11. $\mathbf{u} = \langle 7, -6 \rangle$, $\mathbf{v} = \langle -4, -10 \rangle$

In Exercises 12–16, compute as indicated and draw the indicated vectors.

12. Let $\mathbf{u} = \langle 4, 2 \rangle$. Find $2\mathbf{u}$ and $-2\mathbf{u}$, and then draw **u**, $2\mathbf{u}$, and $-2\mathbf{u}$.

13. Let $\mathbf{v} = \langle -9, 3 \rangle$. Find $-\frac{1}{3}\mathbf{v}$, and then draw **v** and $-\frac{1}{3}\mathbf{v}$.

14. Let $\mathbf{v} = \langle -2, 5 \rangle$. Find $-1\mathbf{v}$, and then draw **v** and $-1\mathbf{v}$.

15. In a certain parallel *RC* circuit, the current through the capacitance in amps is $\mathbf{I}_C = \langle 0, 5 \rangle$ and the current through the resistance in amps is $\mathbf{I}_R = \langle 9, 0 \rangle$.

   (a) Draw $\mathbf{I}_C$ and $\mathbf{I}_R$ on the same axes.

   (b) Draw $\mathbf{I} = \mathbf{I}_C + \mathbf{I}_R$ on the axes used in (a).

   (c) Determine the component form of **I**.

   (d) Determine the scalar value of **I**.

16. In a certain *RLC* series circuit, the resistance in kilohms is $\mathbf{R} = \langle 5, 0 \rangle$, the capacitive reactance is $\mathbf{X}_C = \langle 0, -3 \rangle$, and the inductive reactance in amps is $\mathbf{X}_L = \langle 0, 7 \rangle$.

   (a) Draw $\mathbf{R}$, $\mathbf{X}_C$, and $\mathbf{X}_L$ on the same axes.

   (b) Draw $\mathbf{Z} = \mathbf{R} + \mathbf{X}_C + \mathbf{X}_L$ on the axes used in (a). *Hint:* First draw $\mathbf{X}_C + \mathbf{X}_L$ and then draw $\mathbf{Z} = \mathbf{R} + (\mathbf{X}_C + \mathbf{X}_L)$.

   (c) Determine the component form of $\mathbf{X}_C + \mathbf{X}_L$.

   (d) Determine the component form of **Z**.

   (e) Determine the scalar value of **Z**.

17. **Challenge:** Let $\mathbf{u} = \langle 4, 6 \rangle$ and $\mathbf{v} = \langle 3, 2 \rangle$. What is $2\mathbf{u} - \mathbf{v}$?

FINDING THE MAGNITUDE AND DIRECTION OF A VECTOR

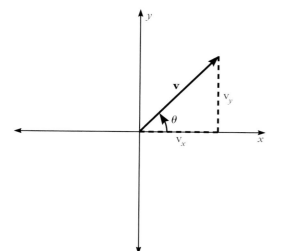

A vector has both magnitude and direction. If we know the component form of a vector we can find:

• the magnitude of the vector by the Pythagorean Theorem, and

• the direction of the vector by trigonometry.

For any vector $\mathbf{v} = \langle v_x, v_y \rangle$, we will use $|\mathbf{v}|$ to represent the magnitude of $\mathbf{v}$. By drawing $\mathbf{v}$ in standard position, we can describe the direction of $\mathbf{v}$ by θ, the counterclockwise angle that $\mathbf{v}$ forms with the positive x-axis. We will call θ the *direction angle*.

Using the x- and y-components, the magnitude of $\mathbf{v}$ can be found by the Pythagorean Theorem:

$$|\mathbf{v}|^2 = v_x^2 + v_y^2$$

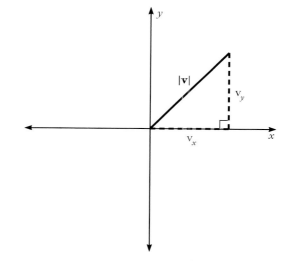

The angle θ can be found by using the inverse tangent function:

$$\tan \theta = \frac{v_y}{v_x}$$

$$\theta = \tan^{-1} \frac{v_y}{v_x}$$

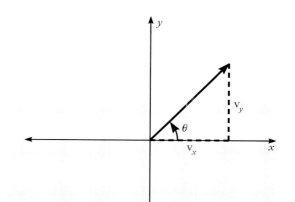

Find the magnitude and direction angle of v = $\langle 4, 3 \rangle$.

Because the vector is given in component form, we can see that $v_x = 4$ and $v_y = 3$. We need to find the magnitude $|\mathbf{v}|$ and the angle θ.

Given v_x and v_y:

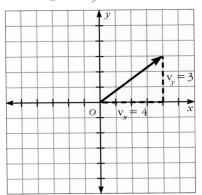

Find $|\mathbf{v}|$ and θ:

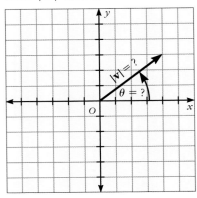

Use the Pythagorean Theorem to find the magnitude $|\mathbf{v}|$.

$|\mathbf{v}|^2 = v_x^2 + v_y^2$ ← Write the Pythagorean Theorem.

$|\mathbf{v}|^2 = 4^2 + 3^2$ ← Substitute the known values.

$|\mathbf{v}|^2 = 16 + 9$ ← Simplify.

$|\mathbf{v}|^2 = 25$

$|\mathbf{v}| = 5$ ← Find the positive square root.

Use the inverse tangent function to find the direction angle θ.

$\tan \theta = \dfrac{v_y}{v_x}$ ← Definition of tangent ratio

$\tan \theta = \dfrac{3}{4}$ ← Substitute the known values.

$\theta = \tan^{-1} \dfrac{3}{4}$ ← Definition of inverse tangent function

$\theta \approx 37°$ ← Use your calculator to approximate $\tan^{-1} \dfrac{3}{4}$.

Vector **v** is 5 units long and forms a counterclockwise angle of about 37° with the positive x-axis.

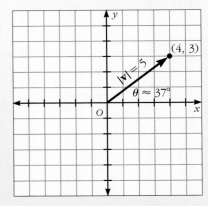

The magnitude of **v** is 5, and the direction angle is about 37°.

EXAMPLE A

Find the magnitude and direction angle of $\mathbf{v} = \langle 5, 2 \rangle$.

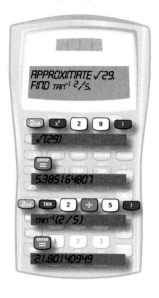

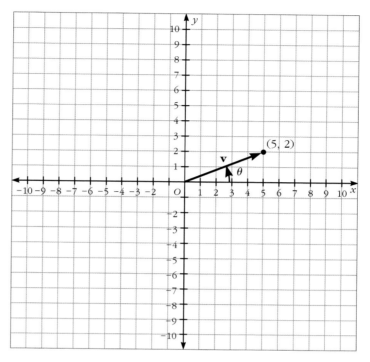

From the component form of the vector, we know that $v_x = 5$ and $v_y = 2$.

Use the Pythagorean Theorem to find the magnitude $|\mathbf{v}|$.

$|\mathbf{v}|^2 = v_x{}^2 + v_y{}^2$ ◄── Write the Pythagorean Theorem.

$|\mathbf{v}|^2 = 5^2 + 2^2$ ◄── Substitute the known values.

$|\mathbf{v}|^2 = 25 + 4$ ◄── Simplify.

$|\mathbf{v}|^2 = 29$

$|\mathbf{v}| \approx 5.4$ ◄── Use your calculator to find an approximation of $\sqrt{29}$.

Use the inverse tangent function to find the direction angle θ.

$\tan \theta = \dfrac{v_y}{v_x}$ ◄── Definition of tangent ratio

$\tan \theta = \dfrac{2}{5}$ ◄── Substitute the known values.

$\theta = \tan^{-1} \dfrac{2}{5}$ ◄── Definition of inverse tangent function

$\theta \approx 22°$ ◄── Use your calculator to find an approximation of $\tan^{-1} \dfrac{2}{5}$.

The magnitude of $\mathbf{v}$ is about 5.4, and the direction angle is about 22°.

When a vector in standard position is in Quadrant II, III, or IV, we will need to use information about reference angles to find the direction angle θ.

EXAMPLE B

Find the magnitude and direction angle of $\mathbf{v} = \langle -12, 5 \rangle$.

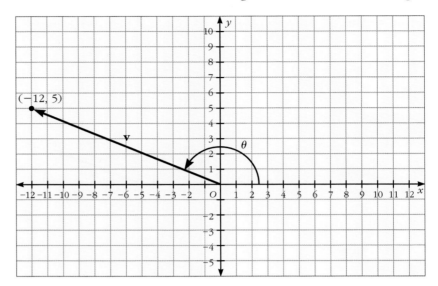

From the component form of the vector, we know that $v_x = -12$ and $v_y = 5$.

Use the Pythagorean Theorem to find the magnitude $|\mathbf{v}|$.

$|\mathbf{v}|^2 = v_x^2 + v_y^2$ ← Write the Pythagorean Theorem.

$|\mathbf{v}|^2 = (-12)^2 + 5^2$ ← Substitute the known values.

$|\mathbf{v}|^2 = 144 + 25$ ← Simplify.

$|\mathbf{v}|^2 = 169$

$|\mathbf{v}| = 13$ ← Find the positive square root.

Use the tangent function to find the direction angle, θ, of $\mathbf{v}$.

$\tan \theta = \dfrac{v_y}{v_x}$ ← Definition of tangent ratio

$\tan \theta = \dfrac{5}{-12}$ ← Substitute the known values.

$\tan^{-1} \dfrac{5}{-12} \approx -23°$ ← Find the inverse tangent of $\dfrac{5}{-12}$.

From the diagram, we can see that θ is not $-23°$. By definition, inverse tangent function values are between $-90°$ and $90°$; that is why $\tan^{-1}\frac{5}{-12} \approx -23°$. However, we now know that the reference angle is $23°$.

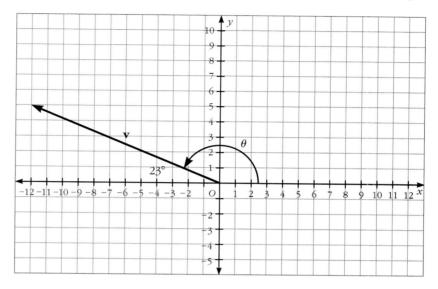

So, $\theta \approx 180° - 23° = 157°$.

The magnitude of **v** is 13, and the direction angle is about $157°$.

In the following examples, we will see that vectors can also lie directly on the coordinate axes. In these cases, it is easy to find the magnitude and direction angle of the vectors.

EXAMPLE C

Find the magnitude and direction angle of **v** = $\langle 4, 0 \rangle$.

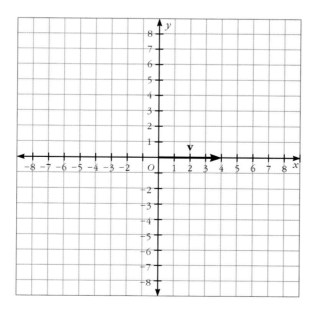

The vector is horizontal, so we can simply count the units to find that it has a length of 4. Because it lies on the positive x-axis, the direction angle is $0°$.

The magnitude of **v** is 4, and the direction angle is $0°$.

EXAMPLE D

Find the magnitude and direction angle of $\mathbf{v} = \langle 0, -5 \rangle$.

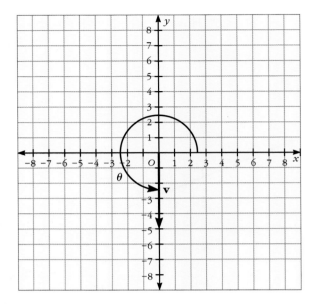

The vector is vertical, so we can simply count the units to find that it has a length of 5. Because it lies on the negative y-axis, the direction angle is 270°.

The magnitude of $\mathbf{v}$ is 5, and the direction angle is 270°.

EXERCISES 18-3

For Exercises 1–4, find the magnitude and direction angle of each vector. Round the magnitude to the nearest tenth, and round the direction angle to the nearest degree, if rounding is necessary.

1.

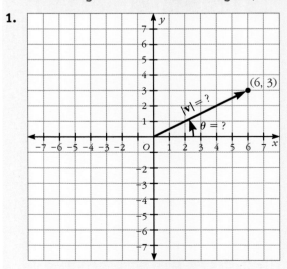

2.

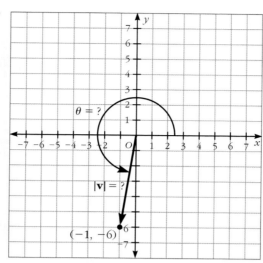

$\theta = ?$

$|\mathbf{v}| = ?$

$(-1, -6)$

3.

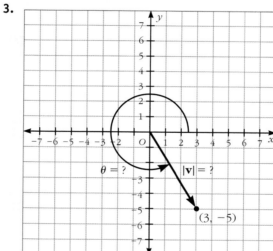

$\theta = ?$

$|\mathbf{v}| = ?$

$(3, -5)$

4.

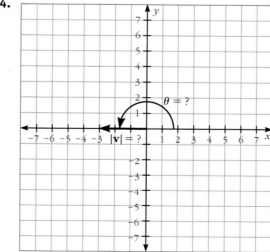

$\theta = ?$

$|\mathbf{v}| = ?$

For Exercises 5–8, draw the vector in standard position and then find its magnitude and direction angle. Round the magnitude to the nearest tenth, and round the direction angle to the nearest degree, if rounding is necessary.

5. $\mathbf{v} = \langle 6, 8 \rangle$

6. $\mathbf{v} = \langle 2, -7 \rangle$

7. $\mathbf{v} = \langle -4, 5 \rangle$

8. $\mathbf{v} = \langle 0, 7 \rangle$

In Exercises 9–10, draw the indicated vectors and compute as indicated. Round the magnitudes to the nearest hundredth and the direction angles to the nearest degree.

9. An inductive reactance, $\mathbf{X}_L$, is described by the vector $\mathbf{X}_L = \langle 0, 42 \rangle$ in kilohms. The resistance, R, in the same circuit is described by the vector $\mathbf{R} = \langle 33, 0 \rangle$ in kilohms.

(a) Draw $\mathbf{X}_L$ and $\mathbf{R}$ on the same axes.

(b) Determine $|\mathbf{X}_L|$.

(c) Determine the direction angle of $\mathbf{X}_L$.

(d) Determine $|\mathbf{R}|$.

(e) Determine the direction angle of $\mathbf{R}$.

10. Three voltages connected in series are described by the vectors $\mathbf{V}_1 = \langle 52, 80 \rangle$, $\mathbf{V}_2 = \langle -104, -60 \rangle$, and $\mathbf{V}_3 = \langle 70, -30 \rangle$.

(a) Draw $\mathbf{V}_1$, $\mathbf{V}_2$, and $\mathbf{V}_3$ on the same axes.

(b) Determine $|\mathbf{V}_1|$.

(c) Determine the direction angle of $\mathbf{V}_1$.

(d) Determine $|\mathbf{V}_2|$.

(e) Determine the direction angle of $\mathbf{V}_2$.

(f) Determine $|\mathbf{V}_3|$.

(g) Determine the direction angle of $\mathbf{V}_3$.

11. Challenge: Let $\mathbf{v} = \langle 2, 8 \rangle$. Find the magnitude and direction angle of $-3\mathbf{v}$.

FINDING AND USING THE COMPONENTS OF A VECTOR

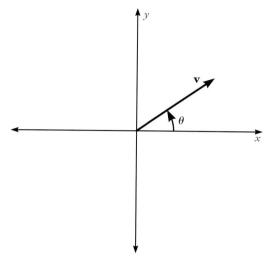

Earlier, we saw that a vector, **v,** can be written in component form as $\mathbf{v} = \langle v_x, v_y \rangle$. When a vector is drawn on a coordinate plane, it is sometimes easy to find its components by counting spaces on the grid. However, if a vector is given in terms of its magnitude and direction, we will need to use trigonometry to find its components.

We can use the cosine and sine ratios to find expressions for components v_x and v_y:

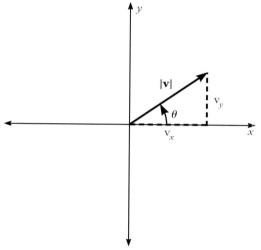

$$\cos \theta = \frac{\text{adj}}{\text{hyp}} \quad \longleftarrow \quad \text{Definition of cosine ratio}$$

$$\cos \theta = \frac{v_x}{|\mathbf{v}|} \quad \longleftarrow \quad \text{Substitute lengths of sides.}$$

$$|\mathbf{v}| \cdot \cos \theta = v_x \quad \longleftarrow \quad \text{Multiply both sides by } |\mathbf{v}|.$$

$$\sin \theta = \frac{\text{opp}}{\text{hyp}} \quad \longleftarrow \quad \text{Definition of sine ratio}$$

$$\sin \theta = \frac{v_y}{|\mathbf{v}|} \quad \longleftarrow \quad \text{Substitute lengths of sides.}$$

$$|\mathbf{v}| \cdot \sin \theta = v_y \quad \longleftarrow \quad \text{Multiply both sides by } |\mathbf{v}|.$$

The rules for finding the components of a vector are given in the box that follows.

Components of a Vector

For vector $\mathbf{v} = \langle v_x, v_y \rangle$,

$v_x = |\mathbf{v}| \cos \theta$
$v_y = |\mathbf{v}| \sin \theta$

where $|\mathbf{v}|$ is the magnitude of **v** and θ is the direction angle.

Using these expressions for v_x and v_y, we can find the component form of a vector that is given in terms of its magnitude and direction.

In the figure below, v is 6 units long and forms a 35° angle with the positive x-axis. Write v in component form.

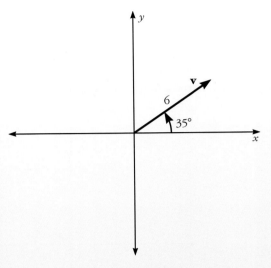

$v_x = |\mathbf{v}| \cos \theta$ ← Write the expression for v_x.

$ = 6 \cos 35°$ ← Substitute known values.

$ \approx 4.9$ ← Use a calculator.

$v_y = |\mathbf{v}| \sin \theta$ ← Write the expression for v_y.

$ = 6 \sin 35°$ ← Substitute known values.

$ \approx 3.4$ ← Use a calculator.

In component form, $\mathbf{v} \approx \langle 4.9, 3.4 \rangle$

EXAMPLE A

Vector **v** has magnitude 9 and direction angle 125°. Write **v** in component form.

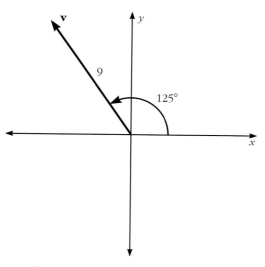

$v_x = |\mathbf{v}| \cos \theta$ ← Write the expression for v_x.

$\quad = 9 \cos 125°$ ← Substitute known values.

$\quad \approx -5.2$ ← Use a calculator.

$v_y = |\mathbf{v}| \sin \theta$ ← Write the expression for v_y.

$\quad = 9 \sin 125°$ ← Substitute known values.

$\quad \approx 7.4$ ← Use a calculator.

$\mathbf{v} \approx \langle -5.2, 7.4 \rangle$

To add two vectors that are given in terms of magnitude and direction, we must first write the vectors in component form. For vectors **u** and **v** below, we will use $\theta_\mathbf{u}$ and $\theta_\mathbf{v}$ to represent their direction angles.

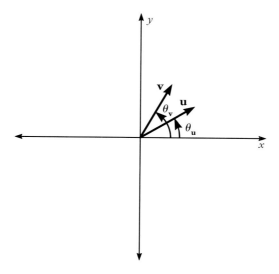

After we have the component forms of **u** and **v**, we can find the resultant vector by adding their components. Then, we will write the resultant vector in terms of its magnitude and direction. In Example B, we will use **w** to represent the resultant vector **u** + **v.**

EXAMPLE B

A series circuit has two voltages, V_1 and V_2. Vector V_1 has a magnitude of 12 and a direction angle of 27°, and vector V_2 has magnitude of 8 and a direction angle of 77°. Find the magnitude and direction of $V = V_1 + V_2$.

STEP 1

Draw V_1 and V_2 in standard position.

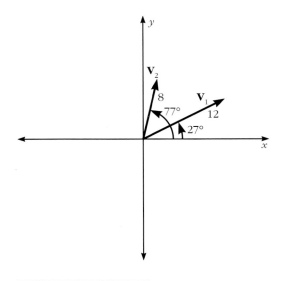

Helpful Hint

Remember that for vectors in standard position, the components of the vector are the same as the coordinates of the terminal point.

STEP 2

Write V_1 and V_2 in component form.

$V_{1x} = 12 \cos 27° \approx 10.7$

$V_1\gamma = 12 \sin 27° \approx 5.4$

So, $V_1 \approx \langle 10.7, 5.4 \rangle$

$V_{2x} = 8 \cos 77° \approx 1.8$

$V_2\gamma = 8 \sin 77° \approx 7.8$

So, $V_2 \approx \langle 1.8, 7.8 \rangle$

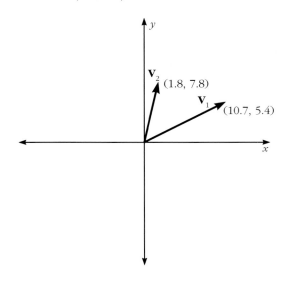

STEP 3

Find V by adding the components of V_1 and V_2.

$$V = V_1 + V_2$$

$$\approx \langle 10.7, \, 5.4 \rangle + \langle 1.8, \, 7.8 \rangle$$

$$\approx \langle 10.7 + 1.8, \, 5.4 + 7.8 \rangle$$

$$\approx \langle 12.5, \, 13.2 \rangle$$

STEP 4

Find the magnitude $|V|$.

$$|V|^2 = V_x^2 + V_y^2 \qquad \longleftarrow \text{Write the Pythagorean Theorem.}$$

$$|V|^2 \approx 12.5^2 + 13.2^2 \qquad \longleftarrow \text{Substitute the known values.}$$

$$|V|^2 \approx 156.3 + 174.2 \qquad \longleftarrow \text{Simplify.}$$

$$|V|^2 \approx 330.5$$

$$|V| \approx 18.2 \qquad \longleftarrow \text{Use your calculator to find an approximation of } \sqrt{330.5}.$$

STEP 5

Find the direction angle θ_V.

The components of **V** are both positive, so **V** is in Quadrant I. Because **V** is in Quadrant I, θ_V can be found directly by using the inverse tangent function.

$$\tan \theta_V = \frac{V_y}{V_x} \qquad \longleftarrow \text{Definition of tangent ratio.}$$

$$\tan \theta_V \approx \frac{13.2}{12.5} \qquad \longleftarrow \text{Substitute the known values.}$$

$$\theta_V \approx \tan^{-1} \frac{13.2}{12.5} \qquad \longleftarrow \text{Definition of inverse tangent function}$$

$$\theta_V \approx 47° \qquad \longleftarrow \text{Use a calculator and round your answer.}$$

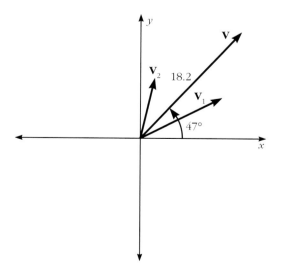

The resultant vector, **V,** has a magnitude of about 18.2 and a direction angle of about 47°.

EXERCISES 18-4

For Exercises 1–4, find the component form of the vector.

1.

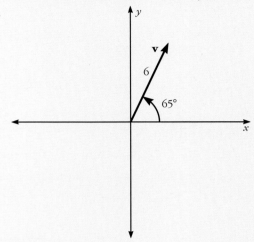

$$|\mathbf{v}| = 6, \theta = 65°$$

2.

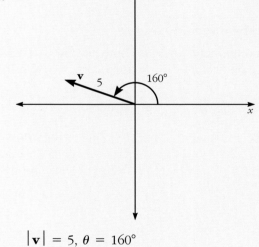

$$|\mathbf{v}| = 5, \theta = 160°$$

3.

$$|\mathbf{v}| = 8, \theta = 193°$$

4.

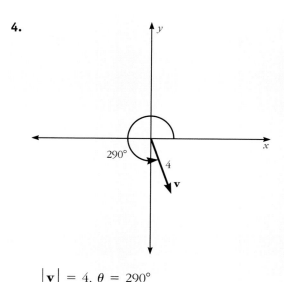

$$|\mathbf{v}| = 4, \theta = 290°$$

For Exercises 5–8, find the magnitude and direction of the resultant vector, w = u + v.

5. $|\mathbf{u}| = 7, \theta_{\mathbf{u}} = 31°$
   $|\mathbf{v}| = 4, \theta_{\mathbf{v}} = 85°$

6. $|\mathbf{u}| = 16, \theta_{\mathbf{u}} = 45°$
   $|\mathbf{v}| = 6, \theta_{\mathbf{v}} = 20°$

7. $|\mathbf{u}| = 4, \theta_{\mathbf{u}} = 38°$
   $|\mathbf{v}| = 10, \theta_{\mathbf{v}} = 295°$

8. $|\mathbf{u}| = 9, \theta_{\mathbf{u}} = 283°$
   $|\mathbf{v}| = 16, \theta_{\mathbf{v}} = 13°$

Answer Exercises 9–10. Round the magnitude to the nearest hundredth and the direction angle to the nearest degree.

9. A series circuit has two voltages, V_1 and V_2. Vector $\mathbf{V}_1$ has a magnitude of 45 V and direction angle 125°, and vector $\mathbf{V}_2$ has magnitude 28 V and direction angle of −35°. Find the magnitude and direction of $\mathbf{V} = \mathbf{V}_1 + \mathbf{V}_2$.

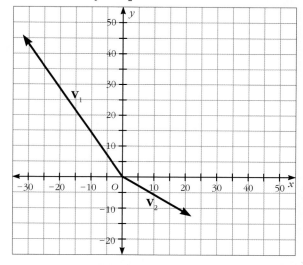

10. In a series-parallel circuit, a current of 20 A in branch *A* has a direction angle of 45° and a current of 35 A in branch *B* has a direction angle of −15°. Find the magnitude and direction of $\mathbf{I} = \mathbf{I}_A + \mathbf{I}_B$.

11. **Challenge:** Using vectors **u** and **v** below, find the magnitude and direction of $\mathbf{w} = 2(\mathbf{u} + \mathbf{v})$.

$$|\mathbf{u}| = 3, \theta_{\mathbf{u}} = 110°$$

$$|\mathbf{v}| = 5, \theta_{\mathbf{v}} = 195°$$

CHAPTER 18 REVIEW EXERCISES

Use the coordinate plane below for Exercises 1–3.

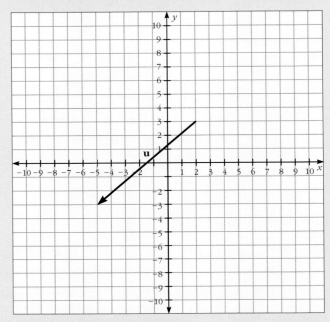

1. Write **u** in component form.

2. Write **−u** in component form.

3. Write 2**u** in component form, and then draw **u**, **−u**, and 2**u** in standard form.

Use the coordinate plane below for Exercises 4–5.

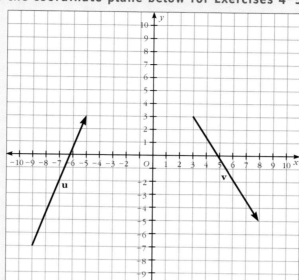

4. Draw **u** and **v**, and use the parallelogram method to draw **u** + **v**.

5. Give the components of **u, v,** and **u** + **v**.

In Exercises 6–7, find the magnitude and direction angle of the given vectors. For the magnitude give the exact value and, if necessary, round to the nearest hundredth. Give the direction angle to the nearest degree.

6. $\mathbf{r} = \langle -2, 7 \rangle$

7. $\mathbf{w} = \langle 65, -72 \rangle$

8. A series circuit has two voltages, V_1 and V_2. Vector $\mathbf{V}_1$ has a magnitude of 54 V and direction angle of 75°, and vector $\mathbf{V}_2$ has magnitude of 38 V and direction angle of −5°. Find the magnitude and direction of $\mathbf{V} = \mathbf{V}_1 + \mathbf{V}_2$.

   (a) Determine the components of $\mathbf{V}_1$.

   (b) Determine the components of $\mathbf{V}_2$.

   (c) Determine the components of $\mathbf{V} = \mathbf{V}_1 + \mathbf{V}_2$.

   (d) Determine the magnitude of **V**.

   (e) Determine the direction of **V**.

9. An inductive capacitance, X_C, is described by the vector $\mathbf{X}_C = (0, -76.9)$ in kilohms. The resistance, R, in the same circuit is described by the vector $\mathbf{R} = (72, 0)$ in kilohms. The impedance, Z, is defined by the vector $\mathbf{Z} = \mathbf{X}_C + \mathbf{R}$.

   (a) Determine $|\mathbf{X}_C|$.

   (b) Determine the direction angle of $\mathbf{X}_C$.

   (c) Determine $|\mathbf{R}|$.

   (d) Determine the direction angle of **R**.

   (e) Determine the components of **Z**.

   (f) Determine $|\mathbf{Z}|$.

   (g) Determine the direction angle of **Z**.

Building a Foundation in Mathematics

Converting between Base-10 and Binary Numbers

Adding Binary Numbers

Complements of Binary Numbers

Subtracting Binary Numbers

Multiplying Binary Numbers

Converting between Base-10 and Octal Numbers

Converting between Binary and Octal Numbers

Converting between Base-10 and Hexadecimal Numbers

Converting between Binary and Hexadecimal Numbers

Overview

There are many programming languages. Among the ones you may have heard of are BASIC, FORTRAN, PASCAL, COBOL Lisp, Ada, and C++. All of these languages have to be converted to numbers in order to communicate with the computer. Each number in a computer is represented by a binary switch. Much like a lightbulb, a binary switch is either on or off.

Because a binary switch is either on or off, or a diode is either conducting or not, it is convenient to represent these options by the binary numbers 0 and

1. It requires many binary digits, or *bits*, to represent large numbers. Octal and hexadecimal numbers are therefore often used because they require much less space. In this chapter, you will learn the basics of binary, octal, and hexadecimal numbers; how to convert numbers from one system to the other and to the decimal system; how to perform addition, subtraction, and multiplication in binary; and ways in which these numbers are used in information and computer technology.

Binary, Octal, and Hexadecimal Numbers

chapter 19

Objectives

After completing this chapter, you will be able to:

- Express binary numbers as decimal numbers
- Express decimal numbers as binary numbers
- Determine the 2's complement of a binary number
- Add, subtract, and multiply binary numbers
- Express octal numbers as decimal numbers
- Express decimal numbers as octal numbers
- Express hexadecimal numbers as decimal numbers
- Express decimal numbers as hexadecimal numbers

CONVERTING BETWEEN BASE-10 AND BINARY NUMBERS

Early computer systems, such as the one shown, used magnetic tape to store data, all in binary form.

The numbers that we use in everyday life are written using the 10 digits 0 through 9. Because this number system uses 10 digits, it is called a **decimal,** or base-10, system. However, there are other types of numbering systems used in various applications. For example, computers perform calculations using a **binary,** or base-2, number system.

A binary number system uses only the digits 0 and 1. Because computers use systems of electrical switches that can only be "off" or "on," a number system with only two digits is a convenient way for computers to store data and perform calculations.

It will be useful to review the meaning of place value in the base-10 system. The number 7,236 is shown in a place value chart below. Recall that a number to the first power is the number itself; a number to the zero power is 1.

Base-10 Place Value System

100,000	10,000	1,000	100	10	1
10^5	10^4	10^3	10^2	10^1	10^0
		7	2	3	6

We can write the decimal number 7,236 in expanded form as shown:

$$7{,}236 = (7 \times 10^3) + (2 \times 10^2) + (3 \times 10^1) + (6 \times 10^0)$$
$$= (7 \times 1{,}000) + (2 \times 100) + (3 \times 10) + (6 \times 1)$$
$$= 7{,}000 + 200 + 30 + 6$$

We can indicate that 7,236 is a base-10 number by using the subscript 10 and writing $7{,}236_{10}$ (read "seven thousand, two hundred thirty-six, base-10"). However, base-10 is understood when no subscript is used.

In the binary, or base-2, system, the place values are expressed as powers of 2. The base-2 number 1001 is shown in a place value chart below.

Base-2 Place Value System

32	16	8	4	2	1
2^5	2^4	2^3	2^2	2^1	2^0
		1	0	0	1

We indicate that 1001 is a base-2 number by using the subscript 2 and writing 1001_2. To convert a base-2 number to a base-10 number, write it in expanded form and evaluate it.

Language Box

Read the binary number 1001_2 as "one zero zero one, base two."

Convert the base-2 number 1001_2 to a base-10 number.

$$1001_2 = (1 \times 2^3) + (0 \times 2^2) + (0 \times 2^1) + (1 \times 2^0)$$
$$= (1 \times 8) + (0 \times 4) + (0 \times 2) + (1 \times 1)$$
$$= \quad 8 \quad + \quad 0 \quad + \quad 0 \quad + \quad 1$$
$$= 9$$

The base-2 number 1001_2 is equivalent to the base-10 number 9 or the number 9_{10}.

EXAMPLE A

Convert the base-2 number 10010_2 to a base-10 number.

Write the digits of the number in a base-2 place value chart.

16	8	4	2	1
2^4	2^3	2^2	2^1	2^0
1	0	0	1	0

Write the base-2 number in exponential form. Then find the value of each digit and add the results.

$$10010_2 = (1 \times 2^4) + (0 \times 2^3) + (0 \times 2^2) + (1 \times 2^1) + (0 \times 2^0)$$
$$= (1 \times 16) + (0 \times 8) + (0 \times 4) + (1 \times 2) + (0 \times 1)$$
$$= \quad 16 \quad + \quad 0 \quad + \quad 0 \quad + \quad 2 \quad + \quad 0$$
$$= 18$$

The base-2 number 10010_2 is equivalent to the base-10 number 18.

The powers of 2 can be used to convert a decimal number (base-10) to a binary number (base 2).

Table of Powers of 2

2^8	2^7	2^6	2^5	2^4	2^3	2^2	2^1	2^0
256	128	64	32	16	8	4	2	1

To convert a decimal number to a binary number, follow these steps:

- Write the decimal number as a sum of powers of 2. The first number in the sum is the highest power of 2 that is less than or equal to the number. Then add the next power of 2 as long as you don't exceed the original number; add zero if it does. Continue in this way until your last number is either 1 (which is 2^0) or 0.

- Rewrite the sum, using multipliers and powers of 2.

- Write the binary number by writing only the multipliers.

This method is shown in Example B.

EXAMPLE B

What is 13 written in binary?

$13 = 8 + 4 + 0 + 1$ ← Write 13 as a sum of powers of 2.

$= (1 \times 2^3) + (1 \times 2^2) + (0 \times 2^1) + (1 \times 2^0)$ ← Rewrite the sum using multipliers (shown in green) and powers of 2.

$= 1101_2$ ← The multipliers are the digits of the binary number.

In binary, 13 is 1101_2.

EXAMPLE C

What is 18 written in binary?

$18 = 16 + 0 + 0 + 2 + 0$ ← Write 18 as a sum of powers of 2.

$= (1 \times 2^4) + (0 \times 2^3) + (0 \times 2^2) + (1 \times 2^1) + (0 \times 2^0)$ ← Rewrite the sum using multipliers and powers of 2.

$= 10010_2$ ← The multipliers are the digits of the binary number.

In binary, 18 is 10010_2.

The method of writing a decimal number as a sum of powers of 2 is not always convenient. Another method uses repeated division. This method is shown in Examples D and E.

EXAMPLE D

What is 52 written in binary?

To convert 52 to binary using repeated division:

- Divide 52 by 2 and record the quotient with its remainder.
- Then divide the quotient by 2 and record the new quotient with its remainder.
- Repeat this process until you reach the last division, which will *always* be 1 ÷ 2 = 0 R1. This is because 2 does not divide into 1, giving you a quotient of 0 and a remainder of 1.

52 ÷ 2 =	26	R 0
26 ÷ 2 =	13	R 0
13 ÷ 2 =	6	R 1
6 ÷ 2 =	3	R 0
3 ÷ 2 =	1	R 1
1 ÷ 2 =	0	R 1

- The binary representation of the decimal number is given by the remainders. The digits of the binary number—as read from *left to right*—are the remainders as read from *bottom to top*.

The number 52 written in binary is 110100_2.

EXAMPLE E

Convert 100 to binary using repeated division.

$100 \div 2 =$	50	R 0	↑
$50 \div 2 =$	25	R 0	
$25 \div 2 =$	12	R 1	
$12 \div 2 =$	6	R 0	
$6 \div 2 =$	3	R 0	
$3 \div 2 =$	1	R 1	
$1 \div 2 =$	0	R 1	

The digits of the binary number (from left to right) are the remainders, read from bottom to top.

The number 100 written in binary is 1100100_2.

IP Addresses

As we mentioned, each digit is called a bit, which stands for *binary digit*. A group of eight bits is called a *byte;* a nibble is half a byte, or four bits.

An Internet Protocol (IP) address is a numerical identification assigned to devices participating in a computer network using the Internet Protocol for communication. An IP address is a binary address consisting of four 8-bit numbers, or bytes. These bytes can be converted back and forth between decimal and binary notation. You may have seen an IP address like 192.168.5.1, but the computer sees this address as 11000000.10101000.00000101.00000001.

All IP addresses have a total of 32 bits, or digits that can have either a 1 or 0 value. Even if all eight bits were ones, a byte can only add up to 255. An IP address can never go above 255 because 1111 1111 = 255.

EXAMPLE F

What is the decimal version of the IP address 11110110. 10001001.10100111. 10100110?

We will work with each byte separately, beginning with the byte on the left.

Converting 11110110_2 to base 10 we get 246. Next, we see that 10001001_2 = 137, then 10100111_2 = 167, and finally 10100110_2 = 166.

The IP address 11110110.10001001.10100111.10100110 is equivalent to 246.137.167.166.

EXERCISES 19-1

In Exercises 1–9, write each binary number in base-10.

1. 111 **2.** 11010 **3.** 110011

4. 1100101 **5.** 11001 **6.** 100001

7. 11111 **8.** 111111 **9.** 1000100

In Exercises 10–18, write each base-10 number in binary.

10. 6 **11.** 14 **12.** 19

13. 27 **14.** 39 **15.** 22

16. 121 **17.** 102 **18.** 87

19. Determine the decimal version of the IP address 10110110.00001001. 10100101.10101110.

20. What is the decimal version of the IP address 01101001.00011001. 10110101.10100010?

21. What is the binary version of the IP address 150.72.180.140?

22. Determine the binary version of the IP address 75.95.125.165.

23. Challenge: What is the only nonzero number that can be written the same way, except for subscripts, in either the base-10 or base-2 number system?

ADDING BINARY NUMBERS

A handheld calculator uses a base-2 number system to perform operations on the numbers you enter in base-10. Of course, it gives the results in base-10.

Computers and calculators are two types of devices that perform calculations using the binary number system. This means that when a computer or calculator adds two numbers, it adds two binary numbers.

Adding binary numbers using pencil and paper is similar to adding numbers in base-10. That is, the numbers are written with the digits aligned by place value. Then, the digits are added in columns from right to left.

Find the sum of 1001_2 and 100_2.

Align the digits; add in columns from right to left.

$$
\begin{array}{r}
1001 \\
+\ 100 \\
\hline
1101
\end{array}
$$

You can check your answer by converting the binary numbers to base-10 numbers, adding the base-10 numbers, and then converting back to a binary number.

$$
\begin{array}{rcr}
1001 &=& 9 \\
+\ 100 &=& +\ 4 \\
\hline
1101 &=& 13
\end{array}
$$ The sums are equal because $1101_2 = 13$.

The sum of 1001_2 and 100_2 is 1101_2.

Helpful Hint

You can drop the subscript 2 during the calculations, but be sure to include it in your answer.

In the previous example, it was not necessary to add $1 + 1$ in a single column. However, when $1 + 1$ occurs in a binary addition, you need to write the binary form of 2, which is 10_2. This fact is shown below, where 2_{10} is converted to 10_2.

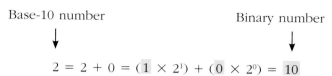

Base-10 number   Binary number

$$2 = 2 + 0 = (1 \times 2^1) + (0 \times 2^0) = 10$$

So, when $1 + 1$ occurs in a column, we will write the result, 10, by writing the 0 in that same column and carrying the 1 to the next column. This is necessary because we cannot write two digits in a single column. The process of regrouping and carrying a digit is shown in Example A.

EXAMPLE A

Find the sum of 101_2 and 1_2.

$$\begin{array}{r} {\scriptstyle 1} \\ 101 \\ + \quad 1 \\ \hline 0 \end{array}$$

Align the digits by place value, and add the rightmost digits: $1 + 1 = 10$. Write the 0 and carry the 1.

$$\begin{array}{r} {\scriptstyle 1} \\ 101 \\ + \quad 1 \\ \hline 110 \end{array}$$

Add the digits in the remaining columns.

Check your answer.

$$\begin{array}{r} 101 = \quad 5 \\ + \quad 1 = +1 \\ \hline 110 = \quad 6 \end{array}$$

The sums are equal because $110_2 = 6$.

The sum of 101_2 and 1_2 is 110_2.

In Example B, regrouping and carrying will be required more than once.

EXAMPLE B

Find the sum of 1101_2 and 101_2.

In this example, it will be necessary to regroup three times. To start, align the digits by place value. Then add by columns from right to left.

Add in column 1.	Add in column 2.	Add in column 3.	Add in column 4.	Add in column 5.
$\begin{array}{r}{\scriptstyle 1}\\1101\\+\ 101\\\hline 0\end{array}$	$\begin{array}{r}{\scriptstyle 1}\\1101\\+\ 101\\\hline 10\end{array}$	$\begin{array}{r}{\scriptstyle 1\ 1}\\1101\\+\ 101\\\hline 010\end{array}$	$\begin{array}{r}{\scriptstyle 1\ 1\ 1}\\1101\\+\ \ 101\\\hline 0010\end{array}$	$\begin{array}{r}{\scriptstyle 1\ 1\ 1}\\1101\\+\ \ 101\\\hline 10010\end{array}$
$1 + 1 = 10$; write the 0 and carry the 1.	$1 + 0 + 0 = 1$	$1 + 1 = 10$; write the 0 and carry the 1.	$1 + 1 = 10$; write the 0 and carry the 1.	The leftmost column has a 1 that was carried.

The sum of 1101_2 and 101_2 is 10010_2.

We know that $1 + 1 + 1 = 3$ in base 10. But in binary, $1 + 1 + 1 = 11$. When you add three 1s in binary, you have to regroup in the next place value to the left because you can only write a 0 or a 1 in the sum. This is shown in Example C.

EXAMPLE C

Find the sum of 11_2 and 11_2.

Write the problem vertically, and then add in the rightmost column.

$$\begin{array}{r} \scriptstyle 1 \\ 11 \\ +11 \\ \hline 0 \end{array} \qquad \begin{array}{r} \scriptstyle 11 \\ 11 \\ +\ 11 \\ \hline 10 \end{array} \qquad \begin{array}{r} \scriptstyle 11 \\ 11 \\ +11 \\ \hline 110 \end{array}$$

$1 + 1 = 10$; write the 0 and carry the 1.

$1 + 1 + 1 = 11$; write the 1 and carry the 1.

The third column has the 1 that was carried. Write 1 in the sum.

The sum of 11_2 and 11_2 is 110_2.

EXERCISES 19-2

In Exercises 1–9, find the sum of the binary numbers. Check your answer by converting the binary numbers to base-10 numbers, adding the base-10 numbers, and converting back to base 2.

1.	101	2.	10010	3.	101010
	+ 10		+ 1000		+ 100

4.	10001	5.	1011	6.	10110
	+ 1001		+ 10		+10100

7.	110	8.	10011	9.	1110
	+110		+10011		+1000

10. **Challenge:** Use addition of binary numbers to find the sum of the numbers: 110_2, 110_2, 101_2, and 111_2.

Helpful Hint

When adding $1 + 1$, write a 0 and carry a 1.

When adding $1 + 1 + 1$, write a 1 and carry a 1.

COMPLEMENTS OF BINARY NUMBERS

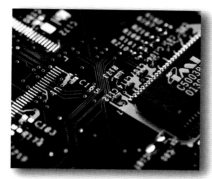

Computers use the 2's complement of an integer to represent the opposite of the integer.

Before discussing subtraction of binary numbers, it is necessary to discuss complements of binary numbers. In this lesson, we will use eight digits for a number. Some examples of binary numbers with eight digits are shown below.

$$2_{10} = 0000\ 0010_2 \qquad 6_{10} = 0000\ 0110_2 \qquad 16_{10} = 0001\ 0000_2$$

Computer programmers often use the leftmost digit of a binary number, called the **most significant digit (MSD),** to indicate the sign of the binary number. If the MSD is 0, the number is positive. If the MSD is 1, the number is negative. In the examples above, the most significant digit is 0 in every case, indicating that the numbers are positive.

To subtract a number, we add its opposite. So, we will need a way to represent opposites in binary form. *The opposite of a binary number is the 2's complement of that number.* Finding the 2's complement of a binary number is a two-step process. First, change every 0 to 1 and every 1 to 0. This step is called finding the **1's complement.** Then add 1 to the 1's complement. The result is the **2's complement.** The steps are shown in the example below.

Use complements to write -3_{10} in binary form.

Base-10 Number	Binary Number	Steps
3 →	0000 0011	Write 3 in binary.
	1111 1100	Write the 1's complement of 3.
	1111 1100 + 1	Add 1 to the 1's complement.
−3 →	1111 1101	This is the 2's complement of 3. It is the opposite of 3.

In binary form, -3_{10} is 1111 1101.

From here on, subscripts may be omitted from some numbers. It will be clear from the context whether a number is a base-10 number or a binary number.

EXAMPLE A

Use complements to write −17 in binary form.

Language Box

Finding the 1's complement is sometimes called *flipping the digits* because every digit "flips" from 0 to 1 or from 1 to 0.

Base-10 Number	Binary Number	Steps
17 →	0001 0001	Write 17 in binary.
	1110 1110	Write the 1's complement of 17.
	1110 1110 + 1	Add 1 to the 1's complement.
−17 →	1110 1111	This is the 2's complement of 17. It is the opposite of 17.

In binary form, −17 is 1110 1111.

KEY IDEA

The opposite of a binary number is the 2's complement of that number.
To find the 2's complement of a binary number:
• Find the 1's complement by changing every 1 to 0 and every 0 to 1.
• Add 1 to the 1's complement.

A binary number and its 2's complement form a **complementary pair.**
So far, we have seen the complementary binary pairs for 3 and −3, as
well as 17 and −17.

We know that the opposite of the opposite of a number is that same
number. For example, −(−5) = 5. It is useful to verify that this property
is also true for binary numbers. For binary numbers, we could state the
property by saying that the 2's complement of the 2's complement of a
number is that same number. This property is illustrated in Example B for
the number 5. Notice that the first and last binary numbers are identical.

EXAMPLE B

Show that the
2's complement of
the 2's complement
of 5 is 5.

5 →	0000 0101	Write 5 in binary.
	1111 1010	1's complement of 5
	+ 1	
−5 →	1111 1011	2's complement of 5
	0000 0100	1's complement of −5
	+ 1	
5 →	0000 0101	2's complement of −5

The 2's complement of −5 is 5, so the 2's complement of the 2's complement of 5 is 5.

EXERCISES 19-3

For Exercises 1–3, find the 1's complement of the number.

1. 0110 1100   **2.** 1101 0001   **3.** 0100 1101

For Exercises 4–6, find the 2's complement of the number.

4. 0000 0101   **5.** 0001 1100   **6.** 1011 1101

For Exercises 7–12, use complements to write each decimal
number in binary form.

7. −41   **8.** −110   **9.** −81

10. −37   **11.** −72   **12.** −102

13. Challenge: If the MSD in an eight-digit binary number is used to
indicate the sign of the number, what is the greatest base-10 number
that can be written using eight binary digits? Explain how you found
your answer.

Helpful Hint

Refer to the Table of
Powers of 2 at the
beginning of the
chapter when convert-
ing decimal numbers
to binary.

SUBTRACTING BINARY NUMBERS

Basic Subtraction

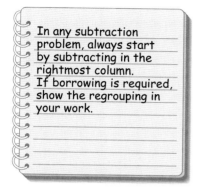

In any subtraction problem, always start by subtracting in the rightmost column. If borrowing is required, show the regrouping in your work.

In this lesson, we will first look at a "paper-and-pencil" method of subtracting eight-digit binary numbers. Then we will look at the method of adding a 2's complement, a method that is actually used in programming computers and calculators.

Subtract 0000 1001 from 0001 1101.

Align the numbers vertically by place value. Then subtract the binary numbers the same way that you subtract base-10 numbers; that is, subtract the digits in each column from right to left.

```
  0001 1101
 −0000 1001
  0001 0100
```

You can check your answer by converting the binary numbers to base-10 numbers, subtracting the base-10 numbers, and converting back to binary numbers.

```
  0001 1101   ⟶      29
 −0000 1001   ⟶     − 9
  0001 0100   ⟶      20
```

The difference is 0001 0100.

In the previous example, there was no need for borrowing. In binary, we need to borrow when we try to subtract 1 from 0 in a column. To do this, borrow 1 from the next column to the left, and use the binary subtraction fact: $10 - 1 = 1$. Remember, 10 in binary is equivalent to 2 in base 10. A subtraction problem that requires borrowing is shown in Example A.

EXAMPLE A

Subtract 0000 0010 from 0000 0101.

Align the numbers vertically and subtract in the first column.

```
  0000 0101
 −0000 0010
          1
```

Now subtract in the second column. To subtract 1 from 0, we must borrow from the next column to the left.

```
        0 10
  0000 0101
 −0000 0010
         11   ⟵   Use the subtraction fact 10 − 1 = 1.
```

Complete the subtractions in the columns to the left.

```
        0 10
  0000 0101
 −0000 0010
  0000 0011
```

Check your answer.

$$
\begin{array}{rcr}
0000\ 0101 & \longrightarrow & 5 \\
-0000\ 0010 & \longrightarrow & -2 \\
\hline
0000\ 0011 & \longrightarrow & 3
\end{array}
$$

The difference is 0000 0011.

Just as in base-10 subtraction, sometimes it is necessary to borrow several times. In Example B, borrowing is needed twice.

EXAMPLE B

Subtract 0001 0110 from 0011 1100.

Align the numbers by place value and subtract in the first column.

$$
\begin{array}{r}
0011\ 1100 \\
-0001\ 0110 \\
\hline
0
\end{array}
$$

Now, to subtract in the second column, borrow from the next column to the left.

$$
\begin{array}{r}
0011\ 11\overset{0\ 10}{\cancel{1}}00 \\
-0001\ 0110 \\
\hline
10
\end{array}
$$

To subtract in the third column, again borrow from the next column to the left. Then complete the subtractions in the remaining columns.

$$
\begin{array}{r}
0011\ \overset{\overset{10}{0\ \cancel{\not{1}}\ 10}}{\cancel{1}\cancel{1}}00 \\
-0001\ 0110 \\
\hline
0010\ 0110
\end{array}
$$

The difference is 0010 0110.

The paper-and-pencil method shown above is not actually used in programming computers and calculators. However, the method of subtraction by adding the 2's complement, shown below, is used.

Subtraction by Adding the 2's Complement

Language Box

The most significant digit (MSD) is the left-most digit of a binary number.

Language Box

The **magnitude** of a number is the absolute value of the number. For example, the magnitude of -9 is 9; the magnitude of 9 is 9.

You can subtract a number by adding its opposite. For example, in base 10, the expression $4 - 3$ can be rewritten as $4 + (-3)$. In the binary number system, we can think of subtraction in the same way. Recall that the 2's complement of a binary number is the opposite of that number. So, *to subtract a binary number, add the 2's complement of that number.*

When we subtract by adding a 2's complement, it is important that we know how to interpret the answer. There are two things to consider—the sign of the answer and the magnitude of the answer. Remember, for binary numbers the most significant digit (MSD) indicates whether the number is positive or negative. A 0 as the MSD indicates that the number is positive; a 1 as the MSD indicates that the number is negative. Subtracting binary numbers by adding the 2's complement and the interpretation of the answer is shown in the examples that follow.

Subtract 0001 0100 from 0001 1010.

Write the subtraction problem vertically.

$$\begin{array}{r} 0001\ 1010 \\ -0001\ 0100 \\ \hline \end{array}$$

Find the 2's complement of the number you are subtracting.

0001 0100   The number you are subtracting

$$\begin{array}{r} 1110\ 1011 \\ +\qquad\quad 1 \\ \hline 1110\ 1100 \end{array}$$ 1's complement

        2's complement

Now, perform the subtraction by adding the 2's complement.

$$\begin{array}{r} 0001\ 1010 \\ -0001\ 0100 \\ \hline \end{array} \longrightarrow \begin{array}{r} 0001\ 1010 \\ +1110\ 1100 \\ \hline 1\ 0000\ 0110 \end{array}$$

Notice the red digit in the sum shown above. This digit is the "carry digit." It is the result of regrouping in the leftmost place. The carry digit can be ignored, because it is not part of the eight digits. The MSD is a zero and the number is positive. The sum is the eight-digit binary number 0000 0110, which equals 6 in base-10. So, we say that the answer is positive and its magnitude is 6.

You can check your answer by converting the binary numbers to base-10 numbers and subtracting:

$$\begin{array}{r} 0001\ 1010 \\ -0001\ 0100 \\ \hline 0000\ 0110 \end{array} \longrightarrow \begin{array}{r} 26 \\ -20 \\ \hline 6 \end{array}$$

The difference is 0000 0110.

Using the 2's complement makes it possible to subtract a greater integer from a lesser integer. This means that the answer will be a negative number, and the MSD will be 1. When the answer to a subtraction problem has a 1 as the MSD, you must find the 2's complement of that answer to determine the magnitude of the answer. This is shown in Example C.

EXAMPLE C

Subtract 0010 1101 from 0010 0100.

Write the subtraction problem vertically.

$$
\begin{array}{r}
0010\ 0100 \\
-0010\ 1101 \\
\hline
\end{array}
$$

Find the 2's complement of the number you are subtracting.

0010 1101   The number you are subtracting

$$
\begin{array}{r}
1101\ 0010 \\
+\qquad 1 \\
\hline
1101\ 0011
\end{array}
$$
        1's complement

        2's complement

Now, perform subtraction by adding the 2's complement.

$$
\begin{array}{r}
0010\ 0100 \\
-0010\ 1101 \\
\hline
\end{array}
\quad\longrightarrow\quad
\begin{array}{r}
0010\ 0100 \\
+1101\ 0011 \\
\hline
1111\ 0111
\end{array}
$$

Notice that there is no carry digit. Also, the MSD (leftmost digit) is 1, indicating that the sum is negative. The magnitude of a negative number is the opposite of that number, so we must find the opposite of 1111 0111 to find its magnitude. Recall that the opposite of a binary number is the 2's complement of that number. Therefore, we need to find the 2's complement of 1111 0111 to find its magnitude.

$$
\begin{array}{r}
0000\ 1000 \\
+\qquad 1 \\
\hline
0000\ 1001
\end{array}
$$
        1's complement of 1111 0111

        2's complement of 1111 0111

We already know that the answer to our subtraction problem is negative. The number in red, (9_{10}), is the magnitude of the answer. Putting the two facts together, the result of the subtraction problem in base-10 is -9.

Check your answer by converting the binary numbers to base-10 numbers and subtracting:

$$
\begin{array}{r}
0010\ 0100 \\
-0010\ 1101 \\
\hline
\end{array}
\quad\longrightarrow\quad
\begin{array}{r}
36 \\
-45 \\
\hline
-9
\end{array}
$$

The answer to the subtraction problem is 1111 0111. This is the binary equivalent of -9.

The interpretation of answers to subtraction problems that are done by adding the 2's complement is summarized in the box that follows.

KEY IDEA

When the difference of two binary numbers has 0 as the MSD, the number is positive. The magnitude of the number can be found by converting it directly to a base-10 number.

When the difference of two binary numbers has 1 as the MSD, the number is negative. The magnitude of the number can be found by finding its 2's complement, and then converting the 2's complement to a base-10 number.

EXERCISES 19-4

In Exercises 1–6, subtract using the pencil-and-paper method. Check your answers by converting the binary numbers to base-10 numbers, subtracting the base-10 numbers, and converting back to binary.

1. 0010 1111
   −0000 1101

2. 0110 1011
   −0010 1010

3. 0011 0010
   −0000 1010

4. 0011 0101
   −0001 0010

5. 0000 1110
   −0000 1011

6. 0110 1100
   −0100 0110

In Exercises 7–12, subtract by adding the 2's complement. Check your answers by converting the binary numbers to base-10 numbers, subtracting the base-10 numbers, and converting back to binary.

7. 0001 0101
   −0001 1111

8. 0111 0111
   −0100 1010

9. 0101 1011
   −0110 1100

10. 0001 0111
    −0001 1011

11. 0010 0000
    −0011 1011

12. 0000 1000
    −0001 1110

13. **Challenge:** Find the solution to the following subtraction problem:
   (1001 1110 − 0100 1101) − 0110 1110.

MULTIPLYING BINARY NUMBERS

In the early 1800s, Charles Babbage designed his "difference engine" to perform calculations and store the results. The photo shown here is only a fragment of the machine.

Because binary numbers use only the digits 0 and 1, there are only four facts you need to know for binary multiplication. These facts are identical to multiplication facts in base 10:

Binary Multiplication	Base-10 Multiplication
$0 \times 0 = 0$	$0 \times 0 = 0$
$1 \times 0 = 0$	$1 \times 0 = 0$
$0 \times 1 = 0$	$0 \times 1 = 0$
$1 \times 1 = 1$	$1 \times 1 = 1$

The process of multiplying two binary numbers is the same as the process of multiplying two base-10 numbers. However, multiplication with binary numbers is simpler than multiplication with base-10 numbers because you multiply only by 0 or 1.

For convenience, the examples of multiplication in this lesson are shown without the leftmost zero digits that would make up the eight-digit length we used in the prior lessons. Omitting the zeros does not affect the product.

Find the product of 101 and 11.

STEP 1

Multiply by the rightmost digit of the multiplier to get the first partial product.

$$
\begin{array}{r}
101 \quad \leftarrow \text{Multiplicand} \\
\times\ 11 \quad \leftarrow \text{Multiplier} \\
\hline
101 \quad \leftarrow 1 \times 101 = 101
\end{array}
$$

STEP 2

Multiply by the next digit of the multiplier to get the second partial product.

$$
\begin{array}{r}
101 \\
\times\ 11 \\
\hline
101 \\
101 \leftarrow 1 \times 101 = 101
\end{array}
$$

STEP 3

Add the partial products.

$$
\begin{array}{r}
101 \\
\times\ 11 \\
\hline
101 \\
+101 \\
\hline
1111
\end{array}
$$

The product of 101 and 11 is 1111.

EXAMPLE A

Find the product of 101 and 101.

STEP 1

Multiply by the rightmost digit of the multiplier to get the first partial product.

$$
\begin{array}{r}
101 \quad \leftarrow \text{Multiplicand} \\
\times 101 \quad \leftarrow \text{Multiplier} \\
\hline
101 \quad \leftarrow 1 \times 101 = 101
\end{array}
$$

STEP 2

Multiply by the next digit to the left in the multiplier to get the second partial product.

$$
\begin{array}{r}
101 \\
\times 101 \\
\hline
101 \\
000 \quad \leftarrow 0 \times 101 = 0
\end{array}
$$

STEP 3

Multiply by the leftmost digit of the multiplier to get the third partial product.

```
    101
  ×101
    101
   000
  101    ◄—  1 × 101 = 101
```

STEP 4

Add the partial products to find the product. The empty spaces represent zeros.

```
    101
  ×101
    101
   000
 +101
  11001
```

Check your answer by converting the binary numbers to base-10 numbers.

```
    101    —►      5
  × 101    —►    ×5
  11001    —►     25
```

The product of 101 and 101 is 1 1001.

In binary multiplication, every partial product is either the **multiplicand** (which may be shifted to the left) or 0. It is a good idea to check that you have written each partial product correctly. Notice in Example B that two partial products contain all zeros and one partial product is the multiplicand shifted two places to the left.

EXAMPLE B

Find the product of 1111 and 100.

Multiply the multiplicand by each digit of the multiplier and then add the partial sums.

```
     1111
   × 100
    0000
   0000
 +1111
  111100
```

Check your answer by converting the binary numbers to base-10 numbers.

```
   1111    —►     15
 × 100    —►    ×4
 111100    —►     60
```

The product of 1111 and 100 is 11 1100.

EXERCISES 19-5

Find the product of the binary numbers. Check your answers by converting all numbers to base 10, multiplying, and converting the answer to binary.

1. 10
×11

2. 110
×110

3. 1010
× 101

4. 101
×100

5. 1010
× 11

6. 1101
×1000

7. 111
×111

8. 1010
× 110

9. 1001
× 101

10. Challenge: In base 10, $2^3 = 8$. Show the work to demonstrate this fact using binary multiplication.

**COMMON
ERROR**

ALERT
If a nonzero partial product does not match the digits of the multiplicand, an error has been made. Check your work.

CONVERTING BETWEEN BASE-10 AND OCTAL NUMBERS

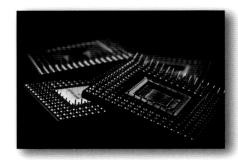

The octal number system was used in the early days of microprocessors.

Just as the binary (base-2) number system uses only the two digits 0 and 1, and the decimal (base-10) number system uses only the ten digits 0 through 9, the **octal number system** uses only the eight digits 0 through 7. The octal number system is a base-8 system. Because the octal number system is a base-8 system, the place values are powers of 8. A subscript of 8 is used to designate an octal number. For example, 632_8 is an octal number.

By writing an octal number in expanded form, we can convert numbers from octal to base 10. The octal number 632_8 is shown in the table below.

Base-8 Place Value System

32,768	4,096	512	64	8	1
8^5	8^4	8^3	8^2	8^1	8^0
			6	3	2

Convert the octal number 632_8 to a base-10 number.

First, write 632_8 in expanded form. Then find the value of each digit of the octal number, and add the results.

$632_8 = (6 \times 8^2) + (3 \times 8^1) + (2 \times 8^0)$

$= (6 \times 64) + (3 \times 8) + (2 \times 1)$

$= 384 + 24 + 2$

$= 410$

So, the octal number 632_8 is equivalent to the base-10 number 410 or the number 410_{10}.

EXAMPLE A

What is the octal number 342_8 in base 10?

Write the digits of the octal number in their place value columns.

64	8	1
8^2	8^1	8^0
3	4	2

Write the octal number in expanded form.

$$342_8 = (3 \times 8^2) + (4 \times 8^1) + (2 \times 8^0)$$
$$= (3 \times 64) + (4 \times 8) + (2 \times 1)$$
$$= 192 + 32 + 2$$
$$= 226$$

The octal number 342_8 is equivalent to the base-10 number 226_{10}.

The conversion of an octal number with zero digits is shown in Example B.

EXAMPLE B

What is the octal number 2007_8 in base 10?

Write the number, using the place value chart for base 8 as a reference.

512	64	8	1
8^3	8^2	8^1	8^0
2	0	0	7

Write the octal number in expanded form.

$$2007_8 = (2 \times 8^3) + (0 \times 8^2) + (0 \times 8^1) + (7 \times 8^0)$$
$$= (2 \times 512) + (7 \times 1)$$
$$= 1,024 + 7$$
$$= 1,031$$

The octal number $2,007_8$ is equivalent to the base-10 number $1,031_{10}$.

Now that we can convert octal numbers to base-10, how can we convert base-10 numbers to octal? We can use a method of repeated division, similar to the method that we used to convert base-10 numbers to binary. However, because we want to write the number in base 8, we need to divide repeatedly by 8, rather than by 2. This is shown in Example C.

EXAMPLE C

What is the number $7,346_{10}$ written in octal?

To convert the base-10 number $7,346_{10}$ to an octal number:

- Divide 7,346 by 8 and record the quotient with its remainder.
- Then divide the quotient by 8 and record the new quotient with its remainder.
- Repeat this process until you reach the last division, which will *always* be a number from 1 to 7 divided by 8.

$7{,}346 \div 8 =$	918	R 2
$918 \div 8 =$	114	R 6
$114 \div 8 =$	14	R 2
$14 \div 8 =$	1	R 6
$1 \div 8 =$	0	R 1

- The octal representation of the decimal number is given by the remainders. The digits of the octal number—as read from *left to right*—are the remainders as read from *bottom to top.*

The number $7{,}346_{10}$ written in octal is $16{,}262_8$.

EXAMPLE D

What is the number $8{,}880_{10}$ written in octal?

Convert $8{,}880_{10}$ to octal using repeated division.

$8{,}880 \div 8 =$	1,110	R 0
$1{,}110 \div 8 =$	138	R 6
$138 \div 8 =$	17	R 2
$17 \div 8 =$	2	R 1
$2 \div 8 =$	0	R 2

The digits of the octal number (from left to right) are the remainders, read from bottom to top.

The number $8{,}880_{10}$ written in octal is $21{,}260_8$.

EXERCISES 19-6

In Exercises 1–9, write each number in base 10.

1. 7_8

2. 26_8

3. 302_8

4. 10_8

5. 413_8

6. 101_8

7. $1{,}036_8$

8. $3{,}221_8$

9. $10{,}015_8$

In Exercises 10–18, write each number in octal.

10. 27_{10}

11. 43_{10}

12. 82_{10}

13. 125_{10}

14. 988_{10}

15. $4{,}389_{10}$

16. 826_{10}

17. 222_{10}

18. $1{,}007_{10}$

19. Challenge: Which nonzero numbers are represented the same way (except for subscripts) in octal and in decimal?

CONVERTING BETWEEN BINARY AND OCTAL NUMBERS

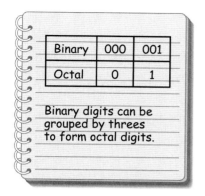

Binary	000	001
Octal	0	1

Binary digits can be grouped by threes to form octal digits.

In the binary number system, each place value is a power of 2. Because of this, binary numbers can become quite long and cumbersome. For example, the nine-digit binary number 100101110_2 is equivalent to the three-digit decimal number 302_{10}. Using the octal number system, we can create a type of "shorthand" system for working with long strings of binary digits. The binary-to-octal conversion is a good choice because the octal numbers are easier to work with, and the conversion is easily accomplished. The conversion method is shown below.

Remember that the octal number system uses the digits 0 through 7. The table below shows the binary equivalents of each of the octal digits, 0 through 7. In the table, notice that a number written with three binary digits can be written with one octal digit.

Equivalent Binary and Octal Numbers

Binary	000	001	010	011	100	101	110	111
Octal	0	1	2	3	4	5	6	7

A binary number can be converted to octal by separating it into groups of three digits each, starting from the right, and then finding the octal digit that each group of three binary digits represents.

Convert 010011110_2 from binary to octal.

First, separate the binary number into groups of three digits each, starting from the right.

010 011 110

Use the table above to determine which octal digit each group of three binary digits represents.

010 011 110 ← Binary

↓ ↓ ↓

2   3   6

← Octal

We can verify this result by converting both numbers to base-10.

Convert 010011110_2 to base-10.

0 1 0   0 1 1 1 1 0

↓

128   +   16 + 8 + 4 + 2   = 158

Convert 236_8 to base-10.

2   3   6

↓   ↓   ↓

$(2 \times 64) + (3 \times 8) + (6 \times 1) = 158$

Both the binary and the octal numbers are equal to 158.

So, 010011110_2 is equivalent to 236_8.

If the number of digits in a binary number is not a multiple of 3, add zeros to the left side of the number, as needed. Adding one or two zeros to the left side of the number does not change its value, but it ensures that groups of three binary digits are converted to one octal digit. This is shown in Example A.

EXAMPLE A

Convert 1011010_2 from binary to octal.

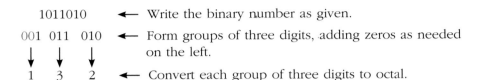

1011010	← Write the binary number as given.
001 011 010	← Form groups of three digits, adding zeros as needed on the left.
↓ ↓ ↓	
1 3 2	← Convert each group of three digits to octal.

The number 1011010_2 is equivalent to 132_8.

> ### Language Box
>
> Adding zeros to the left side of a number is sometimes called *padding* the number with zeros.

To convert an octal number to binary, simply find the binary equivalent of each digit. This is shown in Example B.

EXAMPLE B

Convert $1,723_8$ from octal to binary.

First, find the binary equivalent of each octal digit. Then combine the binary representations of each digit to form the binary number.

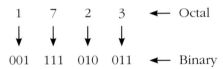

1	7	2	3	← Octal
↓	↓	↓	↓	
001	111	010	011	← Binary

It is not necessary to write the leading zeros in your answer.

The number $1,723_8$ is equivalent to 1111010011_2.

EXERCISES 19-7

In Exercises 1–9, write each number in octal.

1. 111011_2 **2.** 100101_2 **3.** 110111_2

4. 1000101_2 **5.** 10010_2 **6.** 11111_2

7. 1010110_2 **8.** 10010000_2 **9.** 101000010_2

In Exercises 10–18, write each number in binary.

10. 14_8 **11.** 76_8 **12.** 170_8

13. 621_8 **14.** 100_8 **15.** 737_8

16. 513_8 **17.** $1,261_8$ **18.** $2,554_8$

A computer displays error messages as binary numbers using a series of lights. If a light is turned on, it represents a 1, if a light is turned off, it represents a 0. Light on: ◯; Light off: ◯.

19. Consider the computer display shown below:

(a) What is the decimal value of this display?

(b) What is the octal number represented by this display?

20. Consider the computer display shown below:

(a) What is the decimal value of this display?

(b) What is the octal number represented by this display?

21. What's the Error? Howard converted 10_8 to binary with the work shown below:

1	0
100	000
$1 \times 32 + 0 = 32$	

What error did Howard make? What is the binary equivalent of 10_8?

CONVERTING BETWEEN BASE-10 AND HEXADECIMAL NUMBERS

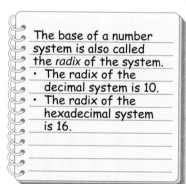

The base of a number system is also called the *radix* of the system.
• The radix of the decimal system is 10.
• The radix of the hexadecimal system is 16.

Another number system that is used by computers is the **hexadecimal,** or base-16, number system. Although the appearance of this number system is slightly different than the decimal, binary, and octal number systems, it is based on the same principles as the other number systems. That is, each place value is a power of the base.

The hexadecimal system is a base-16 system, so it requires 16 digits. The 16 hexadecimal digits and their decimal equivalents are shown below.

Hexadecimal	0	1	2	3	4	5	6	7	8	9	A	B	C	D	E	F
Decimal	0	1	2	3	4	5	6	7	8	9	10	11	12	13	14	15

Notice that the digits 0 through 9 in the two systems are the same and that the letters A through F in the hexadecimal system are equivalent to the numbers 10 through 15 in the decimal system.

Because the hexadecimal system is a base-16 system, the place values are expressed as powers of 16. To convert a number from hexadecimal to base 10, write the number in expanded form. Then find the value of each digit of the hexadecimal number, and add the results. This is the same method that was used to convert numbers from binary and octal to base 10. The hexadecimal number $3D2_{16}$ is shown in the place value table on next page.

Base-16 Place Value System

65,536	4,096	256	16	1
16^4	16^3	16^2	16^1	16^0
		3	D	2

Find the base 10 equivalent of 3D2$_{16}$.

First, write 3D2$_{16}$ in expanded form. Find the value of each digit of the hexadecimal number, and add the results.

$$3D2_{16} = (3 \times 16^2) + (D \times 16^1) + (2 \times 16^0)$$
$$= (3 \times 16^2) + (13 \times 16^1) + (2 \times 16^0)$$
$$= (3 \times 256) + (13 \times 16) + (2 \times 1)$$
$$= \quad 768 \quad + \quad 208 \quad + \quad 2$$
$$= 978$$

So, 3D2$_{16}$ is equivalent to 978$_{10}$.

Language Box

The name *hexadecimal* is derived from the Greek *hexa* for "six," and the Latin *decimal* for "ten."

EXAMPLE A

What is F9BA$_{16}$ written in base-10?

Write the hexadecimal number, and above the digits of the number, from right to left, write the powers of 16, starting with 16^0.

4,096	256	16	1
16^3	16^2	16^1	16^0
F	9	B	A

Now, multiply each digit of the hexadecimal number by the power of 16 above it, and add the results.

$$F9BA = (F \times 16^3) + (9 \times 16^2) + (B \times 16^1) + (A \times 16^0)$$
$$= (15 \times 16^3) + (9 \times 16^2) + (11 \times 16^1) + (10 \times 16^0)$$
$$= (15 \times 4,096) + (9 \times 256) + (11 \times 16) + (10 \times 1)$$
$$= \quad 61,440 \quad + \quad 2,304 \quad + \quad 176 \quad + \quad 10$$
$$= 63,930$$

F9BA$_{16}$ written in base 10 is 63,930$_{10}$.

The method of repeated division can be used to convert base-10 numbers to hexadecimal numbers. Use a divisor of 16 because the base of the hexadecimal system is 16. When there is a remainder greater than 9 in the division process, the remainder must be converted to its hexadecimal equivalent, a letter A through F. This is shown in Example B.

EXAMPLE B

What is the number 650_{10} written in hexadecimal?

To use the method of repeated division:

- Divide 650 by 16 and record the quotient with its remainder.

- Then divide the quotient by 16 and record the new quotient with its remainder.

- Repeat this process until you reach the last division, which will *always* be a number from 1 to 15 divided by 16.

$650 \div 16 =$	40	R 10
$40 \div 16 =$	2	R 8
$2 \div 16 =$	0	R 2

- The hexadecimal representation of the decimal number is given by the remainders. Note that the remainder 10 in the first division must be converted to its hexadecimal equivalent, A.

The number 650_{10} written in hexadecimal is $28A_{16}$.

Helpful Hint

The last division will always be of the form $b \div 16 = 0\ Rb$, where b is an integer from 1 to 15.

Helpful Hint

Refer to the chart below for the values of the letters A through F in hexadecimal.

A = 10
B = 11
C = 12
D = 13
E = 14
F = 15

EXERCISES 19-8

In Exercises 1–9, write each number in base 10.

1. 98_{16}

2. $A1_{16}$

3. ABC_{16}

4. $9D2_{16}$

5. $1B8E_{16}$

6. ACE_{16}

7. $459A3_{16}$

8. $FD45_{16}$

9. $FADED_{16}$

10. Why does the number BB_{16} have a greater value than the number 99_{16}? Explain your answer.

In Exercises 11–19, write each number in base 16.

11. 51_{10}

12. 260_{10}

13. 61_{10}

14. 331_{10}

15. 526_{10}

16. $1{,}616_{10}$

17. $1{,}000_{10}$

18. $2{,}180_{10}$

19. $4{,}100_{10}$

20. **Challenge:** Convert BAD_{16} to its octal equivalent. Explain your steps.

CONVERTING BETWEEN BINARY AND HEXADECIMAL NUMBERS

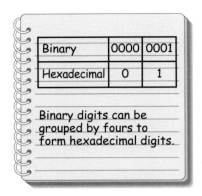

Binary	0000	0001
Hexadecimal	0	1

Binary digits can be grouped by fours to form hexadecimal digits.

Earlier, we converted binary numbers to octal numbers by grouping the binary digits into groups of three. In this lesson, we will convert binary numbers to hexadecimal numbers by grouping the binary digits into groups of four.

The table below shows binary and hexadecimal equivalents. Notice that there are 16 hexadecimal "digits," starting with 0 and ending with F.

Equivalent Binary and Hexadecimal Numbers

Binary	0000	0001	0010	0011	0100	0101	0110	0111
Hexadecimal	0	1	2	3	4	5	6	7

Binary	1000	1001	1010	1011	1100	1101	1110	1111
Hexadecimal	8	9	A	B	C	D	E	F

A binary number can be converted to a hexadecimal number by separating the binary number into groups of four digits each, starting from the right, and then finding the hexadecimal equivalent of each group of four digits.

Convert 100101111011_2 from binary to hexadecimal.

First, separate the binary number into groups of four digits each, starting from the right.

1001   0111   1011

Then use the table above to determine which hexadecimal digit each group of four binary digits represents.

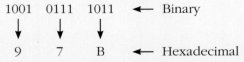

1001   0111   1011   ← Binary

9   7   B   ← Hexadecimal

The number 100101111011_2 is equivalent to $97B_{16}$.

If the number of digits in the binary number is not a multiple of 4, add zeros to the **left** side of the number until the digits can be separated into groups of four digits. This is shown in Example A.

EXAMPLE A

Convert 111100011_2 from binary to hexadecimal.

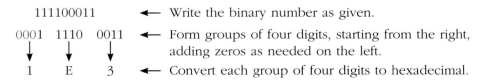

111100011   ← Write the binary number as given.

0001   1110   0011   ← Form groups of four digits, starting from the right, adding zeros as needed on the left.

1   E   3   ← Convert each group of four digits to hexadecimal.

The number 111100011_2 is equivalent to $1E3_{16}$.

EXAMPLE B

Convert $8E4_{16}$ from hexadecimal to binary.

To convert from hexadecimal to binary, find the binary equivalent of each hexadecimal digit. This is shown in Example B.

First, find the binary equivalent of each hexadecimal digit. Then combine the binary representations of each digit to form the binary number.

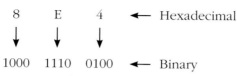

The number, $8E4_{16}$ is equivalent to 100011100100_2.

COMMON ERROR

ALERT

When converting binary to hexa-decimal, an error will result if the binary digits are not grouped by fours, starting from the right side.

EXERCISES 19-9

In Exercises 1–9, write each number in hexadecimal form.

1. 10100110_2 **2.** 10000000_2 **3.** 10110001_2

4. 111011010010_2 **5.** 11111110101_2 **6.** 110010101_2

7. 1000010010_2 **8.** 101100101_2 **9.** 1111001011_2

In Exercises 10–18, write each number in binary.

10. $F2_{16}$ **11.** $98A_{16}$ **12.** ABE_{16}

13. $1ED_{16}$ **14.** $C07_{16}$ **15.** $4D5_{16}$

16. $C5A_{16}$ **17.** $32B_{16}$ **18.** $3DA_{16}$

19. Consider the computer display shown below:

(a) What is the decimal value of this display?

(b) What is the hexadecimal number represented by this display?

20. Consider the computer display shown below:

(a) What is the decimal value of this display?

(b) What is the hexadecimal number represented by this display?

21. Challenge: To convert a number from octal to hexadecimal, convert the number from octal to binary first, and then convert from binary to hexadecimal. What is the hexadecimal equivalent of $4,217_8$?

CHAPTER 19 REVIEW EXERCISES

Answer Exercises 1–12.

1. Write the decimal number 8,531 in expanded form.

2. Write the binary number 10011_2 in expanded form.

3. Write the octal number $5,073_8$ in expanded form.

4. Write the hexadecimal number $A72_{16}$ in expanded form.

5. Convert the decimal number 5,716 to the following

   (a) Its binary equivalent

   (b) Its octal equivalent

   (c) Its hexadecimal equivalent

6. Convert the binary number 10101101_2 to its hexadecimal equivalent.

7. Convert the binary number 100111101_2 to its octal equivalent.

8. Convert the hexadecimal number $F4_{16}$ to its binary equivalent.

9. Determine the indicated sum.

   $$\begin{array}{r} 110101_2 \\ + \ 11001_2 \\ \hline \end{array}$$

10. In the following subtraction problem:

    (a) Determine the 2's complement of each subtrahend.

    (b) Use the 2's complement to solve the subtraction exercise.

    $$\begin{array}{r} 1101_2 \\ - \ 1010_2 \\ \hline \end{array}$$

11. Determine the decimal version of the IP address 11110100.01001001.10101101.10001100.

12. Determine the binary version of the IP address 250.3.27.133.

Building a Foundation in Mathematics

Logic and Truth Tables

Introducing Boolean Algebra

Laws of Boolean Algebra

Using DeMorgan's Law

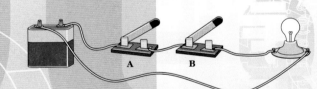

Overview

Electronic machinery, including your calculator and computer, rely on electronic devices such as transistors and integrated circuits. Understanding these devices requires electricians to understand the logic operations of electronic machinery.

These logic operations are known as Boolean algebra and were developed by George Boole in the mid-19th century. The laws of Boolean algebra are

somewhat similar to those of arithmetic and algebra. However, there are some ways in which these laws differ from those of arithmetic and algebra.

In this chapter, we will explore the basic ideas of Boolean algebra to help us use logic to (1) describe circuits mathematically after they have been designed or assembled and (2) design circuits mathematically before they are assembled.

Boolean Algebra Fundamentals

Objectives

After completing this chapter, you will be able to:

- Use the truth table for an AND operation to determine the truth value of an AND statement

- Use the truth table for an OR operation to determine the truth value of an OR statement

- Use the truth table for the negation of a statement to determine the truth value of that statement

- Use the truth table for a NAND operation to determine the truth value of an NAND statement

- Use the truth table for an NOR operation to determine the truth value of an NOR statement

- Use the basic laws and operations of Boolean algebra

- Use truth tables to prove Boolean theorems

- Use the logic gates of AND, OR, NOT, NAND, and NOR

- Use DeMorgan's Law to simplify expressions

LOGIC AND TRUTH TABLES

A system of logic with only two values has direct applications to circuit design. In fact, it is sometimes called *switching logic.*

The lessons of this chapter will deal with the fundamentals of two-valued Boolean algebra. The two values, or elements, in the two-valued Boolean algebra we will study are 0 and 1. Later, we will examine how this type of Boolean algebra can be applied to circuit design, in which every switch has exactly one of two possible states (such as *on* and *off*). To be better prepared to understand these topics later, we will first study a topic in logic that deals with statements and **truth tables.**

The branch of logic in this lesson deals with statements, each of which has exactly one of two possible values, true or false. These values are called truth values. Statements are represented by variables, such as A and B. If two or more statements are joined by a connector, such as AND or OR, the resulting statement is called a **composite statement.** Every composite statement also has exactly one of two possible truth values, either true or false. The truth value of a composite statement is determined by the truth values of the statements that are connected, as well as the type of connector joining them. A convenient way to show all possible truth values of a composite statement is by a **truth table.**

Consider the variables A and B, which represent statements that are either true or false. The **conjunction** of A and B (A AND B) is true only when both statements are true. The truth table for A AND B is shown below. Let T = true and F = false.

The truth table for the statement A AND B is shown below.

A	B	A AND B	
T	T	T	(row 1)
T	F	F	(row 2)
F	T	F	(row 3)
F	F	F	(row 4)

The following examples will illustrate why the statement A AND B has the truth values shown in the table above. Remember that A and B are variables; they may represent many different statements, just as variables like x and y can represent many different numbers in mathematics.

EXAMPLE A

Consider the statements given at the right. Which row of the truth table is illustrated by the statements?

Let A be *A rose is a plant.*

Let B be *Detroit is in Michigan.*

A is true because it is true that *A rose is a plant.*

B is true because it is true that *Detroit is in Michigan.*

A AND B is true because, if it is true that *A rose is a plant* and it is also true that *Detroit is in Michigan*, then it is certainly true that *A rose is a plant and Detroit is in Michigan.*

So, find the row in the truth table that shows the value T for A, T for B, and T for A AND B.

Row 1 of the truth table is illustrated.

EXAMPLE B

Consider the statements given at the right. Which row of the truth table is illustrated by the statements?

Let *A* be *A rose is an animal.*

Let *B* be *Detroit is in Michigan.*

A is false because it is false that *A rose is an animal.*

B is true because it is true that *Detroit is in Michigan.*

A AND *B* is false because, if it is false that *A rose is an animal,* then it is also false that *A rose is an animal and Detroit is in Michigan.*

So, find the row in the truth table that shows the value F for *A*, T for *B*, and F for *A* AND *B*.

Row 3 of the truth table is illustrated.

The truth table for the AND connector indicates that the composite statement *A* AND *B* is true only when both *A* and *B* represent true statements. Another connector in logic is the OR connector. The **disjunction** *A* OR *B* is true when one of the variables or both variables represent a true statement. The truth table for *A* OR *B* is shown next. Again, let T = true and F = false.

A	*B*	*A* OR *B*	
T	T	T	(row 1)
T	F	T	(row 2)
F	T	T	(row 3)
F	F	F	(row 4)

EXAMPLE C

Consider the statements given at the right. Which row of the truth table is illustrated by the statements?

Let *A* be *A dog is an animal.*

Let *B* be 2 + 2 = 5.

Statement *A* is true; statement *B* is false.

A OR *B* is true because at least one of the variables (*A*) represents a true statement.

So, find the row in the truth table that shows the value T for *A*, F for *B*, and T for *A* OR *B*.

This is shown in row 2 of the truth table.

EXAMPLE D

Consider the statements given at the right. Which row of the truth table is illustrated by the statements?

Let *A* be *A dog is a plant.*

Let *B* be 2 + 2 = 5.

Statement *A* is false; statement *B* is false.

A OR *B* is false because neither of the variables represents a true statement.

So, find the row in the truth table that shows the value F for *A*, F for *B*, and F for *A* OR *B*.

This is shown in row 4 of the truth table.

For any statement, *A*, in logic, it is also possible to write another statement, called the **negation of A.** The negation of *A* is also called NOT *A*. The truth table for the negation of a variable is shown below. Again, let T = true and F = false.

A	NOT *A*	
T	F	(row 1)
F	T	(row 2)

The following examples illustrate how the negation of a variable changes the truth value of the variable to its complement. In these examples, *true* and *false* are complements.

EXAMPLE E

Consider the statement given at the right. Which row of the truth table is illustrated by the statement?

Let *A* be *Detroit is in Michigan.*

Statement *A* is true; NOT *A* is false.

This is shown in row 1 of the truth table.

EXAMPLE F

Consider the statement given at the right. Which row of the truth table is illustrated by the statement?

Let *A* be *A rose is an animal.*

Statement *A* is false; NOT *A* is true.

This is shown in row 2 of the truth table.

EXAMPLE G

Let *A* be the statement *4 + 4 = 16* and *B* be the statement *25 is an odd number.* Rewrite the symbolic statement *A* OR $\overline{B}$ as a composite statement and construct a truth table for *A* OR $\overline{B}$.

Statement *A* is given in the problem. $\overline{B}$ is the statement *25 is NOT an odd number.*

So, *A* OR $\overline{B}$ is *4 × 4 = 16* OR *25 is NOT an odd number.*

The truth table for *A* OR $\overline{B}$ is shown below.

A	*B*	$\overline{B}$	*A* OR $\overline{B}$
T	T	F	T
T	F	T	T
F	T	F	F
F	F	T	T

Recall that a logic statement can have only one of two possible values. If statement *A* is true, then NOT *A* must be false; if statement *A* is false, then NOT *A* must be true.

So far in this lesson, we have seen the truth tables for *A* AND *B*, *A* OR *B*, and the negation of *A*. Next, we will see how these same truth tables can be applied to the input and output of circuits using Boolean algebra.

EXERCISES 20-1

Use the truth table for *A* AND *B* for Exercises 1–3. Indicate the truth value for *A, B,* and *A* AND *B*. Identify the row in the truth table represented by each statement.

1. Let *A* be *A square is a rectangle.*
Let *B* be *Five is an even number.*

2. Let *A* be *Paris is in England.*
Let *B* be *Ducks ride bicycles.*

3. Let *A* be *Six is an even number.*
Let *B* be *3 × 3 = 9.*

Use the truth table for *A* OR *B* for Exercises 4–6. Indicate the truth value for *A, B,* and *A* OR *B*. Identify the row in the truth table represented by the statements.

4. Let *A* be *Circles are round.*
Let *B* be *Birds fly.*

5. Let *A* be *Circles are square.*
Let *B* be *Birds fly.*

6. Let *A* be *A car has 18 wheels.*
Let *B* be *A square has 6 sides.*

Use the truth table for the negation of *A* for Exercises 7–8. Identify the row in the truth table illustrated by the statement and its negation.

7. Let *A* be *A carnation is a flower.*

8. Let *A* be *T is a vowel.*

Answer Exercises 9–10.

9. Consider these two statements:

A is the statement *The square root of 121 is 11.*

B is the statement *$14\frac{1}{2}$ is an even number.*

(a) What is the truth value for *A*?

(b) What is the truth value for *B*?

(c) Write the statement *A* AND *B.*

(d) What is the truth value of *A* AND *B*?

(e) Write the statement *A* OR *B.*

(f) What is the truth value of *A* OR *B*?

10. Consider these two statements:

A is the statement *The total voltage in a series circuit with two voltages is $V_T = V_1 + V_2$.*

B is the statement *The total resistance in a parallel circuit with two resistances is $R_T = R_1 + R_2$.*

(a) What is the truth value for *A*?

(b) What is the truth value for *B*?

(c) What is the statement for $\overline{B}$?

(d) What is the truth value for $\overline{B}$?

(e) Write the statement *A* AND $\overline{B}$.

(f) What is the truth value of *A* AND $\overline{B}$?

11. Challenge: Construct a truth table for the conjunction *A* AND *B* AND *C*. Use the headings shown below for your truth table.

A	*B*	*C*	*A* AND *B* AND *C*

Show all the possible combinations of true and false. When is *A* AND *B* AND *C* true?

INTRODUCING BOOLEAN ALGEBRA

Boolean algebra, developed by George Boole, shown above, was later adapted to switching circuits.

The study of Boolean algebra provides a foundation for the design and analysis of logic circuits. Although Boolean algebra was developed in the mid-19th century, it was adapted to circuit design in the 1940s. The adaptation of Boolean algebra to circuit design was done by Claude Shannon when he noticed the similarity between Boolean logic (with its two values) and telephone switching systems.

Earlier we examined the truth tables for statements with the AND and OR connectors, as well as the truth table for the negation of a variable. In this lesson, we will examine how these same truth tables can be applied to the input and output of logic circuits. We will use variables to represent the input and output conditions. Remember that these variables can take on only one of two possible values, designated by 1 or 0.

Although we are working with only two possible values, these two values can take on different interpretations. For example, we earlier used the interpretations *true* and *false*. The table below shows several possible interpretations for the truth values of 1 and 0.

1	0
On	Off
True	False
High input	Low input
Present	Absent
Current	No current

In Boolean algebra, there are three basic operations that can be performed on the input values: *AND, OR,* and *NOT.* These are called **Boolean operators.** The output of the operation is also a discrete value, 1 or 0, which represents the outcome of a logical statement.

The **AND** operation is performed on two input variables. There are two rules for finding the output from an AND operation:

- The output is 1 if both input variables have a value of 1.
- The output is 0 if either input variable has a value of 0.

The AND operation is also known as the **conjunction** of A and B. The symbol for the AND operation is a dot, "·" The statement A AND B is written as $A \cdot B$. The truth table for the AND operation is shown below.

A	B	$A \cdot B$
0	0	0
0	1	0
1	0	0
1	1	1

The OR operation is performed on two input variables. There are two rules for finding the output from an OR operation:

- The output is 1 if either input variable has a value of 1.
- The output is 0 if both variables have a value of 0.

The OR operation is also known as the **disjunction** of A and B. The symbol for the OR operation is a plus sign, "+". The statement A OR B is written as $A + B$. The truth table for the OR operation is shown below.

A	B	$A + B$
0	0	0
0	1	1
1	0	1
1	1	1

At this point, it may be helpful to take special note of the symbols for the connectors AND and OR.

- When two variables are joined by the connector AND, we say they are AND'ed, and we use a dot to join them.
- When two variables are joined by the connector OR, we say they are OR'ed, and we use a plus sign to join them.

The NOT operation is performed on a single input variable. The NOT operation simply changes the value from 0 to 1, or from 1 to 0.

- The output is 1 if the input is 0.
- The output is 0 if the input is 1.

The symbol for the NOT operation is a bar written above the input variable. The statement NOT A is written as $\overline{A}$. The truth table for the NOT operation is shown below:

A	$\overline{A}$
0	1
1	0

Examples A and B demonstrate how the AND and the OR operators can be applied to circuits. In these examples, the variables A and B represent switches that are either off (no current flowing) or on (current flowing).

EXAMPLE A

If a value of 1 indicates that a switch is on and a value of 0 indicates that a switch is off, what input values are necessary for the bulb to light?

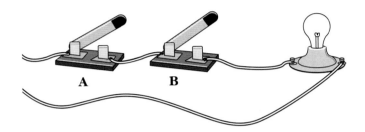

In this circuit, both switches must be on for the current to flow to the lightbulb. If either switch is off, the bulb will not light.

Both input variables must have a value of 1 for the current to flow to the lightbulb.

EXAMPLE B

If a value of 1 indicates that a switch is on and a value of 0 indicates that a switch is off, under what conditions will the bulb light?

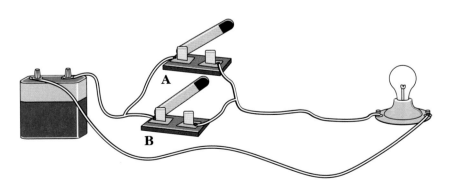

In the circuit, the lightbulb will be lit if one switch, the other switch, or both switches are turned on. The only way for the bulb not to light is for both switches to be off.

If either input variable has a value of 1, the current will flow to the bulb and it will light.

Take a moment to compare the results of Example A to the truth table for the AND operator and the results of Example B to the truth table for the OR operator. You will notice that the result of Example A is given by row 1 of the truth table for the AND operator. Similarly, notice that the result of Example B is given by rows 1, 2, and 3 of the truth table for the OR operator.

When Boolean algebra is applied to circuit design, the Boolean operators are called gates. A gate is a fundamental component of a digital circuit that takes input conditions, performs a logical operation (such as AND, OR, or NOT), and produces an output. The symbols used for AND, OR, and NOT gates are shown below.

The AND gate performs a logical AND operation on two input variables, A and B, to produce the output, $A \cdot B$.

A
B
$A \cdot B$
AND gate

The OR gate performs a logical OR operation on two input variables, A and B, to produce the output, $A + B$.

A
B
$A + B$
OR gate

The NOT gate performs a logical NOT operation on a single variable, A, to produce the output, $\overline{A}$.

A
$\overline{A}$
NOT gate

Given a diagram of logic gates and input variables, we can write an expression for the output of those gates. In Example C, the input variables are A and B. Read the diagram from left to right.

EXAMPLE C

Write an expression for the output of the logic gates shown. Follow the text below the diagram to interpret the gates.

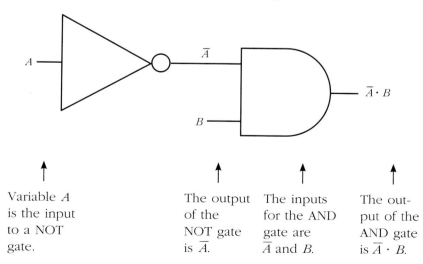

Variable A is the input to a NOT gate.

The output of the NOT gate is $\overline{A}$.

The inputs for the AND gate are $\overline{A}$ and B.

The output of the AND gate is $\overline{A} \cdot B$.

The expression for the output of the gates $\overline{A} \cdot B$.

A truth table can be created to find the value (1 or 0) of the output for every combination of input values. The truth table for $\overline{A} \cdot B$ is shown in Example D.

EXAMPLE D

Construct a truth table for $\overline{A} \cdot B$.

Notice that there is a column for each input variable. The output $\overline{A} \cdot B$ appears in the rightmost column of the truth table.

A	$\overline{A}$	B	$\overline{A} \cdot B$
0	1	0	0
0	1	1	1
1	0	0	0
1	0	1	0

Language Box

The statement $\overline{A} \cdot B$ is read as "NOT A and B."

The truth table for $\overline{A} \cdot B$ is shown above.

EXAMPLE E

Construct a truth table for $A \cdot B + C$.

As in the previous example, there is a column for each input variable and for the output $A \cdot B$. Notice that we entered the column for $A \cdot B$ before the column for C. The output $A \cdot B + C$ appears in the rightmost column of the truth table.

A	B	$A \cdot B$	C	$A \cdot B + C$
0	0	0	0	0
0	1	0	0	0
1	0	0	0	0
1	1	1	0	1
0	0	0	1	1
0	1	0	1	1
1	0	0	1	1
1	1	1	1	1

EXAMPLE F

Draw the logic gates for $A \cdot B + C$.

As in Example C, we first draw the logic gate for $A \cdot B$ and attach that output to the OR logic gate with C. The result is shown in the figure below.

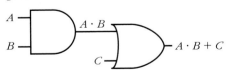

The symbols and rules for using the logical operators are summarized below.

KEY IDEA

The **AND** operator is denoted by the "·" symbol. The symbol for the AND gate is ⊐D.
The output of an AND operation is:
- 1 if both input variables have a value of 1.
- 0 if either input variable has a value of 0.

The **OR** operator is denoted by the "+" symbol. The symbol for the OR gate is ⊐D .
The output of an OR operation is:
- 1 if either input variable has a value of 1.
- 0 if both input variables have a value of 0.

The **NOT** operation is denoted by a bar, "¯", over the variable. The symbol for the NOT gate is ▷∘ . The output of a NOT operation is:
- 1 if the input is 0.
- 0 if the input is 1.

Two other gates are the NAND and NOR gates. An NAND gate is an AND gate followed by a NOT gate. An NOR gate is an OR gate followed by a NOT gate. These gates and their truth tables are shown below. Notice that the small circle on the NAND and NOR gates represents the NOT operation.

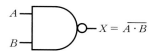

$X = \overline{A \cdot B}$

$X = \overline{A + B}$

NAND gate

NOR gate

A	B	$A \cdot B$	$\overline{A \cdot B}$
0	0	0	1
0	1	0	1
1	0	0	1
1	1	1	0

A	B	$A + B$	$\overline{A + B}$
0	0	0	1
0	1	1	0
1	0	1	0
1	1	1	0

EXAMPLE G

Write an expression for the output of the logic gates shown.

Follow the text below the diagram to interpret the gates.

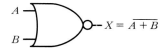

$\overline{A \cdot B}$

$\overline{A \cdot B} + B$

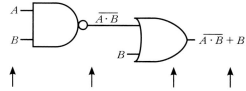

Variables A and B are input in a NAND gate.

The output of the NAND gate is $\overline{A \cdot B}$

The inputs for the OR gate are $\overline{A \cdot B}$ and B.

The output of the OR gate is $\overline{A \cdot B} + B$.

The expression for the output of the gates is $\overline{A \cdot B} + B$.

EXAMPLE H

Construct a truth table for $\overline{A \cdot B} + B$.

A	B	$A \cdot B$	$\overline{A \cdot B}$	$\overline{A \cdot B} + B$
0	0	0	1	1
0	1	0	1	1
1	0	0	1	1
1	1	1	0	1

The truth table for $\overline{A \cdot B} + B$ is shown above.

EXERCISES 20-2

In Exercises 1–4, the input variables are *A* and *B*. Use a truth table to determine the value of the output expression.

1. $A \cdot B$ when
$A = 1, B = 0$

2. $A + B$ when
$A = 1, B = 1$

3. $\overline{A} \cdot B$ when
$A = 0, B = 0$

4. $\overline{A} + \overline{B}$ when
$A = 1, B = 0$

In Exercises 5–7, write an expression for the output, *X*, of the gates shown. Then create a truth table for the output expression.

5.

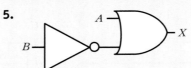

6.

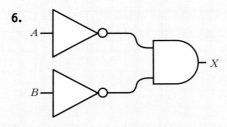

7.

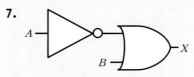

In Exercises 8–9, construct a truth table and a logic gate for each of the statements.

8. $\overline{A} + B \cdot A$

9. $A \cdot B + A \cdot B$

10. Challenge: Write an expression for the output, X, of the gates. Then create a truth table for the output.

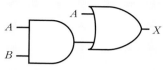

LAWS OF BOOLEAN ALGEBRA

> The laws of logic are composed of postulates, which are assumed, and theorems, which can be proven using the postulates. The postulates and theorems, along with the AND, OR, and NOT operators, make up Boolean Algebra.

In this lesson, we will see that expressions in Boolean algebra follow a set of laws. These laws, or properties as they are sometimes called, govern the way complex expressions are simplified. Some laws have names you are familiar with, for example, the associative and commutative laws. These laws involve grouping and order, just as they do in regular algebra. Other laws have names that are unique to Boolean algebra.

An important characteristic of the laws of logic is the principle of duality. That is, the laws are written in pairs, called *duals*.
To write the dual of a statement, replace all AND operators with OR operators, all OR operators with AND operators, all 0s with 1s, and all 1s with 0s.

Because Boolean algebra has only two possible values, the result of any operation on two input variables is always a 0 or a 1. In mathematics, this property is called *closure*. It is an important property of Boolean algebra. The list that follows contains other important laws of Boolean algebra.

Laws of Boolean Algebra

Commutative Law	$A + B = B + A$ $A \cdot B = B \cdot A$	
Associative Law	$(A + B) + C = A + (B + C)$ $(A \cdot B) \cdot C = A \cdot (B \cdot C)$	
Distributive Law	$A + (B \cdot C) = (A + B) \cdot (A + C)$ $A \cdot (B + C) = (A \cdot B) + (A \cdot C)$	
Identity Laws	$A + 0 = A$ $A \cdot 1 = A$	$A + 1 = 1$ $A \cdot 0 = 0$
Complementation Law	$A + \bar{A} = 1$ $A \cdot \bar{A} = 0$	
Idempotent Law (Tautology)	$A + A = A$ $A \cdot A = A$	
Law of Involution (Double Complementation)	$\bar{\bar{A}} = A$	
Law of Absorption	$A + (A \cdot B) = A$ $A \cdot (A + B) = A$	
DeMorgan's Law	$\overline{A \cdot B} = \bar{A} + \bar{B}$ $\overline{A + B} = \bar{A} \cdot \bar{B}$	

Helpful Hint

The double bar over the variable A is read *NOT (NOT A)*. The expressions $\bar{\bar{A}}$, NOT $\bar{A}$, and NOT (NOT A) are all equivalent.

An important application of Boolean logic is in digital design. A digital design can be described using expressions with variables and Boolean operators. Simplification of expressions using the laws of logic is important because simpler circuits can be built using simpler expressions.

Simplify the expression $(\overline{A} \cdot B) + (A \cdot B)$.

$(\overline{A} \cdot B) + (A \cdot B) =$

$\qquad (\overline{A} + A) \cdot B =$ ← Apply the Distributive Law.

$\qquad\qquad\qquad 1 \cdot B =$ ← Apply the Complementation Law.

$\qquad\qquad\qquad\qquad B$ ← Apply the appropriate Identity Law.

The expression $(\overline{A} \cdot B) + (A \cdot B)$ has been simplified to the single variable B.

$(\overline{A} \cdot B) + (A \cdot B) = B$

EXAMPLE A

Simplify the expression $\overline{\overline{A}} \cdot A + B$.

$A \cdot A + B =$

$A \cdot A + B =$ ← Apply the Law of Involution.

$\qquad A + B$ ← Apply the Idempotent Law.

The expression $A + B$ is simpler than the expression $\overline{\overline{A}} \cdot A + B$ because it has only two variable terms and one operator.

$\overline{\overline{A}} \cdot A + B = A + B$

EXAMPLE B

Simplify the expression $A \cdot B \cdot \overline{A}$.

$A \cdot B \cdot \overline{A} =$

$A \cdot \overline{A} \cdot B =$ ← Apply the Commutative Law.

$\qquad 0 \cdot B =$ ← Apply the Complementation Law.

$\qquad\qquad 0$ ← Apply the appropriate Identity Law.

The expression $A \cdot B \cdot \overline{A}$ has been simplified to the constant 0.

$A \cdot B \cdot \overline{A} = 0$

When a value of 0 or 1 appears in an expression, it represents a constant input rather than a variable. In Example C, the value 1 in the expression represents a constant input of 1.

EXAMPLE C

Simplify the expression $A + (\overline{B} + 1)$.

$A + (\overline{B} + 1) =$

$\qquad A + 1 =$ ← Apply the appropriate Identity Law.

$\qquad\qquad 1$ ← Apply the appropriate Identity Law.

$A + (\overline{B} + 1) = 1$

EXAMPLE D

Write the dual of
the statement
$A + (\overline{B} + 1) = 1.$

To write the dual of the statement $A + (\overline{B} + 1) = 1$, replace each "+" sign
with a "·"; replace each 1 with a 0:

$A + \overline{B} + 1 = 1$

$$A \cdot \overline{B} \cdot 0 = 0$$

Both the original statement and its dual are true.

The dual of the statement $A + (\overline{B} + 1) = 1$ is $A \cdot \overline{B} \cdot 0 = 0$.

EXERCISES 20-3

**Use the laws of Boolean algebra to simplify each expression. Give
an explanation for each step of your work.**

1. $A + (\overline{A} \cdot C)$

2. $A + \overline{A} + \overline{\overline{A}}$

3. $A + B + C + \overline{\overline{A}} + \overline{\overline{B}} + \overline{\overline{C}}$

4. $A \cdot (A \cdot B)$

5. $A \cdot (A \cdot (A + B))$

6. $(\overline{A} + 1) \cdot (B + C)$

7. $A + (\overline{A} \cdot B \cdot C)$

8. $(A \cdot A) + A + 1$

9. $A + \overline{A} \cdot B$

**Use truth tables to verify the laws of Boolean algebra in Exercises
10–11.**

10. Prove the Law of Absorption: $A \cdot (A + B) = A$.

11. Prove the Associative Law: $(A + B) + C = A + (B + C)$.

Write the dual of each statement in Exercises 12–13.

12. $(\overline{A} + 0) \cdot (A + 1) = 0$

13. $(A \cdot A) + (1 + A) = 1$

14. **Challenge:** Simplify the expression $A \cdot B + (B \cdot C) \cdot (B + C)$ so that
    the result contains three variable terms and two operators.

USING DeMORGAN'S LAW

DeMorgan's Law is named after English mathematician Augustus DeMorgan (1806–1871), a friend of George Boole.

The final law of logic that we will discuss is **DeMorgan's Law.** DeMorgan's Law is composed of dual statements that tell us how to find the negation, or complement, of an expression.

Just as a short bar over a variable indicates the negation of that variable, a long bar over an entire expression indicates the negation of the entire expression. For example, $\overline{A}$ indicates the negation of variable A, and $\overline{A + B}$ indicates the negation of $A + B$. DeMorgan's Law states that the complement of an AND or OR expression is found by changing the operator (from AND to OR, or from OR to AND) and writing the complement of each individual variable. Compare this explanation to the statement of the law given below.

DeMorgan's Law

$$(1)\ \overline{A \cdot B} = \overline{A} + \overline{B}$$
$$(2)\ \overline{A + B} = \overline{A} \cdot \overline{B}$$

DeMorgan's Law is sometimes referred to as *breaking the bar and changing the operation.* Comparing the left side of each equation to the right side, you can see how this description applies. In both statements, the long bar is "broken" into smaller bars and the operation is changed.

Statement (1) of DeMorgan's Law states that the complement of A AND B is equal to $\overline{A}$ OR $\overline{B}$. We can show that this is true by looking at the truth tables for $\overline{A \cdot B}$ and $\overline{A} + \overline{B}$. Two logical expressions are equivalent if their truth tables are the same.

Construct truth tables for $\overline{A \cdot B}$ and $\overline{A} + \overline{B}$.

A	B	$A \cdot B$	$\overline{A \cdot B}$
0	0	0	1
0	1	0	1
1	0	0	1
1	1	1	0

A	B	$\overline{A}$	$\overline{B}$	$\overline{A} + \overline{B}$
0	0	1	1	1
0	1	1	0	1
1	0	0	1	1
1	1	0	0	0

The last columns of both truth tables are the same; $\overline{A \cdot B}$ and $\overline{A} + \overline{B}$ are equivalent.

Statement (2) of DeMorgan's Law states that the complement of $A + B$ is equal to the complement of A AND the complement of B.

We can show that this is true by looking at the truth tables for $\overline{A + B}$ and $\overline{A} \cdot \overline{B}$:

A	B	$A + B$	$\overline{A + B}$
0	0	0	1
0	1	1	0
1	0	1	0
1	1	1	0

A	B	$\overline{A}$	$\overline{B}$	$\overline{A} \cdot \overline{B}$
0	0	1	1	1
0	1	1	0	0
1	0	0	1	0
1	1	0	0	0

The last columns of both truth tables are the same; $\overline{A + B}$ and $\overline{A} \cdot \overline{B}$ are equivalent.

Apply DeMorgan's Law to the expression $\overline{A \cdot B} + C$.

$\overline{A \cdot B} + C =$

$\overline{A} + \overline{B} + C$ ⟵ Apply DeMorgan's Law to $\overline{A \cdot B}$.

So, $\overline{A \cdot B} + C = \overline{A} + \overline{B} + C$.

DeMorgan's Law may be applied more than once to an expression. This is shown in Example A.

EXAMPLE A

Apply DeMorgan's
Law to the
expression
$\overline{A \cdot B + C}$.

$\overline{A \cdot B + C} =$

$\overline{A \cdot B} \cdot \overline{C} =$ ⟵ Apply DeMorgan's Law to $\overline{A \cdot B + C}$. $A \cdot B \cdot C$

$\overline{A} + \overline{B} \cdot \overline{C}$ ⟵ Apply DeMorgan's Law to $\overline{A \cdot B}$.

$\overline{A \cdot B + C} = \overline{A} + \overline{B} \cdot \overline{C}$

In Example B, you will see how DeMorgan's Law can be used along with the other laws of logic to simplify an expression.

EXAMPLE B

Simplify the
expression $\overline{\overline{A} + B}$.

Apply DeMorgan's Law to the longest bar in an expression first.

$\overline{\overline{A} + B} =$

$\overline{\overline{A}} \cdot \overline{B} =$ ⟵ Apply DeMorgan's Law to $\overline{\overline{A} + B}$.

$A \cdot \overline{B}$ ⟵ Apply the Law of Involution to $\overline{\overline{A}}$.

$\overline{\overline{A} + B} = A \cdot \overline{B}$

DeMorgan's Law can also be applied to expressions with more than two variables. This is done by changing the operations and complementing each term. For example, the dual statements of the law with three variables are shown below.

$$\overline{A \cdot B \cdot C} = \overline{A} + \overline{B} + \overline{C}$$
$$\overline{A + B + C} = \overline{A} \cdot \overline{B} \cdot \overline{C}$$

The first of these statements is used in Example C.

EXAMPLE C

Simplify the expression $\overline{A \cdot B \cdot C} + C$.

$$\overline{A \cdot B \cdot C} + C =$$
$$\overline{A} + \overline{B} + \overline{C} + C = \quad \longleftarrow \text{Apply DeMorgan's Law to } \overline{A \cdot B \cdot C}.$$
$$\overline{A} + \overline{B} + 1 \quad \longleftarrow \text{Apply the Complementation Law to } \overline{C} + C.$$

$$\overline{A \cdot B \cdot C} + C = \overline{A} + \overline{B} + 1$$

EXERCISES 20-4

Simplify each expression. Give an explanation for each step of your work.

1. $\overline{A + B \cdot C}$ 2. $A + \overline{B \cdot C}$

3. $\overline{A + B + C} \cdot A$ 4. $\overline{\overline{A \cdot B}}$

5. $\overline{\overline{A} + B + C}$ 6. $\overline{A} \cdot \overline{B} \cdot C$

7. **Challenge:** $\overline{A + B} \cdot \overline{A \cdot B}$

CHAPTER 20 REVIEW EXERCISES

In Exercises 1–4, use the following two statements.

A is the statement *The total resistance in a parellel circuit with two resistances is* $\dfrac{1}{R_T} = \dfrac{1}{R_1} + \dfrac{1}{R_2}$
B is the statement *Some rectangles have five sides.*

1. What is the truth value for A?

2. What is the truth value for B?

3. What is the truth value for A AND B?

4. What is the truth value for A NOR B?

In Exercises 5–6, write an expression for the output, *X*, of the gates shown. Then create a truth table for the output expression.

5.
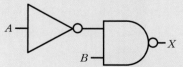

6.
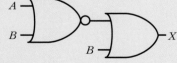

Answer Exercises 7–10.

7. Use a truth table to prove the Law of Involution: $\overline{\overline{A}} = A$.

8. Use the laws of Boolean algebra to simplify the expression $\overline{A} + 0 + A \cdot 1$. Give an explanation for each step of your work.

9. Use the laws of Boolean algebra to simplify the expression $A \cdot B + \overline{A} + B$. Give an explanation for each step of your work.

10. Write the dual of the statement $\overline{A} + 0 + A = 1$.

APPENDIX A

Mathematical Symbols

Operation	Symbol	Phrases That Might Be Used
addition	$+$	sum, added to, plus, increased by, more than
subtraction	$-$	difference, subtracted from, minus, decreased by, less than
multiplication	$\times, \cdot$	product, times
division	$\div, \overline{)\,}$	quotient, divided by

Symbol	Meaning
$=$	is equal to, equals, is the same as
$\neq$	is not equal to, is not the same as
$\approx$	is approximately equal to
$<$	is less than
$>$	is greater than
$\leq$	is less than or equal to
$\geq$	is greater than or equal to
$[\ ]$	brackets, a grouping symbol
$(\ )$	parentheses, a grouping symbol
$1.\overline{3}$	repeating decimal 1.33333 . . .
$|\ |$	absolute value; $|x|$ means absolute value of x
$\sqrt{}$	radical sign, used to indicate nonnegative square root; $\sqrt{x}$ means nonnegative square root of a nonnegative number x
$\sqrt[3]{}$	radical sign with an index of 3, used to indicate cube root; $\sqrt[3]{x}$ means cube root of x
$\%$	percent
$\ldots$	continues on
$^{\circ}$	degree(s)
π	pi, approximately 3.14
$\parallel$	is parallel to
$\perp$	is perpendicular to
$\cong$	is congruent to
$\sim$	is similar to
$\overset{\bullet}{P}$	point P

Symbol	Meaning
$\overline{AB}$	line segment AB
AB	length of $\overline{AB}$
$\overrightarrow{AB}$	ray AB
$\overleftrightarrow{AB}$	line AB
$\angle A$	angle with vertex point A
$m\angle A$	measure of angle A
	right angle indicated by small square at vertex
θ	Greek letter *theta*, often used to name an angle
α	Greek letter *alpha*, often used to name an angle
β	Greek letter *beta*, often used to name an angle
	directed line segment AB
$\langle v_x, v_y \rangle$	component form of a vector $\boldsymbol{v}$
$|\boldsymbol{v}|$	magnitude of vector $\boldsymbol{v}$
	AND gate
	OR gate
	NOT gate
	NAND gate
	NOR gate

A P P E N D I X B

Addition Facts

0 + 0 0	0 + 1 1	0 + 2 2	0 + 3 3	0 + 4 4	0 + 5 5	0 + 6 6	0 + 7 7	0 + 8 8	0 + 9 9
1 + 0 1	1 + 1 2	1 + 2 3	1 + 3 4	1 + 4 5	1 + 5 6	1 + 6 7	1 + 7 8	1 + 8 9	1 + 9 10
2 + 0 2	2 + 1 3	2 + 2 4	2 + 3 5	2 + 4 6	2 + 5 7	2 + 6 8	2 + 7 9	2 + 8 10	2 + 9 11
3 + 0 3	3 + 1 4	3 + 2 5	3 + 3 6	3 + 4 7	3 + 5 8	3 + 6 9	3 + 7 10	3 + 8 11	3 + 9 12
4 + 0 4	4 + 1 5	4 + 2 6	4 + 3 7	4 + 4 8	4 + 5 9	4 + 6 10	4 + 7 11	4 + 8 12	4 + 9 13
5 + 0 5	5 + 1 6	5 + 2 7	5 + 3 8	5 + 4 9	5 + 5 10	5 + 6 11	5 + 7 12	5 + 8 13	5 + 9 14
6 + 0 6	6 + 1 7	6 + 2 8	6 + 3 9	6 + 4 10	6 + 5 11	6 + 6 12	6 + 7 13	6 + 8 14	6 + 9 15
7 + 0 7	7 + 1 8	7 + 2 9	7 + 3 10	7 + 4 11	7 + 5 12	7 + 6 13	7 + 7 14	7 + 8 15	7 + 9 16
8 + 0 8	8 + 1 9	8 + 2 10	8 + 3 11	8 + 4 12	8 + 5 13	8 + 6 14	8 + 7 15	8 + 8 16	8 + 9 17
9 + 0 9	9 + 1 10	9 + 2 11	9 + 3 12	9 + 4 13	9 + 5 14	9 + 6 15	9 + 7 16	9 + 8 17	9 + 9 18

Subtraction Facts

1	2	3	4	5	6	7	8	9	10
-1	-1	-1	-1	-1	-1	-1	-1	-1	-1
0	1	2	3	4	5	6	7	8	9

2	3	4	5	6	7	8	9	10	11
-2	-2	-2	-2	-2	-2	-2	-2	-2	-2
0	1	2	3	4	5	6	7	8	9

3	4	5	6	7	8	9	10	11	12
-3	-3	-3	-3	-3	-3	-3	-3	-3	-3
0	1	2	3	4	5	6	7	8	9

4	5	6	7	8	9	10	11	12	13
-4	-4	-4	-4	-4	-4	-4	-4	-4	-4
0	1	2	3	4	5	6	7	8	9

5	6	7	8	9	10	11	12	13	14
-5	-5	-5	-5	-5	-5	-5	-5	-5	-5
0	1	2	3	4	5	6	7	8	9

6	7	8	9	10	11	12	13	14	15
-6	-6	-6	-6	-6	-6	-6	-6	-6	-6
0	1	2	3	4	5	6	7	8	9

7	8	9	10	11	12	13	14	15	16
-7	-7	-7	-7	-7	-7	-7	-7	-7	-7
0	1	2	3	4	5	6	7	8	9

8	9	10	11	12	13	14	15	16	17
-8	-8	-8	-8	-8	-8	-8	-8	-8	-8
0	1	2	3	4	5	6	7	8	9

9	10	11	12	13	14	15	16	17	18
-9	-9	-9	-9	-9	-9	-9	-9	-9	-9
0	1	2	3	4	5	6	7	8	9

Multiplication Facts

0	1	2	3	4	5	6	7	8	9
×0	×0	×0	×0	×0	×0	×0	×0	×0	×0
0	0	0	0	0	0	0	0	0	0

0	1	2	3	4	5	6	7	8	9
×1	×1	×1	×1	×1	×1	×1	×1	×1	×1
0	1	2	3	4	5	6	7	8	9

0	1	2	3	4	5	6	7	8	9
×2	×2	×2	×2	×2	×2	×2	×2	×2	×2
0	2	4	6	8	10	12	14	16	18

0	1	2	3	4	5	6	7	8	9
×3	×3	×3	×3	×3	×3	×3	×3	×3	×3
0	3	6	9	12	15	18	21	24	27

0	1	2	3	4	5	6	7	8	9
×4	×4	×4	×4	×4	×4	×4	×4	×4	×4
0	4	8	12	16	20	24	28	32	36

0	1	2	3	4	5	6	7	8	9
×5	×5	×5	×5	×5	×5	×5	×5	×5	×5
0	5	10	15	20	25	30	35	40	45

0	1	2	3	4	5	6	7	8	9
×6	×6	×6	×6	×6	×6	×6	×6	×6	×6
0	6	12	18	24	30	36	42	48	54

0	1	2	3	4	5	6	7	8	9
×7	×7	×7	×7	×7	×7	×7	×7	×7	×7
0	7	14	21	28	35	42	49	56	63

0	1	2	3	4	5	6	7	8	9
×8	×8	×8	×8	×8	×8	×8	×8	×8	×8
0	8	16	24	32	40	48	56	64	72

0	1	2	3	4	5	6	7	8	9
×9	×9	×9	×9	×9	×9	×9	×9	×9	×9
0	9	18	27	36	45	54	63	72	81

Division Facts

$2 \overline{)2}^{\,1}$ $2 \overline{)4}^{\,2}$ $2 \overline{)6}^{\,3}$ $2 \overline{)8}^{\,4}$ $2 \overline{)10}^{\,5}$ $2 \overline{)12}^{\,6}$ $2 \overline{)14}^{\,7}$ $2 \overline{)16}^{\,8}$ $2 \overline{)18}^{\,9}$

$3 \overline{)3}^{\,1}$ $3 \overline{)6}^{\,2}$ $3 \overline{)9}^{\,3}$ $3 \overline{)12}^{\,4}$ $3 \overline{)15}^{\,5}$ $3 \overline{)18}^{\,6}$ $3 \overline{)21}^{\,7}$ $3 \overline{)24}^{\,8}$ $3 \overline{)27}^{\,9}$

$4 \overline{)4}^{\,1}$ $4 \overline{)8}^{\,2}$ $4 \overline{)12}^{\,3}$ $4 \overline{)16}^{\,4}$ $4 \overline{)20}^{\,5}$ $4 \overline{)24}^{\,6}$ $4 \overline{)28}^{\,7}$ $4 \overline{)32}^{\,8}$ $4 \overline{)36}^{\,9}$

$5 \overline{)5}^{\,1}$ $5 \overline{)10}^{\,2}$ $5 \overline{)15}^{\,3}$ $5 \overline{)20}^{\,4}$ $5 \overline{)25}^{\,5}$ $5 \overline{)30}^{\,6}$ $5 \overline{)35}^{\,7}$ $5 \overline{)40}^{\,8}$ $5 \overline{)45}^{\,9}$

$6 \overline{)6}^{\,1}$ $6 \overline{)12}^{\,2}$ $6 \overline{)18}^{\,3}$ $6 \overline{)24}^{\,4}$ $6 \overline{)30}^{\,5}$ $6 \overline{)36}^{\,6}$ $6 \overline{)42}^{\,7}$ $6 \overline{)48}^{\,8}$ $6 \overline{)54}^{\,9}$

$7 \overline{)7}^{\,1}$ $7 \overline{)14}^{\,2}$ $7 \overline{)21}^{\,3}$ $7 \overline{)28}^{\,4}$ $7 \overline{)35}^{\,5}$ $7 \overline{)42}^{\,6}$ $7 \overline{)49}^{\,7}$ $7 \overline{)56}^{\,8}$ $7 \overline{)63}^{\,9}$

$8 \overline{)8}^{\,1}$ $8 \overline{)16}^{\,2}$ $8 \overline{)24}^{\,3}$ $8 \overline{)32}^{\,4}$ $8 \overline{)40}^{\,5}$ $8 \overline{)48}^{\,6}$ $8 \overline{)56}^{\,7}$ $8 \overline{)64}^{\,8}$ $8 \overline{)72}^{\,9}$

$9 \overline{)9}^{\,1}$ $9 \overline{)18}^{\,2}$ $9 \overline{)27}^{\,3}$ $9 \overline{)36}^{\,4}$ $9 \overline{)45}^{\,5}$ $9 \overline{)54}^{\,6}$ $9 \overline{)63}^{\,7}$ $9 \overline{)72}^{\,8}$ $9 \overline{)81}^{\,9}$

APPENDIX C

Method for Computing the Square Root of a Number

To compute the square root of a number, such as $\sqrt{2916}$, the following major steps are necessary.

STEP 1

Pair the digits of the radicand, 2916, proceeding away from the decimal point. (Remember, a number like 2916 is the decimal number 2916 without the decimal point.)

$$\sqrt{2916}$$

STEP 2

Determine what number **squared** will be the closest to the value of the first two digits, without going over their value. In this case, the number would be **5**, since 5×5 equals 25 and 25 is less than 29; 6×6 would be too great a value. Place the **5** above the **first two** digits of the radicand.

$$\overset{5}{\sqrt{2916}}$$

STEP 3

Under the first pair (29), square the value 5. Place that result, 25, under the 29. The value 25 is the largest perfect square that is less than 29.

$$\begin{array}{c} 5 \\ \sqrt{2916} \\ 25 \end{array}$$

STEP 4

Subtract the 25 from 29 and obtain the difference of 4.

$$\begin{array}{c} 5 \\ \sqrt{2916} \\ 25 \\ \hline 4 \end{array}$$

STEP 5

Bring down the next pair of digits from the radicand, which in this case is 16.

$$\begin{array}{c} 5 \\ \sqrt{2916} \\ 25 \\ \hline 416 \end{array}$$

STEP 6

Double the **5** (the partial answer so far), and place that product (10), on the outside of the 416, as shown below.

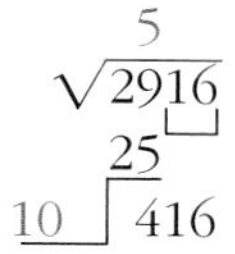

$$
\begin{array}{r}
5 \\
\sqrt{2916} \\
25 \\
\underline{10}\ \rfloor\ 416
\end{array}
$$

STEP 7

Disregard (or cover up for the moment) the *last digit* in 416. Divide 10 into the 41 that is showing, to obtain **4.** Place this **4** on the top above the 16 in the radicand.

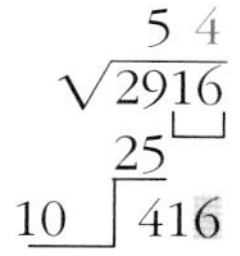

$$
\begin{array}{r}
5\ 4 \\
\sqrt{2916} \\
25 \\
\underline{10}\ \rfloor\ 416
\end{array}
$$

STEP 8

This **4** must also be placed just after the 10, as shown below.

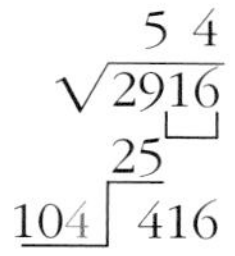

$$
\begin{array}{r}
5\ 4 \\
\sqrt{2916} \\
25 \\
\underline{104}\ \rfloor\ 416
\end{array}
$$

STEP 9

Multiply the 104 by **4** (as in long division) to obtain 416.

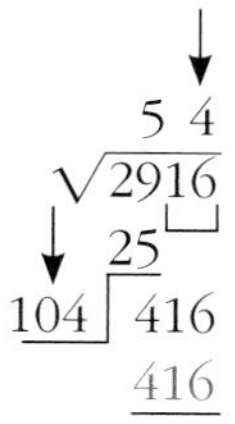

$$
\begin{array}{r}
5\ 4 \\
\sqrt{2916} \\
25 \\
\underline{104}\ \rfloor\ 416 \\
416
\end{array}
$$

STEP 10

Subtract the 416 to obtain a remainder of zero. In this illustration, 54 is found to be the **exact** square root of 2916.

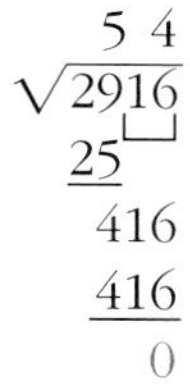

$$
\begin{array}{r}
5\ 4 \\
\sqrt{2916} \\
25 \\
416 \\
\underline{416} \\
0
\end{array}
$$

APPENDIX D

Common Percents, Fractions, and Decimal Equivalents

Percent	Decimal	Fraction
10%	0.1	$\frac{1}{10}$
20%	0.2	$\frac{1}{5}$
25%	0.25	$\frac{1}{4}$
30%	0.3	$\frac{3}{10}$
$33\frac{1}{3}\%$	≈ 0.333	$\frac{1}{3}$
40%	0.4	$\frac{2}{5}$
50%	0.5	$\frac{1}{2}$
60%	0.6	$\frac{3}{5}$
$66\frac{2}{3}\%$	≈ 0.667	$\frac{2}{3}$
70%	0.7	$\frac{7}{10}$
75%	0.75	$\frac{3}{4}$
80%	0.8	$\frac{4}{5}$
90%	0.9	$\frac{9}{10}$
100%	1.0	$\frac{1}{1}$

APPENDIX E

Formulas in Geometry

Rectangle

Area

$A = bh$, where b is the base and h is the height of the rectangle, or $A = lw$, where l is the length and w is the width of the rectangle.

Triangle

Area

$A = \frac{1}{2}bh$, where b is the base and h is the height of the triangle.

Parallelogram

Area

$A = bh$, where b is the base and h is the height of the parallelogram.

Trapezoid

Area

$A = \frac{1}{2}(b_1 + b_2)h$, where b_1 and b_2 are bases 1 and 2 and h is the height of the trapezoid.

Circle

Circumference

$C = \pi d$ or $C = 2\pi r$, where d is the diameter and r is the radius of the circle.

Area

$A = \pi r^2$, where r is the radius of the circle.

Right Prism

Lateral Area

$LA = ph$, where p is the perimeter of the base and h is the height of the prism.

Surface Area

$SA = LA + 2B$, where B is the area of the base.

Volume

$V = Bh$, where B is the area of the base and h is the height of the prism.

Right Cylinder

Lateral Area

$LA = 2\pi rh$, where r is the radius of the cylinder and h is the height of the cylinder.

Surface Area

$SA = LA + 2B = 2\pi rh + 2\pi r^2$, where B is the area of the base.

Volume

$V = Bh = \pi r^2 h$, where B is the area of the base, h is the height, and r is the radius of the cylinder.

Right Cone

Lateral Area

$LA = \pi r\ell$, where r is the radius of the base and ℓ is the slant height of the cone.

Surface Area

$SA = LA + B = \pi r\ell + \pi r^2$, where B is the area of the base of the cone.

Volume

$V = \frac{1}{3}Bh = \frac{1}{3}\pi r^2 h$, where B is the area of the base, r is the radius, and h is the height of the cone.

Regular Pyramid

Lateral Area

$LA = \frac{1}{2}p\ell$, where p is the perimeter of the base and ℓ is the slant height of the pyramid.

Surface Area

$SA = LA + B$, where B is the area of the base of the pyramid.

Volume

$V = \frac{1}{3}Bh$, where B is the area of the base and h is the height of the pyramid.

Sphere

Surface Area

$SA = 4\pi r^2$, where r is the radius of the sphere.

Volume

$V = \frac{4}{3}\pi r^3$, where r is the radius of the sphere.

APPENDIX F

Values of the Trigonometric Functions

Angle	Sin	Cos	Tan	Angle	Sin	Cos	Tan	Angle	Sin	Cos	Tan
0°	0.0000	1.0000	0.0000								
1°	0.0175	0.9998	0.0175	31°	0.5150	0.8572	0.6009	61°	0.8746	0.4848	1.8040
2°	0.0349	0.9994	0.0349	32°	0.5299	0.8480	0.6249	62°	0.8829	0.4695	1.8807
3°	0.0523	0.9986	0.0524	33°	0.5446	0.8387	0.6494	63°	0.8910	0.4540	1.9626
4°	0.0698	0.9976	0.0699	34°	0.5592	0.8290	0.6745	64°	0.8988	0.4384	2.0503
5°	0.0872	0.9962	0.0875	35°	0.5736	0.8192	0.7002	65°	0.9063	0.4226	2.1445
6°	0.1045	0.9945	0.1051	36°	0.5878	0.8090	0.7265	66°	0.9135	0.4067	2.2460
7°	0.1219	0.9925	0.1228	37°	0.6018	0.7986	0.7536	67°	0.9205	0.3907	2.3559
8°	0.1392	0.9903	0.1405	38°	0.6157	0.7880	0.7813	68°	0.9272	0.3746	2.4751
9°	0.1564	0.9877	0.1584	39°	0.6293	0.7771	0.8098	69°	0.9336	0.3584	2.6051
10°	0.1736	0.9848	0.1763	40°	0.6428	0.7660	0.8391	70°	0.9397	0.3420	2.7475
11°	0.1908	0.9816	0.1944	41°	0.6561	0.7547	0.8693	71°	0.9455	0.3256	2.9042
12°	0.2079	0.9781	0.2126	42°	0.6691	0.7431	0.9004	72°	0.9511	0.3090	3.0777
13°	0.2250	0.9744	0.2309	43°	0.6820	0.7314	0.9325	73°	0.9563	0.2924	3.2709
14°	0.2419	0.9703	0.2493	44°	0.6947	0.7193	0.9657	74°	0.9613	0.2756	3.4874
15°	0.2588	0.9659	0.2679	45°	0.7071	0.7071	1.0000	75°	0.9659	0.2588	3.7321
16°	0.2756	0.9613	0.2867	46°	0.7193	0.6947	1.0355	76°	0.9703	0.2419	4.0108
17°	0.2924	0.9563	0.3057	47°	0.7314	0.6820	1.0724	77°	0.9744	0.2250	4.3315
18°	0.3090	0.9511	0.3249	48°	0.7431	0.6691	1.1106	78°	0.9781	0.2079	4.7046
19°	0.3256	0.9455	0.3443	49°	0.7547	0.6561	1.1504	79°	0.9816	0.1908	5.1446
20°	0.3420	0.9397	0.3640	50°	0.7660	0.6428	1.1918	80°	0.9848	0.1736	5.6713
21°	0.3584	0.9336	0.3839	51°	0.7771	0.6293	1.2349	81°	0.9877	0.1564	6.3138
22°	0.3746	0.9272	0.4040	52°	0.7880	0.6157	1.2799	82°	0.9903	0.1392	7.1154
23°	0.3907	0.9205	0.4245	53°	0.7986	0.6018	1.3270	83°	0.9925	0.1219	8.1443
24°	0.4067	0.9135	0.4452	54°	0.8090	0.5878	1.3764	84°	0.9945	0.1045	9.5144
25°	0.4226	0.9063	0.4663	55°	0.8192	0.5736	1.4281	85°	0.9962	0.0872	11.4301
26°	0.4384	0.8988	0.4877	56°	0.8290	0.5592	1.4826	86°	0.9976	0.0698	14.3007
27°	0.4540	0.8910	0.5095	57°	0.8387	0.5446	1.5399	87°	0.9986	0.0523	19.0811
28°	0.4695	0.8829	0.5317	58°	0.8480	0.5299	1.6003	88°	0.9994	0.0349	28.6363
29°	0.4848	0.8746	0.5543	59°	0.8572	0.5150	1.6643	89°	0.9998	0.0175	57.2900
30°	0.5000	0.8660	0.5774	60°	0.8660	0.5000	1.7321	90°	1.0000	0.0000	undefined

GLOSSARY

WORD	DEFINITION	EXAMPLE OR EXPLANATION
1's complement (of a binary number)	A binary number in which every digit is the inverse of the original digit.	Example: The 1's complement of 1011_2 is 0100_2.
2's complement (of a binary number)	The sum of the 1's complement of the binary number and 1.	The opposite of a binary number is the 2's complement of that number.
Absolute value	The distance that a number is from zero on the number line. It is always positive or zero.	The absolute value of a number is written: $\lvert-5\rvert = 5$, $\lvert5\rvert = 5$.
Absolute zero	The temperature at which, in theory, all movement stops.	On the Kelvin scale, 0 K represents absolute zero.
Acute angle	An angle that measures less than 90°.	
Addends	In an addition problem, the numbers being added are called addends.	In $32 + 51 = 83$, the addends are 32 and 51.
Additive inverse	The opposite of a number.	Examples: The additive inverse of 9 is -9; the additive inverse of -6.4 is 6.4.
Adjacent angles	Two angles in the same plane that have a common vertex and a common side.	
Algebra	The study of operations and relationships among numbers, often using symbols to represent numbers.	
Algebraic expression	An expression that contains one or more variables.	
Alternate exterior angles	Angles that are located outside a set of parallel lines and on opposite sides of a transversal.	In the diagram, angles 1 and 8 are alternate exterior angles; angles 2 and 7 are alternate exterior angles. 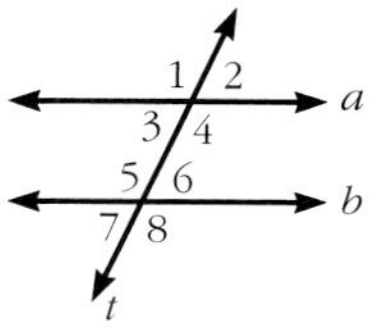
Alternate interior angles	Angles that are located between a set of parallel lines and on opposite sides of a transversal.	In the diagram, angles 3 and 6 are alternate interior angles; angles 4 and 5 are alternate interior angles. 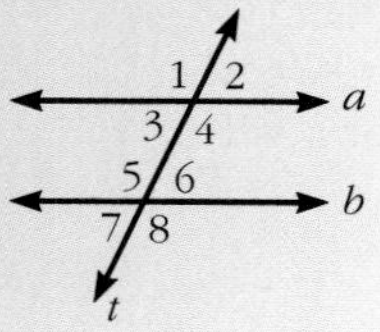
Altitude	A line segment that shows the height of a figure.	
Angle	A geometric figure formed by two rays that have the same endpoint.	

WORD	DEFINITION	EXAMPLE OR EXPLANATION
Antilogarithm	If the logarithm of N is p, then the antilogarithm of p is N.	Examples: log 100 = 2 and antilog 2 = 100; log 1,000 = 3 and antilog 3 = 1,000
Area	The number of square units it takes to cover a two-dimensional space.	
Base (of a power)	A number used as a factor in repeated multiplication.	In the power 2^3, the base is 2. In the power 3^{-5}, the base is 3.
Bel	The basic unit of gain or loss of power.	
Binary number system	A number system that uses only two digits, 0 and 1.	The binary number system is also called the base 2 system.
Boolean operators	The three basic operators in Boolean algebra: AND, OR, and NOT.	
Celsius scale	A system of temperature measurement used in the metric system in which the freezing point of water is 0°C and the boiling point of water is 100°C.	
Center (of a circle)	The point within the circle that is the same distance from all points on the circle.	
Characteristic	The integer part of a logarithm.	Example: log 1,125 = 3.0512; the characteristic of 3.0512 is 3.
Circle	The set of all points in a plane that are the same distance from a fixed point, called the center.	
Circular mil	The area of a circle that has a diameter of 1 mil.	
Circumference	The distance around a circle.	
Coefficient	The numerical factor in a term with a variable.	Examples: In the term $12x^3$, the coefficient is 12. In the term xy, a coefficient of 1 is understood.
Common logarithm	A logarithm with base 10.	When a logarithm is written without a base, it is understood to have a base of 10: $\log_{10} 100 = 2$ or log 100 = 2.
Complementary angles	Two angles whose measures have a sum of 90°.	
Complementary pair	A binary number and its 2's complement.	
Component form (of a vector)	The form $\langle v_x, v_y \rangle$, where v_x and v_y are the horizontal and vertical distances from the initial point to the terminal point of the vector, **v**.	
Composite statement	Two or more statements joined by a connector, such as AND or OR, in Boolean algebra.	
Cone	A solid figure that has one circular base and one vertex.	

WORD	DEFINITION	EXAMPLE OR EXPLANATION
Congruent angles	Angles with the same measure.	The symbol $\cong$ is used to indicate that two angles have the same measure. $\angle A \cong \angle B$ means that angles A and B are congruent.
Congruent polygons	Polygons that have the same size and shape.	The symbol $\cong$ is used to indicate that two polygons are congruent.
Conjunction	The joining of variables with the AND operator in Boolean algebra.	The conjunction of A and B may be written A AND B or $A \cdot B$.
Constant of variation	In the equation $\frac{y}{x} = k$, $(k \neq 0)$, describing a direct proportion, k is called the constant of variation.	Example: In the equation $\frac{y}{x} = 4$, the constant of variation is 4.
Conversion factor	A fraction that is equal to 1, with the numerator expressed in the units that you want to convert *to*, and the denominator expressed in the units that you want to convert *from*.	Example: $\frac{1 \text{ gallon}}{4 \text{ quarts}}$ is a conversion factor used to convert quarts to gallons.
Convex polygon	A polygon with the property that any line segment connecting two points in the polygon lies entirely in the interior of the polygon.	
Coordinate plane	A coordinate system formed by two number lines, one horizontal and one vertical, intersecting at right angles.	The number lines in a coordinate plane are commonly referred to as the x-axis (horizontal) and the y-axis (vertical).
Corresponding angles	Angles that hold corresponding positions when a line intersects two or more parallel lines in a plane.	
Coterminal angles	Two angles in standard position with the same terminal side.	Example: In standard position, angles of 120° and −240° are coterminal.
Customary system	A system of measurement used in the United States.	In the customary system, some units of measure are the inch, foot, cup, quart, ounce, and pound.
Cylinder	A solid figure with two circular congruent bases that lie in parallel planes.	
Decibel	One tenth of a bel, the basic unit of gain or loss of power.	The abbreviation for decibel is dB.
Decimal	A number that contains a decimal point.	Examples: 2.6, 0.063, 7.0, and 8.05 are decimals.
Decimal part	The digits found in the decimal places of a number.	The decimal part of 45.125 is 125.
Decimal places	The places to the right of the decimal point in a decimal.	The decimal, 5.025, has tenths, hundredths, and thousandths places.
Decimal point	The dot in a decimal that separates the whole number from the decimal part.	In 12.375, the decimal point separates the whole number, 12, from the decimal part, 375.
Decimal system	The system of numbers composed of the digits 0, 1, 2, . . ., 9, along with the operations of addition, subtraction, multiplication, and division.	The decimal system is also called the base 10 system.

WORD	DEFINITION	EXAMPLE OR EXPLANATION
Degree	A unit of measure for an angle.	
DeMorgan's Law	The law of logic, given in dual statements, which explains how to find the negation of an expression.	
Denominator	In a fraction, the number or expression below the fraction bar line.	In the fraction $\frac{3}{7}$, the denominator is 7; in the fraction $\frac{3}{(x-5)}$, the denominator is $x + 5$.
Diameter	The length of any line segment that connects two points on a circle and passes through the center.	Diameter is also used to name any line segment that connects two points on a circle and passes through its center.
Difference	The result of subtraction.	In $100 - 70 = 30$, the difference is 30.
Digits	With reference to the base 10 number system, any of the numbers 0, 1, 2, 3, 4, 5, 6, 7, 8, or 9.	
Directed line segment	The representation of a vector on a coordinate plane.	The length of the directed line segment represents the magnitude of the vector; the arrowhead shows the direction of the vector.
Directly proportional	Two variables x and y are directly proportional if the ratio of y to x is a nonzero constant.	When two variables are directly proportional, as one variable increases, so does the other.
Disjunction	The joining of variables with the OR operator in Boolean algebra.	The disjunction of A and B may be written A OR B or $A + B$.
Dividend	A number that is divided by another number.	In $32 \div 4 = 8$, the dividend is 32.
Divisor	The number by which another number is divided.	In $32 \div 4 = 8$, the divisor is 4.
Edge (of a polyhedron)	A line segment where two faces of the polyhedron meet.	
Elimination method	An algebraic method of solving systems of equations, which involves adding like terms in order to solve for one variable at a time.	
Engineering notation	A number in the form of $c \times 10^n$, where c is greater than or equal to 1 and less than 1,000, and n is an integer that is divisible by 3.	Examples: 2.5×10^3, 25×10^6, 90×10^{-3}, and 900×10^{-6} are written in engineering notation.
Equal vectors	Directed line segments that have the same length and same direction.	
Equation	A mathematical sentence that states that two expressions are equal.	Examples: $2 \times 5 = 10$, $3n = 12$, and $\sin 30° = 0.5$ are equations.
Equiangular polygon	A polygon in which all angles are congruent (have the same measure).	
Equilateral polygon	A polygon in which all sides are congruent (have the same length).	
Equivalent decimals	Decimals that represent the same part of a whole.	Example: 0.5, 0.50, 0.500, and 0.5000 are equivalent decimals.
Equivalent fractions	Fractions that represent the same part of a whole.	Examples: $\frac{1}{3}$, $\frac{2}{6}$, and $\frac{4}{12}$ are equivalent fractions.

WORD	DEFINITION	EXAMPLE OR EXPLANATION
Expanded form (of a number)	The expression of a number as a sum of the products of each digit and its place value.	In expanded form, 6,085 is written $(6 \times 1{,}000) + (8 \times 10) + (5 \times 1)$.
Exponent	The number that represents how many times the base is to be used in repeated multiplication.	In the power 2^3, the exponent is 3.
Expression	A number, a variable, or a combination of numbers, variables, and operations.	Examples: 7, $7n$, $2n + 7$, $2mn^2p$, and $\frac{2m}{7n^2}$ are all expressions.
Extremes	In a proportion, the extremes are the numerator of the first ratio and the denominator of the second ratio.	When a proportion is written in the form $a{:}b = c{:}d$, a and d are the extremes.
Face (of a polyhedron)	A polygon that is the side of a polyhedron.	
Factor	A number in a multiplication statement.	In $4 \times 5 \times 12 = 240$, the factors are 4, 5, and 12.
Fahrenheit scale	A system of temperature measurement used in the customary system in which the freezing point of water is 32°F and the boiling point of water is 212°F.	
Formula	An equation that describes a relationship among quantities, usually containing more than one variable.	Example: $d = rt$, where d represents distance, r represents rate, and t represents time, is called the distance formula.
Graduations	The equal sections between tick marks on a ruler.	If one inch is divided into 8 sections on a ruler, each graduation is $\frac{1}{8}$ inch.
Greatest common factor (GCF)	The greatest factor of two or more nonzero numbers.	The GCF of 9 and 15 is 3; the GCF of 25 and 125 is 25.
Hexadecimal number system	A number system that uses the sixteen digits 0 through 9 and the letters A, B, C, D, E, and F.	The hexadecimal number system is also called the base 16 system.
Hypotenuse	In a right triangle, the side that is opposite the right angle.	The hypotenuse is the longest side of a right triangle.
Identity element	Zero is the identity element for addition, and 1 is the identity element for multiplication.	For any number a, $0 + a = a + 0 = a$ $1 \cdot a = a \cdot 1 = a$
Improper fraction	A fraction in which the numerator is greater than or equal to the denominator.	Examples: $\frac{7}{4}$, $\frac{20}{15}$, and $\frac{8}{8}$ are improper fractions.
Inequality	A statement that compares two quantities using one of the following symbols: $<, >, \leq, \geq, \neq$.	Examples: $7 < 8$, $2n \leq 14$, and $4 \neq 5$ are examples of inequalities.
Integers	The counting numbers 1, 2, 3, 4, . . . and their opposites, together with zero.	Examples: -15, -4, 0, 4, and 15 are integers.
Inversely proportional	Two variables x and y are inversely proportional if the product of x and y is a nonzero constant.	When two variables are inversely proportional, as one variable increases, the other variable decreases.
Irrational number	A real number that cannot be written as the quotient of two integers.	Examples: $\sqrt{2}$, $\sqrt{5}$, $\sqrt{8}$, and π are irrational numbers.

WORD	DEFINITION	EXAMPLE OR EXPLANATION
Kelvin scale	A system of temperature measurement used in the metric system in which zero kelvin (0 K) represents absolute zero.	
Lateral area (LA)	The area of a solid figure that does not include the area of the base(s).	
Least common multiple (LCM)	The least whole number that is a multiple of two or more numbers.	The LCM of 6 and 9 is 18; the LCM of 3, 5, and 30 is 30.
Like terms	Terms that have the same variables raised to the same powers.	Examples: $4x^3$ and $-2x^3$ are like terms; 3 and 5 are like terms.
Line	A set of points in a straight path that extends endlessly in both directions.	A line may be named by any two points that lie on the line.
Line segment	A portion of a line consisting of two points on the line and all the points between them.	Line segments are named by their endpoints.
Linear equation	An equation whose graph is a line.	Examples: $y = 2x$, $y = 2x + 5$, and $y = 2x - 5$ are linear equations.
Logarithm	When a base and a power of that base are known, the logarithm is the exponent needed to achieve that power.	Example: exponent $5^2 = 25$ base power; logarithm $\log_5 25 = 2$ base power
Magnitude	The absolute value of a number.	The magnitude of -5 is 5; the magnitude of 5 is 5.
Mantissa	The decimal part of a logarithm.	Example: log 1,125 = 3.0512; the mantissa of 3.0512 is 0.0512.
Means	In a proportion, the means are the denominator of the first ratio and the numerator of the second ratio.	When a proportion is written in the form $a:b = c:d$, b and c are the means.
Metric system	The system of measurement using the meter, liter, and gram as basic units.	
Mil	A unit of measure equal to one thousandth of an inch.	
Minuend	The number from which you are subtracting.	In $10 - 7 = 3$, the minuend is 10.
Mixed number	A whole number together with a fraction.	Examples: $1\frac{2}{3}$, $5\frac{7}{8}$, and $9\frac{1}{2}$ are mixed numbers.
Most significant digit (MSD)	The leftmost digit of a number.	Examples: In the number 0111_2, the MSD 0 can be used to indicate that the number is positive; in the number 1001_2, the MSD 1 can be used to indicate that the number is negative.
Multiplicand	The number that is being multiplied in multiplication.	In the problem shown, 9 $\times$ 4, 36, the multiplicand is 9.

WORD	DEFINITION	EXAMPLE OR EXPLANATION
Multiplicative inverse	The reciprocal of any nonzero number.	Examples: The multiplicative inverse of 5 is $\frac{1}{5}$; the multiplicative inverse of $\frac{7}{8}$ is $\frac{8}{7}$.
Multiplier	The number by which you are multiplying.	In the problem shown, $$\begin{array}{r} 9 \\ \times\,4 \\ \hline 36 \end{array}$$ the multiplier is 4.
Natural logarithm	A logarithm whose base is the irrational number e.	The natural logarithm is denoted by the symbol ln.
Negation (of a statement or expression)	The complement of the statement or expression.	The negation of A is written $\overline{A}$ and is read *not A*.
Net (of a solid figure)	A two-dimentional drawing that shows the shapes that will form the solid figure.	
Numerator	In a fraction, the number or expression above the fraction bar.	In the fraction $\frac{5}{9}$, the numerator is 5; in the fraction $\frac{a-3}{4}$, the numerator is $a - 3$.
Obtuse angle	An angle that measures more than 90° and less than 180°.	
Octal number system	A number system that uses eight digits, 0 through 7.	The octal number system is also called the base 8 system.
Opposites	A pair of numbers that are the same distance, but in opposite directions, from zero on the number line.	Examples: 4 and -4 are opposites; 2.5 and -2.5 are opposites.
Order of operations	A list of steps to be followed for finding the value of an expression.	
Ordered pair (x, y)	Every point in a coordinate plane can be identified by an ordered pair of coordinates, (x, y).	Example: To locate the point given by the ordered pair $(3, -4)$, begin at the origin, go 3 units right, and 4 units straight down.
Ordered triple (x, y, z)	A solution of a system of equations with three variables x, y, and z is the point given by (x, y, z).	
Origin	The point of intersection of the horizontal axis and the vertical axis of the coordinate plane.	The origin is represented by the ordered pair $(0, 0)$.
Parallel lines	Lines in the same plane that do not intersect.	
Parallelogram	A quadrilateral with both pairs of opposite sides parallel.	
Parallelogram method	A method of vector addition in which the resultant vector is the diagonal of a parallelogram.	
Partial product	When the multiplier has 2 or more digits, the product of each digit and the multiplicand is called a partial product.	In the problem shown, the partial products are 344 and 43. $$\begin{array}{r} 43 \\ \times\,18 \\ \hline 344 \\ +\,43 \\ \hline 774 \end{array}$$

WORD	DEFINITION	EXAMPLE OR EXPLANATION
Percent	A special ratio that compares a number to 100 using the symbol %.	Examples: $\frac{52}{100}$ = 52%, read as *fifty-two percent*; $\frac{125}{100}$ = 125%, read as *one hundred twenty-five percent*.
Perfect square	A number that is the square of an integer.	Examples: 16 is a perfect square because $4^2 = 16$; 81 is a perfect square because $(-9)^2 = 81$.
Perimeter (of a polygon)	The sum of the lengths of all its sides.	
Perpendicular lines	Lines that form a right angle.	
Plane	A flat surface that extends endlessly in all directions.	A plane can be named by a single letter or by three points in the plane.
Point	A specific location represented by a dot and named by a single letter.	
Point-slope form (of a linear equation)	A linear equation written in the form $y - y_1 = m(x - x_1)$, where m is the slope and (x_1, y_1) represents one point on the line.	Example: $y - 2 = \frac{1}{2}(x + 4)$ is the point-slope form of the equation of a line through $(2, -4)$ with slope of $\frac{1}{2}$.
Polygon	A geometric figure made up of three or more line segments (sides) that intersect only at their endpoints.	A polygon is named by giving its vertices in consecutive order. A polygon is named by the number of sides, such as hexagon (6 sides).
Polyhedron	A solid that is formed by polygons.	Examples: Cubes, pyramids and prisms are polyhedrons.
Power	A power is an expression with an exponent and a base.	The power 5^2 has a base of 5 and an exponent of 2.
Principal square root	The positive square root of a number.	Examples: $\sqrt{9} = 3$, $\sqrt{100} = 10$, $\sqrt{625} = 25$
Prism	A solid figure with two parallel congruent bases and rectangles or parallelograms as faces.	
Product	The result of multiplication.	In $4 \times 9 = 36$, the product is 36.
Proper fraction	A fraction in which the numerator is less than the denominator.	Examples: $\frac{1}{4}$, $\frac{6}{9}$, and $\frac{13}{16}$ are proper fractions.
Proportion	An equation that states that two ratios are equal.	A proportion may contain all numbers, such as $\frac{1}{3} = \frac{20}{60}$, or it may contain a variable, such as $\frac{1}{3} = \frac{n}{60}$.
Protractor	A tool used to measure and draw angles in degrees.	
Pyramid	A solid figure whose base can be any polygon and whose faces are triangles.	A pyramid is named by the shape of its base. For example, a square pyramid has a square as its base.
Pythagorean Theorem	For any right triangle, the square of the length of the hypotenuse c is equal to the sum of the squares of the lengths of the legs, a and b; $a^2 + b^2 = c^2$.	
Quadrantal angle	An angle in standard position that terminates on an axis.	Examples: In standard position, angles of 90°, 270°, $-180°$, and $-360°$ are quadrantal angles.
Quadrants	The horizontal axis and the vertical axis separate the coordinate plane into four sections, called quadrants.	

WORD	DEFINITION	EXAMPLE OR EXPLANATION
Quadrilateral	A polygon with four sides.	Examples: Squares, rectangles, and parallelograms are quadrilaterals.
Quotient	The result of division.	In $32 \div 4 = 8$, the quotient is 8.
Radius	The distance from the center of a circle to any point on the circle.	Radius is also used to name any line segment that goes from the center of a circle to a point on the circle. The plural of radius is radii.
Radix	A term used to indicate the base of a number system.	The radix of the binary system is 2; radix of the decimal system is 10.
Rate	A type of ratio that compares two unlike quantities.	Examples: Speed is a rate that is often expressed in miles per hour, or *mph*.
Ratio	The comparison of two quantities by division.	A ratio can be expressed with a colon, 7:5, as a fraction, $\frac{7}{5}$, or written 7 to 5.
Rational numbers	Numbers that can be expressed as the ratio of two integers, where the denominator is not zero.	Rational numbers include fractions, decimals, and integers.
Ray	A portion of a line that starts at one point and extends endlessly in one direction.	Rays are named by their endpoint and another point on the ray, in that order.
Reciprocal (of a number)	Formed by writing the number as a fraction and then inverting the fraction.	Examples: The reciprocal of $\frac{5}{6}$ is $\frac{6}{5}$; the reciprocal of 8 is $\frac{1}{8}$; the reciprocal of $\frac{1}{10}$ is 10.
Reciprocals	Two nonzero numbers that have a product of 1.	Examples: $\frac{3}{4}$ and $\frac{4}{3}$ are reciprocals, 6 and $\frac{1}{6}$ are reciprocals, 10 and 0.1 are reciprocals.
Regular polygon	A polygon that is both equilateral and equiangular.	Example: A square is a regular polygon because the sides are the same length and the angles are all 90°.
Regular pyramid	A right pyramid whose base is a regular polygon.	
Remainder	In a division problem, the result of the last subtraction is the *remainder*.	In $33 \div 4 = 8$ R1, there is a remainder of 1.
Resultant vector	The result of vector addition.	
Right angle	An angle with a measure of exactly 90°.	A right angle is indicated by a small square at its vertex.
Scalar	A number that describes a physical quantity; the magnitude.	
Scalar quantity	A physical quantity that can be described by a single number.	
Scale	A ratio that compares the dimensions in a drawing or model to the actual dimensions.	Example: The scale of a map may be 1 inch = 25 miles.
Scientific notation	A number written in the form $c \times 10^n$, where c is greater than or equal to 1 and less than 10, and n is an integer.	Examples: 2.3×10^3, 6.5×10^{-4} and 8.05×10^6 are written in scientific notation.
Similar polygons	Polygons that have the same shape, but not necessarily the same size.	The symbol $\sim$ is used to indicate that two polygons are similar.

WORD	DEFINITION	EXAMPLE OR EXPLANATION
Simplest form (of a fraction)	A fraction in which the greatest common factor of the numerator and the denominator is 1.	The fractions $\frac{2}{5}$, $\frac{8}{15}$, and $\frac{24}{25}$ are in simplest form.
Slant height ℓ (of a regular pyramid or a right cone)	The height of any face other than the base of a regular pyramid, or the distance from the vertex to a point on the edge of the base of a right cone.	
Slope (of a line)	The ratio of the vertical change (rise) to the horizontal change (run).	
Slope-intercept form (of a linear equation)	A linear equation written in the form $y = mx + b$, where the slope of the line is m and the y-intercept is b.	Example: In the equation $y = -4x + 3$, the slope is -4 and the y-intercept is 3.
Solid figure	A figure that has three dimensions and occupies space.	Examples: Cubes, prisms, and spheres are solid figures.
Solution	A value that makes an equation or inequality true.	
Solution set	The set of all values that makes an equation or an inequality true.	
Sphere	A solid figure formed by the set of all points that are equidistant from a given point, called the center.	
Square mil	The area of a square measuring 1 mil on each side.	
Square root	A number that can be squared to get the given number.	The two square roots of 36 are 6 and -6.
Square unit	A square that measures 1 unit on each side.	Examples: Area may be measured in square units, such as square centimeters, square feet, or square miles.
Standard form (of a number)	A number written using only digits.	In standard form, four hundred thousand, fifty is written 400,050.
Straight angle	An angle that measures 180°.	An angle that measures 180° is a straight line. It is denoted by a semicircle.
Substitution method	An algebraic method of solving systems of equations.	To use the substitution method with two equations and two variables, solve one equation for one variable, and then substitute in the other equation to solve for the other variable.
Subtrahend	The number being subtracted.	In $10 - 7 = 3$, the subtrahend is 7.
Sum	The result of addition.	In $32 + 51 = 83$, the sum is 83.
Supplementary angles	Two angles whose measures have a sum of 180°.	
Surface area (SA)	The area of a solid figure that includes the lateral area plus the area of the base(s).	
System of linear equations	Two or more equations whose graphs are lines.	
Term	A number, a variable, or a product of numbers and variables, separated by addition and/or subtraction signs.	The expression $5x$ consists of one term; the expression $5x + 2y + 4$ consists of three terms.

WORD	DEFINITION	EXAMPLE OR EXPLANATION
Transversal	A line that intersects two or more lines that lie in the same plane.	
Triangle	A three-sided polygon.	
Triangle method	A method of vector addition in which the resultant vector is the third side of a triangle.	
Trigonometric ratios	The six ratios formed by using the lengths of two of the sides of a right triangle.	The six trigonometric ratios are: sine (sin), cosine (cos), tangent (tan), secant (sec), cosecant (csc), and cotangent (cot).
Truth table	A method of displaying all the possible truth values of a composite statement in Boolean algebra.	
Unit circle	A circle whose radius is 1.	
Unit cost	The cost of one item.	Example: The cost of grocery items may be given in dollars per pound or dollars per quart.
Unit rate	A rate in which the denominator is 1.	Example: The crop yield for 1 acre may be expressed as 60 bushels per acre, or $\frac{60 \text{ bu}}{1 \text{ acre}}$.
Variable	A symbol that represents a number.	Example: In the expressions $\frac{3ab}{5c}$ and $3a + 5b - c$, the variables are a, b, and c.
Vector	The numbers that describe a vector quantity.	A vector is a measure of "how much" and "in what direction."
Vector quantity	A physical quantity that has both magnitude and direction.	
Vertex	The common endpoint of the sides of an angle or the common endpoint of two sides of a polygon.	
Vertex (of a polyhedron)	A point where three or more edges of the polyhedron meet.	Examples: A pyramid has one vertex; a cube has 8 vertices.
Vertical angles	Angles formed by two intersecting lines with a common vertex, but no common side.	
Volume	The amount of space that a solid figure occupies.	Volume is measured in cubic units.
Whole numbers	The numbers in the set 0, 1, 2, 3, 4, 5, ...	Examples: 9, 19, 395, 1,000, and 480,035 are whole numbers.
Word form (of a number)	A number written using only words.	In word form, 10,549 is written *ten thousand, five hundred forty-nine*.
x-axis	The horizontal number line on a coordinate plane.	
x-intercept	The x-coordinate of the point where a line crosses the x-axis.	If a line crosses the x-axis at (5, 0), the x-intercept is 5.
y-axis	The vertical number line on a coordinate plane.	
y-intercept	The y-coordinate of the point where a line crosses the y-axis.	If a line crosses the y-axis at (0, −3), the y-intercept is −3.

Chapter 1

Exercises 1-1

1. $(5 \times 10{,}000) + (1 \times 1{,}000) + (3 \times 10) + (1 \times 1)$; fifty-one thousand, thirty-one

3. $(4 \times 100{,}000) + (1 \times 10{,}000) + (6 \times 1{,}000) + (5 \times 100) + (1 \times 1)$; four hundred sixteen thousand, five hundred one

5. 306,018; three hundred six thousand, eighteen

7. 49,220

9. 51,300

11. 17,800

13. (a) October: 9,470 feet; November: 8,750 feet; and December: 6,950 feet.
(b) October: 9,500 feet; November: 8,700 feet; and December: 7,000 feet.
(c) October: 9,000 feet; November: 9,000 feet; and December: 7,000 feet.

Exercises 1-2

1. 35

3. 114

5. 4,570

7. 1,038

9. 1,366,175

11. (a) 21,000 feet
(b) 20,942 feet

13. 53,435 feet

Exercises 1-3

1. 26

3. 75

5. 182

7. 3,512

9. 6,764

11. (a) 300
(b) 550
(c) 548

13. (a) $1,817
(b) $157

Exercises 1-4

1. 150

3. 2,600

5. 2,100

7. 1,080

9. 29,952

11. 97,792

13. (a) 19,200 turns
(b) 19,908 turns

15. 2,279 outlets

Exercises 1-5

1. 12

3. 15 R3

5. 35

7. 34 R2

9. 120 R10

11. 3,491

13. (a) 123
(b) 170,478
(c) 1,705 watts
(d) 1,386 watts

15. (a) $722
(b) $19

Review Exercises

1. $(7 \times 10{,}000) + (2 \times 1000) + (3 \times 100) + (5 \times 10) + (6 \times 1)$; seventy-two thousand three hundred fifty-six

3. 105

5. 108

7. 518

9. 14

11. 79Ω

13. 272

15. (a) 18,000 feet
(b) 72 rolls

17. (a) 37 hours
(b) $555
(c) $15 per hour

19. (a) $354
(b) $13,452
(c) $14,481

Chapter 2

Exercises 2-1

1. $\frac{6}{18}$

3. $\frac{1}{2}$

5. $\frac{6}{24}$

7. $\frac{4}{5}$

9. $\frac{5}{11}$

11. $\frac{5}{13}$

13. $\frac{3}{14}$

15. Lisa, Trent, José

17. Possible answer:

$$\frac{5}{16} \qquad \frac{5}{8} = \frac{10}{16}$$

Exercises 2-2

1. $1\frac{1}{5}$

3. $5\frac{1}{2}$

5. $1\frac{1}{4}$

7. $2\frac{1}{25}$

9. $\frac{29}{4}$

11. $\frac{16}{3}$

13. $\frac{11}{4}$

15. $\frac{101}{2}$

17. 31

19. (a) $\frac{60}{16} = \frac{15}{4}H$

(b) $3\frac{3}{4}H$

21. $14\frac{1}{2} = \frac{29}{2} = \frac{58}{4}$, so $14\frac{1}{2} > \frac{57}{4}$.

Exercises 2-3

1. $\frac{3}{5}$

3. $1\frac{1}{5}$

5. $1\frac{7}{12}$

7. $11\frac{2}{3}$

9. $\frac{5}{6}$

11. $1\frac{1}{4}$

13. $21\frac{3}{8}$ feet

15. $1\frac{3}{4}$ pints

17. Those students added the numerators and the denominators. They should have added the numerators only and used the common denominator, 10, to find the sum, $\frac{6}{10}$, and reduced this to $\frac{3}{5}$.

Exercises 2-4

1. $\frac{3}{8}$

3. $\frac{1}{2}$

5. $\frac{5}{16}$

7. $6\frac{5}{16}$

9. $8\frac{3}{4}$

11. $11\frac{7}{12}$

13. $4\frac{5}{8}$

15. $8\frac{1}{6}$

17. $9\frac{3}{5}$ miles

19. $2\frac{5}{8}$ A

Exercises 2-5

1. $\frac{6}{25}$

3. $\frac{5}{6}$

5. $1\frac{2}{3}$

7. $14\frac{7}{16}$

9. $36\frac{1}{4}$

11. $\frac{1}{49}$

13. 25 hours

15. $736

17. The student did not convert the mixed numbers to improper fractions.
$3\frac{3}{5} = \frac{18}{5}$ and $1\frac{1}{8} = \frac{9}{8}$
$\frac{18}{5} \times \frac{9}{8} = \frac{162}{40} = 4\frac{2}{40} = 4\frac{1}{20}$

Exercises 2-6

1. $1\frac{1}{3}$

3. 100

5. 1

7. 2

9. $\frac{22}{45}$

11. $1\frac{5}{6}$

13. $\frac{30}{31}$ Ω

15. (a) 7

(b) 7

(c) $2\frac{1}{8}$ inches

17. The quotient is less than $\frac{1}{4}$ because $\frac{1}{4} \div 4 = \frac{1}{4} \times \frac{1}{4} = \frac{1}{16}$. The fraction $\frac{1}{16} < \frac{1}{4}$.

Exercises 2-7

1. $\frac{3}{8}$ in.

3. $1\frac{1}{4}$ in.

5. $3\frac{1}{4}$ in.

7. $75\frac{1}{4}$ in.; 6 ft $3\frac{1}{4}$ in.

9. $76\frac{1}{8}$ in.; 6 ft $4\frac{1}{8}$ in.

11. (a) 7 feet 6 inches

(b) $7\frac{1}{2}$ feet

(c) $2\frac{1}{2}$ yards

13. $\frac{54}{16}$ inches

Review Exercises

1. (a) 3

(b) $\frac{11}{15}$

3. $\frac{14}{3}$

5. $\frac{17}{15}$

7. (a) $\frac{79}{60}$

(b) $1\frac{19}{60}$

9. (a) $\frac{23}{12}$ or $1\frac{11}{12}$

(b) $\frac{5}{24}$

(c) $\frac{3}{2}$ or $1\frac{1}{2}$

(d) 6

(e) $12\frac{7}{12}$

(f) $7\frac{7}{8}$

(g) $29\frac{1}{4}$

(h) $2\frac{10}{17}$

11. (a) $\frac{495}{4} = 123\frac{3}{4}$ A

(b) $\frac{99}{280}$ A

13. $8\frac{32}{101}$ Ω

15. (a) 24 feet

(b) $23\frac{19}{24}$ feet = 23 feet $9\frac{1}{2}$ inches

(c) 7 yards 2 feet $9\frac{1}{2}$ inches

17. (a) $38\frac{3}{4}$ feet = 38 feet 9 inches

(b) 12 yards 2 feet 9 inches

19. $\frac{40}{33}$ $\Omega = 1\frac{7}{33}$ Ω

Chapter 3

Exercises 3-1

1. sixty-three thousandths;

$0.063 = \left(\frac{6}{100}\right) + \left(\frac{3}{1,000}\right)$

3. twenty and seven hundredths;

$20.07 = 20 + \left(\frac{7}{100}\right)$

5. three and four ten-thousandths;

$3.0004 = 3 + \left(\frac{4}{10,000}\right)$

7. no

9. no

11. no

13. $2.16 < 2.61$
15. $0.01 < 0.5$
17. $8.405 < 8.504$
19. 6.6
21. 75.7
23. 25.34
25. 1785 meters

Exercises 3-2

1. 5.56
3. 68.0
5. 6.7

7. 13.938
9. 14.1 g
11. 4.583 Ω

13. The sum in the tenths place is 16. The digit 6 should be written in the tenths place and the digit 1 should be written in the ones place.

Exercises 3-3

1. 1.14
3. 192.82
5. 4.721

. **7.** 0.246
9. 23.9
11. 235.2 V

13. maximum width: 2.308 cm; minimum width: 2.292 cm

Exercises 3-4

1. 3.968
3. 0.006
5. 0.36

7. 0.1914
9. 0.07
11. 667.5 g

13. (a) 0.002209375 Ω

(b) 0.002 Ω

15. 0.000512

Exercises 3-5

1. 21
3. 900
5. 6.8

7. 20
9. 837.5
11. 20.9

13. 106.9 mph
15. 2.225 V

Exercises 3-6

1. $\frac{16}{25}$

3. $11\frac{7}{100}$

5. $3\frac{2}{25}$

7. 1.75

9. 8.07
11. 10.1
13. 0.44
15. 0.83
17. 4.06

19. $\frac{1}{80}$

21. $3\frac{2}{25}$ kg

23. 35.25 lb

Exercises 3-7

1. 5 mm; 0.5 cm
3. 40 mm; 4.0 cm

5. 85 mm; 8.5 cm
7. 9000 cm

9. (a) 7045 mm
(b) 234 mm
(c) 29 or 30
(d) 29
(e) 21.413 cm

Review Exercises

1. twenty-seven and fifty-three thousandths
$$27.053 = (2 \times 10) + (7 \times 1)$$
$$+ (\tfrac{5}{100}) + (\tfrac{3}{1,000})$$
$$= 20 + 7 + 0.05 + 0.003$$
3. (a) 28.51
(b) 2.289

(c) 10.5
(d) 6.4
(e) 1,176.311
5. (a) $\tfrac{7}{8}$
(b) $17\tfrac{1}{4}$
7. (a) 0.81
(b) 12.77
9. 7.104 Ω

11. 4.8 A
13. 3.41 Ω
15. 6.67 Ω
17. (a) 4350 mm
(b) 185 mm
(c) 22
(d) 23
(e) 60.0 mm

Chapter 4

Exercises 4-1

1. -9
3. -1
5. 15
7. 19
9. 30

11. -9
13. -8
15. $=$
17. $>$
19. $>$

21. $-26, 0, 45$
23. -7 A
25. The absolute value of a number is never negative.

Exercises 4-2

1. -6
3. -4
5. -8
7. 28

9. -12
11. 3
13. 8
15. 67

17. 0 V
19. (a) 28 V
(b) -26 V
(c) -0 V

Exercises 4-3

1. 6
3. -14
5. 4

7. 4
9. 25
11. -16

13. 22
15. -21
17. 397°C higher

Exercises 4-4

1. -6
3. 12
5. 77
7. 210
9. -108

11. -44
13. (a) $-\$55$
(b) $-\$42$
(c) $\$141$
(d) $-\$104$

15. The product is positive.
$[(\text{pos} \times \text{neg}) \times (\text{pos} \times \text{neg})]$
$\times$ pos
$= (\text{neg} \times \text{neg}) \times \text{pos}$
$= \text{pos} \times \text{pos}$
$= \text{pos}$

Exercises 4-5

1. 5
3. -2
5. -14

7. 2
9. 3
11. 2

13. $-\$1,262$
15. $-1,000$ ft per min

Review Exercises

1. (a) 137
(b) 43
(c) 6

3. (a) -14
(b) 20
(c) -28

(d) -7
(e) 1
(f) -40
(g) 19
(h) 48
(i) -33

5. (a) $-37\ \Omega$
(b) $45\ \Omega$
(c) $-51\ \Omega$

7. (a) $-\$72$
(b) $\$156$
(c) $-\$156$

Chapter 5

Exercises 5-1

1. $-\dfrac{2}{3}$

3. $-3\dfrac{3}{5}$

5. $1\dfrac{1}{8}$

7. 10.7

9. 8.3

11. $-6\dfrac{7}{12}$

13. $\$15.00$

15. (a) -9.775 V

(b) $25\dfrac{9}{40}$ V $= 25.23$ V

(c) -11.75 V

Exercises 5-2

1. 10.6

3. $-2\dfrac{1}{16}$

5. -3.85

7. $14\dfrac{1}{4}$

9. -18.7

11. -2.56

13. $2\dfrac{17}{40}$

15. $-4\dfrac{17}{24}$

17. $1{,}629.33$ V

19. The fraction $\left(\dfrac{-1}{-3}\right)$ is positive; it is equal to $\dfrac{1}{3}$.

$6\dfrac{2}{3} - \dfrac{1}{3} = 6\dfrac{1}{3}$

Exercises 5-3

1. -90

3. $\dfrac{1}{6}$

5. $-51\dfrac{1}{5}$

7. $\dfrac{1}{25}$

9. $\dfrac{9}{250}$

11. -0.27

13. The product will be positive. Working from left to right, the signs of the factors are:
(neg) $\times$ (pos) $\times$ (pos) $\times$ (neg) =

(neg) $\times$ (pos) $\times$ (neg) =

(neg) $\times$ (neg) =
(pos)

15. (a) $\$253.56$
(b) $-\$105.68$
(c) $\$505.80$
(d) $-\$58.56$

17. $-20°$C

Exercises 5-4

1. $-1\dfrac{1}{4}$

3. 0

5. -9

7. $26\dfrac{2}{3}$

9. $-8\dfrac{3}{4}$

11. $-2\dfrac{1}{5}$

13. 3.6

15. -2.3

17. $\$142.25$

19. $\left[-\dfrac{1}{2} \div \left(-\dfrac{1}{2}\right)\right] \div \left(-\dfrac{1}{2}\right) =$
$1 \div \left(-\dfrac{1}{2}\right) = -2$

Review Exercises

1. (a) $15\dfrac{1}{2}$

(b) 97.325

(c) $1\dfrac{17}{20}$ or 1.85

3. (a) -9.375 or $-9\dfrac{3}{8}$

(b) -12.28

(c) 17.8595

(d) 0.944

(e) -16.857

(f) $-15\dfrac{25}{66} = -15.3\overline{78}$

(g) -61.75

(h) $32.525 = 32\dfrac{21}{40}$

(i) -72.13

5. (a) $-15\dfrac{1}{30}$

(b) $-5\dfrac{1}{90}$ V ≈ -5.011 V

7. (a) $-\$55.17$
(b) $-\$61.50$
(c) $\$307.45$

Chapter 6

Exercises 6-1

1. 32

3. $\frac{8}{27}$

5. $\frac{1}{81}$

7. $\left(-\frac{1}{32}\right)$

9. 1

11. $\frac{1}{49}$

13. 1

15. 3

17. 14.44

19. (a) $-\frac{1}{1,000}$; $-\frac{1}{1,000}$

    (b) $-\frac{1}{100}$; $\frac{1}{100}$

    (c) $-\frac{1}{10}$; $-\frac{1}{10}$

    (d) -1; 1

    (e) -10; -10

    (f) -100; 100

    (g) $-1,000$; $-1,000$

21. 28.6225 W

23. 1.048 Ω

25. $(-5)^2 = (-5)(-5) = 25$, but $-5^2 = (-1) \times 5 \times 5 = -25$. In the expression $(-5)^2$, -1 is part of the base. In the expression, -5^2, -1 is not part of the base.

Exercises 6-2

1. 0.4^3

3. 3.6^{10}

5. $\frac{1}{(-8)^2} = \frac{1}{64}$

7. $\frac{16}{81}$

9. $\frac{1}{4}$

11. 5.1^2

13. 1,000

15. 1

17. $(-3)^{11}$

19. $\frac{3^{-4}}{4^{-10}} = \frac{4^{10}}{3^4}$

21. True; to evaluate 2^{-3}, use the reciprocal of the base and the opposite of the exponent. So $2^{-3} = \left(\frac{1}{2}\right)^3 = \frac{1}{2^3}$.

23. $\frac{6^6}{6^3} = 6^3$; when the base is the same in a division problem, the exponents are subtracted, not divided.

Exercises 6-3

1. 2.19×10^2

3. 5.05001×10^1

5. 7.16×10^1

7. 1.835×10^3

9. 3.65×10^6

11. 0.000022

13. 0.0707

15. 0.45

17. 9.2

19. 2.389×10^{-4}

21. $4.9 \times 10^{-2} = 0.049$; $9.4 \times 10^{-3} = 0.0094$
Conclusion: $0.049 > 0.0094$

$$4.9 \times 10^{-2} > 9.4 \times 10^{-3}$$

Exercises 6-4

1. 3.24×10^6

3. 4.0×10^3

5. 3.654×10^{10}

7. 2.0×10^{-1}

9. 1.5×10^{-6}

11. 5,000 times heavier

13. 5.0×10^3 meters = 5,000 meters = 5,000,000 millimeters
Error: Multiply meters by 1,000 to find millimeters; do not divide.

Exercises 6-5

1. 78.0×10^{-3}

3. 16.0×10^0

5. 3.75×10^3

7. 4.7×10^3 meters; 4.7 kilometers; 4.7 km

9. 18.0×10^{-3} watts; 18.0 milliwatts; 18.0 mW

11. 468.75×10^6

Exercises 6-6

1. 7

3. $\frac{2}{3}$

5. $4\frac{1}{2}$

7. $\frac{1}{3}$

9. 1

11. $-3\sqrt{10}$

13. $3\sqrt{2}$

15. $5\sqrt{10}$

17. 2

19. $\frac{3}{4}$

21. 7

23. 2.5

25. 1

27. 2

29. 0.45 A

Review Exercises

1. -49

3. $\frac{81}{121}$

5. 1

7. 1.5

9. $\frac{2}{3}$

11. 0.5^8

13. $4.2^0 = 1$

15. $5\sqrt{3}$

17. 1.296×10^{13}

19. 37.5×10^9

21. 25.6 W

23. 3600 C

25. (a) 7.5×10^{-7} W
(b) 750×10^{-9} W
(c) $750n$ W

Chapter 7

Exercises 7-1

1. $\log_5 125 = 3$

3. $\log_4 16 = 2$

5. $\log_7 2{,}401 = 4$

7. $8^2 = 64$

9. $3^5 = 243$

11. $9^{-2} = \frac{1}{81}$

13. $6^x = 1;\ x = 0$

15. $11^x = 121;\ x = 2$

17. $3^x = 9;\ x = 2$

19. $2^x = 8;\ x = 3$

21. $5^x = \frac{1}{25};\ x = -2$

23. 0.01

25. $7{,}500 = 7.5 \times 10^3$. To get
log 7,500, add 3 to log 7.5.
$\log 7.5 = 0.8751$
$\log 7{,}500 = 3.8751$

27. $\log_2 16 = 4;\ \log_4 16 = 2$

Exercises 7-2

1. $\log 3 - \log 8$

3. $3 \cdot \log 5$

5. $2 \cdot \log 7.5 + \log 9.4$

7. $\log_2 4 + \log_2 7.5$

9. $\log (2 \times 5{,}000) =$
$\log 10{,}000 = 4$

11. $\log_5 \frac{50}{2} = \log_5 25 = 2$

13. $2 \log_4 16 - \log_4 4 =$
$2 \cdot 2 - 1 = 3$

15. $\log \left(\frac{1}{10^{-3}}\right) = \log 1 - \log 10^{-3} =$
$\log 1 + 3 \cdot \log 10 =$
$0 + 3 \cdot 1 = 3$

17. (a) $10 \log \frac{500}{5} = 10 \log 100$
(b) 20 dB

Exercises 7-3

1. $4.5 \times 0.03 = \text{antilog } (\log 4.5 + \log 0.03)$
$= \text{antilog } (0.6532 + (-1.5229))$
$= \text{antilog } (-0.8697)$
≈ 0

3. $\frac{21}{18} = \text{antilog } (\log 21 - \log 18)$
$= \text{antilog } (1.3222 - 1.2553)$
$= \text{antilog } (0.0669)$
≈ 1

5. $1{,}000 \times 1{,}000 = \text{antilog } (\log 1{,}000 + \log 1{,}000)$
$= \text{antilog } (3 + 3)$
$= \text{antilog } (6)$
$= 1{,}000{,}000$

7. (a) $2 \log 80.25 - \log 225$
(b) 1.4567
(c) 28.6

9. (a) $\log 28 + \log 32 - \log(28 + 32)$
(b) 1.1742
(c) 14.93

Exercises 7-4

1. 3.64

3. -1.27

5. 10.72

7. 4.39

9. -0.63

11. $5 \ln 0.3$

13. $\ln 5 - \ln 6$

15. $3 \ln 6.7$

17. (a) 5.1 years
(b) 15.7 years (almost 15
years, 9 months)

Exercises 7-5

1. 10 dB

3. Possible answer: A gain of 3 dB indicates that the power has been doubled; a loss of 3 dB indicates that the power has been cut in half.

5. 8 dB

Review Exercises

1. (a) $\log_{12} 144 = 2$
(b) $\log_5 0.008 = -3$

3. (a) 2.1584
(b) 3.1584
(c) 5.4161
(d) 7.7187

5. (a) $\log 2 - \log 3$
(b) $\log 5 + 7 \log 3$
(c) i$\ln 56 - 3$ in 7
(d) in $125 - $ (in $18 - 3$ In 4)

7. (a) 0.9638
(b) 5.9638

9. -3.4 dB

11. (a) 0.001 W
(b) 0.125 W
(c) 21 dB

Chapter 8

Exercises 8-1

1. $\dfrac{2 \text{ pt}}{1 \text{ qt}}$

3. $\dfrac{1 \text{ lb}}{16 \text{ oz}}$

5. $\dfrac{1\text{T}}{2{,}000 \text{ lb}}$

7. 10 yd

9. 8 pt

11. 3 days

13. 31,680 ft

15. 41.7 days

17. 304.8 in.

19. 219 feet 9 inches

21. 75 lb

Exercises 8-2

1. $\dfrac{1 \text{ kg}}{1000 \text{ g}}$

3. $\dfrac{1000 \text{ mL}}{1 \text{ L}}$

5. $\dfrac{1 \text{ m}}{1000 \text{ mm}}$; $\dfrac{100 \text{ cm}}{1 \text{ m}}$ or $\dfrac{1 \text{ cm}}{10 \text{ mm}}$

7. 2 cm

9. 3 A

11. 60,000 mg

13. 27.5 mm

15. 0.97155 m

17. (a) 110 000 V
(b) 0.11 MV

Exercises 8-3

1. 17.5 yd

3. 133.2 g

5. 3.3 ft

7. 6.3 qt

9. 1.1 yd

11. 6.9 m

13. (a) 166.5 lb
(b) No, the motor is 16 lb too heavy.

15. (a) about 73.9 mL
(b) about 4 cents

Exercises 8-4

1. 257°F

3. 112.1°F

5. 37.8°C

7. 122°F

9. 20°C

11. 232.8°C

13. 39.7°C

15. -40°F

17. The error in the work shown is that subtraction inside the parentheses was not done first. The correct order of steps is to subtract inside the parentheses first, and then multiply by $\frac{5}{9}$. The correct answer is 12.2°C.

Exercises 8-5

1. 1,920 acres

3. 12,000 cm²

5. 900 sq ft

7. 1.2 cu ft

9. 450 000 mm³

11. 8,640 cu in.

13. 1.2 cu in.

15. 9.3 m²

17. 76.5 m³

19. 90 cm³

21. 532.6 cm³

Exercises 8-6

1. 3 in.
3. 12,375 mils
5. 0.85 in.
7. 0.02 in.

9. 62,500 sq mil
11. 2,000,000 sq mils
13. 6,400 sq mils; 8,149 cmil

15. 750,000 sq mils; 954,927 cmil
17. 35,156 sq mils; 44,762 cmil
19. 7.5 sq mils

Review Exercises

1. $\frac{1 \text{ m}}{100 \text{ cm}}$

3. $\frac{25.4 \text{ mm}}{1 \text{ in.}}$

5. 18 in.

7. 324 in.^{2}

9. 0.775 L

11. 650 A

13. 40 250 kV

15. 3.0 m^2

17. 4,750 mils

19. 6250 µA

21. (a) 1.350 m
  (b) 135.0 cm

23. (a) 252 770 km
  (b) 230 000 V

25. −46°F, 22°F, and 66°F

27. 33.0 sq mil

Chapter 9

Exercises 9-1

1. $2 \cdot 3^2 - 9 = 2 \cdot 9 - 9 =$
  $18 - 9 = 9$

3. $\frac{2 \times 9}{20 - 2} - 5 = \frac{18}{18} - 5 =$
  $1 - 5 = -4$

5. $8^2 \div 4(9 - 7)^2 = 8^2 \div 4(2)^2$
  $= 64 \div 4(4) = 64 \div 16 = 4$

7. $54 + 14 - 5 \cdot 2^3 = 54 + 14$
  $- 5 \cdot 8 = 54 + 14 - 40 =$
  $68 - 40 = 28$

9. $5^2 + 9^2 - 6 = 25 + 81 - 6$
  $= 106 - 6 = 100$

11. $45 \div [20 - (10 + 1)] = 45 \div$
  $[20 - 11] = 45 \div 9 = 5$

13. 28
15. −4
17. −40
19. 7
21. 10
23. Perform the subtraction inside the parentheses first. Then multiply the result by $\frac{5}{9}$.
25. $P = 2.5^2 \cdot 1.6 = 10$
27. Possible answers: $9 - 6 - 3 \times 1$;
  $9 - 6 - 3 \div 1$

Exercises 9-2

1. 6; Commutative Property of Addition
3. 6; Associative Property of Multiplication
5. 1; Identity Property for Multiplication
7. $\left(2\frac{3}{4}\right)^{-1} = \frac{4}{11}$
9. $\frac{1}{12}$; Property of Multiplicative Inverses

11. 1; Associative Property of Addition
13. x; Associative Property of Multiplication
15. a; Associative Property of Addition
17. y; Distributive Property

19. 7; Commutative Property of Addition
21. Associative and Commutative Properties of Addition
23. $36I_a^2 + 48I_b^2$
25. $I(V - RI)$

Exercises 9-3

1. $2n - 5$
3. $a^2 + 8a + 16$
5. $8a + 36$

7. $-ab - 6a$
9. $3b^2 + 20b + 12$

11. $2t^2 + 7t - 15$
13. 0

Review Exercises

1. 26.75
3. 26
5. (−5), Distributive Property

7. $18x - 6$
9. 10

11. 2.5
13. $3(5I_a^2 + 4I_b^2)$

Chapter 10

Exercises 10-1

1. $\frac{x}{6} = 7$

3. $\frac{x}{7} = -35$

5. $12 - x = 8$

7. $A = \pi r^2$

9. $6x + 1 = 55$

11. $x - 8 = -2$

13. $\sqrt{V - 9} = 4$

15. $15.3 = \sqrt{\frac{P}{12.25}}$

17. Choice b: *One number squared is equal to a second number squared plus the square of a third number.* In the equation $x^2 = y^2 + z^2$, the variables y and z are squared, and then the squares of y and z are added together.

Exercises 10-2

1. $n = 12$

3. $d = 64$

5. $x = \frac{2}{7}$

7. $b = 9$ or $b = -9$

9. $c = -60$

11. $x = 10$

13. $n = 9$

15. $w = 625$

17. $a = -30$

19. $n = 24$

21. $n = \frac{16}{81}$

23. $w = \frac{1}{10}$

25. $x = \frac{1}{4}$

27. $x = 49$

29. 4 A

31. Multiplying both sides of an equation by $\frac{1}{3}$ is the same as dividing both sides by 3. You can use either operation to solve this equation. ($n = 7$)

33. The student took the square root of both the numerator and denominator instead of squaring the numerator and denominator. The solution is $x = \frac{81}{256}$.

Exercises 10-3

1. $b = -6$

3. $x = 9$

5. $x = 4$ or $x = -4$

7. $x = -3$

9. $b = 14$

11. $b = 25$

13. $q = -1$

15. $x = 7$ or $x = -7$

17. $a = 6$ or $a = -6$

19. 3.25 A

21. (a) $4.25 = \frac{V^2}{2850}$

(b) 110 V

23. $c = 1$

Exercises 10-4

1. $l = \frac{A}{w}$

3. $r = \sqrt{\frac{V}{\pi h}}$

5. $I = \frac{P}{E}$

7. $y = \frac{x - t}{a}$

9. $V = \frac{q}{C}$

11. $I = \sqrt{\frac{P}{R}}$

13. $L = \frac{1}{(2\pi f)^2 C}$

15. $K = \frac{A \cdot R}{L}$; $K = 125$

17. $h = \frac{V}{lw}$; $h = 15$ meters

19. $R_3 = R - \frac{1}{\frac{1}{R_1} + \frac{1}{R_2}}$, $R_3 = 10\,\Omega$

21. $r = \frac{A - p}{pt}$; $r = 0.04$

Exercises 10-5

1. The number is 13.

3. Rita is 38 years old.

5. The two integers are -12 and -10.

7. The integers are 33, 34, and 35.

9. The two odd integers are -1 and 1.

11. 6 hours

Exercises 10-6

1. $\frac{n}{6} < 24$

3. $-4n < 16$

5. $3(n + 5) \geq 20$

7. $p > 250{,}000$

9. $s \leq 875$

11. Yes, both statements have the same meaning. If a number is not negative, it must be equal to zero or greater than zero.

Exercises 10-7

1. $x > -8$

3. $w < 27$

5. $x \geq -5$

7. $n < -1$

9. $n \leq 3$

11. $n > -60$

13. $x > -24$

15. $x \leq 8$

17. (a) $32.50m + 145 \geq 475.95$
(b) 11 months

19. Possible lengths for the side of the square include lengths greater than 16 inches and less than 38 inches.

Review Exercises

1. (a) $R - 25 = 37$
(b) $R = 62 \ \Omega$

3. (a) $40 = I^2(250)$
(b) $I = 0.4$ A

5. $r = \frac{E}{I} - R$

7. $1{,}672.5$ ft

9. $30 \ \Omega$

11. $c \leq 1{,}750$

Chapter 11

Exercises 11-1

1. 1 to 3, 1:3, $\frac{1}{3}$

3. 47 to 702, 47:702, $\frac{47}{402}$

5. 1 to 1,000, 1:1,000, $\frac{1}{1{,}000}$

7. 1 cm:50 km; 1 cm = 50 km

9. 1:58, 1 inch = 4 feet 10 inches

Exercises 11-2

1. 32 miles per gallon

3. $12.25 per shirt

5. 100 people per square mile

7. 27 cubic feet per cubic yard

9. $2.35/ft

11. 1.24 Canadian dollars per U.S. dollar

13. The unit cost for product A (48 ounces for $3.84 = $0.08 per ounce) is lower than the unit cost for product B (64 ounces for $5.44 = $0.085 per ounce).

Exercises 11-3

1. $n = 60$

3. $x = 6$

5. $w = 6$

7. $x = 14$

9. $w = 12.8$

11. $n = 2$

13. $\frac{6}{15} = \frac{x}{45}$; $x = 18$

15. $1.05 \ \Omega$

17. $ad = bc$; the equation represents the product of the extremes equal to the product of the means.

Exercises 11-4

1. $x_2 = 1$

3. $y_2 = 28$

5. $y_1 = 180$

7. $x_1 = 28$

9. 7.8 ohms

11. 18 gallons

13. 90 volts

Exercises 11-5

1. $y_1 = 12$

3. $x_1 = 5$

5. $x_2 = \frac{1}{3}$

7. $y_1 = 17$

9. The constant of variation is 20.

11. $s_2 = 8$ cm

13. (a) $I_1 R_1 = I_2 R_2$
(b) 122.5 ohms

Review Exercises

1. 1.5 to 42.6, 1.5 : 42.6 and $\frac{1.5}{42.6}$

3. 32 miles per gallon

5. 4.5

7. 13.7

9. (a) $\frac{P_1}{V_1^2} = \frac{P_2}{V_2^2}$

(b) 172.8 W

11. (a) $R_1 \cdot I_1 = R_2 \cdot I_2$

(b) 3.125 Ω

Chapter 12

Exercises 12-1

1. $\frac{23}{100}$

3. $\frac{3}{10}$

5. $\frac{2}{25}$

7. $\frac{12}{25}$

9. $\frac{3}{100}$

11. 0.82

13. 1.28

15. 0.002

17. 0.31

19. 37%

21. 65%

23. 80%

25. 77%

27. 11.1%

29. $\frac{3}{100}$

31. $\frac{1}{1,000}$

33. $2\frac{1}{10}$

35. $\frac{4}{5}$

37. 55%

39. 12.5%

41. 62.5%

43. 7.5%

45. 7.5%

47. 0.85, $\frac{17}{20}$

49. 2.9, $\frac{29}{10} = 2\frac{9}{10}$

Exercises 12-2

1. 24.7

3. 87.5%

5. $1.16

7. 0.5%

9. 500

11. 63

13. $33\frac{1}{3}$%

15. (a) $17.81

(b) $81.14

17. Resistors: 47.7%, capacitors: 28.8%, transistors: 10.8%, integrated circuits: 12.6%

19. The first ratio in the proportion is incorrect; it should be $\frac{2.5}{100}$. The correct answer is 0.45.

Exercises 12-3

1. 6.4

3. 15%

5. 340

7. 150

9. 40%

11. 5,000

13. (a) 5.5 V

(b) 214.5 V

15. $77.15

Exercises 12-4

1. 1,561 families

3. 3,000 shingles

5. 5,500 ohms

7. About 3.6%

9. 29%; possible explanations: (1) Let p represent the percent of shirts in size *Large* and solve the proportion $\frac{p}{100} = \frac{25}{85}$, $p \approx 29$; 29% or (2) Let x represent the unknown percent and solve the equation $25 = 85x$; $x \approx 0.29 = 29\%$.

11. $552,000

13. 199.9 miles

Review Exercises

1. (a) $\frac{3}{20}$

(b) $\frac{21}{400}$

(c) $\frac{9}{8} = 1\frac{1}{8}$

3. (a) 25%

(b) 150%

(c) 0.63%

5. (a) 53%

(b) 2.1%

(c) 125%

7. 32.4 A

9. $17.50

Chapter 13

Exercises 13-1

1. (−6, −4), Quadrant III

3. (4, −5), Quadrant IV

5. (7, 6), Quadrant I

7. The order in which the coordinates are given in the ordered pair makes a difference. To go to (2, −5) from (0, 0), go 2 units to the right and then 5 units down into Quadrant IV. To go to (−5, 2) from (0, 0), go 5 units to the left and then 2 units up into Quadrant II.

9. (a) *C, E;* (b) *A, H;* (c) *B, G;* (d) *D*

11. Point *B*(−4, 0) is plotted incorrectly. Point *B* should be located on the *x*-axis, 4 units to the left of (0, 0).

Exercises 13-2

1.

x	y = 2x − 4	(x, y)
−2	2(−2) − 4 = −8	(−2, −8)
−1	2(−1) − 4 = −6	(−1, −6)
0	2(0) − 4 = −4	(0, −4)
1	2(1) − 4 = −2	(1, −2)
2	2(2) − 4 = 0	(2, 0)
3	2(3) − 4 = 2	(3, 2)

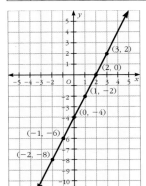

3.

x	y = x − 1	(x, y)
−1	−2	(−1, −2)
0	−1	(0, −1)
1	0	(1, 0)

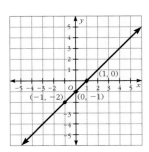

5.

x	y = $\frac{1}{3}$x + 2	(x, y)
−6	0	(−6, 0)
0	2	(0, 2)
6	4	(6, 4)

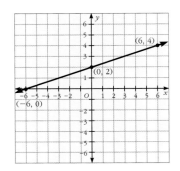

7.

x	y = −$\frac{3}{4}$x	(x, y)
−4	3	(−4, 3)
0	0	(0, 0)
4	−3	(4, −3)

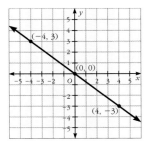

9.

x	y = x − 8	(x, y)
0	−8	(0, −8)
4	−4	(4, −4)
8	0	(8, 0)

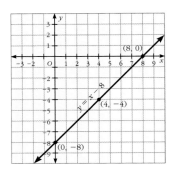

11.

x	y = −3x − 5	(x, y)
−1	−2	(−1, −2)
0	−5	(0, −5)
1	−8	(1, −8)

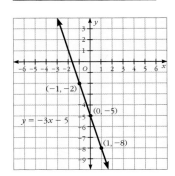

Exercises 13-2 (continued)

13.

V	I = 2V	(V, I)
0	2(0) = 0	(0, 0)
2	2(2) = 4	(2, 4)
4	2(4) = 8	(4, 8)
6	2(6) = 12	(6, 12)
8	2(8) = 16	(8, 16)
10	2(10) = 20	(10, 20)

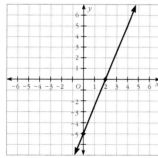

15.

n	C = 6.50n + 8.76	(n, C)
1	6.50(1) + 8.76 = 15.26	(1, 15.26)
2	6.50(2) + 8.76 = 21.76	(2, 21.76)
5	6.50(5) + 8.76 = 41.26	(5, 41.26)
8	6.50(8) + 8.76 = 60.76	(8, 60.76)

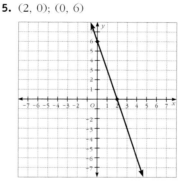

17.

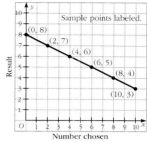

Exercises 13-3

1. x-intercept: -4; y-intercept: 3

3.

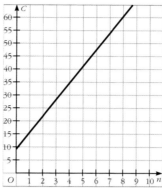

5. (2, 0); (0, 6)

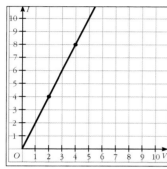

7. Sample answer: The y-coordinate of the point where a line crosses the x-axis is 0. So, substituting 0 for y in the equation gives the x-intercept. The x-intercept of the line is -3.

9. The x-intercept is $-\dfrac{b}{m}$.

Exercises 13-4

1. $\dfrac{3}{2}$

3. $\dfrac{2}{3}$

5. $\dfrac{1}{5}$

7. $\dfrac{10}{7}$

9. slope = 3; y-intercept = 9

11. slope = $\dfrac{1}{2}$; y-intercept = $-\dfrac{1}{2}$

13. slope = 0; y-intercept = 3

15. $y = 3x + 2$; slope = 3; y-intercept = 2

17. $y = 0x - 1$; slope = 0; y-intercept = -1

19. $y = -x + 8$; slope = -1; y-intercept = 8

21. (a) 2
(b) 0

23. (a) 6.5
(b) 8.76

Exercises 13-5

1. slope $= \frac{4}{5}$, y-intercept $= 1$

3. slope $= -8 = \frac{-8}{1}$ or $\frac{8}{-1}$,
y-intercept $= 2$

5. slope $= 1 = \frac{1}{1}$,
y-intercept $= -5$

7.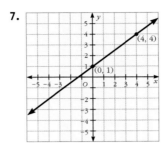

9.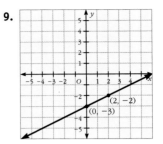

11. (a) 0.0045

(b) 8.54

Exercises 13-6

1. $y = -\frac{1}{2}x + 1$

3. $y = \frac{2}{3}x + \frac{7}{9}$

5. $y = -\frac{5}{6}x - 2$

7. slope $= -\frac{3}{2}$; $y = -\frac{3}{2}x + 1$

9. $y = 2x - 8$

11. $y = -2$

13. $y = x + 5$

15. $y = \frac{2}{7}x$

17. $y = -4x - 2$

19. $R = 0.039T + 4.251$

21. $y = -\frac{3}{4}x - 3$

Review Exercises

1.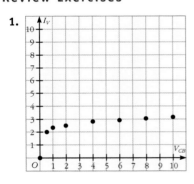

3. (a)

V	$I = \dfrac{V}{1.6}$	(I, V)
0	$\dfrac{0}{1.6} = 0$	(0, 0)
2	$\dfrac{2}{1.6} = 1.25$	(2, 1.25)
4	$\dfrac{4}{1.6} = 2.5$	(4, 2.5)
6	$\dfrac{6}{1.6} = 3.75$	(6, 3.75)
8	$\dfrac{8}{1.6} = 5$	(8, 5)
10	$\dfrac{10}{1.6} = 6.25$	(10, 6.25)

(b)

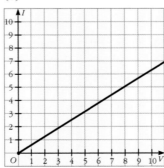

(c) 0.625

(d) 0

5. (a) 19.65

(b) 29,625

(c) $C = 19.65t + 29,625$

Chapter 14

Exercises 14-1

1.
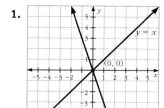
Solution: $(0, 0)$

3.
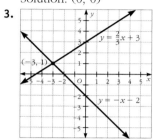
Solution: $(-3, 1)$

5.

Solution: $(2, -2)$

7. $(V, I) = (-4, 4)$
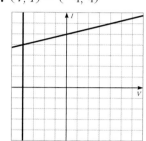

9. (a)
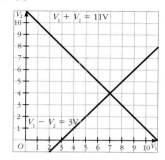

(b) $V_1 = 7$ V, $V_2 = 4$ V

Exercises 14-2

1. $(-2, -3)$
3. $(8, 2)$
5. $(-4, 4)$
7. $(3, 2)$

9. $(4, 1)$
11. $(1, 1)$
13. $(-6, -1)$
15. $V_1 = 20$, $V_2 = 3$

17. Sample answer: I do not agree. The line $x = -5$ is vertical; the line $y = 6$ is horizontal. The lines intersect at the point $(-5, 6)$, so the solution is $(-5, 6)$.

Exercises 14-3

1. $(11, -6)$
3. all ordered pairs that satisfy the equation $x - y = 3$
5. $\left(-1, \frac{2}{3}\right)$
7. no solution

9. $(-9, -7)$
11. $(10, -2)$
13. $(2, -1)$
15. $I_1 = 1$ A, $I_2 = 1$ A, $I_3 = 0$ A

17. Sample answer: Multiply both sides of Equation 1 by -6 and then add similar terms of the equations. The solution of the system is $(-6, -8)$.

Exercises 14-4

1. $(3, -2, 2)$
3. $(2, -1, 3)$
5. $(10, 1, 5)$
7. $(5, 4, 3)$

9. $(-5, 4, -2)$
11. $I_1 = 9$ A, $I_2 = 5$ A, $I_3 = 14$ A
13. Sample answer: Substitute $2x$ for y in Equation 2: $z = 2y = 2(2x) = 4x$. Then substitute $4x$ for z in Equation 3: $x = 2z = 2(4x) = 8x$.

Solve the equation $x = 8x$: $x = 0$. Substitute 0 for x in Equations 1 and 3 to find that $y = z = 0$. The solution is $(0, 0, 0)$.

Exercises 14.5

1. lesser number: 2; greater number: 7

3. 5 hours, $175
5. 35 V, 17 V

7. Resistor: $0.40, transistor: $1.02

Review Exercises

1. $(-9, 18)$

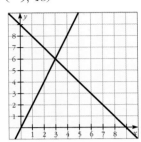

3. (a)

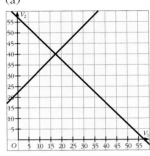

(b) $V_1 = 17$ V, $V_2 = 40$ V

5. $(5, 3, 7)$
7. $I_1 = 17$ A, $I_2 = 31$ A,
   $I_3 = 14$ A

Chapter 15

Exercises 15-1

1. Possible answers: $\overrightarrow{AB}$ or $\overrightarrow{BA}$
3. Possible answers: $\overline{AC}$ or $\overline{CA}$;
   $\overline{AB}$ or $\overline{BA}$; $\overline{AD}$ or $\overline{DA}$

5. point M
7. $\angle Y$, $\angle 1$, $\angle XYZ$, and $\angle ZYX$

9. Intersection of planes R and
   S: $\overrightarrow{AX}$; intersection of line ℓ
   and plane S: point T

Exercises 15-2

1. $60°$
3. $125°$
5.

7.

$150°$ (diagram with points G, H, I)

9. $45°$

11. $150°$
13. $m\angle B = 75°$
15. Rodney used the wrong scale
   on his protractor;
   $m\angle RST = 100°$.

Exercises 15-3

1. right
3. straight
5. $\angle AEC$, $\angle CEB$, $\angle BED$, $\angle DEA$
7. $\angle ABD$ and $\angle DBC$ are complementary; $m\angle ABD = 19°$.
9. $x = 74$, $y = 106$

11. $m\angle 1 = m\angle 4 = m\angle 5 =$
   $m\angle 8 = 110°$, $m\angle 2 = m\angle 3$
   $= m\angle 6 = m\angle 7 = 70°$
13. If two parallel lines are cut
   by a transversal that is perpendicular to both lines, then
   each of the 8 angles will
   measure $90°$.

15. Two right angles are a pair
   of supplementary angles, but
   right angles are neither acute
   nor obtuse.

Exercises 15-4

1. nonagon; none of these
3. hexagon; equilateral, but not
   equiangular or regular
5. heptagon; none of these

7. $m\angle D = 62°$
9. $m\angle B = 125°$
11. $m\angle C = 102°$

13. No; a rhombus does not
   always have all right angles,
   but a square always has all
   right angles. So, not every
   rhombus is a square.
15. $x = 80$

Exercises 15-5

1. 20 m
3. 45 m
5. 32 in.

7. 26 m
9. 420 ft

11. Width: 4.3 mm,
   length: 7.4 mm
13. $x = 30$ in.

Exercises 15-6

1. 48 ft²
3. 72 yd²
5. 30 in.²

7. 75 cm²
9. 15 in.²
11. 45 in.²

13. 16 ft²
15. 60 m²

Exercises 15-7

1. 18 in.
3. 18.8 yd
5. 19.1 m

7. 201.0 ft²
9. 176.6 cm²
11. 5.73 in.
13. 1.86 in.²

15. One 16-inch pizza; three 8-inch pizzas is about 151 in.² of pizza, and one 16-inch pizza is about 201 in.² of pizza.

Exercises 15-8

1. $m\angle D = 26°$, $DE = 4$ m
3. $m\angle Y = 94°$, $XY = 10$ m

5. $x = 6$ in.
7. $x = 3$ ft

9. (a) 44 mm
  (b) 8 mm

Review Exercises

1. (a) 120°
  (b) 30°

3. 7 lights

5. 155.06 cm²

Chapter 16

Exercises 16-1

1. triangular prism
3. pentagonal pyramid
5. rectangular prism
7. triangular pyramid
9. Sample answer: Use Euler's Formula to find the number of vertices: $F + V = E + 2$,

$8 + V = 18 + 2$, $V = 12$. Two faces of every prism are the bases, so there are 6 faces joining the bases. That means the bases each have 6 sides and the prism is a hexagonal prism.

11. a) $\overline{AE} \parallel \overline{BF} \parallel \overline{CG} \parallel \overline{DH}$,
  b) $\overline{AD} \parallel \overline{EH} \parallel \overline{FG} \parallel \overline{BC}$,
  c) $\overline{AB} \parallel \overline{EF} \parallel \overline{HG} \parallel \overline{DC}$

Exercises 16-2

1. $LA = 136$ in.², $SA = 280$ in.²
3. $LA \approx 138.16$ in.²,
  $SA \approx 163.28$ in.²
5. $LA = 90$ in.²

7. $LA \approx 65.94$ in.², $SA \approx 78.50$ in.²
9. $LA \approx 41.45$ cm²,
  $SA \approx 56.65$ cm²

11. $SA \approx 50.24$ ft²
13. (a) About 1.98 in.²
  (b) About 2.79 in.²
15. About 2,385 in.²

Exercises 16-3

1. 121.5 in.³
3. 2400 cm³
5. 280 cm³
7. 94.2 in.³

9. 113.04 in.³
11. 261.67 in.³
13. 5.96 in.³

Review Exercises

1. (a) Triangular prism
  (b) 520 cm²
  (c) 640 cm²
  (d) 780 cm³
3. 1,253.75 in³

5. $LA \approx 447.61$ cm²
  $SA \approx 617.51$ cm²
  $V \approx 1163.80$ cm³
7. $LA \approx 28\ 512$ mm²
  $SA \approx 33\ 696$ mm²
  $V \approx 312\ 768$ mm³

9. $SA \approx 14\ 957.12$ mm²
  $V \approx 172\ 006.91$ mm³

Chapter 17

Exercises 17-1

1. $c = 20$
3. $b = 9$
5. $c = 4.2$
7. 8 cm

9. yes
11. yes
13. 24.9 feet
15. 10.9 feet or
10 feet-10.8 inches

17. 5.7 inches
19. $\overline{BC} = 0.5$, $\overline{DE} = 0.75$,
$\overline{FG} = 1.5$

Exercises 17-2

1. $\csc \theta = \frac{29}{20}$, $\cos \theta = \frac{21}{29}$,
$\sec \theta = \frac{29}{21}$, $\tan \theta = \frac{20}{21}$,
$\cot \theta = \frac{21}{20}$

3. (a) 0.9397, (b) 0.9962,
(c) 1.1918, (d) 0.9877,
(e) 0.2419, (f) 0.3249
5. $\sin \theta = \frac{2}{\sqrt{13}} \approx 0.5547$,
$\cos \theta = \frac{3}{\sqrt{13}} \approx 0.8321$,
$\tan \theta = \frac{2}{3} \approx 0.6667$

7. $\sin \theta = 0.6000$, $\cos \theta = 0.8000$, $\tan \theta = 0.7500$,
$\csc \theta = 1.6667$, $\sec \theta = 1.2500$, $\cot \theta = 1.3333$
9. 4.9 V

Exercises 17-3

1. II
3. IV
5. I
7. II
Sample answers are given for Exercises 9, 11, and 13.
9. $405°$, $-315°$
11. $545°$, $-175°$
13. $720°$, $-360°$
15. $120°$

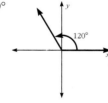

17. $\cos 60° = \frac{1}{2}$
19. $\sin (-30°) = -\frac{1}{2}$, $\cos (-30°) = \frac{\sqrt{3}}{2}$
21. $\sin (-90°) = -1$, $\csc (-90°) = -1$, $\cos (-90°) = 0$,
$\sec (-90°)$ is undefined,
$\tan (-90°)$ is undefined,
$\cot (-90°) = 0$
23. (b) $45°$, $\frac{1}{\sqrt{2}}$, $-\frac{1}{\sqrt{2}}$;
(c) $60°$, $\frac{\sqrt{3}}{2}$, $\frac{\sqrt{3}}{2}$

25. (a) $45°$, 1, -1; (b) $30°$; $\frac{1}{\sqrt{3}}$, $\frac{1}{\sqrt{3}}$;
(c) $60°$, $\sqrt{3}$, $-\sqrt{3}$
27. 24.3 mA.
29. Sample answer: In a unit circle, $\sin \theta = y$ and $\cos \theta = x$. Because y and x represent the lengths of the legs of a right triangle with hypotenuse 1, these values may be substituted into the Pythagorean equation. So, $y^2 + x^2 = 1$, or $(\sin \theta)^2 + (\cos \theta)^2 = 1$.

Exercises 17-4

1. Sample answer: $45°$, $405°$, $225°$
3. $-45°$
5. $56°$

7. $63°$
9. $-49°$
11. $\theta \approx 37°$
13. $39.5°$

15. (a) $32.8°$
(b) $147.2°$
17. $\alpha \approx 65°$

Exercises 17-5

1. $A = 22°$, $a \approx 1.5$
3. $B = 16°$, $b = 7$

5. $A = 66°$, $a \approx 22.5$, $c \approx 24.6$
7. 222.8 feet

9. $Z = 219.9$ kΩ, $R = 170.4$ kΩ, and $\alpha = 50.8°$
11. $A \approx 62°$, $B \approx 28°$

Exercises 17-6

1. 1,205 feet
3. 15 feet

5. 11.3 inches
7. $Z = 219.9$ kΩ, $R = 170.4$ kΩ, and $= 50.8°$

9. $51.3°$

Review Exercises

1. $c = 26.2$
$\alpha = 16.4°$
$\beta = 73.6°$
3. $b = 42.2$
$c = 50.3$
$\alpha = 32.9°$

5. Reference angle: $68°$
$\sin \alpha = -0.9272$
$\cos \alpha = -0.3746$
$\tan \alpha = 2.4751$
7. (a) 10 m = 1000 cm
(b) 8.80 m = 880 cm

(c) 26.27 m = 2627 cm
(d) 3.73 m = 373 cm
9. (a) about 14 ft 4 in.
(b) $59°$

Chapter 18

Exercises 18-1

1. scalar
3. vector
5. $\mathbf{v}_4$: initial (3, 4), terminal (2, 0)
7. $\mathbf{v}_2$ and $\mathbf{v}_4$ are equal.

9. initial point: (0, 0); terminal point: (5, 7)
11. $-\mathbf{v}_1 = \langle -5, -3 \rangle$

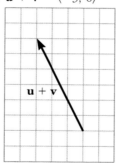

13. (a) $\mathbf{v}_4 = \langle 8, 2 \rangle$
(b) $\mathbf{v}_5 = \langle 3, 3 \rangle$
(c) $\mathbf{v}_6 = \langle 11, 5 \rangle$
15. $\mathbf{v} = \langle 3, 2 \rangle$

Exercises 18-2

1.

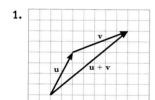

3. $\mathbf{u} + \mathbf{v} = \langle 8, 6 \rangle$

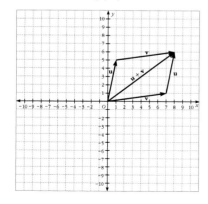

5. $\mathbf{u} + \mathbf{v} = \langle -3, 6 \rangle$

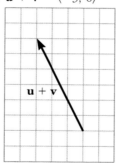

7. $\mathbf{u} + \mathbf{v} = \langle 1, -14 \rangle$
9. $\mathbf{u} - \mathbf{v} = \langle -8, -3 \rangle$

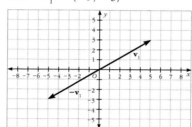

11. $\mathbf{u} - \mathbf{v} = \langle 11, 4 \rangle$
13. $-\frac{1}{3}\mathbf{v} = \langle 3, -1 \rangle$

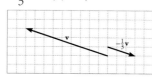

15. (a)

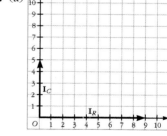

(b)

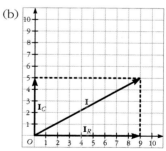

(c) (9, 5)
(d) $\sqrt{106} \approx 10.3\text{A}$
17. $\langle 5, 10 \rangle$

Exercises 18-3

1. $|\mathbf{v}| \approx 6.7$ units, $\theta \approx 27°$
3. $|\mathbf{v}| \approx 5.8$ units, $\theta \approx 301°$
5.

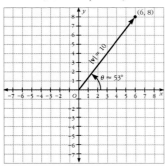

magnitude $|\mathbf{v}| = 10$ units
direction angle $\theta \approx 53°$

7.

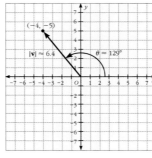

magnitude $|\mathbf{v}| \approx 6.4$ units
direction angle $\theta \approx 129°$

9. (a)

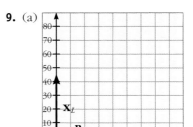

(b) 42 kΩ
(c) 90°
(d) 33 kΩ
(e) 0°
11. $|-3\mathbf{v}| \approx 24.7$ units, $\theta \approx 256°$

Exercises 18-4

1. $\mathbf{v} \approx \langle 2.5, 5.4 \rangle$
3. $\mathbf{v} \approx \langle -7.8, -1.8 \rangle$

5. $|\mathbf{w}| \approx 9.9$, $\theta_\mathbf{w} \approx 50°$
7. $|\mathbf{w}| \approx 9.9$, $\theta_\mathbf{w} \approx 318°$

9. $|\mathbf{V}| \approx 21.00$ V, $\theta_\mathbf{v} = 98°$
11. $|\mathbf{w}| \approx 12.1$, $\theta_\mathbf{w} \approx 165°$

Review Exercises

1. $\langle -7, -6 \rangle$
3. $\langle -14, -12 \rangle$

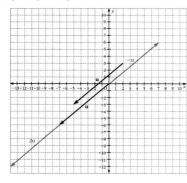

5. $\mathbf{u} = \langle 4, 10 \rangle$, $\mathbf{v} = \langle 5, -8 \rangle$,
$\mathbf{u} + \mathbf{v} = \langle 9, 2 \rangle$
7. $|\mathbf{w}| = 97$, $\theta_\mathbf{w} \approx -48°$ or
312°

9. (a) $|\mathbf{x}_c| = 76.9$
(b) $\theta_{\mathbf{x}_C} \approx -90°$ or 270°
(c) $|\mathbf{R}| = 72$
(d) $\theta_\mathbf{R} = 0°$
(e) Z = $\langle 72, -76.9 \rangle$
(f) $|\mathbf{Z}| = \sqrt{11,097.61} \approx$
105.35
(g) $\theta_\mathbf{Z} \approx -47°$ or 313°

Chapter 19

Exercises 19-1

1. 7
3. 51
5. 25
7. 31
9. 68

11. 1110
13. 11011
15. 10110
17. 1100110
19. 182.9.165.174

21. 10010110.01001000.10110100.
10001100
23. 1; $1_{10} = 1_2$

Exercises 19-2

1. 111_2; $101_2 = 5$, $10_2 = 2$;
$5 + 2 = 7$; $7 = 111_2$
3. 101110_2; $101010_2 = 42$, 100_2
$= 4$; $42 + 4 = 46$; $46 =$
101110_2

5. 1101_2; $1011_2 = 11$, $10_2 = 2$;
$11 + 2 = 13$; $13 = 1101_2$
7. 1100_2; $110_2 = 6$, $110_2 = 6$;
$6 + 6 = 12$; $12 = 1100_2$

9. 10110_2; $1110_2 = 14$, $1000_2 =$
8; $14 + 8 = 22$; $22 = 10110_2$

Exercises 19-3

1. 1001 0011

3. 1011 0010

5. 1110 0100

7. 1101 0111

9. 1010 1111

11. 1011 1000

13. The greatest number that can be written is 127. This number is written with a 0 in the MSD (to indicate it is positive) and 1's in all the other places: 0111 1111.

Exercises 19-4

1. 0010 0010; 0010 1111$_2$ = 47; 0000 1101$_2$ = 13; 47 − 13 = 34; 34 = 0010 0010$_2$

3. 0010 1000; 0011 0010$_2$ = 50; 0000 1010$_2$ = 10; 50 − 10 = 40; 40 = 0010 1000$_2$

5. 0000 0011; 0000 1110$_2$ = 14; 0000 1011$_2$ = 11; 14 − 11 = 3; 3 = 0000 0011$_2$

7. 1111 0110; 0001 0101$_2$ = 21; 0001 1111$_2$ = 31; 21 − 31 = −10. To find the magnitude of the negative result,

1111 0110$_2$, find its 2's complement. Add 0000 1001 + 1 = 0000 1010$_2$ = 10.

9. 1110 1111; 0101 1011$_2$ = 91; 0110 1100$_2$ = 108; 91 − 108 = −17. To find the magnitude of the negative result, 1110 1111$_2$, find its 2's complement. Add 0001 0000 + 1 = 0001 0001$_2$ = 17.

11. 1110 0101; 0010 0000$_2$ = 32; 0011 1011$_2$ = 59; 32 − 59 = −27. To find the magnitude of the negative result, 1110 0101$_2$, find its 2's complement. Add 0001 1010 + 1 = 0001 1011$_2$ = 27.

13. 1110 0011

Exercises 19-5

1. 110; 11$_2$ = 3, 10$_2$ = 2; 3 × 2 = 6; 6 = 110$_2$

3. 11 0010; 101$_2$ = 5, 1010$_2$ = 10; 5 × 10 = 50; 50 = 11 0010$_2$

5. 1 1110; 11$_2$ = 3, 1010$_2$ = 10; 3 × 10 = 30; 30 = 1 1110$_2$

7. 11 0001; 111$_2$ = 7, 111$_2$ = 7; 7 × 7 = 49; 49 = 11 0001$_2$

9. 10 1101; 101$_2$ = 5, 1001$_2$ = 9; 5 × 9 = 45; 45 = 10 1101$_2$

Exercises 19-6

1. 7_{10}

3. 194_{10}

5. 267_{10}

7. 542_{10}

9. $4,109_{10}$

11. 53_8

13. 175_8

15. 10445_8

17. 336_8

19. 1, 2, 3, 4, 5, 6, and 7

Exercises 19-7

1. 73_8

3. 67_8

5. 22_8

7. 126_8

9. 502_8

11. 111110_2

13. 110010001_2

15. 111011111_2

17. 1010110001_2

19. (a) 3,258

  (b) 6272_8

21. Sample answer: Howard converted the 1 in the octal number incorrectly. A 1 in octal is 001 in binary, so Howard should have written $10_8 = 001000_2$.

Exercises 19-8

1. 152_{10}

3. $2,748_{10}$

5. $7,054_{10}$

7. $285,091_{10}$

9. $1,027,565_{10}$

11. 33_{16}

13. $3D_{16}$

15. $20E_{16}$

17. $3E8_{16}$

19. 1004_{16}

Exercises 19-9

1. $A6_{16}$
3. $B1_{16}$
5. $7F5_{16}$
7. 212_{16}

9. $3CB_{16}$
11. 100110001010_2
13. 111101101_2
15. 10011010101_2

17. 1100101011_2
19. (a) 2,746
(b) ABA_{16}
21. $88F_{16}$

Review Exercises

1. $8531 = (8 \times 10^3) + (5 \times 10^2) + (3 \times 10^1) + (1 \times 10^0)$
$= (8 \times 1000) + (5 \times 100) + (3 \times 10) + (1 \times 1)$

3. $5073_8 = (5 \times 8^3) + (7 \times 8^1) + (3 \times 8^0)$
$= (5 \times 512) + (7 \times 8) + (3 \times 1)$

5. (a) $1\ 0110\ 0101\ 0100_2$,
(b) $13\ 124_8$
(c) 1654_{16}

7. 475_8

9. $100\ 1110_2$
11. 244.73.173.140

Chapter 20

Exercises 20-1

1. Statement A is true; statement B is false, A AND B is false; row 2

3. Statement A is true; statement B is true, A AND B is true; row 1

5. Statement A is false; statement B is true, A OR B is true; row 3

7. Statement A is true; NOT A is false; row 1

9. (a) A is true.
(b) B is false.
(c) A AND B is the statement *The square root of 121 is 11 AND $14\frac{1}{2}$ is an even number.*
(d) A AND B is false
(e) A OR B is the statement *The square root of 121 is 11 OR $14\frac{1}{2}$ is an even number.*
(f) A OR B is true.

11.

A	B	C	A AND B AND C
T	T	T	T
T	T	F	F
T	F	T	F
T	F	F	F
F	T	T	F
F	T	F	F
F	F	T	F
F	F	F	F

A AND B AND C is only true when all three statements A, B, and C are true. Otherwise it is false.

Exercises 20-2

1. $A \cdot B = 0$
3. $\overline{A} \cdot B = 0$
5. output: $A + \overline{B}$;

A	B	$\overline{B}$	$A + \overline{B}$
0	0	1	1
0	1	0	0
1	0	1	1
1	1	0	1

7. output: $\overline{A} + B$;

A	$\overline{A}$	B	$\overline{A} + B$
0	1	0	1
0	1	1	1
1	0	0	0
1	0	1	1

9.

A	B	$A \cdot B$	$A \cdot B + A \cdot B$
0	0	0	0
0	1	0	0
1	0	0	0
1	1	1	1

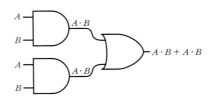

Exercises 20-3

1. $A + (\overline{A} \cdot C) =$
$(A + \overline{A}) \cdot (A + C) =$ ◄— Apply the Distributive Law.
$1 \cdot (A + C) =$ ◄— Apply the Complementation Law.
$A + C$ ◄— Apply the appropriate Identity Law.

3. $A + B + C + \overline{\overline{A}} + \overline{\overline{B}} + \overline{\overline{C}} =$
$A + B + C + A + B + C =$ ◄— Apply the Law of Involution.
$(A + A) + (B + B) + (C + C) =$ ◄— Apply the Commutative and Associative Laws.
$A + B + C$ ◄— Apply the Idempotent Law.

5. $A \cdot (A \cdot (A + B)) =$
$A \cdot (A) =$ ◄— Apply the Absorption Law.
A ◄— Apply the Idempotent Law.

7. $A + (\overline{A} \cdot B \cdot C) =$
$A \cdot \overline{A} + A \cdot B + A \cdot C =$ ◄— Apply the Distributive Law.
$0 + A \cdot B + A \cdot C =$ ◄— Apply the Complementation Law.
$A \cdot (B + C)$ ◄— Apply the Distributive Law.

9. $A + (\overline{A} \cdot B) =$
$(A + \overline{A}) \cdot (A + B) =$ ◄— Apply the Distributive Law.
$0 + (A \cdot B) =$ ◄— Apply the Law of Complementation.
$A \cdot B$ ◄— Apply the appropriate Law of Identity.

11.

A	B	C	$A + B$	$B + C$	$(A + B) + C$	$A + (B + C)$
0	0	0	0	0	0	0
0	0	1	0	1	1	1
0	1	0	1	1	1	1
0	1	1	1	1	1	1
1	0	0	1	0	1	1
1	0	1	1	1	1	1
1	1	0	1	1	1	1
1	1	1	1	1	1	1

13. $(A + B) \cdot (0 \cdot A) = 0$

Exercises 20-4

1. $\overline{A + B} \cdot C =$

$\overline{A} \cdot \overline{B} \cdot C =$ ← Apply DeMorgan's Law.

3. $\overline{A + B + C} \cdot A =$

$\overline{A} \cdot \overline{B} \cdot \overline{C} \cdot A =$ ← Apply DeMorgan's Law.

$A \cdot \overline{A} \cdot \overline{B} \cdot \overline{C} =$ ← Apply the Commutative Law.

$0 \cdot \overline{B} \cdot \overline{C} =$ ← Apply the Complementation Law.

0 ← Apply the Identity Law: $A \cdot 0 = 0$.

5. $\overline{\overline{A} + B + C} =$

$\overline{\overline{A}} \cdot \overline{B} \cdot \overline{C} =$ ← Apply DeMorgan's Law.

$A \cdot \overline{B} \cdot \overline{C}$ ← Apply the Law of Involution.

7. $\overline{A + B} \cdot \overline{A \cdot B} =$

$(\overline{A} \cdot \overline{B}) \cdot (\overline{A} + \overline{B}) =$ ← Apply DeMorgan's Law to $\overline{A + B}$ and $\overline{A \cdot B}$.

$(\overline{A} \cdot \overline{B} \cdot \overline{A}) + (\overline{A} \cdot \overline{B} \cdot \overline{B}) =$ ← Apply the Distributive Law.

$(\overline{A} \cdot \overline{A} \cdot \overline{B}) + (\overline{A} \cdot \overline{B} \cdot \overline{B}) =$ ← Apply the Commutative Law to $\overline{A} \cdot \overline{B} \cdot \overline{A}$.

$(\overline{A} \cdot \overline{B}) + (\overline{A} \cdot \overline{B}) =$ ← Apply the Idempotent Law to $\overline{A} \cdot \overline{A}$ and $\overline{B} \cdot \overline{B}$.

$\overline{A} \cdot \overline{B}$ ← Apply the Idempotent Law to $\overline{A} \cdot \overline{B}$.

Review Exercises

1. A is true

3. A AND B is false

5. $\overline{\overline{A} \cdot B}$

A	$\overline{A}$	B	$\overline{A} \cdot B$	$\overline{\overline{A} \cdot B}$
0	1	0	0	1
0	1	1	1	0
1	0	0	0	1
1	0	1	0	1

7.

A	$\overline{A}$	$\overline{\overline{A}}$
0	1	0
1	0	1

9. $\overline{A \cdot B} + \overline{A + B} = (\overline{A} + \overline{B}) + (\overline{A} \cdot \overline{B})$ ← Apply DeMorgan's Law to $\overline{A \cdot B}$ and $\overline{A + B}$.

$= (\overline{A} + \overline{B} + \overline{A}) \cdot (\overline{A} + \overline{B} + \overline{B})$ ← Apply the Distributive Law.

$= (\overline{A} + \overline{A} + \overline{B}) \cdot (\overline{A} + \overline{B} + \overline{B})$ ← Apply the Commutative Law to $\overline{A} + \overline{B} \cdot \overline{A}$

$= (\overline{A} + \overline{B}) \cdot (\overline{A} + \overline{B})$ ← Apply the Idempotent Law to $\overline{A} + \overline{A}$ and $\overline{B} \cdot \overline{B}$

$= \overline{A} + \overline{B}$ ← Apply the Idempotent Law

INDEX

PHOTO CREDITS

Notes

Notes

Notes

Notes

Notes